Vitamin D and Cancer

Donald L. Trump • Candace S. Johnson
Editors

Vitamin D and Cancer

 Springer

Editors
Donald L. Trump, MD
President & CEO
Roswell Park Cancer Institute
Elm & Carlton Streets
Buffalo, NY 14263, USA
donald.trump@roswellpark.org

Candace S. Johnson, PhD
Deputy Director
Chair, Pharmacology & Therapeutics
Roswell Park Cancer Institute
Elm & Carlton Streets
Buffalo, NY 14263, USA
candace.johnson@roswellpark.org

ISBN 978-1-4419-7187-6   e-ISBN 978-1-4419-7188-3
DOI 10.1007/978-1-4419-7188-3
Springer New York Dordrecht Heidelberg London

Library of Congress Control Number: 2010938476

© Springer Science+Business Media, LLC 2011
All rights reserved. This work may not be translated or copied in whole or in part without the written permission of the publisher (Springer Science+Business Media, LLC, 233 Spring Street, New York, NY 10013, USA), except for brief excerpts in connection with reviews or scholarly analysis. Use in connection with any form of information storage and retrieval, electronic adaptation, computer software, or by similar or dissimilar methodology now known or hereafter developed is forbidden.
The use in this publication of trade names, trademarks, service marks, and similar terms, even if they are not identified as such, is not to be taken as an expression of opinion as to whether or not they are subject to proprietary rights.

Printed on acid-free paper

Springer is part of Springer Science+Business Media (www.springer.com)

Preface

Over the past 30 years numerous provocative studies have provided clues suggesting that vitamin D may play an important role in cancer. In vitro studies have shown that cancer cells metabolize vitamin D and that vitamin D compounds can induce differentiation, inhibit cellular proliferation, and induce cell death. In addition, epidemiologic data suggest that vitamin D compounds may play a role in the prevention of cancer. In the past few years the understanding of the molecular effects of vitamin D has expanded substantially and investigators have begun to delineate the role of genetic factors that influence the response to vitamin D.

With this considerable history of development of vitamin D and cancer, it is timely and appropriate to summarize the current "state of the art" in the study of vitamin D and cancer. Scientists who have made many of the seminal contributions to this field of study have contributed to this volume. These collected data describe the foundation and current state for this important domain of cancer research – a domain that the coeditors of this book believe will yield important advances in cancer prevention and therapy.

Vitamin D Analogues as Antineoplastics: A Prologue Long Overdue?

Numerous investigators have drawn attention to the high prevalence of the vitamin D receptor (VDR) in human and murine cancer cells, the frequent evidence of intact vitamin D signaling pathways in such cells, and the ability of high concentrations of vitamin D analogues to inhibit the replication of cancer cells, induce apoptosis, and even inhibit angiogenesis. These data are cited in preceding and following chapters. Had such studies been completed with a new molecule – e.g., a new "targeted agent" – it is very likely that the following steps would have been undertaken promptly:

(a) Careful in vivo delineation of schedule and dose dependencies of these anticancer activities
(b) Careful determination of the maximum tolerated dose of analogues and exploration of optimal biologic dose

(c) *Direct* comparisons of toxicity and antitumor efficacy of analogues and parent compound (calcitriol) using the apparently most active drug schedules

Unfortunately, for vitamin D based studies in cancer, very little of this rudimentary work has been carried out. Studies with most analogues of vitamin D (paricalcitol, seocalcitol, inecalcitol) have employed continuous dosing schedules even though practically all in vitro and in vivo studies which have shown anticancer activity of vitamin D have exposed cells and tumors to intermittent, high-pulse doses. Many have been encouraged by the study of daily dosing of analogues and parent compound (calcitriol) and finding the analogue causes less hypercalcemia. Often and not surprisingly, the analogue binds less avidly to VDR. Such studies have led to small to medium sized studies using daily dosing algorithms which have shown no antitumor effects and been halted without any toxicity remotely resembling those defensible in patients with advanced cancer.

Further limiting work with calcitriol has been the absence of a formulation suitable for high-dose therapy. This limitation is due primarily to the lack of an economic motivation for the development of such formulations.

The following chapters provide excellent and comprehensive discussions of the potential role of vitamin D based therapies in breast, colorectal, prostate cancer, and leukemia and myelodysplastic syndromes. These chapters also point out that the focus on these diseases is largely determined by the interests and expertise of the outstanding scientists who have chosen to pursue vitamin D based cancer therapeutics. To our knowledge, every tumor type ever evaluated has shown some biochemical and antiproliferative response to vitamin D. Similarly, vitamin D analogues, especially calcitriol, potentiate almost every cytotoxic agent with which combination therapies have been tested. In our view the slow and halting development of vitamin D based cancer therapeutics could be greatly accelerated by following standard principles of anticancer drug development:

(a) Development of a standardized formula.
(b) Determination of MTD (current data indicate the MTD of calcitriol on an intermittent [weekly] schedule is ≥ 100 mcg). No reliable oral MTD have ever been determined and few data on the optimal biologic dose developed in the laboratory, much less in the clinic.
(c) Conduct of carefully designed clinical trials.

The field of vitamin D based cancer therapeutics has very few such data items available. Perhaps the extensive preclinical data on the antitumor effectiveness of high-dose vitamin D analogue therapy are misleading or in fact wrong. But until the agent is examined in the fashion one would follow for an antineoplastic – we will never know. The following chapters point out what is known and the direction that can be followed in clinical development of vitamin D as an anticancer agent.

Buffalo, NY
Buffalo, NY

Donald L. Trump
Candace S. Johnson

Contents

1 **Vitamin D: Synthesis and Catabolism – Considerations for Cancer Causation and Therapy**.. 1
Heide S. Cross

2 **The Molecular Cancer Biology of the VDR**... 25
James Thorne and Moray J. Campbell

3 **Anti-inflammatory Activity of Calcitriol in Cancer** 53
Aruna V. Krishnan and David Feldman

4 **The Epidemiology of Vitamin D and Cancer Risk**............................... 73
Edward Giovannucci

5 **Vitamin D and Angiogenesis** 99
Yingyu Ma, Candace S. Johnson, and Donald L. Trump

6 **Vitamin D: Cardiovascular Function and Disease**............................... 115
Robert Scragg

7 **Induction of Differentiation in Cancer Cells by Vitamin D: Recognition and Mechanisms** 143
Elzbieta Gocek and George P. Studzinski

8 **Vitamin D and Cancer Chemoprevention** 175
Sarah A. Mazzilli, Mary E. Reid, and Barbara A. Foster

9 **Molecular Biology of Vitamin D Metabolism and Skin Cancer** 191
Florence S.G. Cheung and Juergen K.V. Reichardt

10 **Vitamin D and Prostate Cancer**.. 221
Christine M. Barnett and Tomasz M. Beer

| 11 | **Vitamin D and Hematologic Malignancies** | 251 |

Ryoko Okamoto, Tadayuki Akagi, and H. Phillip Koeffler

| 12 | **The Vitamin D Signaling Pathway in Mammary Gland and Breast Cancer** | 279 |

Glendon M. Zinser, Carmen J. Narvaez, and JoEllen Welsh

| 13 | **Vitamin D and Colorectal Cancer** | 295 |

Marwan Fakih, Annette Sunga, and Josephia Muindi

| 14 | **Unique Features of the Enzyme Kinetics for the Vitamin D System, and the Implications for Cancer Prevention and Therapeutics** | 315 |

Reinhold Vieth

| 15 | **Assessment of Vitamin D Status in the 21st Century** | 327 |

Bruce W. Hollis

Index 339

Contributors

Tadayuki Akagi Ph.D.
Division of Hematology and Oncology, Cedars-Sinai Medical Center, UCLA
School of Medicine, Los Angeles, CA, USA

Christine M. Barnett M.D.
Division of Hematology and Medical Oncology, Knight Cancer Institute,
Oregon Health & Science University, CH-14R, 3303 SW Bond Ave, Portland,
OR 97239, USA

Tomasz M. Beer M.D.
Division of Hematology & Medical Oncology, Oregon Health & Science
University, Knight Cancer Institute 3303 SN Bond Avenue, CH14R Portland,
OR 97239–3098, USA

Moray J. Campbell Ph.D.
Department of Pharmacology & Therapeutics, Roswell Park Cancer Institute,
Elm & Carlton Streets, Buffalo, NY 14263, USA

Florence SG. Cheung M.D., Ph.D.
Plunkett Chair of Molecular Biology (Medicine), Bosch Institute, The University
of Sydney, Camperdown, NSW 2006, Australia

Heidi S. Cross Ph.D.
Department of Pathophysiology, Medical University of Vienna,
Waehringer Guertel 18–20, A-1090 Vienna, Austria

Marwan Fakih M.D.
Roswell Park Cancer Institute, Elm & Carlton Streets, Buffalo, NY 14263, USA

David Feldman M.D.
Department of Medicine, Division of Endocrinology, Stanford University School
of Medicine, 300 Pasteur Drive, Room S-025, Stanford, CA 94305–5103, USA

Barbara A. Foster Ph.D.
Pharmacology & Therapeutics, Roswell Park Cancer Institute, Elm & Carlton
Streets, Buffalo, NY 14263, USA

Edward Giovannucci M.D., ScD
Department of Nutrition, 2–371, Harvard School of Public Health,
665 Huntington Avenue, Boston, MA 02115, USA
and
Department of Epidemiology, Harvard School of Public Health,
Boston, MA 02115, USA
and
Channing Laboratory, Department of Medicine,
Brigham and Women's Hospital and Harvard Medical School,
181 Longwood Avenue, Boston, MA 02115, USA

Elzbieta Gocek
Faculty of Biotechnology, University of Wroclaw, Tamka 2, 50–137 Wroclaw,
Poland

Bruce W. Hollis Ph.D.
Department of Pediatrics, Darby Children's Research Institute,
Medical University of South Carolina, 173 Ashley Ave., Room 313,
Charleston, SC 29425, USA

Candace S. Johnson Ph.D.
Deputy Director, Chair, Pharmacology & Therapeutics, Roswell Park Cancer Institute,
Elm & Carlton Streets, Buffalo, NY 14263, USA

H. Phillip Koeffler M.D.
Director, Hematology/Oncology Division, Cedar-Sinai Medical Center,
8700 Beverly Blvd, Los Angeles, CA 90048, USA

Aruna V. Krishnan
Stanford University School of Medicine, 300 Pasteur Drive, Room S-025,
Stanford, CA 94305–5103, USA

Yingyu Ma M.D., Ph.D.
Pharmacology & Therapeutics, Roswell Park Cancer Institute,
Elm & Carlton Streets, Buffalo, NY 14263, USA

Josephia Muindi M.D., Ph.D.
Associate Professor of Oncology, Roswell Park Cancer Institute,
Elm and Carlton Streets, Buffalo, NY 14263, USA

Carmen J. Narvaez Ph.D.
GenNYsis Center for Excellence in Cancer Genomics,
122H Cancer Research Center, 1 Discovery Drive, Rensselaer, NY 12144, USA

Ryoko Okamoto Ph.D.
Division of Hematology and Oncology, Cedars-Sinai Medical Center, UCLA
School of Medicine, 8700 Beverly Blvd, Los Angeles, CA 90048, USA

Contributors   xi

Juergen K.V. Reichardt
Plunket Chair of Molecular Biology (Medicine),
Bosch Institute, The University of Sydney,
Medical Foundation Building (K25), 92–94 Parramatta Road,
Camperdown, NSW 2006, Australia
and
School of Pharmacy and Molecular Sciences, James Cook University,
Townsville, Qld 4811, Australia

Robert Scragg Ph.D.
Associate Professor, School of Population Health, University of Auckland,
Auckland Mail Centre, Private Bag 92019, Auckland 1142, New Zealand

George P. Studzinski M.D., Ph.D.
Professor, Department of Pathology & Laboratory Medicine, UMDNJ-New Jersey
Medical School, 185 So. Orange Avenue, MSB C-540, Newark, NJ 07101-1709, USA

Annette Sunga M.D., MPH
Assistant Professor of Oncology, Roswell Park Cancer Institute, Elm & Carlton
Streets, Buffalo, NY 14263, USA

James Thorne
Division of Experimental Haematology, University of Leeds, Leeds Institute
of Molecular Medicine, Wellcome Trust Brenner Building, St James's University
Hospital, Leeds LS9 7TF, UK

Donald L. Trump M.D.
President & CEO, Roswell Park Cancer Institute, Elm & Carlton Streets,
Buffalo, NY 14263, USA

Reinhold Vieth Ph.D.
Departments of Nutritional Sciences, and Laboratory Medicine and Pathobiology,
University of Toronto, Ontario, M5G 1X5, Canada

JoEllen Welsh Ph.D.
GenNYsis Center for Excellence in Cancer Genomics, University of Albany,
Rensselaer, NY 12144, USA

Glendon M. Zinser Ph.D.
Department of Cancer & Cell Biology, Vontz Center for Molecular Studies,
3125 Eden Avenue, Cincinnati, OH 45267–0521, USA

Chapter 1
Vitamin D: Synthesis and Catabolism – Considerations for Cancer Causation and Therapy

Heide S. Cross

Abstract Protection from sporadic malignancies by vitamin D can be traced to the role of its hormonally active metabolite, 1,25-dihdroxyvitamin D_3 (1,25-$(OH)_2D_3$) which, by binding to the nuclear vitamin D receptor (VDR), can maintain cellular homeostasis. Human colonic, prostatic, and breast cells express the CYP27B1-encoded 25-(OH)D-1α-hydroxylase, the enzyme responsible for conversion of 25-$(OH)D_3$ to 1,25-$(OH)_2D_3$. In vitamin D insufficiency, availability of 25-(OH) D_3 is low, so that extrarenal CYP27B1 activity may not be high enough to achieve tissue concentrations of 1,25-$(OH)_2D_3$ necessary to control growth and prevent neoplastic transformation of colonocytes.

While adequate supply of the vitamin D precursor 25-$(OH)D_3$ is essential for prevention of tumor progression, activity of the extrarenal synthesizing CYP27B1 is of paramount importance especially in view of the fact that 1,25-$(OH)_2D_3$ catabolism is progressively elevated during tumor progression. To counteract catabolism, enhancement of 1,25-$(OH)_2D_3$ synthesis is discussed. Early during cancer progression growth factors and sex hormones may elevate CYP27B1 expression and suppress that of CYP24A1. Also, genetic variations and epigenetic regulation of vitamin D hydroxylases could determine actual accumulation of 1,25-$(OH)_2D_3$ in mammary, prostate, and colonic tissue and are considered both for prevention of progression as well as for potential therapy.

Primarily in the colon as part of the digestive system, the chemopreventive potential of vitamin D can also be augmented by nutrient factors that induce appropriate changes in CYP27B1 and/or CYP24A1 expression. Among these factors are calcium, the phytoestrogen genistein and potentially also folate. Adequate intake levels of these nutrients could augment effectiveness of 1,25-$(OH)_2D_3$ for prevention of cancers in humans. Especially folate, as a methyl donor, could affect epigenetic regulation of CYP27B1 and of CYP24A1, and could therefore play a central role in vitamin D-mediated inhibition of tumor progression.

H.S. Cross (✉)
Department of Pathophysiology, Medical University of Vienna,
Waehringer Guertel 18–20, A-1090 Vienna, Austria
e-mail: heide.cross@meduniwien.ac.at

D.L. Trump and C.S. Johnson (eds.), *Vitamin D and Cancer*,
DOI 10.1007/978-1-4419-7188-3_1, © Springer Science+Business Media, LLC 2011

Keywords Expression of extrarenal vitamin D hydroxylases • Cancer prevention • Regulation of colonic vitamin D synthesis • Calcium • Estrogens

1.1 Introduction

The enzyme 25-hydroxyvitamin D3-1α-hydroxylase (CYP27B1) plays a central role in calcium homeostasis [1], but alternative physiological actions have been suspected for decades. The enzyme catalyzes the conversion of 25-hydroxyvitamin D_3 (25-(OH)D_3) to the hormone 1,25-dihydroxyvitamin D_3 (1α,25-(OH)$_2D_3$) that is known to regulate calcium and phosphate transport in intestine, bone, and kidney. While initially it was thought that only proximal tubule kidney cells express CYP27B1, it became evident in the mid-1980s that extrarenal cells, for instance, bone cells, macrophages, and keratinocytes (see, e.g., [2]) could also express CYP27B1 enzymatic activity in vitro. Mawer et al. [3] demonstrated that certain lung cells had measurable CYP27B1 activity. Apparently, this particular 25-hydroxyvitamin D3-1α-hydroxylase was not up-regulated by PTH and was not down-regulated by plasma calcium, a hallmark of the renal enzyme. In addition, while in renal cells sufficiency of serum 1,25-(OH)$_2D_3$ concentration leads to induction of the vitamin D-inactivating enzyme 1,25-(OH)$_2D_3$–24-hydroxylase (CYP24A1) [4], the extrarenal CYP27B1 is not necessarily inversely correlated with CYP24A1 expression, a fact that will be enlarged upon later in this chapter.

While extrarenal CYP27B1 activity in macrophages might be the reason for the hypercalcemia associated with sarcoidosis and lymphomas, there was also the possibility that it might be coded by a gene different from the renal one, and this could lead to alternative regulatory mechanisms. The renal CYP27B1 is a combination of three proteins: a cytochrome P450 as well as two other proteins, ferredoxin and ferredoxin reductase. Purified preparations of these proteins possess the CYP27B1 enzyme activity in vitro [5]. These enzyme complexes were cloned from rodents and human renal cells and response elements were found in promoter regions that allow up-regulation by PTH. Proof was provided that extrarenal CYP27B1 is a product of the same gene as the renal form. However, regulation of the newly discovered CYP27B1 suggested existence of a paracrine loop in extrarenal tissues for the modification of cellular proliferation and differentiation, though subsequent conversion of the active vitamin D metabolite into a C-24 oxidation product by CYP24A1 was similar to renal catabolism [6].

In the last few decades, there has been growing appreciation for the multitude of physiological roles that vitamin D has in many body tissues. As early as in 1979, Stumpf et al. demonstrated that cells from heart, stomach, pancreas, colon, brain, skin, gonads, etc., have the nuclear receptor for 1,25-(OH)$_2D_3$ [7], the so-called vitamin D receptor (VDR), and such tissues are potential targets for 1,25-(OH)$_2D_3$ activity. Many of these VDR-positive tissues are also positive for CYP27B1, i.e., the enzyme that can convert 25-(OH)D_3 to the active metabolite [8], and many of these tissues are known to be targets for development of malignancies.

1 Vitamin D: Synthesis and Catabolism   3

As mentioned above, regulation of CYP27B1 in these non-renal tissues differs from that observed in the kidney and, importantly and in contrast to the renal enzyme, may be dependent on substrate concentration for activity. This led to the novel concept that maintenance of adequate serum 25-(OH)D$_3$ levels would be essential for providing the substrate for the synthesis of the active metabolite at extrarenal sites, which in turn would have physiological functions apart from those involved in bone mineral metabolism. This concept will be enlarged upon in the following. Evidence will be provided for the function and regulation of vitamin D synthesizing and catabolic hydroxylases, i.e., CYP27B1 and CYP24A1, respectively, in colorectal, prostate, and mammary gland-derived cells that are from organs particularly affected by sporadic malignancies during advancing age.

1.1.1 1,25-(OH)$_2$D$_3$ Synthesis

7-Dehydrocholesterol, the immediate precursor in the cholesterol biosynthetic pathway, is produced in rather large quantities in the skin of most vertebrates, also humans. Ageing decreases the capacity of skin to produce 7-dehydrocholesterol by as much as 75% [9] and this is of particular relevance when considering that sporadic cancers occur primarily in the elderly. When exposed to sunlight, skin cells absorb UVB radiation with wavelengths of 290–315 nm leading to a rearrangement of the molecular structure of 7-dehydrocholesterol to form the more thermodynamically stable previtamin D$_3$. Protection of the skin by topical sunscreens will reduce previtamin D$_3$ production by almost 100%. Persons that have greater amounts of melanin in their epidermis require much higher exposure to sunlight than whites to avoid vitamin D deficiency. Living at geographic latitudes above 35° will not provide enough UVB photons for sufficient production of vitamin D$_3$ in skin during winter time (for further reading see, e.g., [10]). Very few foods naturally contain vitamin D. Cod liver oil and oily fish are the best dietary source which, in some Scandinavian countries, can provide a positive balance to the lack of dermal vitamin D production.

Vitamin D$_3$ is first hydroxylated in the liver by CYP27A1, a cytochrome P450 25-hydroxylase, to the precursor 25-(OH)D$_3$. To be fully active, 25-(OH)D$_3$ must be converted to 1,25-(OH)$_2$D$_3$ by CYP27B1, a mitochondrial cytochrome P450 enzyme present primarily in proximal renal tubule cells but also in many extrarenal cells [11]. While the hormone regulates calcium and phosphate metabolism in intestine, bone, and kidney, at extrarenal sites it has a wide range of biological effects that are essentially noncalcemic in nature. The most surprising one is its ability to suppress hyperproliferative growth of cells and to support differentiation. In 1982, Tanaka et al. [12] provided the first evidence that 1,25-(OH)$_2$D$_3$ was able to promote differentiation of HL-60 leukemia cells. This, and a pronounced antimitotic effect, has subsequently been shown for many types of cancer cells in vitro (see, e.g., [13–18]), though only at nanomolar concentrations. However, serum 1,25-(OH)$_2$D$_3$ never exceeds picomolar concentrations, regardless of whether

sunlight exposure is increased or whether there is increased oral uptake of vitamin D [19], since its synthesis in renal cells is tightly regulated by PTH, calcium, and phosphate.

As early as 1980, Garland et al. raised the question whether sunlight and vitamin D can protect against colon cancer [20]. Strong support for this hypothesis was obtained when Garland et al. [21] in 1985 published the results of a 19-year prospective trial, showing that low dietary intakes of vitamin D and of calcium are associated with a significant risk of colorectal cancer. In the following decades, a compromised vitamin D status as indicated by low 25-(OH)D_3 serum levels has been associated with pathogenesis of diverse types of malignancy (for review see, e.g., [22, 23]). This, and the realization that there was vitamin D synthesis at extrarenal sites potentially enhancing $1,25\text{-(OH)}_2\text{D}_3$ concentrations in certain tissues without contributing to serum levels of $1,25\text{-(OH)}_2\text{D}_3$, suggested a hypothesis on how decreased sunlight exposure and low serum 25-(OH)D_3 could contribute to tumor pathogenesis.

1.2   Regulation of $1,25\text{-(OH)}_2\text{D}_3$ Synthesis in Extrarenal Cells

Regulation of $1,25\text{-(OH)}_2\text{D}_3$ production at multiple levels is a crucial determinant of nonclassical aspects of $1,25\text{-(OH)}_2\text{D}_3$ function. When we showed that normal and neoplastic human colon epithelial cells are endowed with a functional 25-hydroxyvitamin D-1α-hydroxylase and can thus convert 25-(OH)D_3 to $1,25\text{-(OH)}_2\text{D}_3$ [24–26], we hypothesized that adequate accumulation of the active metabolite could slow down or inhibit progression of malignant disease by promoting differentiation and apoptosis and by suppressing antimitotic activity locally. Renal CYP27B1 activity is tightly regulated by serum Ca^{++} and parathyroid hormone (PTH), as well as by feedback inhibition from $1,25\text{-(OH)}_2\text{D}_3$. In contrast, CYP27B1 expression, at least in colonocytes and prostate cells, is relatively insensitive to modulation via the $PTH/[Ca^{++}]_o$ axis [27, 28]. Intracellular synthesis of $1,25\text{-(OH)}_2\text{D}_3$ at extrarenal sites depends largely on ambient 25-(OH)D_3 levels and is not influenced by plasma levels of $1,25\text{-(OH)}_2\text{D}_3$ [29]. This may explain why the incidence of vitamin D-dependent cancers, e.g., of the colorectum [30], breast [31], and prostate gland [32], is correlated with low serum 25-(OH)D_3 rather than with serum concentrations of $1,25\text{-(OH)D}_3$. Strong support for the importance of intracellularly produced over circulating $1,25\text{-(OH)}_2\text{D}_3$ for regulation of cell functions comes from a study by Rowling et al. [33] who have shown that in mammary gland cells VDR-mediated actions depended more on megalin-mediated endocytosis of 25-(OH)D_3 than on ambient $1,25\text{-(OH)}_2\text{D}_3$. Also Lechner et al. [34] could induce the characteristic antimitogenic effect of $1,25\text{-(OH)}_2\text{D}_3$ when human colon carcinoma cells were treated with 25-(OH)D_3, though only when they were CYP27B1-positive. Similar observations were made in prostate [35] and mammary cells [36].

1   Vitamin D: Synthesis and Catabolism   5

However, at low serum levels of 25-(OH)D$_3$, CYP27B1 activity in extrarenal cells may be not high enough (in normal colonic mucosa without hyperproliferative signaling, positivity for CYP27B1 is extremely low [26]) to achieve those steady-state tissue concentrations of 1,25-(OH)$_2$D$_3$ necessary to maintain normal cellular growth and differentiation during hyperproliferation. In addition, 1,25-(OH)$_2$D$_3$ itself is an important regulator of CYP27B1 gene expression. Down-regulation of the CYP27B1 gene involves a negative vitamin D response element and cell specificity for this could be due to differential expression of protein complexes associated with the CYP27B1 promoter [37, 38].

1.2.1   Expression of CYP27B1 and of VDR During Hyperproliferation and Tumor Progression

The relevance of 1,25-(OH)$_2$D$_3$ to maintain normal epithelial cell turnover in the large intestine was demonstrated by studies with mice, which were genetically altered to block 1,25-(OH)$_2$D$_3$/VDR signaling: The colon mucosa of VDR null (VDR$^{-/-}$) mice show a pattern of increased DNA damage and cell division, the former probably due to formation of reactive oxygen species [39]. Interestingly, the large intestine reacts to inflammatory and hyperproliferative conditions with up-regulation of the VDR and of its ligand-synthesizing enzyme, CYP27B1: Liu et al. [40] reported that in a mouse model of ulcerative colitis, a disease considered to be a precursor lesion to colorectal cancer, expression of CYP27B1 was increased four-fold compared with controls. With respect to human colon cancer, we have shown that expression of CYP27B1 rises about fourfold in the course of progression from adenomas to well and moderately differentiated (G1 and G2) tumors, and then substantially declines during further progression [41]. Expression of the VDR showed the same dependence on tumor cell differentiation [41, 42]. However, cells from poorly differentiated (G3) colonic lesions, are frequently devoid of immuno-reactivity for VDR and CYP27B1, while, at the same time, epidermal growth factor (EGF) receptor mRNA can be detected by in situ hybridization in almost any cancer cell [43]. Statistical evaluation actually showed an inverse expression of EGF receptor positivity compared to that of VDR. We suggested therefore that the 1,25-(OH)$_2$D$_3$/VDR system can be activated in colon epithelial cells in response to mitogenic stimulation, e.g., by EGF, respectively, TGF-α [43, 44]. A strong auto-crine/paracrine antimitogenic action of 1,25-(OH)$_2$D$_3$ would retard further tumor growth as long as cancer cells retain a certain degree of differentiation and high levels of CYP27B1 activity and of VDR expression. However, during progression to high grade malignancy, signaling from the 1,25-(OH)$_2$D$_3$/VDR system would be too weak to effectively counteract proliferative effects from, e.g., EGF-R activation [43]. We confirmed these hypotheses by demonstrating that, in differentiated colon cell lines, EGF stimulates expression of VDR and CYP27B1, whereas in a primary culture derived from a G2 tumor, expression of VDR and of CYP27B1 was actually reduced [45]. Palmer et al. [46] demonstrated that induction of the adhesion protein

6   H.S. Cross

E-cadherin by vitamin D enhanced differentiation of colon cancer cells. This in turn opposed hyperproliferation and thus indicates the importance of vitamin D activity for normal maintenance of the wnt pathway. It is significant that repression of E-cadherin and of VDR, and parallel enhanced expression of the transcription factor SNAIL, was found in patients with aggressive tumor characteristics [47].

CYP27B1 and VDR expression is present also in some prostate and mammary gland-derived cells, since growth inhibition by 25-(OH)D$_3$ occurs with concomitant upregulation of CYP24A1. If mammary cells are negative for CYP27B1, there is no mitotic inhibition, and no induction of CYP24A1 expression [48]. When the antimitotic potencies of 25-(OH)D$_3$ and of 1,25-(OH)$_2$D$_3$, both in the nanomolar range, were studied in prostate cells, they were quite similar as long as cells expressed CYP27B1 [49]. However, it was suggested that during tumor progression, prostate cells no longer express CYP27B1 [35], though the biological grade of cells was not established in these studies. Quite similar to colon cells, EGF stimulated CYP27B1 promoter activity in prostate cell lines via involvement of the MAPK pathway, at least in those cancer cells that are still differentiated [50]. In normal human prostatic epithelial cells mitogen-activated protein kinase phosphatase 5 was induced by 1,25-(OH)$_2$D$_3$ leading to deactivation of protein kinase p38 [51]. Activation of p38 and downstream production of interleukin-6 are proinflammatory. Inflammation as well as interleukin-6 overproduction have been implicated in initiation and progression of prostate as well as of colon cancer [52]. Similar regulatory networks appear to exist in mammary gland cells (for review see [53]).

1.2.2   Expression of CYP24A1 During Hyperproliferation and Tumor Progression

It must be taken into account that the effective tissue concentration of 1,25-(OH)$_2$D$_3$ is determined not only by substrate availability but by additional regulatory factors that may govern also renal vitamin D synthesis: (i) in colonocytes, in prostate and mammary gland cells, 1,25-(OH)$_2$D$_3$ downregulates CYP27B1 and the VDR (see, e.g., [34]); (ii) 1,25-(OH)$_2$D$_3$ at the same time induces CYP24A1-encoded 25-(OH)D$_3$-24-hydroxylase, the enzyme that initiates stepwise degradation of the hormone; and (iii) at least in colon tumors, expression of CYP24A1 increases dramatically during progression to a poorly differentiated state (G3-G4) though CYP27B1 expression is diminished [54].

Therefore, one major mechanism for vitamin D resistance or reduced sensitivity in VDR-positive cancer cells is 1,25-(OH)$_2$D$_3$ catabolism via the C-24 hydroxylation pathway. An inverse relation between cellular metabolism of 1,25-(OH)$_2$D$_3$ via 24-hydroxylation and growth inhibition of prostate cancer cells by vitamin D has been suggested [55]. A 1,25-(OH)$_2$D$_3$ resistant prostate cell line was growth-inhibited when cultured with the active vitamin D metabolite combined with the CYP24A1 inhibitor liarozole [56]. Colon cells isolated from well-advanced (G3)

1 Vitamin D: Synthesis and Catabolism   7

tumors express extremely high levels of CYP24A1, and cannot be growth-inhibited by 1,25-(OH)$_2$D$_3$. Actually, when these CYP27B1-negative cells are exposed to 16.6 nmol 25-(OH)D$_3$, they will efficiently use up the precursor within 12 h for 24,25-(OH)$_2$D$_3$ production and further degradation [34]. Androgen-independent prostate cell lines also tend to express high levels of CYP24A1, whereas CYP27B1 expression is negligible (see, e.g., [57]). These few examples clearly demonstrate an uncoupling of 1,25(OH)$_2$D$_3$ action from expression of CYP24A1 during advancing malignancy: whereas, in differentiated colon and prostate cancer cells, 1,25-(OH)$_2$D$_3$ will induce CYP24A1 expression, undifferentiated cells express basally extremely high levels of CYP24A1 that can no longer be enhanced by treatment with the active metabolite [38, 58]. Therefore, such basally high expression of CYP24A1 during advanced malignancy will not permit effective treatment of patients with vitamin D or vitamin D analogs that can be degraded via the C-24 pathway. However, this also clearly shows that inhibition of CYP24A1 activity in tumor cells could be of primary importance for cancer therapy. This aspect will be discussed further in the section on epigenetic regulation of CYP24A1 (see section 1.2.5.)

1.2.3   *Regulation of CYP27B1 and CYP24A1 Expression by Sex Hormones*

Although men and women suffer from similar rates of colorectal cancer deaths in their lifetime, the age-adjusted risk for colorectal cancer is less for women than for men [59]. This strongly indicates a protective role of female sex hormones, particularly of estrogens, against colorectal cancer (see, e.g., [60, 61]). Observational studies have further suggested that postmenopausal hormone therapy is associated with a lower risk for colorectal cancer and a lower death rate in women [62]. A meta-analysis of studies showed a 34% reduction in the incidence of this tumor in postmenopausal women receiving hormone replacement therapy [63]. A mechanism of action for estrogens in lowering colon cancer risk is not known yet. Since estrogen receptors are present in both normal intestinal epithelium and in colorectal cancers, the hormone is probably protective through these receptors and resultant post-receptor cellular activities.

While the colon cannot be considered an estrogen-dependent tissue, it must be defined as an estrogen-responsive organ. Expression of estrogen receptor (ER) subtypes α and β have been detected in cancer cell lines. Whereas human colon mucosa expresses primarily the ER-β type regardless of gender [64], ER-α is mainly expressed in the breast and the urogenital tract [65]. Both receptors bind estrogen, but they activate promoters in different modes. Studies of breast and prostate carcinogenesis suggested opposite roles for ER-α and ER-β in proliferation and differentiation [66]. Therefore, the ER-α/ER-β ratio has been suggested as a possible determinant of the susceptibility of a tissue to estrogen-induced carcinogenesis: in some cells, binding of estrogen to ER-α induces cancer-promoting effects, whereas binding to ER-β exerts a protective action. With respect to colon

cancer, the concept of a protective role of ER-β gained support recently: decreasing levels of the receptor were reported during colonic tumorigenesis compared with expression in the adjacent normal mucosa from the same patient [67].

Estrogens may indirectly oppose progression of malignancies by changing VDR expression or vitamin D metabolism in colonic epithelial cells. As early as in 1986, a study on the effect of endogenous estrogen fluctuation with respect to 25-(OH)D$_3$ metabolism was published [68]. This study in healthy premenopausal women suggested that 25-(OH)D$_3$ was metabolized predominantly to 24,25-(OH)$_2$D$_3$ at low estrogen, but to 1,25-(OH)$_2$D$_3$ at higher serum estrogen concentrations. While this, in 1986, primarily concerned renal synthesis of vitamin D metabolites, it was the first suggestion that estrogen elevates CYP27B1 expression.

Liel et al. [69] reported that estrogen increased VDR activity in epithelial cells of the gastrointestinal tract. In the colon adenocarcinoma-derived cell line Caco-2, which is ER-β positive but negative for ER-α, we demonstrated an increase of CYP27B1 mRNA expression and also of enzymatic activity after treatment with 17β-estradiol [70]. Based on these findings a clinical pilot trial was designed, in which postmenopausal women with a past history of rectal adenomas were given 17β-estradiol daily for 1 month to reach premenopausal serum levels. Rectal biopsies were obtained at the beginning and end of trial. A predominant result was the elevation of VDR mRNA [71]. We also observed significant induction of CYP27B1 mRNA in parallel to a decrease in COX-2 mRNA expression in those patients who had particularly high levels of the inflammatory marker at the beginning of the trial (Cross HS, The vitamin D system and colorectal cancer prevention. In: Vitamin D, 3rd edition. D. Feldman ed. Elsevier 2010).

To study modification of vitamin D hydroxylase activity by 17β-estradiol further, we used a mouse model to measure actual 1,25-(OH)$_2$D$_3$ synthesis and accumulation in colonic mucosa. In female compared with male mice, CYP27B1 mRNA was doubled and 1,25-(OH)$_2$D$_3$ concentration in the mucosa was increased by more than 50%. This occurred in the proximal colon only and suggested that there may be site-specific action of 17β-estradiol [127]. In this respect it is significant, that the estrogen receptor ESR1 is more methylated (inactivated) in the human distal than in the proximal colon [72].

There is equivocal evidence for the role of estrogen receptors (ER)-α and (ER)-β, and therefore for estrogenic activity, during mammary tumor progression. It has been suggested that higher ER-α expression in normal breast epithelium increases breast cancer risk. Since 1,25-(OH)$_2$D$_3$ synthesis, not only in colonocytes but also in mammary cells, may in part be regulated by 17β-estradiol [70], and since epidemiological evidence points to a correlation between breast cancer incidence and low levels of the precursor 25-(OH)D$_3$ [73], evaluation of the vitamin D system during progression of mammary carcinogenesis could be important. When normal breast tissue was compared with that derived from cancer patients, CYP27B1 mRNA was found in both tissues. In one study it was claimed, that expression was higher during early malignancy similar to colonic tissue [74]. Primary cultures established from human mammary tissue expressed CYP27B1 and were growth-inhibited by physiologically relevant concentrations of 25-(OH)D$_3$ [48], while

1 Vitamin D: Synthesis and Catabolism 9

established breast cancer cell lines showed a wide range of vitamin D hydroxylase expression. In general, however, CYP27B1 mRNA expression is relatively low and that of CYP24A1 is rather high. For example, hydroxylation of 25-(OH)D$_3$ in MCF-7 cells occurred primarily on the C-24 pathway [38], though we were able to demonstrate that 17β-estradiol elevates CYP27B1 mRNA expression and activity in these cells as well [70]. Kemmis and Welsh [36] recently showed that oncogenic transformation of human mammary epithelial cells was associated with reduced 1,25-(OH)$_2$D$_3$ synthesis and decreased sensitivity to its antimitotic action. This suggests enhanced expression of the catabolic CYP24A1 during progression.

Growth and function of the prostate is dependent on androgens. Initial endocrine therapy in prostate cancer aims to eliminate androgenic activity from cells. However, cells invariably become refractory to this therapy and grow androgen-independently. During this progression, estrogen influence appears to increase and oxidative and reductive 17β-hydroxysteroid dehydrogenase activities are modified [75]. In another report, 17β-hydroxysteroid dehydrogenase subtypes 2, 4, and 5 were up-regulated in prostatic cell lines treated with 1,25-(OH)$_2$D$_3$ [76]. Interestingly, aromatase enzymatic activity was enhanced by 1,25-(OH)$_2$D$_3$ in prostate cancer cell lines suggesting synthesis of estradiol from testosterone, whereas 5α-reductase was not modified [77]. On the other hand, 1,25-(OH)$_2$D$_3$ apparently inhibited androgen glucuronidation and thus androgen inactivation [78], while it stimulated androgen receptor expression [79]. Quantification of CYP27B1 mRNA [80] and of enzymatic activity in prostate cancer compared with normal cells suggested deficiency during progression [35], which would result in reduced dependence on 25-(OH)D$_3$ for growth control.

1.2.4 Regulation of CYP27B1 and of CYP24A1 Expression by Splicing Mechanisms and Polymorphisms

Alternative gene splicing affects up to 70% of human genes and enhances genetic diversity by generating proteins with distinct new functions. In line with many cytochrome P450s, CYP27B1 is known to exhibit alternative splicing and, in kidney cells, this led to modified 1,25-(OH)$_2$D$_3$ synthesis [81]. There have been several reports on differential expression of splice variants for CYP27B1 also in cancerous cells derived from diverse tissues suggesting a role for gene splicing in tissue-specific regulation of 1,25-(OH)$_2$D$_3$ production [82–85]. In MCF-7 mammary cells, and several subclones of this cell line, at least six splice variants of CYP27B1 were detected resulting in at least six protein variants present in Western blots at varying band intensity [85]. It is yet unknown whether some of these splice variants present during breast tumor progression lack 1α-hydroxylation activity.

Splice variants of CYP24A1 could lead to abnormal vitamin D catabolism respectively reduced or enhanced 1,25-(OH)$_2$D$_3$ accumulation (see, e.g., [86, 128]). In prostate tumor-derived cell lines, constitutive CYP24A1 was expressed as a splice variant in some cells, whereas others had CYP24A1 splice variants after

treatment with 1,25-$(OH)_2D_3$ only [87]. In colon tumors, a CYP24A1 splice variant at 754 bp was much more prominent in differentiated (G1) tumors than in undifferentiated ones [25]. In colon cells derived from a G2 tumor, the normal CYP24A1 band as well as the variant were present with similar intensity, but the variant was not found in differentiated Caco-2 cells. This particular splice variant also disappeared after treatment with 1,25-$(OH)_2D_3$ [45].

Studies of genetic polymorphisms with respect to vitamin D hydroxylases are rare. In colon cancer patients, genetic variants of several markers, among them the VDR, were investigated to explore associations with microsatellite instability (MSI) or the CpG Island methylator phenotype (CIMP). Fok1 VDR polymorphism was associated with CIMP-positive tumors [88]. When investigating prostate tumors in a group of Caucasian and African American patients, several non-coding SNPs were identified in the CYP27B1 gene. However, these SNPs probably do not enhance susceptibility to tumors since they were found also in an unaffected control group [89]. Novel SNPs were detected in the human CYP24A1 promoter that did result in reduced expression of CYP24A1. This variant was found primarily in the African American population [90]. Since this population group is recognized to suffer from vitamin D insufficiency and to present with prostate tumors more frequently than Caucasian Americans, a relevance of this variant for protection against tumor incidence by the vitamin D system appears questionable.

1.2.5 *Epigenetic Regulation of CYP27B1 and of CYP24A1 Expression*

DNA methylation of cytosine residues of CpG islands in the promoter region of genes is associated with transcriptional silencing of gene expression in mammalian cells, while decreased methylation of CpG islands enhances gene activity. The CpG island methylator phenotype (CIMP) is a distinct phenotype in sporadic colorectal cancer. For instance, a CIMP-high status is significantly associated with tumors of the proximal colon. Also relative survival can be associated with methylation status [91]. While these studies certainly are not definitive yet, it seems unlikely that methylation/demethylation processes in general can be associated with colon tumor incidence; though CIMP status coupled with other information such as microsatellite instability could be used as a prognostic factor. However, methylation/demethylation processes concerning promoters of certain genes may predispose to, or protect against, sporadic malignancies.

In the normal colon, methylation is age- and also apparently site-related. When evaluating the promoter region of the estrogen receptor (ESR1), it was found to be more highly methylated (inactivated) in the human distal than in the proximal colon [72]. Since estrogen apparently enhances 1,25-$(OH)_2D_3$ synthesis in mucosal cells, this suggests that in women the distal colon is less protected by vitamin D against tumor incidence (see Sect. 1.2.3.).

1 Vitamin D: Synthesis and Catabolism

Other genes modified by epigenetic events could be those coding for the vitamin D system. Kim et al. [92] demonstrated that the negative response element in the CYP27B1 promoter is regulated by the ligand-activated vitamin D receptor through recruitment of histone deacetylase, a critical step for chromatin structure remodeling in suppression of the CYP27B1 gene. In addition, this transrepression by VDR requires DNA methylation in the CYP27B1 gene promoter. However, this study was done in kidney cells and not in tumor-derived cells. Another study highlighted the relevance of different microenvironments (tumor versus normal) for the regulation of CYP24A1: CYP24A1 promoter hypermethylation was present in endothelial cells derived from tumors, but not from normal tissue [93].

In a mouse model of chemically induced colon cancer, protection against tumor incidence by estrogen was associated with decreased CpG island methylation of the VDR promoter and enhanced VDR expression [94]. When we tested colon cancer cell lines derived from moderately differentiated G2 tumors (COGA-1 cells) and from undifferentiated G3 tumors (COGA-13 cells) for expression of vitamin D hydroxylases and compared results with the differentiated colon cancer cell line Caco-2, it became evident that Caco-2 cells had high levels of CYP27B1 mRNA, while COGA-1 and COGA-13 had low expression or none. In contrast, constitutive CYP24A1 expression was extremely high in COGA-13, and not apparent in COGA-1 and Caco-2 cells (Fig. 1.1). Addition of the methyltransferase inhibitor 5-aza-2'-deoxycytidine induced CYP24A1 mRNA expression significantly in Caco-2 and also in COGA-1 cells. In COGA-13 cells, however, the methyltransferase inhibitor did not further raise the already high basal CYP24A1 expression. Interestingly, CYP27B1 appeared to be under epigenetic control as well, since COGA-1 and COGA-13 cells showed a distinct elevation of CYP27B1 mRNA after treatment with 5-aza-2'-deoxycytidine (Fig. 1.1) (Khorchide et al., manuscript in preparation).

Differences in expression of vitamin D hydroxylases in the course of tumor progression as observed in colon cancer patients [41, 54] could be caused by epigenetic regulation of gene activity via methylation/demethylation processes as well as histone acetylation/deacetylation. In low-grade cancerous lesions, CYP27B1 expression is exceedingly high compared to normal mucosa in non-cancer patients [26].

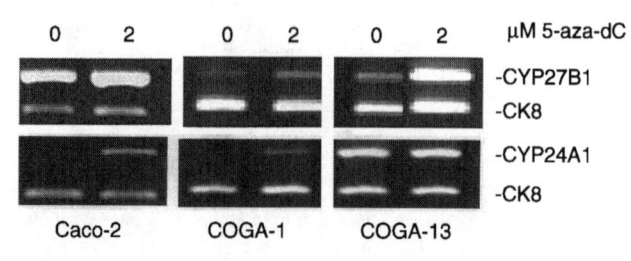

Fig. 1.1 Evaluation of CYP27B1 and CYP24A1 mRNA expression by RT-PCR in colon cancer cells. Cells were treated for 3 days with 2 μM 5-aza-deoxycytidine treatment. Caco-2, differentiated cells; COGA-1, established from a moderately differentiated tumor; COGA-13, established from an undifferentiated tumor. Reference mRNA was cytokeratin 8 (CK8)

Enhanced synthesis and accumulation of 1,25-$(OH)_2D_3$ in the colon mucosa would be responsible for up-regulation of transcriptional activity of CYP24A1 [34] and also for autocrine/paracrine inhibition of tumor cell growth. We suggest that this enhanced expression of CYP27B1 could be due, at least in part, to epigenetic regulation, i.e., demethylation, while raised CYP24A1 expression probably results from the normal regulatory loop following accumulation of 1,25-$(OH)_2D_3$ in colonic mucosa. However, in highly malignant tumors, an efficient antimitogenic effect by 1,25-$(OH)_2D_3$ is unlikely, because expression of the catabolic vitamin D hydroxylase by far exceeds that of CYP27B1. Our hypothesis, therefore, is that during cancer progression CYP27B1 would be inactivated by epigenetic mechanisms, whereas that of CYP24A1 would be activated. To test this, we studied expression of vitamin D hydroxylases in 105 colon tumor patients entering a Viennese hospital for tumor resection. Uncoupling of CYP24A1 expression from regulation by colonic 1,25-$(OH)_2D_3$ would lead to vitamin D hydroxylase expression in opposite directions during progression to a highly malignant state. This is actually the case: Transition from low- to high-grade cancers is associated with a further highly significant rise in CYP24A1 mRNA expression and a simultaneous decline of CYP27B1 activity (Fig. 1.2). Analysis of a selected (small) number of tumor biopsies

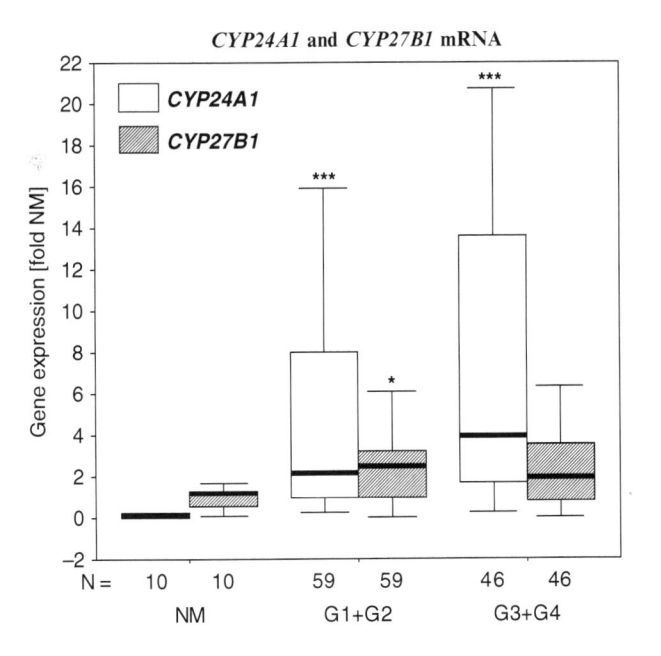

Fig. 1.2 CYP24A1 and CYP27B1 mRNA expression in 105 colon cancer patients. $n=59$ patients with G1/G2 (highly to moderately differentiated) tumors; $n=46$ patients with G3/G4 (low and undifferentiated) tumors. Cancer patient data were compared with those derived from non-cancer (NM) patients (tissue from stoma reoperations after diverticulitis surgery) $n=10$. Densitometric data of tumor patients were expressed in fold increase compared to NM. Significant differences are expressed as: $*p \leq 0.05$; $***p \leq 0.001$

1   Vitamin D: Synthesis and Catabolism   13

suggested that in poorly differentiated cancerous lesions, regions of the CYP24A1 promoter were demethylated and those of CYP27B1 were methylated (Khorchide et al., manuscript in preparation).

In prostate cells, Khorchide et al. [95] demonstrated that human normal prostate cells possess CYP27B1 expression, but are devoid of CYP24A1, whereas DU-145 prostate cancer cells display high CYP24A1 and very low CYP27B1 mRNA expression. Treatment with the methylation inhibitor 5-aza-2'-deoxycytidine together with the histone deacetylation (HDAC) inhibitor trichostatin A, elevated both CYP27B1 as well as CYP24A1 mRNA expression in the normal cell line. In DU-145 cells, 5-aza-2'-deoxycytidine plus trichostatin A elevated CYP27B1 mRNA and, importantly, also its activity as measured by HPLC [95]. Another HDAC inhibitor, SAHA, induced CYP27B1 mRNA expression in prostate cells as well, however at the high dose of 15 µM only [96]. In contrast, Banwell et al. were able to demonstrate that vitamin D-insensitive prostate and breast cells when treated with $1,25$-$(OH)_2D_3$ together with nanomolar doses of HDAC inhibitors, were growth-inhibited synergistically. They suggest that insensitivity to vitamin D could be due to epigenetic mechanisms involving the VDR [97].

1.3   Regulation of CYP27B1 and of CYP24A1 Expression by Nutrition

The colorectum, as part of the digestive system, clearly is particularly affected by nutritional components. Therefore, this section will address nutrient regulation of vitamin D hydroxylases primarily in colorectal malignancies. However, there is some indication that also prostate as well as mammary cancer cells might be affected, though mechanistic evidence for this is more difficult to obtain.

It is clear that, for prevention of sporadic malignancies, average 25-(OH)D$_3$ levels at or above at least 40 nM need to be achieved in the general population, though there is still some discussion about the exact amount. However, optimization of extrarenal production of $1,25$-$(OH)_2D_3$ is essential as well. Experimental proof is accumulating that nutrient factors such as calcium, phytoestrogens, and folate could regulate expression of vitamin D hydroxylases.

1.3.1   Regulation of Vitamin D Metabolism in the Gut Mucosa by Calcium

It is intriguing that vitamin D in combination with high intake of calcium from dietary sources or supplements, apparently is much more effective in reducing the risk of colorectal cancer than when given alone [98–100]. To investigate this further, we availed ourselves of a mouse model. Feeding male and female mice an AIN76 minimal diet containing 0.04% calcium led to enhanced positivity for

PCNA (proliferating cell nuclear antigen) and for cyclin D1, while that for p21, a cyclin-dependent kinase inhibitor, was diminished. Mice on a calcium-deficient diet also expressed CYP24A1 mRNA at a six- to eightfold higher level than their counterparts on a 0.9% calcium diet [27]. Interestingly, CYP27B1 mRNA was significantly up-regulated in animals on 0.04% compared to 0.9% calcium as well, though in female mice only [129]. Importantly, measurement of 1,25-(OH) D_3 concentrations in mucosal homogenates by a newly developed assay [127] indicated that up-regulation of CYP27B1 by low calcium is translated into increased CYP27B1 protein activity causing accumulation of 1,25-(OH)D_3 in colonic mucosal cells. In parallel, in these cells apoptotic pathways, i.e., expression of the downstream effector proteases, caspase-3 and of caspase 7, are stimulated. This strongly suggests that enhanced synthesis of 1,25-(OH)D_3 in females overrides the gender-independent stimulatory effect of low calcium on CYP24A1-mediated vitamin D catabolism, thereby providing protection against incipient hyperproliferation induced by inadequate calcium nutrition. This enhanced synthesizing activity occurred in the proximal colon only and suggests that there may be site-specific action of 17β-estradiol. As mentioned previously, the estrogen receptor ESR1 is more methylated (inactivated) in the human distal than in the proximal colon [72] (see also Sect. 1.2.3).

At present it is not clear whether signals from low luminal calcium are transduced by the calcium sensing receptor (CaR). Alternatively, a lack of calcium is known to increase concentrations of free bile acids in the gut lumen. Of these, lithocholic acid by binding to the VDR can induce expression of CYP24A1 [101]. Our results suggest that in humans also calcium supplementation could lower the risk of colorectal cancer because high dietary calcium suppresses vitamin D catabolism and this would favor accumulation of 1,25-(OH)D_3 in the colon mucosa. Furthermore, 1,25-(OH)D_3 would increase expression of the CaR by binding to a vitamin D responsive element in its promoter region [102].

1.3.2 Regulation of the Vitamin D System by Phytoestrogens

It can be inferred that in human colonocytes, estrogenic compounds have positive effects on endogenous synthesis of 1,25-(OH)$_2$D$_3$ and consequently on VDR-mediated anti-inflammatory and antimitogenic actions (see Sect. 1.2.3). In this context, it is of interest that in East Asian populations the risk of cancers of sex hormone-responsive organs, viz., breast and prostate gland, as well as of the colorectum is clearly lower than elsewhere. This has been traced to the typical diet in this part of the world, which is rich in soy products and therefore contains high amounts of phytoestrogens. Of these, genistein induced CYP27B1 and reduced CYP24A1 expression and activity in a mouse model and in human colon adenocarcinoma-derived cell lines [103], while daidzein, another phytoestrogen prominent in soy and, importantly, its metabolite equol, which is strongly active in other biological systems, did not affect any of the colonic vitamin D hydroxylases [70].

1   Vitamin D: Synthesis and Catabolism   15

Genistein could also have anti-inflammatory properties in the colon: When mice were fed 0.04% dietary calcium, COX-2 mRNA and protein were increased two-fold in the female colon mucosa and to a lesser extent in males. Supplementation of genistein to the diet lowered COX-2 expression to control levels (0.5% dietary calcium) in both genders [104]. This suggests that genistein could have a beneficial effect on colonic inflammation similar to that seen with 17β-estradiol in the human pilot study described before (Sect. 1.2.3). Since genistein preferentially activates ER-β [105, 106], which is equally expressed in the colon of women and men, low rates of colorectal cancer incidence in both genders in soy-consuming populations could be due to appropriate modulation of the anti-inflammatory and anticancer potential of vitamin D by phytoestrogens.

Also the human prostate is frequently affected by inflammatory disease, which could predispose to development of malignancies. Since the inflammation-related prostaglandin pathway is negatively affected in prostate cancer cells by genistein [107], this suggests a potential mechanism of prostate cancer prevention in soy-consuming countries. Experimental data from Farhan et al. indicated that genistein very efficiently reduced the activity of CYP24A1 in human prostate cancer cells [57, 108], probably by direct binding to the CYP24A1 protein [58]. In contrast to the colon, genistein inhibited CYP27B1 mRNA expression in prostate cells, and this may involve histone deacetylation since trichostatin A rescues CYP27B1 from transcriptional inactivation [58] (see also [95]). Treatment of prostate cancer cells with $1,25\text{-}(OH)_2D_3$ together with genistein potentiated the antimitotic activity of the active metabolite. This suggests an increased half-life of $1,25\text{-}(OH)_2D_3$ due to inhibition of CYP24A1 activity [109], as already indicated in previous studies [58].

1.3.3   Effect of Folate on CYP24A1 Expression

Folate, a water-soluble vitamin of the B family, is essential for synthesis, repair, and methylation of DNA. As a methyl donor, folate could play an important role in epigenetic regulation of gene expression. While folic acid was supplemented to foods in the USA in the late 1990s to curb incidence of neural tube defects, and blood folate concentrations increased in the survey period shortly thereafter, there has been a decline since and its causes are unknown [110].

Sporadic cancers evolve over a lifetime and could therefore be at least equally affected by low folic acid intake as neural tube development. Older age and inadequate folate intake lead to altered methylation patterns [111]. Evidence is increasing that a low folate status predisposes to development of several common malignancies including colorectal cancer [112]. Giovannucci et al. [113] and others demonstrated that prolonged intake of folate above currently recommended levels significantly reduced the risk of colorectal cancer.

To investigate the relevance of folate for regulation of the vitamin D system, we used C57/BL6 mice on the semisynthetic AIN76A diet, which contained, among others, 5% fat, 0.025 µg/g vitamin D_3, 5 mg/g calcium, and 2 µg/g folic acid [114, 115].

When this basal diet was modified to contain high fat, low calcium, low vitamin D_3, and low folic acid, mice exhibited signs of hyperplasia and hyperproliferation in the colon mucosa [115], which were accompanied by a more than 2.5-fold elevated CYP24A1 mRNA expression [116]. When calcium and vitamin D_3 in the diet were optimized while fat was still high and folic acid low, CYP24A1 mRNA expression fell by 50%, but was still higher than in the colon mucosa of mice fed the basal (control) diet. Finally, when the diet contained high fat, low calcium, and low vitamin D, but folic acid content was optimized, only then any increment in colonic CYP24A1 due to dietary manipulations was completely abolished [116].

1.4 Can Regulation of Vitamin D Hydroxylases Be Implemented for Therapy?

The high levels of $1,25\text{-}(OH)_2D_3$ respectively of its analogs initially used for cancer therapy invariably caused hypercalcemia. However, it was observed that doses of the active metabolite could be reduced without loss of activity when given as combination therapy.

$1,25\text{-}(OH)_2D_3$ and vitamin D analogs can enhance, either synergistically or additively, the antitumor activities of several classes of antineoplastic agents (see, e.g., [117–119]). This has led to several clinical studies with drugs such as docetaxel in combination with $1,25\text{-}(OH)_2D_3$ in the treatment of androgen-independent prostate cancer, though mechanisms of action are poorly understood yet. It was observed that the antimitotic action of $1,25\text{-}(OH)_2D_3$ associated with G0/G1 arrest, enhanced apoptosis, and differentiation could be achieved with lower concentrations of vitamin D substances when they were given to patients in combination therapy with cytotoxic agents such as carboplatin and taxanes. Even an intermittent $1,25\text{-}(OH)_2D_3$ schedule was possible in this treatment regimen. It was also attempted to use ketoconazole, an unspecific cytochrome P450 inhibitor, for combination treatment. Very low doses of $1,25\text{-}(OH)_2D_3$ could be used under such conditions since degradation of vitamin D was attenuated [120]. Recently it was demonstrated that antineoplastic agents themselves can target CYP24A1 for degradation by decreasing stability of CYP24A1 mRNA. When kidney cells positive for CYP27B1 were treated with $25\text{-}(OH)D_3$, they synthesized $1,25\text{-}(OH)_2D_3$ as expected. Treatment with daunorubicin, etoposide, and vincristine caused enhanced accumulation of $1,25\text{-}(OH)_2D_3$. While CYP27B1 mRNA expression was not altered by cytotoxic drug treatment, that of CYP24A1 was reduced highly significantly [121]. Since mitogen-activated protein (MAP) kinases play an important role in mediating the stimulatory effect of $1,25\text{-}(OH)_2D_3$ on CYP24A1 expression [122], and antineoplastic agents apparently stimulate activity of MAP kinases [123], this seems a likely mechanism of action.

Enhancing apoptotic activity of malignant cells could be another approach to cancer patient therapy. Pretreatment with a high dose of $1,25\text{-}(OH)_2D_3$ augmented the antitumor activity of docetaxel, which manifested itself by an increased

1 Vitamin D: Synthesis and Catabolism   17

population of apoptotic cells, raised Bax (a pro-apoptotic protein), and also reduced expression of a multidrug resistance-associated protein [124]. In an animal model for squamous cell carcinoma a combination of only 10 nM $1,25\text{-}(OH)_2D_3$ together with cisplatin resulted in greater caspase-3 activation than either substance given alone. It was suggested that increased cytotoxicity resulting from a $1,25\text{-}(OH)_2D_3$/ cisplatin treatment could be due to raised $1,25\text{-}(OH)_2D_3$-induced apoptotic signaling through the MEKK-1 pathway [118]. Also the anti-EGFR drug cetuximab applied together with $1,25\text{-}(OH)_2D_3$ seems to provide increased cell cycle arrest and apoptosis in prostate cancer cell cultures [125].

Another valid approach to cancer therapy with $1,25\text{-}(OH)_2D_3$ would be the use of vitamin D analogs to block CYP24A1 activity directly. A 24-phenylsulfone analog of vitamin D raised CYP24A1 mRNA expression in colon, prostate, and mammary cancer cells, but inhibited its activity very rapidly in a dose-dependent manner. This analog apparently binds to the VDR to stimulate transactivation, but also directly interacts with and inhibits CYP24A1 protein [126].

These few examples suggest that there are various options for the use of vitamin D for patient therapy. Most approaches are concerned with reducing activity of the catabolic hydroxylase CYP24A1. This is based on the hypothesis that reduced degradation of the active metabolite in combination therapy will allow the use of much lower concentrations of $1,25\text{-}(OH)_2D_3$.

1.5 Conclusion

It is well-recognized that sporadic malignancies have a multifactorial etiology. While there is strong evidence that serum $25\text{-}(OH)D_3$ levels are inversely related to tumor incidence, there are other factors equally important that will determine the optimal concentration of $1,25\text{-}(OH)_2D_3$ synthesized from the precursor in extrarenal tissues. A person's genetic background with respect to VDR, CYP27B1 and CYP24A1 expression caused by specific splicing mechanisms and polymorphisms will determine production in kidney as well as in extrarenal cells. Growth factors and sex hormones regulate expression of vitamin D hydroxylases and of the VDR in several tissues known to be affected by sporadic cancers. Hyperproliferative cells early during tumor progression may express CYP27B1 strongly as a defense against progression, resulting in enhanced apoptosis and reduced mitosis. High concentrations of $1,25\text{-}(OH)_2D_3$ in such tissues will invariably result in raised expression of the catabolic hydroxylase and this necessitates the use of potent CYP24A1 inhibitors to maintain tissue levels of the active metabolite. This highlights the need for reliable methods to measure tissue concentrations of $1,25\text{-}(OH)_2D_3$. However, functional analysis of vitamin D metabolism in cancer is complicated by the heterogeneous composition of tumors, not only with respect to cell types but also to biological grade of cells. In at least 50% of G3 undifferentiated colon tumors, expression of CYP24A1 mRNA is extremely high whereas that of CYP27B1 is very low. This is probably because of epigenetic mechanisms and

could be age- and colonic-site-related. Activation of the CYP24A1 gene during progression could potentially be halted by a combination of methyltransferase inhibitors and histone deacetylase inhibitors.

When considering prevention of cancer by vitamin D, we speculate that nutritional folate as a methyl donor for epigenetic control, as well as enhanced consumption of calcium and phytoestrogens could optimize expression of vitamin D hydroxylases during the decades that it takes a sporadic tumor to develop. Maintaining high extrarenal tissue concentrations of $1,25\text{-}(OH)_2D_3$, by whatever means, could prove to be a most effective cancer-preventive approach.

References

1. Jones G, Strugnell SA, DeLuca HF (1998) Current understanding of the molecular actions of vitamin D. Physiol Rev 78(4):1193–1231
2. Adams JS, Singer FR, Gacad MA et al (1985) Isolation and structural identification of 1, 25-dihydroxyvitamin D3 produced by cultured alveolar macrophages in sarcoidosis. J Clin Endocrinol Metab 60(5):960–966
3. Mawer EB, Hayes ME, Heys SE et al (1994) Constitutive synthesis of 1, 25-dihydroxyvitamin D3 by a human small cell lung cancer cell line. J Clin Endocrinol Metab 79(2):554–560
4. Knutson JC, DeLuca HF (1974) 25-Hydroxyvitamin D3-24-hydroxylase. Subcellular location and properties. Biochemistry 13(7):1543–1548
5. Ghazarian JG, Jefcoate CR, Knutson JC, Orme-Johnson WH, DeLuca HF (1974) Mitochondrial cytochrome p450. A component of chick kidney 25-hydrocholecalciferol-1alpha-hydroxylase. J Biol Chem 249(10):3026–3033
6. Jones G, Ramshaw H, Zhang A et al (1999) Expression and activity of vitamin D-metabolizing cytochrome P450s (CYP1alpha and CYP24) in human nonsmall cell lung carcinomas. Endocrinology 140(7):3303–3310
7. Stumpf WE, Sar M, Reid FA, Tanaka Y, DeLuca HF (1979) Target cells for 1, 25-dihydroxyvitamin D3 in intestinal tract, stomach, kidney, skin, pituitary, and parathyroid. Science 206(4423):1188–1190
8. Zehnder D, Bland R, Williams MC et al (2001) Extrarenal expression of 25-hydroxyvitamin d(3)-1 alpha-hydroxylase. J Clin Endocrinol Metab 86(2):888–894
9. MacLaughlin J, Holick MF (1985) Aging decreases the capacity of human skin to produce vitamin D3. J Clin Invest 76(4):1536–1538
10. Holick MF (2004) Vitamin D: importance in the prevention of cancers, type 1 diabetes, heart disease, and osteoporosis. Am J Clin Nutr 79(3):362–371
11. Hewison M, Burke F, Evans KN, et al (2007) Extra-renal 25-hydroxyvitamin D3–1alpha-hydroxylase in human health and disease. J Steroid Biochem Mol Biol 103(3–5):316–321
12. Tanaka H, Abe E, Miyaura C et al (1982) 1 Alpha, 25-dihydroxycholecalciferol and a human myeloid leukaemia cell line (HL-60). Biochem J 204(3):713–719
13. Lointier P, Wargovich MJ, Saez S, Levin B, Wildrick DM, Boman BM (1987) The role of vitamin D3 in the proliferation of a human colon cancer cell line in vitro. Anticancer Res 7(4B):817–821
14. Cross HS, Pavelka M, Slavik J, Peterlik M (1992) Growth control of human colon cancer cells by vitamin D and calcium in vitro. J Natl Cancer Inst 84(17):1355–1357
15. Eisman JA, Koga M, Sutherland RL, Barkla DH, Tutton PJ (1989) 1, 25-Dihydroxyvitamin D3 and the regulation of human cancer cell replication. Proc Soc Exp Biol Med 191(3):221–226

1   Vitamin D: Synthesis and Catabolism

16. Pols HA, Birkenhager JC, Foekens JA, van Leeuwen JP (1990) Vitamin D: a modulator of cell proliferation and differentiation. J Steroid Biochem Mol Biol 37(6):873–876
17. Saez S, Falette N, Guillot C, Meggouh F, Lefebvre MF, Crepin M (1993) William L. McGuire Memorial Symposium. 1,25(OH)2D3 modulation of mammary tumor cell growth in vitro and in vivo. Breast Cancer Res Treat 27(1–2):69–81
18. Peehl DM, Skowronski RJ, Leung GK, Wong ST, Stamey TA, Feldman D (1994) Antiproliferative effects of 1, 25-dihydroxyvitamin D3 on primary cultures of human prostatic cells. Cancer Res 54(3):805–810
19. Adams JS, Clemens TL, Parrish JA, Holick MF (1982) Vitamin-D synthesis and metabolism after ultraviolet irradiation of normal and vitamin-D-deficient subjects. N Engl J Med 306(12):722–725
20. Garland CF, Garland FC (1980) Do sunlight and vitamin D reduce the likelihood of colon cancer? Int J Epidemiol 9(3):227–231
21. Garland C, Shekelle RB, Barrett-Connor E, Criqui MH, Rossof AH, Paul O (1985) Dietary vitamin D and calcium and risk of colorectal cancer: a 19-year prospective study in men. Lancet 1(8424):307–309
22. Peterlik M, Cross HS (2005) Vitamin D and calcium deficits predispose for multiple chronic diseases. Eur J Clin Invest 35(5):290–304
23. Deeb KK, Trump DL, Johnson CS (2007) Vitamin D signalling pathways in cancer: potential for anticancer therapeutics. Nat Rev Cancer 7(9):684–700
24. Cross HS, Peterlik M, Reddy GS, Schuster I (1997) Vitamin D metabolism in human colon adenocarcinoma-derived Caco-2 cells: expression of 25-hydroxyvitamin D3-1alpha-hydroxylase activity and regulation of side-chain metabolism. J Steroid Biochem Mol Biol 62(1):21–28
25. Bareis P, Bises G, Bischof MG, Cross HS, Peterlik M (2001) 25-hydroxy-vitamin d metabolism in human colon cancer cells during tumor progression. Biochem Biophys Res Commun 285(4):1012–1017
26. Bises G, Kallay E, Weiland T et al (2004) 25-hydroxyvitamin D3-1alpha-hydroxylase expression in normal and malignant human colon. J Histochem Cytochem 52(7):985–989
27. Kallay E, Bises G, Bajna E et al (2005) Colon-specific regulation of vitamin D hydroxylases – a possible approach for tumor prevention. Carcinogenesis 26(9):1581–1589
28. Young MV, Schwartz GG, Wang L et al (2004) The prostate 25-hydroxyvitamin D-1 alpha-hydroxylase is not influenced by parathyroid hormone and calcium: implications for prostate cancer chemoprevention by vitamin D. Carcinogenesis 25(6):967–971
29. Anderson PH, O'Loughlin PD, May BK, Morris HA (2005) Modulation of CYP27B1 and CYP24 mRNA expression in bone is independent of circulating 1, 25(OH)(2)D(3) levels. Bone 36(4):654–662
30. Feskanich D, Ma J, Fuchs CS et al (2004) Plasma vitamin D metabolites and risk of colorectal cancer in women. Cancer Epidemiol Biomarkers Prev 13(9):1502–1508
31. Bertone-Johnson ER, Chen WY, Holick MF et al (2005) Plasma 25-hydroxyvitamin D and 1, 25-dihydroxyvitamin D and risk of breast cancer. Cancer Epidemiol Biomarkers Prev 14(8):1991–1997
32. Tuohimaa P, Tenkanen L, Ahonen M et al (2004) Both high and low levels of blood vitamin D are associated with a higher prostate cancer risk: a longitudinal, nested case-control study in the Nordic countries. Int J Cancer 108(1):104–108
33. Rowling MJ, Kemmis CM, Taffany DA, Welsh J (2006) Megalin-mediated endocytosis of vitamin d binding protein correlates with 25-hydroxycholecalciferol actions in human mammary cells. J Nutr 136(11):2754–2759
34. Lechner D, Kallay E, Cross HS (2006) 1alpha, 25-Dihydroxyvitamin D(3) downregulates CYP27B1 and induces CYP24A1 in colon cells. Mol Cell Endocrinol 263(1–2):55–64
35. Hsu JY, Feldman D, McNeal JE, Peehl DM (2001) Reduced 1alpha-hydroxylase activity in human prostate cancer cells correlates with decreased susceptibility to 25-hydroxyvitamin D3-induced growth inhibition. Cancer Res 61(7):2852–2856

36. Kemmis CM, Welsh J (2008) Mammary epithelial cell transformation is associated with deregulation of the vitamin D pathway. J Cell Biochem 105(4):980–988

37. Turunen MM, Dunlop TW, Carlberg C, Vaisanen S (2007) Selective use of multiple vitamin D response elements underlies the 1 alpha, 25-dihydroxyvitamin D3-mediated negative regulation of the human CYP27B1 gene. Nucleic Acids Res 35(8):2734–2747

38. Lechner D, Kallay E, Cross HS (2007) 1alpha, 25-dihydroxyvitamin D3 downregulates CYP27B1 and induces CYP24A1 in colon cells. Mol Cell Endocrinol 263(1–2):55–64

39. Kallay E, Pietschmann P, Toyokuni S et al (2001) Characterization of a vitamin D receptor knockout mouse as a model of colorectal hyperproliferation and DNA damage. Carcinogenesis 22(9):1429–1435

40. Liu N, Nguyen L, Chun RF et al (2008) Altered endocrine and autocrine metabolism of vitamin D in a mouse model of gastrointestinal inflammation. Endocrinology 149(10):4799–4808

41. Cross HS, Bareis P, Hofer H et al (2001) 25-Hydroxyvitamin D(3)-1alpha-hydroxylase and vitamin D receptor gene expression in human colonic mucosa is elevated during early cancerogenesis. Steroids 66(3–5):287–292

42. Cross HS, Bajna E, Bises G et al. (eds) (1996) Vitamin D receptor and cytokeratin expression may be progression indicators in human colon cancer. Anticancer Res 16:2333–2337

43. Sheinin Y, Kaserer K, Wrba F et al (2000) In situ mRNA hybridization analysis and immunolocalization of the vitamin D receptor in normal and carcinomatous human colonic mucosa: relation to epidermal growth factor receptor expression. Virchows Arch 437(5):501–507

44. Tong WM, Kallay E, Hofer H et al (1998) Growth regulation of human colon cancer cells by epidermal growth factor and 1, 25-dihydroxyvitamin D3 is mediated by mutual modulation of receptor expression. Eur J Cancer 34(13):2119–2125

45. Bareis P, Kallay E, Bischof MG et al (2002) Clonal differences in expression of 25-hydroxyvitamin D(3)-1alpha- hydroxylase, of 25-hydroxyvitamin D(3)-24-hydroxylase, and of the vitamin D receptor in human colon carcinoma cells: effects of epidermal growth factor and 1alpha,25-dihydroxyvitamin D(3). Exp Cell Res 276(2):320–327

46. Palmer HG, Gonzalez-Sancho JM, Espada J et al (2001) Vitamin D(3) promotes the differentiation of colon carcinoma cells by the induction of E-cadherin and the inhibition of beta-catenin signaling. J Cell Biol 154(2):369–387

47. Pena C, Garcia JM, Silva J et al (2005) E-cadherin and vitamin D receptor regulation by SNAIL and ZEB1 in colon cancer: clinicopathological correlations. Hum Mol Genet 14(22):3361–3370

48. Kemmis CM, Salvador SM, Smith KM, Welsh J (2006) Human mammary epithelial cells express CYP27B1 and are growth inhibited by 25-hydroxyvitamin D-3, the major circulating form of vitamin D-3. J Nutr 136(4):887–892

49. Barreto AM, Schwartz GG, Woodruff R, Cramer SD (2000) 25-Hydroxyvitamin D3, the prohormone of 1, 25-dihydroxyvitamin D3, inhibits the proliferation of primary prostatic epithelial cells. Cancer Epidemiol Biomarkers Prev 9(3):265–270

50. Wang L, Flanagan JN, Whitlatch LW, Jamieson DP, Holick MF, Chen TC (2004) Regulation of 25-hydroxyvitamin D-1alpha-hydroxylase by epidermal growth factor in prostate cells. J Steroid Biochem Mol Biol 89–90(1–5):127–130

51. Nonn L, Peng L, Feldman D, Peehl DM (2006) Inhibition of p38 by vitamin D reduces interleukin-6 production in normal prostate cells via mitogen-activated protein kinase phosphatase 5: implications for prostate cancer prevention by vitamin D. Cancer Res 66(8):4516–4524

52. Brozek W, Bises G, Fabjani G, Cross HS, Peterlik M (2008) Clone-specific expression, transcriptional regulation, and action of interleukin-6 in human colon carcinoma cells. BMC Cancer 8:13

53. Fernandez-Pol JA (2008) Modulation of EGF receptor protooncogene expression by growth factors and hormones in human breast carcinoma cells. Crit Rev Oncog x2(2):173–185

1 Vitamin D: Synthesis and Catabolism

54. Cross H, Bises G, Lechner D, Manhardt T, Kallay E (2005) The Vitamin D endocrine system of the gut - Its possible role in colorectal cancer prevention. J Steroid Biochem Mol Biol 97:121–128
55. Miller GJ, Stapleton GE, Hedlund TE, Moffat KA (1995) Vitamin D receptor expression, 24-hydroxylase activity, and inhibition of growth by 1alpha, 25-dihydroxyvitamin D3 in seven human prostatic carcinoma cell lines. Clin Cancer Res 1(9):997–1003
56. Ly LH, Zhao XY, Holloway L, Feldman D (1999) Liarozole acts synergistically with 1alpha, 25-dihydroxyvitamin D3 to inhibit growth of DU 145 human prostate cancer cells by blocking 24-hydroxylase activity. Endocrinology 140(5):2071–2076
57. Farhan H, Wahala K, Adlercreutz H, Cross HS (2002) Isoflavonoids inhibit catabolism of vitamin D in prostate cancer cells. J Chromatogr B Anal Technol Biomed Life Sci 777(1–2):261–268
58. Farhan H, Wahala K, Cross HS (2003) Genistein inhibits vitamin D hydroxylases CYP24 and CYP27B1 expression in prostate cells. J Steroid Biochem Mol Biol 84(4):423–429
59. Calle EE, Miracle-McMahill HL, Thun MJ, Heath CW Jr (1995) Estrogen replacement therapy and risk of fatal colon cancer in a prospective cohort of postmenopausal women. J Natl Cancer Inst 87(7):517–523
60. DeCosse JJ, Ngoi SS, Jacobson JS, Cennerazzo WJ (1993) Gender and colorectal cancer. Eur J Cancer Prev 2(2):105–115
61. Jemal A, Murray T, Samuels A, Ghafoor A, Ward E, Thun MJ (2003) Cancer statistics, 2003. CA Cancer J Clin 53(1):5–26
62. Grodstein F, Newcomb PA, Stampfer MJ (1999) Postmenopausal hormone therapy and the risk of colorectal cancer: a review and meta-analysis. Am J Med 106(5):574–582
63. Chlebowski RT, Wactawski-Wende J, Ritenbaugh C et al (2004) Estrogen plus progestin and colorectal cancer in postmenopausal women. N Engl J Med 350(10):991–1004
64. Foley EF, Jazaeri AA, Shupnik MA, Jazaeri O, Rice LW (2000) Selective loss of estrogen receptor beta in malignant human colon. Cancer Res 60(2):245–248
65. Nelson LR, Bulun SE (2001) Estrogen production and action. J Am Acad Dermatol 45 (3 Suppl):S116–S124
66. Lindberg MK, Moverare S, Skrtic S et al (2003) Estrogen receptor (ER)-beta reduces ERalpha-regulated gene transcription, supporting a "ying yang" relationship between ERalpha and ERbeta in mice. Mol Endocrinol 17(2):203–208
67. Campbell-Thompson M, Lynch IJ, Bhardwaj B (2001) Expression of estrogen receptor (ER) subtypes and ERbeta isoforms in colon cancer. Cancer Res 61(2):632–640
68. Buchanan JR, Santen R, Cauffman S, Cavaliere A, Greer RB, Demers LM (1986) The effect of endogenous estrogen fluctuation on metabolism of 25-hydroxyvitamin D. Calcif Tissue Int 39(3):139–144
69. Liel Y, Shany S, Smirnoff P, Schwartz B (1999) Estrogen increases 1, 25-dihydroxyvitamin D receptors expression and bioresponse in the rat duodenal mucosa. Endocrinology 140(1): 280–285
70. Lechner D, Bajna E, Adlercreutz H, Cross HS (2006) Genistein and 17beta-estradiol, but not equol, regulate vitamin D synthesis in human colon and breast cancer cells. Anticancer Res 26(4A):2597–2603
71. CH PP, Hopkins ME, Kallay E, Bises G, Dreyhaupt E, Augenlicht L, Lipkin M, Lesser M, Livote E, Holt PR (2009) Chemoprevention of colorectal neoplasia by estrogen: potential role of vitamin D. Cancer Prev Res 2(1):43–51
72. Horii J, Hiraoka S, Kato J et al (2008) Age-related methylation in normal colon mucosa differs between the proximal and distal colon in patients who underwent colonoscopy. Clin Biochem 41(18):1440–1448
73. Grant WB (2002) An ecologic study of dietary and solar ultraviolet-B links to breast carcinoma mortality rates. Cancer 94(1):272–281
74. Friedrich M, Diesing D, Cordes T et al (2006) Analysis of 25-hydroxyvitamin D3-1alpha-hydroxylase in normal and malignant breast tissue. Anticancer Res 26(4A):2615–2620

75. Soronen P, Laiti M, Torn S et al (2004) Sex steroid hormone metabolism and prostate cancer. J Steroid Biochem Mol Biol 92(4):281–286
76. Wang JH, Tuohimaa P (2007) Regulation of 17beta-hydroxysteroid dehydrogenase type 2, type 4 and type 5 by calcitriol, LXR agonist and 5alpha-dihydrotestosterone in human prostate cancer cells. J Steroid Biochem Mol Biol 107(1–2):100–105
77. Lou YR, Murtola T, Tuohimaa P (2005) Regulation of aromatase and 5alpha-reductase by 25-hydroxyvitamin D(3), 1alpha, 25-dihydroxyvitamin D(3), dexamethasone and progesterone in prostate cancer cells. J Steroid Biochem Mol Biol 94(1–3):151–157
78. Kaeding J, Belanger J, Caron P, Verreault M, Belanger A, Barbier O (2008) Calcitrol (1alpha, 25-dihydroxyvitamin D3) inhibits androgen glucuronidation in prostate cancer cells. Mol Cancer Ther 7(2):380–390
79. Zhao XY, Ly LH, Peehl DM, Feldman D (1999) Induction of androgen receptor by 1alpha, 25-dihydroxyvitamin D3 and 9-cis retinoic acid in LNCaP human prostate cancer cells. Endocrinology 140(3):1205–1212
80. Ma JF, Nonn L, Campbell MJ, Hewison M, Feldman D, Peehl DM (2004) Mechanisms of decreased Vitamin D 1alpha-hydroxylase activity in prostate cancer cells. Mol Cell Endocrinol 221(1–2):67–74
81. Wu S, Ren S, Nguyen L, Adams JS, Hewison M (2007) Splice variants of the CYP27b1 gene and the regulation of 1, 25-dihydroxyvitamin D3 production. Endocrinology 148(7):3410–3418
82. Diesel B, Radermacher J, Bureik M et al (2005) Vitamin D(3) metabolism in human glioblastoma multiforme: functionality of CYP27B1 splice variants, metabolism of calcidiol, and effect of calcitriol. Clin Cancer Res 11(15):5370–5380
83. Maas RM, Reus K, Diesel B et al (2001) Amplification and expression of splice variants of the gene encoding the P450 cytochrome 25-hydroxyvitamin D(3) 1, alpha-hydroxylase (CYP 27B1) in human malignant glioma. Clin Cancer Res 7(4):868–875
84. Radermacher J, Diesel B, Seifert M et al (2006) Expression analysis of CYP27B1 in tumor biopsies and cell cultures. Anticancer Res 26(4A):2683–2686
85. Fischer D, Seifert M, Becker S et al (2007) 25-Hydroxyvitamin D3 1alpha-hydroxylase splice variants in breast cell lines MCF-7 and MCF-10. Cancer Genom Proteom 4(4):295–300
86. Ren S, Nguyen L, Wu S, Encinas C, Adams JS, Hewison M (2005) Alternative splicing of vitamin D-24-hydroxylase: a novel mechanism for the regulation of extrarenal 1, 25-dihydroxyvitamin D synthesis. J Biol Chem 280(21):20604–20611
87. Muindi JR, Nganga A, Engler KL, Coignet LJ, Johnson CS, Trump DL (2007) CYP24 splicing variants are associated with different patterns of constitutive and calcitriol-inducible CYP24 activity in human prostate cancer cell lines. J Steroid Biochem Mol Biol 103(3–5):334–337
88. Slattery ML, Wolff RK, Curtin K et al (2009) Colon tumor mutations and epigenetic changes associated with genetic polymorphism: Insight into disease pathways. Mutat Res 660(1–2):12–21
89. Hawkins GA, Cramer SD, Zheng SL et al (2002) Sequence variants in the human 25-hydroxyvitamin D3 1-alpha-hydroxylase (CYP27B1) gene are not associated with prostate cancer risk. Prostate 53(3):175–178
90. Roff A, Wilson RT (2008) A novel SNP in a vitamin D response element of the CYP24A1 promoter reduces protein binding, transactivation, and gene expression. J Steroid Biochem Mol Biol 112(1–3):47–54
91. Barault L, Charon-Barra C, Jooste V et al (2008) Hypermethylator phenotype in sporadic colon cancer: study on a population-based series of 582 cases. Cancer Res 68(20):8541–8546
92. Kim MS, Fujiki R, Kitagawa H, Kato S (2007) 1alpha, 25(OH)2D3-induced DNA methylation suppresses the human CYP27B1 gene. Mol Cell Endocrinol 265–266:168–173
93. Chung I, Karpf AR, Muindi JR et al (2007) Epigenetic silencing of CYP24 in tumor-derived endothelial cells contributes to selective growth inhibition by calcitriol. J Biol Chem 282(12):8704–8714

1 Vitamin D: Synthesis and Catabolism

94. Smirnoff P, Liel Y, Gnainsky J, Shany S, Schwartz B (1999) The protective effect of estrogen against chemically induced murine colon carcinogenesis is associated with decreased CpG island methylation and increased mRNA and protein expression of the colonic vitamin D receptor. Oncol Res 11(6):255–264

95. Khorchide M, Lechner D, Cross HS (2005) Epigenetic regulation of vitamin D hydroxylase expression and activity in normal and malignant human prostate cells. J Steroid Biochem Mol Biol 93(2–5):167–172

96. Wang L, Persons KS, Jamieson D et al (2008) Prostate 25-hydroxyvitamin D-1alpha-hydroxylase is up-regulated by suberoylanilide hydroxamic acid (SAHA), a histone deacetylase inhibitor. Anticancer Res 28(4A):2009–2013

97. Banwell CM, Singh R, Stewart PM, Uskokovic MR, Campbell MJ (2003) Antiproliferative signalling by 1, 25(OH)2D3 in prostate and breast cancer is suppressed by a mechanism involving histone deacetylation. Recent Results Cancer Res 164:83–98

98. Cho E, Smith-Warner SA, Spiegelman D et al (2004) Dairy foods, calcium, and colorectal cancer: a pooled analysis of 10 cohort studies. J Natl Cancer Inst 96(13):1015–1022

99. Grau MV, Baron JA, Sandler RS et al (2003) Vitamin D, calcium supplementation, and colorectal adenomas: results of a randomized trial. J Natl Cancer Inst 95(23):1765–1771

100. Holt PR, Bresalier RS, Ma CK et al (2006) Calcium plus vitamin D alters preneoplastic features of colorectal adenomas and rectal mucosa. Cancer 106(2):287–296

101. Nehring JA, Zierold C, DeLuca HF (2007) Lithocholic acid can carry out in vivo functions of vitamin D. Proc Natl Acad Sci USA 104(24):10006–10009

102. Canaff L, Hendy GN (2002) Human calcium-sensing receptor gene. Vitamin D response elements in promoters P1 and P2 confer transcriptional responsiveness to 1,25-dihydroxyvitamin D. J Biol Chem 277(33):30337–30350

103. Cross HS, Kallay E, Lechner D, Gerdenitsch W, Adlercreutz H, Armbrecht HJ (2004) Phytoestrogens and vitamin D metabolism: a new concept for the prevention and therapy of colorectal, prostate, and mammary carcinomas. J Nutr 134(5):1207S–1212S

104. Bises G, Bajna E, Manhardt T, Gerdenitsch W, Kallay E, Cross HS (2007) Gender-specific modulation of markers for premalignancy by nutritional soy and calcium in the mouse colon. J Nutr 137(1):211S–215S

105. An J, Tzagarakis-Foster C, Scharschmidt TC, Lomri N, Leitman DC (2001) Estrogen receptor beta-selective transcriptional activity and recruitment of coregulators by phytoestrogens. J Biol Chem 276(21):17808–17814

106. Kuiper GG, Lemmen JG, Carlsson B et al (1998) Interaction of estrogenic chemicals and phytoestrogens with estrogen receptor beta. Endocrinology 139(10):4252–4263

107. Swami S, Krishnan AV, Moreno J, Bhattacharyya RB, Peehl DM, Feldman D (2007) Calcitriol and genistein actions to inhibit the prostaglandin pathway: potential combination therapy to treat prostate cancer. J Nutr 137(1 Suppl):205S–210S

108. Cross HS, Kallay E, Farhan H, Weiland T, Manhardt T (2003) Regulation of extrarenal vitamin D metabolism as a tool for colon and prostate cancer prevention. Recent Results Cancer Res 164:413–425

109. Swami S, Krishnan AV, Peehl DM, Feldman D (2005) Genistein potentiates the growth inhibitory effects of 1, 25-dihydroxyvitamin D3 in DU145 human prostate cancer cells: role of the direct inhibition of CYP24 enzyme activity. Mol Cell Endocrinol 241(1–2):49–61

110. Pfeiffer CM, Johnson CL, Jain RB et al (2007) Trends in blood folate and vitamin B-12 concentrations in the United States, 1988–2004. Am J Clin Nutr 86(3):718–727

111. Keyes MK, Jang H, Mason JB et al (2007) Older age and dietary folate are determinants of genomic and p16-specific DNA methylation in mouse colon. J Nutr 137(7):1713–1717

112. Choi SW, Mason JB (2000) Folate and carcinogenesis: an integrated scheme. J Nutr 130(2):129–132

113. Giovannucci E, Stampfer MJ, Colditz GA et al (1998) Multivitamin use, folate, and colon cancer in women in the Nurses' Health Study. Ann Intern Med 129(7):517–524

114. Newmark HL, Lipkin M (1992) Calcium, vitamin D, and colon cancer. Cancer Res 52 (7 Suppl):2067s–2070s
115. Newmark HL, Yang K, Lipkin M et al (2001) A Western-style diet induces benign and malignant neoplasms in the colon of normal C57Bl/6 mice. Carcinogenesis 22(11):1871–1875
116. Cross HS, Lipkin M, Kallay E (2006) Nutrients regulate the colonic vitamin D system in mice: relevance for human colon malignancy. J Nutr 136(3):561–564
117. Beer TM (2005) ASCENT: the androgen-independent prostate cancer study of calcitriol enhancing taxotere. BJU Int 96(4):508–513
118. Hershberger PA, McGuire TF, Yu WD et al (2002) Cisplatin potentiates 1, 25-dihydroxyvitamin D3-induced apoptosis in association with increased mitogen-activated protein kinase kinase kinase 1 (MEKK-1) expression. Mol Cancer Ther 1(10):821–829
119. Trump DL, Hershberger PA, Bernardi RJ et al (2004) Anti-tumor activity of calcitriol: pre-clinical and clinical studies. J Steroid Biochem Mol Biol 89–90(1–5):519–526
120. Trump DL, Muindi J, Fakih M, Yu WD, Johnson CS (2006) Vitamin D compounds: clinical development as cancer therapy and prevention agents. Anticancer Res 26(4A):2551–2556
121. Tan J, Dwivedi PP, Anderson P et al (2007) Antineoplastic agents target the 25-hydroxyvitamin D3 24-hydroxylase messenger RNA for degradation: implications in anticancer activity. Mol Cancer Ther 6(12 Pt 1):3131–3138
122. Dwivedi PP, Hii CS, Ferrante A et al (2002) Role of MAP kinases in the 1,25-dihydroxyvitamin D3-induced transactivation of the rat cytochrome P450C24 (CYP24) promoter. Specific functions for ERK1/ERK2 and ERK5. J Biol Chem 277(33):29643–29653
123. Tang D, Wu D, Hirao A et al (2002) ERK activation mediates cell cycle arrest and apoptosis after DNA damage independently of p53. J Biol Chem 277(15):12710–12717
124. Ting HJ, Hsu J, Bao BY, Lee YF (2007) Docetaxel-induced growth inhibition and apoptosis in androgen independent prostate cancer cells are enhanced by 1alpha, 25-dihydroxyvitamin D3. Cancer Lett 247(1):122–129
125. Belochitski O, Ariad S, Shany S, Fridman V, Gavrilov V (2007) Efficient dual treatment of the hormone-refractory prostate cancer cell line DU145 with cetuximab and 1, 25-dihydroxyvitamin D3. In Vivo 21(2):371–376
126. Lechner D, Manhardt T, Bajna E, Posner GH, Cross HS (2007) A 24-phenylsulfone analog of vitamin D inhibits 1alpha, 25-dihydroxyvitamin D(3) degradation in vitamin D metabolism-competent cells. J Pharmacol Exp Ther 320(3):1119–1126
127. Nittke T, Kallay E, Manhardt T, Cross HS (2009) Parallel elevation of colonic 1,25-dihydroxyvitamin D3 levels and apoptosis in female mice on a calcium-deficient diet. Anticancer Res 29:3727–32
128. Horvath HC, Khabir Z, Nittke T, Gruber S, Speer G, Manhardt T, Bonner E, Kallay E (2010) CYP24A1 splice variants-implications for the antitumorigenic actions of 1,25-(OH)2D3 in colorectal cancer. J Steroid Biochem Mol Biol 121:76–9
129. Nittke T, Selig S, Kallay E, Cross HS (2008) Nutritional calcium modulates colonic expression of vitamin D receptor and pregnane X receptor target genes. Mol Nutr Food Res 52:S45–51

Chapter 2
The Molecular Cancer Biology of the VDR

James Thorne and Moray J. Campbell

Abstract The development of an understanding of the role the vitamin D receptor (VDR) endocrine system plays to regulate serum calcium levels began approximately three centuries ago with the first formal descriptions of rickets. The parallel appreciation of a role for the VDR in cancer biology began approximately 3 decades ago and subsequently a remarkable increase has occurred in the understanding of its actions in normal and malignant systems.

Principally, much of this understanding has focused on understanding the extent and mechanism by which the VDR influences expression of multiple proteins whose combined actions are to govern cell cycle progression, induce differentiation, and contribute to the regulation of programmed cell death, perhaps in response to loss of genomic integrity. Predominantly, although not exclusively, these increases in target proteins reflect the transcriptional control exerted via the VDR. Reflecting the expanding understanding of how chromatin architecture is sensed and altered by transcription factors, the actions of the VDR have been defined through the large transcriptional complexes it is found in. The diversity of these complexes is large, and presumably underpins the pleiotropic biological actions that the VDR is associated with. The VDR is neither mutated nor deleted in malignancy but instead polymorphic variation distorts its ability to function, as indeed does expression of a number of associated cofactors, thereby skewing the ability to transactivate target genes.

Exploitation of this understanding into cancer therapeutic settings may occur through several routes, but perhaps a more systems orientated approach may yield insight by identifying and modeling points where the VDR, and closely related nuclear receptors, exert the most dominant control over cellular processes such as cell cycle control.

M.J. Campbell (✉)
Department of Pharmacology & Therapeutics,
Roswell Park Cancer Institute,
Elm & Carlton Streets, Buffalo, NY 14263, USA
e-mail: Moray.Campbell@RoswellPark.org

D.L. Trump and C.S. Johnson (eds.), *Vitamin D and Cancer*,
DOI 10.1007/978-1-4419-7188-3_2, © Springer Science+Business Media, LLC 2011

Abbreviations

AR	Androgen receptor
bHLH	Bacis helix loop helix
9 cRA	9 *cis* retinoic acid
$1\alpha,25(OH)_2D_3$	$1\alpha,25$DihydroxyvitaminD$_3$
DREAM	Downstream regulatory element antagonist modulator
ER	Estrogen receptor
FXR	Farnesoid X-activated receptor
HDAC	Histone deacetylase
HDACi	Histone deacetylase inhibitor
HSP	Heat shock protein
LCOR	Ligand-dependent nuclear receptor corepressor
LCA	Lithocholic acid
LXR	Liver X receptor
NCOR1	Nuclear receptor corepressor 1
NCOR2/SMRT	Silencing mediator of retinoid and thyroid hormone receptors/Nuclear receptor corepressor 2
NR	Nuclear receptor
PPAR	Peroxisome proliferator activated receptor
RAR	Retinoic acid receptor
RXR	Retinoid X receptor
SLIRP	SRA stem loop-interacting RNA-binding protein
SRC	Steroid receptor coactivator
TRIP2/DRIP205	Thyroid hormone receptor interactor 2
TRIP15/COPS2/Alien	Thyroid hormone receptor interactor 15
VDR	Vitamin D receptor

2.1 Choreography of VDR Signaling

2.1.1 General Findings for VDR Transcriptional Actions

$1\alpha,25(OH)_2D_3$ and its precursor $25(OH)D_3$, in common with most NR ligands, are highly hydrophobic and transported in the aqueous blood stream associated with a specific binding protein (DBP) [1–3]. At the cell membrane they are free to diffuse across the lipid membrane, although the identification of Megalin as an active transport protein for $25(OH)D_3$ suggests that transport into the cell of vitamin D$_3$ metabolites may be more tightly regulated than merely by passive diffusion alone [4]. Once in the cells of the target organ, $1\alpha,25(OH)_2D_3$ associates with the VDR.

In the absence of ligand, the VDR may be distributed throughout the cell, although predominantly located in the nucleus. There is evidence of cytoplasmic expression and cell-membrane-associated VDR that may mediate non-genomic

signal transduction responses [5, 6]. This is a feature of several NRs, such as the ERα, where the NR is cycled through caveolae at the cell membrane to initiate signal transduction pathways [6, 7]. The contribution of these actions to the overall functions of $1\alpha,25(OH)_2D_3$ remains to be clarified fully. Interestingly, there is also evidence for the VDR to be actively trafficked into the nucleus upon ligand activation, in tandem with the heterodimeric partner RXRs [8], each in association with specific importins [9].

The majority of findings to date have addressed a nuclear function for the VDR associated with transcription. Structurally, the VDR is uncommon, compared to other NRs (NRs), as it does not contain an activation domain at its amino terminus (AF1). In most other receptors, this is an important domain for activation, for example, for autonomous ligand-independent AF function domain. The VDR instead relies on a domain in the carboxy terminus (AF-2) for activation and other domains for heterodimerization with RXR [10]. The VDR ligand-binding pocket contains hydrophobic residues such as His-305 and -397 that are important in the binding of $1\alpha,25(OH)_2D_3$. Ligand binding specifically requires interaction of the hydroxyl group of the A ring at carbon 1 of $1\alpha,25(OH)_2D_3$, which is added by the action of the 1α hydroxylase enzyme. The binding of ligand causes an LBD conformational change, which allows the C-terminal helix 12 of the AF2 domain to reposition into an active conformation, exposing a docking surface for transcriptional co-regulators [11–13]. This switch of conformation of the LBD in the presence of ligand is a common feature in all ligand-binding NRs, as is the capacity to undergo receptor–cofactor interactions. Thus, both the unliganded and liganded VDR associates with a large number of different proteins involved with transcriptional suppression and activation, respectively.

When located within the nucleus and in the absence of ligand, the VDR exist in an "apo" state associated with RXR and corepressors (e.g., NCOR1 and NCOR2/SMRT) [14, 15] as part of large complexes (~2.0 MDa) [14, 16] and bound to RE sequences. These complexes in turn actively recruit a range of enzymes that posttranslationally modify histone tails, for example, histone deacetylases (HDACs) and methyltransferases, and thereby maintain a locally condensed chromatin structure around response element sequences [17–20]. Ligand binding induces a so-called *holo* state, facilitating the association of the VDR-RXR dimer with coactivator complexes. A large number of interacting coactivator proteins have been described, which can be divided into multiple families including the p160 family, the non-p160 members, and members of the large "bridging" TRAP/DRIP/ARC complex, which links the receptor complex to the co-integrators CBP/p300 and basal transcriptional machinery [21, 22].

The complex choreography of these events has recently emerged from the study of the VDR [17, 23–28] and other NRs [29–32], and involves cyclical rounds of promoter-specific complex assembly, gene transactivation, complex disassembly, and proteosome-mediated receptor degradation coincident with corepressor binding and silencing of transcription. This gives rise to the characteristic periodicity of NR transcriptional activation and pulsatile mRNA and protein accumulation. However, the periodicity of VDR-induced mRNA accumulation of target genes is not shared, but

rather tends toward patterns that are specific for individual target genes and suggests that promoter-specific complexes combine to determine the precise periodicity [23, 24].

2.1.2 VDR Signal Specificity

Historically, researchers have tended to consider transcription factor actions in a somewhat monochrome view, for example, as illustrated for MYC and AP-1. These views are currently being revised in the light of surveys of genome binding sites and dissection of biological actions in a broader context (for example, reviewed in [33, 34]). These findings suggest that the functions of a given transcription factor superfamily are distilled through interaction with multiple cellular processes such that the normal capacity represents an extremely flexible and integrated signaling module. In malignancy, however, these transcriptional choices and phenotypic outputs generally become restricted [35].

The diversity of VDR expression sites, being detected in virtually all cells of a human, and the disparate phenotypic effects, from regulating calcium transport to sensing redox potential and DNA damage, also suggests that the cell specificity of actions may be distilled in a cell-type-specific manner. Therefore, the questions emerge as to what governs the temporal regulation of VDR-dependent transcritpomes, among different cell types. Recent findings suggest that a high level of specificity of the timing and choice of VDR cofactor interactions may provide a mechanistic basis for signaling specificity. Combined expression and choice of interacting cofactors yield a high degree of NR transcriptional plasticity over choice, and timing of gene regulation [32, 36, 37].

Of the principal corepressors, it remains to be established to what extent specificity and redundancy occur. The expression, localization, and isoforms of NCOR1 and NCOR2/SMRT corepressors strongly influence the spatio-temporal equilibrium between repressing and activating NR complexes and transcriptional outputs [38]. The specificity of these corepressor interactions is beginning to emerge. Ncor1 and Ncor2/Smrt knockouts are embryonically lethal, whereas stem cell components from these mice and conditional approaches are revealing tissue-specific interactions [39–41] with distinct interacting domains being used to distinguish NR recognition [42]. Equally, the list continues to grow of novel corepressor proteins that the VDR interacts with.

Compared to the relatively massive size of the corepressors NCOR1 and NCOR2/SMRT, a number of smaller molecules have emerged as showing corepressor function. TRIP15/COPS2/Alien has been demonstrated to interact with the VDR and act as a corepressor, in an AF-2 independent manner that may not require the same interactions with HDACs that NCOR1 does [43]. Intriguingly, this protein contributes to the lid sub-complex of the 26S proteasome and thereby potentially links VDR function with the regulation of protein stability [44]. Similarly, SLIRP [45] has also emerged as a repressive factor for the VDR, although to date very little is known about the specificity, in terms of tissue and target gene.

2 The Molecular Cancer Biology of the VDR

Other repressors appear to demonstrate more specific phenotypic specificity. Hairless blocks VDR-mediated differentiation of keratinocytes, whereas addition of $1\alpha,25(OH)_2D_3$ displaces Hairless from the promoter of target genes and recruits coactivators to promote differentiation [46–48]. Similarly, DREAM (downstream regulatory element antagonist modulator) usually binds to direct repeat response elements in the promoters of target genes to enhance transcription in VDR and RAR target genes, in a calcium-dependent manner, and suggests that specificity arises from the interactions of VDR with further tissue-specific cofactors [49].

Finally, the Williams syndrome transcription factor (WSTF), contained within WINAC complex, identified by Kato and colleagues, directly interacts with unliganded VDR and mediates binding to promoter sequences and can then bind and recruit other co-regulatory proteins. WINAC has ATP-dependent chromatin-remodeling activity and contains both SWI/SNF components and DNA replication-related factors. WINAC mediates the recruitment of unliganded VDR to its promoter target sites, and may organize local nucleosomal positioning to allow promoters access to co-regulators. This suggests a novel mechanism in transcriptional regulation, in which VDR binds to gene promoters before ligand is present [50, 51].

A similar level of coactivator specificity is also beginning to emerge. Members of the TRAP/DRIP complex were identified independently in association with the VDR and other NRs including the GR [52, 53] and TR [54–56]. The exact specificity of many of the co-regulatory factors remains to be established fully, although there are some suggestions that certain co-activators are VDR-specific, for example, NCoA-62 [57]. Similarly, knockout of TRAP220, which has multiple NR interacting domains, has begun to reveal distinct interactions, and notably disrupts the ability of the VDR to regulate hematopoietic differentiation [58, 59]. In keeping with the skin being a critical target for VDR actions, the specificity of VDR interactions with cofactor complexes has been dissected in detail by Bikle and colleagues who have demonstrated the timing and extent of coactivator binding, and established a role for SRC3 during specific stages of keratinocyte differentiation [60, 61].

Aside from the established co-regulators, some chaperone proteins have been reported to be regulators of VDR-mediated transcription. HSP70 down-regulates VDR to repress transcription [62], whereas BAG1L, an HSP70 binding protein, has been shown to bind to the VDR, and enhances VDR-mediated transcription [63]. Similarly, p23 and HSP90 have been shown to release the VDR/coactivator complex from the promoter of target genes in the presence of $1\alpha,25(OH)_2D_3$ [64]. The association of these HSPs suggests a natural cross-talk with other NRs, such as the AR, that associate with these chaperones in the cytoplasm.

Posttranslational modifications (PTM) possibly confer further VDR specificity of function. PTMs resulting from signal transduction processes, for example, bring about phosphorylation, acetylation, and ubiquitinylation events on the AR [65]. The VDR has been less extensively studied, but crucial roles have emerged for the phosphorylation of serine and threonine residues [66]. Subsequently, several residues have been identified that appear to regulate DNA binding and cofactor recruitment. The zinc finger DNA-binding domain is located at the N terminal of the VDR and

adjacent to this domain is the Serine 51 residue. This residue appears crucial for ligand-induced and phosphorylation-dependent transcriptional activation by the VDR. When Ser51 is mutated, phosphorylation of the VDR, by PKC at least, is all but completely abolished and its transcriptional activity is markedly reduced [67]. It is intriguing that the crucial site of PKC activity is located so close to the DNA-binding domain, but whether there are allosteric or biochemical changes that alter the ability of the VDR to bind DNA remains to be elucidated.

The common NR partner RXR can also be phosphorylated and as a result alters recruitment of cofactors to its *holo*-complexes. Ser260 is located within the ligand-binding domain of the RXR and appears crucial for mediating cofactor binding and ligand-induced transcriptional responses. When phosphorylated, Ser260 allows binding between the RXR and VDR, but presumably through allosteric changes to the complex, limits the recruitment of cofactors to the complex [68].

The recruitment of cofactors to the VDR *holo*-complex also appears to be regulated further by the presence of PTMs, for example, kinase CK-II. The phospho-mimic mutant VDRS208D does not increase or decrease VDR–DNA, VDR–RXR, or VDR–SRC interactions but it does increase the levels of VDR–DRIP205 complexes present. CK-II which specifically phosphorylates Ser208 enhances $1,25(OH)_2D_3$-induced transactivation of VDR targets [69, 70]. In addition, phosphatase inhibitors (okadoic acid) in combination with $1,25(OH)_2D_3$ shifts the cofactor preference from NCOA2/GRIP-1 to TRIP2/DRIP205 [71]. Taken together, these data suggest that the TRIP2/DRIP205 coactivator complex enhances the transcriptional response by VDR and is recruited by CK-II dependent phosphorylation of the VDR at Ser208.

2.1.3 Vitamin D Response Elements

A further level of specificity may arise from the specificity of binding sequence contained within the REs sequences of genomic targets. Simple REs are formed by two recognition motives and their relative distance and orientation contributes to receptor-binding specificity. Thus, the first identified VDRE was the DR3 – an imperfect hexameric direct repeat sequence AGTTCA with a spacer of three nucleotides. In the DR3 configuration, RXR, the heterodimer partner is believed to occupy the upstream half-site and VDR the downstream motif with two half-sites spaced by three nucleotides. Other types of VDREs have since been identified. One such VDRE is a palindromic sequence with a nine base-pair nucleotide spacer (IR9). This sequence was identified in the human calbindin D9K gene and like most VDREs the VDR/RXR binds this sequence in a 5′-RXR-VDR-3′ polarity (reviewed in [72]). More recently, a novel everted repeat sequence with a six base-pair nucleotide spacer (ER6) has been identified in the gene for *CYP3A4* (an enzyme important in xenobiotic metabolism) in addition to the DR3 already known to be present in this gene [73]. An inverted repeat with no spacer (IR0) has also been identified in the *SULT2A1* gene [74].

Similarly, the ability of VDR to display transrepression, that is, ligand-dependent transcriptional repression has received significant interest and reflects emerging themes for other NRs, for example, PPARs [75, 76], and highlights further the hitherto unsuspected flexibility of the VDR to associate with a diverse array of protein factors to adapt function [77, 78]. For example, analysis of the avian PTH gene has revealed a ligand-dependent repression of this gene by VDR [79]. The element mediating this effect was identified as a DR3, and since it resulted in transcriptional repression, the motif was referred to as a negative nVDRE. A similar nVDRE has been identified in the human kidney in the *CYP27b1* gene [80]. Interestingly, the VDR does not bind directly to this sequence; binding has been shown to be mediated by an intermediary factor known as a bHLH-type transcription factor, VDR interacting repressor (VDIR). It has since been shown that liganded VDR binds to the VDIR and indirectly causes repression through HDAC mechanisms [77].

More recently, larger and integrated responsive regions have been identified, suggesting a yet more intricate control involving integration with other transcription factors, for example, p53 and C/EBPα as demonstrated on the promoter/enhancer regions of *CDKN1A* and *SULT2A1*, respectively [23, 81]. Thus, the combinatorial actions of the VDR with other TFs most likely go some way toward explaining the apparent diversity of VDR biological actions. Again, for other NRs (e.g., AR and ERα), more dominant transcription factors, so-called pioneer factors, appear to be highly influential in determining choice and magnitude of transcriptional actions [82]. Recently, C/EBP family members have been demonstrated to act in a similar cooperative manner with the related PPARγ [36] and it remains to be established to what extent the VDR interacts similarly with other transcription factors. The above findings are suggestive of a similar mechanism.

Efforts to understand VDR function have at their basis the antagonism between these *apo* and *holo* receptor complexes and the ability of these complexes to sense and regulate a diverse range of histone modifications. Histone modifications at the level of meta-chromatin architecture appear to form a stable and heritable "histone code," such as in X chromosome inactivation (reviewed in [83]). The extent to which similar processes operate to govern the activity of micro-chromatin contexts, such as gene promoter regions, is an area of debate [84, 85]. The *apo* and *holo* NR complexes initiate specific and coordinated histone modifications [86, 87] to govern transcriptional responsiveness of the promoter. There is good evidence that specific histone modifications also determine the assembly of transcription factors on the promoter, and control individual promoter transcriptional responsiveness [88–90]. It is less clear to what extent complexes containing NRs in general, and VDR specifically, recognize basal histone modifications on target gene promoters; functional studies of the SANT motif contained in the corepressor NCoR2/SMRT support this latter idea [91]. This is a complex and rapidly evolving area and the reader is referred to an excellent recent review [75].

Collectively, these findings support the concept that the VDRs transcriptional actions reflect a convergence of multiple complexes, the details of which are still emerging and reflect the cross-talk, both cooperatively and antagonistically

between different cellular-signaling systems. Furthermore, the arena for VDR actions and interplay extends beyond the nucleus and integrates levels of cytoplasmic signal transduction, genomic and epigenomic regulation. Establishing the specificity of function and selectivity of VDR interactions has to an extent been limited by technical approaches. Unbiased approaches are now required to dissect VDR interactions (in the membrane, cytoplasm, and nucleus) in either individual cells or very pure populations, thereby to generate a comprehensive understanding of the spatial temporal network of its interactions.

2.2 Integrated VDR Actions

2.2.1 Lessons from Murine Models

The VDR plays a well-established endocrine role in the regulation of calcium homeostasis by regulating calcium absorption in the gut and kidney, and bone mineralization. $1\alpha,25(OH)_2D_3$ status is dependent upon cutaneous synthesis initiated by solar radiation and also on dietary intake – a reduction of either one or both sources leads to insufficiency, although UV-initiated cutaneous $1\alpha,25(OH)_2D_3$ synthesis is the principal route in a vitamin D-sufficient individual. The importance of the relationships between solar exposure and the ability to capture UV-mediated energy is underscored by the inverse correlation between human skin pigmentation and latitude. That is, the individual capacity to generate vitamin D_3 in response to solar UV exposure is intimately associated with forebear environmental adaptation. The correct and sufficient level of solar exposure and serum vitamin D_3 are matters of considerable debate. Current recommendations for daily vitamin D_3 intake are in the range of 400–800 IU/day [92]. More recently, reassessment of the $1\alpha,25(OH)_2D_3$ impact on the prevention of osteoporosis has suggested that the correct level may be as high as 2–3,000 IU/day, which may reflect more accurately "ancestral" serum levels [93].

The importance of the relationship between UV exposure and calcium homeostasis has driven the endocrine view of $1\alpha, 25(OH)_2D_3$ synthesis and signaling. In parallel, local generation of $1\alpha, 25(OH)_2D_3$ in target tissues has become apparent and supported a separate autocrine role to regulate cell proliferation and differentiation, and other functions including the modulation of immune responses.

Key insights into these functions have been gained in *Vdr*-deficient mice [94–96]. The Vdr is expressed widely during murine embryonic development in tissues involved in calcium homeostasis and bone development. Vdr disruption results in a profound phenotype in these models, which is principally observed post-weaning and is associated with the alteration of duodenal calcium absorption and bone mineralization, resulting in hypocalcemia, secondary hyperparathyroidism, osteomalacia, rickets, impaired bone formation, and elevated serum levels of $1\alpha,25(OH)_2D_3$. In parallel, a range of more subtle effects are seen more clearly when the animals are rescued with dietary calcium supplementation

2   The Molecular Cancer Biology of the VDR   33

and may represent autocrine and non-calcemic actions. The animals became
growth-retarded, display alopecia, uterine hypoplasia, impaired ovarian follicu-
logenesis, reproductive dysfunction, cardiac hypertrophy, and enhanced
thrombogenicity.

2.2.2   Self-renewing Epithelial Systems

The sporadic, temporal acquisition of a cancer phenotype is compatible with
models of disruption of the self-renewal of epithelial tissues. It has become
increasingly clear that breast, colon, and prostate tissues, in common with other
epithelial tissues and many other cell types in the adult human, are self-renewing
and contain committed stem cell components [97–102].These stem cells are
slowly proliferating and are able to undergo asymmetric divisions to give rise
both to other stem cells and transiently amplifying (TA) populations of progenitor
cells, that in turn give rise to the differentiated cell types, which typify the functions of these tissues and are subsequently lost through programmed cell death
processes and replaced by newly differentiated TA cells. The mechanisms that
control the intricate balance of these processes of division, differentiation, and
programmed cell death are subjects of significant investigations. These studies
have revealed common roles for Wnt and hedgehog signaling and the actions of
other signal transduction processes that govern cell cycle progression, with gene
targets such as the cyclin-dependent kinase inhibitor *CDKN1A* (which encodes
p21$^{(wafl/cip1)}$) emerging as points of criticality upon which numerous signal pathways converge.

Against this backdrop, the Vdr operates in several self-renewing tissues. The
Vdr is readily detected in keratinocytes and co-treatment of calcium and
1α,25(OH)$_2$D$_3$ decreases proliferation and promotes differentiation of cultured
keratinocytes [103]. The Vdr is also detected in outer root sheath and hair follicle
bulb, as well as in the sebaceous glands [104] and the Vdr -/- mice develop hair loss
and ultimately alopecia totalis, associated with large dermal cysts, that is not prevented by the high calcium rescue diet. The alopecia arises due to a complete failure to initiate anagen, which is the first postnatal hair growth phase. Subsequently,
the hair follicles convert into epidermal cysts [105]. Hair follicle formation requires
highly coordinated signaling between different cell types including contributions
from the stem cells components and therefore the alopecia phenotype has attracted
significant research interest as it may represent a role for the VDR in stem cell
maintenance. Subsequent studies have demonstrated that a failure to maintain hair
follicles in Vdr -/- animals does not actually reflect a loss of follicle stem cells but
rather an inability of the primitive progenitor cells to migrate along the follicle at
the onset of anagen [106].

Interestingly, these effects appear independent of ligand binding, in that they
can be rescued even when Vdr is mutated in the LBD, but not completely if
the AF2 domain is interrupted, suggesting that the association with cofactors is

required [107]. Notably, the corepressor, Hairless plays a clear role in hair formation with either knockout or mutation resulting in alopecia strikingly similar to that observed in the Vdr null mice [108, 109].

Wnt signaling is one of the major processes regulating postmorphogenic hair follicle development. Interestingly, the development of dermal cysts and increase in sebaceous glands observed in the Vdr and Hairless -/- mice are also similar to mice expressing a keratinocyte-specific disruption to β-catenin [110, 111]. These findings have raised the possibility that one function of the Vdr may be to co-regulate aspects of Wnt signaling, a concept that is supported further by the physical association of VDR in a complex with β-catenin and other Wnt components [112].

Another unexpected finding of the Vdr -/- animals was the uterine hypoplasia and impaired ovarian function in the females that leads to dramatically reduced fertility. Similarly to the hair phenotype, this was not restored by the rescue diet of high calcium [94]. Estradiol supplementation, however, of the female mice restored uterine function and fertility and suggests the fault lies with an inability to generate estrogen. The mammary gland has also been studied extensively, in a comprehensive series of experiments by Welsh and coworkers [113, 114] and represents an intriguing tissue where endocrine (calcemic8) and autocrine (antimitotic, pro-differentiative, pro-apoptotic) effects of the VDR appear to converge.

These phenotypes underscore the integrated nature of VDR signaling. That is, the biology of hair regeneration and mammary gland function reflects the choreographed actions of VDR, with other NRs, alongside other regulatory processes including Wnt signaling. Dysfunction of multiple aspects of this is seen in many cancer phenotypes.

2.3 VDR Transcriptional Networks in Malignancy

Defining the mechanisms by which the VDR exerts desirable anticancer effects has been an area of significant investigation since the early 1980s. In 1981, $1\alpha,25(OH)_2D_3$ was shown to inhibit human melanoma cell proliferation significantly in vitro at nanomolar concentrations [115], and was subsequently found to induce differentiation in cultured mouse and human myeloid leukemia cells [116, 117]. Following these studies, anti-proliferative effects have been demonstrated in a wide variety of cancer cell lines, including those from prostate, breast, and colon [118–125]. To identify critical target genes that mediate these actions, comprehensive genome-wide *in silico* and transcriptomic screens have analyzed the anti-proliferative VDR transcriptome and revealed broad consensus on certain targets, but has also highlighted variability [118, 126–128]. This heterogeneity may in part reflect experimental conditions, cell line differences, and genuine tissue-specific differences of cofactor expression that alter the amplitude and periodicity of VDR transcriptional actions.

2   The Molecular Cancer Biology of the VDR   35

2.3.1   Cell Cycle Arrest

A common anti-proliferative VDR function is associated with arrest at G_0/G_1 of the
cell cycle, coupled with upregulation of a number of cell cycle inhibitors including
p21$^{(waf1/cip1)}$ and p27$^{(kip1)}$. Promoter characterization studies have demonstrated a
series of VDREs in the promoter/enhancer region of *CDKN1A* [23, 129]. By contrast, the regulation of the related CDKI p27$^{(kip1)}$ is mechanistically enigmatic,
reflecting both transcriptional and translational regulation such as enhanced mRNA
translation, and attenuating degradative mechanisms [130–133].

The up-regulation of p21$^{(waf1/cip1)}$ and p27$^{(kip1)}$ principally mediate G_1 cell cycle
arrest, but $1\alpha,25(OH)_2D_3$ has been shown to mediate a G_2/M cell cycle arrest in a
number of cancer cell lines via direct induction of *GADD45α* [127, 134, 135].
Again, this regulation appears to combine direct gene transcription and a range of
posttranscriptional mechanisms. These studies highlight the difficulty of establishing strict transcriptional effects of the VDR, as a range of posttranscriptional effects
act in concert to regulate target protein levels. Concomitant with changes in the cell
cycle there is some evidence that $1\alpha,25(OH)_2D_3$ also induces differentiation, most
clearly evidenced in myeloid cell lines, but also supported by other cell types and
most likely reflects the intimate links that exist between the regulation of the G_1
transition, the expression of CDKIs such as p21$^{(waf1/cip1)}$, and the induction of cellular
differentiation [136].

Historically, hematological malignancies combined an ease of interrogation with
robust classification of cellular differentiation capacity which was envied by investigators of solid tumors. It is therefore no coincidence that these cell systems
yielded many important insights for cancer cell biologists generally, such as chro-
mosomal translocations and instability, and the role of committed adult stem cells.

Indeed, the capacity to readily differentiate in response to external and internal
signals has fascinated leukemia researchers as they have sought to understand why
leukemia cells appear to fail at certain stages of differentiation. It is within this
context that in the 1980s, investigators [137, 138] considered a role for the VDR
and the related retinoic acid receptor (RAR) to reactivate dormant differentiation
programs in so-called differentiation therapies. Over the following 2 decades,
researchers began to reveal how these receptors instill mitotic restraint and facilitate
differentiation programs and how discord over the control and integration of these
processes is central to leukemogenesis. Despite these efforts, clinical exploitation
of these receptors has largely proved to be equivocal. The one exception to this
translational failure has been the exploitation of RAR signaling in patients with
acute promyelocytic leukemia. Again, understanding the basic signaling behind this
application proved significant to the developing understanding of epigenetic regulation of transcription and the promise of HDAC inhibitors [139].

Against this backdrop, various groups, including that of Studzinski, have worked
consistently exploring mechanisms of resistance to VDR signaling and methods of
exploitation and recently demonstrated, elegantly, a role for VDR to down-regulate
miR181a, which when left unchecked degrades p27$^{(kip1)}$. Thus, indirectly VDR

activation elevates expression of p27$^{(kip1)}$, initiates cell cycle arrest, and commits cells toward differentiation. Transcriptional control of miRNAs and their biological effects are clearly a field of rapid expansion, and members of the NR superfamily are implicated in their regulation [140, 141]. A role for the VDR to govern the expression of this regulatory miRNA and, importantly, place its role in the well-understood map of differentiation is highly novel.

2.3.2 Sensing DNA Damage

An important and emergent area, both in terms of physiology and therapeutic exploitation, is the role the liganded VDR appears to play in maintaining genomic integrity and facilitating DNA repair. There appears to be close cooperation between VDR actions and the p53 tumor suppressor pathway. The maintenance of genomic fidelity against a backdrop of self-renewal is central to the normal development and adult function of many tissues including the mammary and prostate glands, and the colon. For example, in the mammary gland p53 family members play a role in gland development and maintenance. *P63 -/-* animals have an absence of mammary and other epithelial structures, associated with a failure of lineage commitment (reviewed in [142]), whereas *p53 -/-* animals have delayed mammary gland involution, reflecting the *Vdr -/-* animals, and wider tumor susceptibility (reviewed in [143]).

The overlap between p53 and VDR appears to extend beyond cellular phenotypes. The *VDR* is a common transcriptional target of both p53 and p63 [144, 145] and VDR and p53 share a cohort of direct target genes associated with cell cycle arrest, signal transduction, and programmed cell death including *CDKN1A GADD45A, RB1, PCNA, Bax, IGFBP3, TGFB1/2*, and *EGFR* [23, 128, 135, 146–150]. At the transcriptional level, both VDR heterodimers and p53 tetramers associate, for example, with chromatin remodeling factors CBP/p300 and the SWI/SNF to initiate transactivation [51, 151] By contrast, in the gene repressive state VDR and p53 appear to associate with distinct repressor proteins, for example, p53 with SnoN [152], and VDR with NCOR1, suggesting the possibly association with distinct sets of histone deacetylases. Indeed, *CDKN1A* promoter-dissection studies revealed adjacent p53 and VDR-binding sites, suggesting composite responsive regions [23]. Together, these findings suggest that $1\alpha,25(OH)_2D_3$-replete environments enhance p53 signaling to regulate mitosis negatively.

Similarly the role of $1\alpha,25(OH)_2D_3$ in the skin is also suggestive of its chemo-preventive effects. UV light from sun exposure has several effects in the skin; UVA light induces DNA damage through increasing the level of reactive oxygen species (ROS), but importantly UVB light also catalyzes the conversion of 7-dehydroxycholesterol to 25(OH)-D and induces the expression of VDR.

In addition, antimicrobial and anti-inflammatory genes are another subset of VDR targets that are induced by UV radiation. Suppression of the adaptive inflammatory response is thought to be protective for several reasons. Inflamed tissues

contain more ROS, which in turn can damage DNA and prevent proper function of DNA repair machinery. Also the induction of cytokines and growth factors associated with inflammation act to increase the proliferative potential of the cells. NF-κB is a key mediator of inflammation and the VDR attenuates this process by negatively regulating NF-κB signaling [153]. This control by VDR is underscored by studies showing *Vdr-/-* mice are more sensitive to chemicals that induce inflammation than their wild-type counterparts [154]. The normally protective effect of inflammation that occurs under other conditions is lost through VDR-mediated suppression but is compensated for by the induction of a cohort of antimicrobial and antifungal genes [155–157]. The induction of antimicrobials not only prevents infection in damaged tissue but can be cytotoxic for cells with increased levels of anion phospholipids within their membranes, a common feature of transformed cells [158]. Finally, and most recently, network strategies have been used in different strains of mice with altered sensitivity toward skin cancer. Remarkably, in such unbiased screens, the VDR emerges as a key nodal control point in determining sensitivity toward skin tumors as it regulates both turnover of self-renewal and inflammatory infiltrate [159].

The key question, and central to exploiting any therapeutic potential of this receptor, is why should the VDR exert such pleiotropic actions? One possible explanation for this pleiotropism is that it represents an adaptation of the skin to UV exposure, coupling the paramount importance of initiating $1\alpha,25(OH)_2D_3$ synthesis with protection of cell and tissue integrity. Thus, VDR actions are able to maximize UV-initiated synthesis of $1\alpha,25(OH)_2D_3$ production, whilst controlling the extent of local inflammation that can result from sun exposure. To compensate for the potential loss of protection associated with immunosuppression, the VDR mediates a range of antimicrobial actions. Equally, local genomic protection is ensured through the upregulation of target genes which induce G_0/G_1 arrest, cooperation with p53, and the induction of cell differentiation. It remains a tantalizing possibility that the functional convergence between p53 family and VDR signaling, which arose in the dermis as an evolutionary adaptation to counterbalance the conflicting physiological requirements of vitamin D synthesis and genome protection, are sustained in epithelial systems, such as the lining of the mammary gland, to protect against genotoxic insults derived from either the environment or local inflammation.

2.3.3 Programmed Cell Death

VDR actions, notably in MCF-7 breast cancer cells, are associated with a profound and rapid induction of apoptosis, irrespective of p53 content. This may reflect the VDR role in the involution of the post-lactating mammary gland. The direct transcriptional targets which regulate these actions remain elusive, although there is evidence of an involvement of the *BAX* family of proteins [160, 161]. Induction of programmed cell death following $1\alpha,25(OH)_2D_3$ treatment is also associated with increased ROS generation. $1\alpha,25(OH)_2D_3$ treatment up-regulates *VDUP1* encoding

vitamin D up-regulated protein 1, which binds to the disulfide reducing protein thioredoxin and inhibits its ability to neutralize ROS, thereby potentiating stress-induced apoptosis [162, 163]. In other cells, the apoptotic response is delayed and not so pronounced, and probably reflecting less direct effects. Taken together, these data suggest that extent and timing of apoptotic events depend on the integration of VDR actions with other cell signaling systems. This regulation of apoptosis in human cancer cell lines reflects, of course, the absence of apoptosis in chondrocytes in the Vdr -/- animals [7].

2.4 Mechanisms of Resistance Toward the VDR

A major limitation in the therapeutic exploitation of VDR in cancer therapies is the resistance of cancer cells toward $1\alpha,25(OH)_2D_3$. An understanding of the molecular mechanisms of resistance has emerged.

2.4.1 Reduced Local Availability of $1\alpha,25(OH)_2D_3$

Tumors, such as breast cancer appear to distort the VDR signaling axis locally, with reduced *CYP27b1* mRNA and protein levels, and comparative genome hybridization studies have found that *CYP24* is amplified in human breast cancer [164, 165]. Thus, cancer cells maybe associated with low circulating concentrations of $25(OH)D_3$, arising as a result of reduced exposure to sunlight, altered dietary patterns, and exacerbated further by impaired local generation of $1\alpha,25(OH)_2D_3$. In support of these in vitro findings, a large number of epidemiological studies have identified an association between environments of reduced serum 25(OH)D and cancer rates.

2.4.2 Dominant Signal Transduction Events

In terms of distribution, evidence is emerging that the normally dynamic flux of the VDR becomes altered in more transformed and aggressive cancer cells, becoming restricted to the nucleus [166, 167]. These findings that the normal transport rates, such as importin-mediated processes, become distorted in malignancy and may result in a reduced ability for the VDR to sample the cytoplasm for $1\alpha,25(OH)_2D_3$.

Reflecting the cooperative and integrated nature of the VDR to function as a transcription factor, a number of workers have identified mechanisms by which more dominant signaling process are able either to ablate or attenuate VDR

2   The Molecular Cancer Biology of the VDR   39

signaling. For example, Munoz and coworkers have dissected the interrelationships between the VDR, E-cadherin, and the Wnt signaling pathway in colon cancer cell lines and primary tumors. In these studies, the induction of *CDH1* (encodes E-Cadherin) was seen in subpopulations of SW480 colon cancer cells, which express the VDR and respond to $1\alpha,25(OH)_2D_3$. The VDR thereby limits the transcriptional effects of β-catenin by physically and directly binding it in the nucleus, and by upregulating E-cadherin to sequestrate β-catenin in the cytoplasm. In malignancy, these actions are corrupted through downregulation of *VDR* mRNA, which appears to be a direct consequence of binding by the transcriptional repressor SNAIL; a key regulator of the epithelial-mesenchyme transition, which is overexpressed in colon cancer [168–170]. Equally underscoring the central importance of β-catenin, it has recently been shown to be posttranslationally modified to act as VDR coactivator and supports a model of checks and balances between these two signaling processes [168, 171].

2.4.3   Genetic Resistance

In cancer, and outside of the very limited pool of mutations reported in the VDR in type II rickets, the receptor, generally, is neither mutated nor does it appear to be the subject of cytogenetic abnormalities [172]. By contrast, polymorphic variations of the *VDR* have been widely reported. Thus polymorphisms in the 3′ and 5′ regions of the gene have been described and variously associated with risk of breast, prostate, and colon cancer, although the functional consequences remain to be established clearly. For example, a start codon polymorphism in exon II at the 5′ end of the gene, determined using the *fok*-I restriction enzyme, results in a truncated protein. At the 3′ end of the gene, three polymorphisms have been identified that do not lead to any change in either the transcribed mRNA or the translated protein. The first two sequences generate *Bsm*I and *Apa*I restriction sites and are intronic, lying between exons 8 and 9. The third polymorphism, which generates a *Taq*I restriction site, lies in exon 9 and leads to a silent codon change (from ATT to ATC) which both inserts an isoleucine residue at position 352. These three polymorphisms are linked to a further gene variation, a variable length adenosine sequence within the 3′ untranslated region (3′UTR). The poly(A) sequence varies in length and can be segregated into two groups; long sequences of 18–24 adenosines or short ones [173–176]. The length of the poly(A) tail can determine mRNA stability [177–179] so the polymorphisms resulting in long poly(A) tails may increase the local levels of the VDR protein.

Multiple studies have addressed the association between *VDR* genotype and cancer risk and progression. In breast cancer, the *Apa*I polymorphism shows a significant association with breast cancer risk, as indeed have *Bsm*I and the "L" poly(A) variant. Similarly, the *Apa*I polymorphism is associated with metastases to bone [180, 181]. The functional consequences of the *Bsm*I, *Apa*I, and *Taq*I polymorphisms are unclear, but because of genetic linkage may act as a marker for the

poly(A) sequence within the 3′UTR, which in turn determines transcript stability. Interestingly, combined polymorphisms and serum 25OH-D levels have been shown to compound breast cancer risk and disease severity further [182].

Earlier studies suggested that polymorphisms in the VDR gene might also be associated with risk factor of prostate cancer. Ntais and coworkers performed a meta-analysis of 14 published studies with four common gene polymorphisms (*Taq*1, poly A repeat, *Bsm*1, and *Fok*1) in individuals of European, Asian, and African descent. They concluded that these polymorphisms are unlikely to be major determinants of susceptibility to prostate cancer on a wide population basis [183]. Equally, studies in colon cancer have yet to reveal conclusive relationships and may be dependent upon ethnicity of the population studied.

2.4.4 Epigenetic Resistance

In cancer cells, the lack of an antiproliferative response is reflected by a suppression of the transcriptional responsiveness of anti-proliferative target genes such as *CDKN1A CDKNIB, GADD45A* and *IGFBPs, BRCA1* [120, 135, 184, 185]. Paradoxically, VDR transactivation of other targets is sustained or even enhanced, as measured by induction of the highly $1\alpha25(OH)_2D_3$-inducible *CYP24* gene [186, 187]. Together these data suggest that the lack of functional VDR alone cannot explain resistance and instead the VDR transcriptome is skewed in cancer cells to disfavor anti-proliferative target genes. It has been proposed that this apparent $1\alpha,25(OH)_2D_3$-insensitivity is the result of epigenetic events that selectively suppress the ability of the VDR to transactivate target genes [188].

The epigenetic basis for such transcriptional discrepancies has been investigated intensively in prostate cancer. VDR-resistant prostate cancer cells are associated with elevated levels of NCOR2/SMRT [135, 184]; these data indicate that the ratio of *VDR* to corepressor may be critical to determine $1\alpha,25(OH)_2D_3$ responsiveness in cancer cells. An siRNA approach toward *NCoR2/SMRT* demonstrated a role for this corepressor to regulate this response *GADD45α* in response to $1\alpha,25(OH)_2D_3$. By contrast, knockdown of NCOR1 does not restore anti-proliferative responsiveness toward $1\alpha,25(OH)_2D_3$ but does reactivate transcriptional networks governed by PPARs [189]. These data support a central role for elevated NCOR2/SMRT levels to suppress the induction of key target genes, resulting in loss of sensitivity to the anti-proliferative action of $1\alpha,25(OH)_2D_3$; other workers have reinforced these concepts [190, 191].

Parallel studies have demonstrated a similar spectrum of reduced $1\alpha,25(OH)_2D_3$-responsiveness between nonmalignant breast epithelial cells and breast cancer cell lines. Again, this was not determined solely by a linear relationship between the levels of $1\alpha,25(OH)_2D_3$ and *VDR* expression. Rather, elevated corepressor mRNA levels, notably of *NCoR1*, in ERα negative breast cancer cell lines and primary cultures, were associated with $1\alpha,25(OH)_2D_3$ insensitivity [192]. Elevated NCOR1

2 The Molecular Cancer Biology of the VDR

has also been demonstrated to suppress the VDR responsiveness of bladder cancer cell lines [166], notably toward the VDR ligand lithocholic acid (LCA) [193], suggesting a role for epigenetic disruption of the capacity of cells to sense and metabolize potential genotoxic insults.

The epigenetic lesion rising from elevated NCOR1 can be targeted by co-treatment of either $1\alpha,25(OH)_2D_3$ or its analogs, plus the HDAC inhibitors such as trichostatin A, and can restore the $1\alpha,25(OH)_2D_3$-responses of androgen-independent PC-3 cells to levels indistinguishable from control normal prostate epithelial cells. This reversal of $1\alpha,25(OH)_2D_3$ insensitivity was associated with reexpression of gene targets associated with the control of proliferation and induction of apoptosis, notably *GADD45A* [120, 135, 185]. Similarly, targeting in breast cancer cells through co-treatments of $1\alpha,25(OH)_2D_3$ with HDAC inhibitors coordinately regulated VDR targets and restored anti-proliferative responsiveness [192, 194]. Similarly, other workers have used combinatorial chemistry to combine aspects of the structure of $1\alpha,25(OH)_2D_3$ and HDAC inhibitors into a single molecule that demonstrates very significant potency [195].

Together, these data support the concept that altered patterns of corepressors inappropriately sustain histone deacetylation around the VDRE of specific target gene promoter/enhancer regions, and shifts the dynamic equilibrium between *apo* and *holo* receptor conformations, to favor transcriptional repression of key target genes. Furthermore, targeting this epigenetic lesion with co-treatments of vitamin D_3 compounds plus HDAC inhibitors generates a temporal window where the equilibrium point between *apo* and *holo* complexes is shifted to sustain a more transcriptionally permissive environment.

These findings compliment a number of parallel studies that have established cooperativity between $1\alpha,25(OH)_2D_3$ and butyrate compounds, such as sodium butyrate (NaB) [196–201]. These compounds are short-chain fatty acids produced during fermentation by endogenous intestinal bacteria and have the capacity to act as HDAC inhibitors. Stein and coworkers have identified the effects in colon cancer cells of $1\alpha,25(OH)_2D_3$ plus NaB co-treatments to include the coordinate regulation of the VDR itself. Together these studies underscore further the importance of the dietary-derived milieu to regulate epithelial proliferation and differentiation beyond sites of action in the gut.

2.5 Toward an Integrated Understanding of the VDR

A highly conserved VDR is found widely throughout metazoans, even in certain non-calcified chordates such as the lamprey (reviewed in [202]). Within prokaryotes there appears to be the capacity to undertake UV-catalyzed metabolism of cholesterol compounds and suggests that the evolution of vitamin D biochemistry is very ancient. These findings suggest that the VDR system has been adapted to regulate calcium function and retains other functions that are calcium-independent and include the capacity to sense the local environment.

Phylogenetic classification has defined seven NR subfamilies, and within these the VDR is in the group 1 subfamily, sharing homology with the LXRs and FXR, and more distantly the PPARs [203, 204]. The receptors within this subfamily preferentially form homo- or heterodimeric complexes with RXR acting as a common central partner for VDR, PPARs, LXRs, and FXR. Thus, the receptors in the group appear to be all responsive to either bile acid or xenobiotic receptors and, therefore, widely integrated with bile acid homeostasis and detoxification. In keeping with this capacity, the bile acid lithocholic acid (LCA) has recently been shown to be a potent ligand for the VDR all be it with lower millimolar affinity [193].

VDR biology participates in at least three fundamental areas of biology required for human health, and which are disrupted in human disease. It participates in the regulation of serum calcium, and by implication the maintenance of bone integrity; the control of cell proliferation and differentiation; and by implication the disruption of these actions in malignancy; and as a modifier of immune responses and by implication contributes toward auto-immune diseases [205]. The divergence of these actions may make the VDR a particularly challenging receptor to understand in terms of biology and to exploit therapeutically.

Specifically dietary-derived fatty acids and bile acids cycle rapidly in response to dietary intake and work hormonally to coordinate multiple aspects of tissue function in response to changing energetic status. Thus, it is unlikely that the VDR alone plays a key and dominant role in cell and tissue function by acting singularly, but instead is intimately linked to the actions of related NRs (e.g., PPARs, FXR, and LXR) and cofactors. In this manner, the actions of the VDR to regulate cell growth and differentiation, as part of a network of environmental and dietary sensing receptors, may be the central and common function for the VDR. The differentiated phenotype of these cells then participates in diverse biology that range from calcium transport to dermis formation and mammary gland function.

For "next generation" developments to occur it will be necessary to adopt a broader view of VDR signaling. Historically, researchers have studied the abilities of single NRs such as the VDR to regulate a discrete group of gene targets and influence cell function. This has led to substantial knowledge concerning many of these receptors, individually. Cell and organism function, however, depends on the dynamic interactions of a collection of receptors, through the networks that link them, and against the backdrop of intrinsic cellular programs, such as those governing development and differentiation.

In such a view, it is apparent that NRs act as an adaptive homeostatic network in several tissues to sense environmental dietary and xenobiotic lipophilic compounds and sustain the cell, for example, through the diurnal patterns of fast and feeding (reviewed in [204, 206]). The VDR was originally described for a central endocrine role in maintenance of serum calcium levels. Similarly, the FXR and LXRs were described for their central role in cholesterol metabolism and bile acid synthesis in the enterohepatic system. However, their expression in multiple target tissues such a broader role. Examination of the known target genes for VDR, RARs, PPARs, FXR, and LXRs reveals that they share in common the regulation of cell cycle, programmed cell death, differentiation, and xenobiotic and metabolic clearance.

2 The Molecular Cancer Biology of the VDR 43

The challenge is to model the spatio-temporal actions of the NR network and, in particular, the extent to which the VDR exerts critical control over transcription and translation. Such an understanding requires a clear awareness of the chromatin architecture and context of the promoter regions (e.g., histone modifications, DNA methylation), genomic organization, gene regulation hierarchies, and $1\alpha,25(OH)_2D_3$-based metabolomic cascades, all within the context of specific cell backgrounds. The ultimate research goal will be to translate this understanding to strategies that can predict the capacity of subsets of VDR actions to be regulated and targeted in distinct cell types and exploited in discrete disease settings.

The current lack of an integral view of how these interactions bring about function and dysfunction, for example, in the aging human individual, can be attributed to the limitations of previously available techniques and tools to undertake such studies. The implementation of post-genomic techniques together with bioinformatics and systems biology methodology is expected to generate an integral view, thereby revealing and quantifying the mechanisms by which cells, tissues, and organisms interact with environmental factors such as diet [207, 208].

References

1. Deeb KK, Trump DL, Johnson CS (2007) Vitamin D signalling pathways in cancer: potential for anticancer therapeutics. Nat Rev Cancer 7:684–700
2. Imawari M, Kida K, Goodman DS (1976) The transport of vitamin D and its 25-hydroxy metabolite in human plasma. Isolation and partial characterization of vitamin D and 25-hydroxyvitamin D binding protein. J Clin Invest 58:514–523
3. Bouillon R, Van Assche FA, Van Baelen H, Heyns W, De Moor P (1981) Influence of the vitamin D-binding protein on the serum concentration of 1, 25-dihydroxyvitamin D3. Significance of the free 1, 25-dihydroxyvitamin D3 concentration. J Clin Invest 67:589–596
4. Nykjaer A et al (1999) An endocytic pathway essential for renal uptake and activation of the steroid 25-(OH) vitamin D3. Cell 96:507–515
5. Barsony J, Renyi I, McKoy W (1997) Subcellular distribution of normal and mutant vitamin D receptors in living cells. Studies with a novel fluorescent ligand. J Biol Chem 272:5774–5782
6. Huhtakangas JA, Olivera CJ, Bishop JE, Zanello LP, Norman AW (2004) The vitamin D receptor is present in caveolae-enriched plasma membranes and binds 1 alpha, 25(OH)2-vitamin D3 in vivo and in vitro. Mol Endocrinol 18:2660–2671
7. Boyan BD et al (2006) Regulation of growth plate chondrocytes by 1, 25-dihydroxyvitamin D3 requires caveolae and caveolin-1. J Bone Miner Res 21:1637–1647
8. Prufer K, Racz A, Lin GC, Barsony J (2000) Dimerization with retinoid X receptors promotes nuclear localization and subnuclear targeting of vitamin D receptors. J Biol Chem 275:41114–41123
9. Yasmin R, Williams RM, Xu M, Noy N (2005) Nuclear import of the retinoid X receptor, the vitamin D receptor, and their mutual heterodimer. J Biol Chem 280:40152–40160
10. Quack M, Carlberg C (2000) The impact of functional vitamin D(3) receptor conformations on DNA-dependent vitamin D(3) signaling. Mol Pharmacol 57:375–384
11. Renaud JP et al (1995) Crystal structure of the RAR-gamma ligand-binding domain bound to all-trans retinoic acid. Nature 378:681–689

12. Nakabayashi M et al (2008) Crystal structures of rat vitamin D receptor bound to adamantyl vitamin D analogs: structural basis for vitamin D receptor antagonism and partial agonism. J Med Chem 51:5320–5329
13. Carlberg C, Molnar F (2006) Detailed molecular understanding of agonistic and antagonistic vitamin D receptor ligands. Curr Top Med Chem 6:1243–1253
14. Saramaki A et al (2009) Cyclical chromatin looping and transcription factor association on the regulatory regions of the p21 (CDKN1A) gene in response to 1alpha, 25-dihydroxyvitamin D3. J Biol Chem 284(12):8073–82
15. Kim JY, Son YL, Lee YC (2009) Involvement of SMRT corepressor in transcriptional repression by the vitamin D receptor. Mol Endocrinol 23:251–264
16. Li J et al (2000) Both corepressor proteins SMRT and N-CoR exist in large protein complexes containing HDAC3. EMBO J 19:4342–4350
17. Malinen M et al (2007) Distinct HDACs regulate the transcriptional response of human cyclin-dependent kinase inhibitor genes to trichostatin A and 1{alpha}, 25-dihydroxyvitamin D3. Nucleic Acids Res 36(1):121–32
18. Yoon HG et al (2003) Purification and functional characterization of the human N-CoR complex: the roles of HDAC3, TBL1 and TBLR1. EMBO J 22:1336–1346
19. Alenghat T, Yu J, Lazar MA (2006) The N-CoR complex enables chromatin remodeler SNF2H to enhance repression by thyroid hormone receptor. EMBO J 25(17):3966–74
20. Yu C et al (2005) The nuclear receptor corepressors NCoR and SMRT decrease PPARgamma transcriptional activity and repress 3T3-L1 adipogenesis. J Biol Chem 280(14):13600–5
21. Oda Y et al (2003) Two distinct coactivators, DRIP/mediator and SRC/p160, are differentially involved in vitamin D receptor transactivation during keratinocyte differentiation. Mol Endocrinol 17:2329–2339
22. Rachez C et al (2000) The DRIP complex and SRC-1/p160 coactivators share similar nuclear receptor binding determinants but constitute functionally distinct complexes. Mol Cell Biol 20:2718–2726
23. Saramaki A, Banwell CM, Campbell MJ, Carlberg C (2006) Regulation of the human p21(waf1/cip1) gene promoter via multiple binding sites for p53 and the vitamin D3 receptor. Nucleic Acids Res 34:543–554
24. Vaisanen S, Dunlop TW, Sinkkonen L, Frank C, Carlberg C (2005) Spatio-temporal Activation of Chromatin on the Human CYP24 Gene Promoter in the Presence of 1alpha, 25-Dihydroxyvitamin D(3). J Mol Biol 350(1):65–77
25. Zella LA, Kim S, Shevde NK, Pike JW (2006) Enhancers located within two introns of the vitamin D receptor gene mediate transscriptional autoregulation by 1, 25-dihydroxyvitamin D3. Mol Endocrinol 103(3–5):435–9
26. Kim S, Shevde NK, Pike JW (2005) 1, 25-Dihydroxyvitamin D3 stimulates cyclic vitamin D receptor/retinoid X receptor DNA-binding, co-activator recruitment, and histone acetylation in intact osteoblasts. J Bone Miner Res 20:305–317
27. Seo YK et al (2007) Xenobiotic- and vitamin D-responsive induction of the steroid/bile acid-sulfotransferase Sult2A1 in young and old mice: the role of a gene enhancer in the liver chromatin. Gene 386:218–223
28. Meyer MB, Watanuki M, Kim S, Shevde NK, Pike JW (2006) The human TRPV6 distal promoter contains multiple vitamin D receptor binding sites that mediate activation by 1, 25-dihydroxyvitamin D3 in intestinal cells. Mol Endocrinol 20(6):1447–61
29. Reid G et al (2003) Cyclic, proteasome-mediated turnover of unliganded and liganded ERalpha on responsive promoters is an integral feature of estrogen signaling. Mol Cell 11:695–707
30. Metivier R et al (2003) Estrogen receptor-alpha directs ordered, cyclical, and combinatorial recruitment of cofactors on a natural target promoter. Cell 115:751–763
31. Yang X et al (2006) Nuclear receptor expression links the circadian clock to metabolism. Cell 126:801–810
32. Carroll JS et al (2006) Genome-wide analysis of estrogen receptor binding sites. Nat Genet 38:1289–1297

33. Efer R, Wagner EF (2003) AP-1: a double-edged sword in tumorigenesis. Nat Rev Cancer 3:859–868
34. Watt FM, Frye M, Benitah SA (2008) MYC in mammalian epidermis: how can an oncogene stimulate differentiation? Nat Rev Cancer 8:234–242
35. Thorne JL, Campbell MJ, Turner BM (2009) Transcription factors, chromatin and cancer. Int J Biochem Cell Biol 41:164–175
36. Lefterova MI et al (2008) PPAR{gamma} and C/EBP factors orchestrate adipocyte biology via adjacent binding on a genome-wide scale. Genes Dev 22:2941–2952
37. Lupien M et al (2008) FoxA1 translates epigenetic signatures into enhancer-driven lineage-specific transcription. Cell 132:958–970
38. Goodson ML, Jonas BA, Privalsky ML (2005) Alternative mRNA splicing of SMRT creates functional diversity by generating corepressor isoforms with different affinities for different nuclear receptors. J Biol Chem 280(9):7493–503
39. Jepsen K et al (2007) SMRT-mediated repression of an H3K27 demethylase in progression from neural stem cell to neuron. Nature 450:415–419
40. Alenghat T et al (2008) Nuclear receptor corepressor and histone deacetylase 3 govern circadian metabolic physiology. Nature 456(7224):997–1000
41. Astapova I et al (2008) The nuclear corepressor, NCoR, regulates thyroid hormone action in vivo. Proc Natl Acad Sci USA 105(49):19544–9
42. Sutanto MM, Symons MS, Cohen RN (2007) SMRT recruitment by PPARgamma is mediated by specific residues located in its carboxy-terminal interacting domain. Mol Cell Endocrinol 267:138–143
43. Polly P et al (2000) VDR-Alien: a novel, DNA-selective vitamin D(3) receptor-corepressor partnership. FASEB J 14:1455–1463
44. Lykke-Andersen K et al (2003) Disruption of the COP9 signalosome Csn2 subunit in mice causes deficient cell proliferation, accumulation of p53 and cyclin E, and early embryonic death. Mol Cell Biol 23:6790–6797
45. Hatchell EC et al (2006) SLIRP, a small SRA binding protein, is a nuclear receptor corepressor. Mol Cell 22:657–668
46. Hsieh JC et al (2003) Physical and functional interaction between the vitamin D receptor and hairless corepressor, two proteins required for hair cycling. J Biol Chem 278:38665–38674
47. Miller J et al (2001) Atrichia caused by mutations in the vitamin D receptor gene is a phenocopy of generalized atrichia caused by mutations in the hairless gene. J Invest Dermatol 117:612–617
48. Xie Z, Chang S, Oda Y, Bikle DD (2005) Hairless suppresses vitamin D receptor transactivation in human keratinocytes. Endocrinology 147(1):314–23
49. Scsucova S et al (2005) The repressor DREAM acts as a transcriptional activator on Vitamin D and retinoic acid response elements. Nucleic Acids Res 33:2269–2279
50. Fujiki R et al (2005) Ligand-induced transrepression by VDR through association of WSTF with acetylated histones. EMBO J 24:3881–3894
51. Kitagawa H et al (2003) The chromatin-remodeling complex WINAC targets a nuclear receptor to promoters and is impaired in Williams syndrome. Cell 113:905–917
52. Ding XF et al (1998) Nuclear receptor-binding sites of coactivators glucocorticoid receptor interacting protein 1 (GRIP1) and steroid receptor coactivator 1 (SRC-1): multiple motifs with different binding specificities. Mol Endocrinol 12:302–313
53. Eggert M et al (1995) A fraction enriched in a novel glucocorticoid receptor-interacting protein stimulates receptor-dependent transcription in vitro. J Biol Chem 270:30755–30759
54. Zhang J, Fondell JD (1999) Identification of mouse TRAP100: a transcriptional coregulatory factor for thyroid hormone and vitamin D receptors. Mol Endocrinol 13:1130–1140
55. Yuan CX, Ito M, Fondell JD, Fu ZY, Roeder RG (1998) The TRAP220 component of a thyroid hormone receptor- associated protein (TRAP) coactivator complex interacts directly with nuclear receptors in a ligand-dependent fashion. Proc Natl Acad Sci USA 95:7939–7944

56. Lee JW, Choi HS, Gyuris J, Brent R, Moore DD (1995) Two classes of proteins dependent on either the presence or absence of thyroid hormone for interaction with the thyroid hormone receptor. Mol Endocrinol 9:243–254

57. Zhang C et al (2003) Nuclear coactivator-62 kDa/Ski-interacting protein is a nuclear matrix-associated coactivator that may couple vitamin D receptor-mediated transcription and RNA splicing. J Biol Chem 278:35325–35336

58. Urahama N et al (2005) The role of transcriptional coactivator TRAP220 in myelomonocytic differentiation. Genes Cells 10:1127–1137

59. Ren Y et al (2000) Specific structural motifs determine TRAP220 interactions with nuclear hormone receptors. Mol Cell Biol 20:5433–5446

60. Teichert A et al (2009) Quantification of the Vitamin D Receptor-Coregulator Interaction (dagger). Biochemistry 48(7):1254–61

61. Hawker NP, Pennypacker SD, Chang SM, Bikle DD (2007) Regulation of human epidermal keratinocyte differentiation by the vitamin D receptor and its coactivators DRIP205, SRC2, and SRC3. J Invest Dermatol 127:874–880

62. Lutz W, Kohno K, Kumar R (2001) The role of heat shock protein 70 in vitamin D receptor function. Biochem Biophys Res Commun 282:1211–1219

63. Guzey M, Takayama S, Reed JC (2000) BAG1L enhances trans-activation function of the vitamin D receptor. J Biol Chem 275:40749–40756

64. Bikle D, Teichert A, Hawker N, Xie Z, Oda Y (2007) Sequential regulation of keratinocyte differentiation by 1, 25(OH)2D3, VDR, and its coregulators. J Steroid Biochem Mol Biol 103:396–404

65. Blok LJ, de Ruiter PE, Brinkmann AO (1996) Androgen receptor phosphorylation. Endocr Res 22:197–219

66. Hilliard GMT, Cook RG, Weigel NL, Pike JW (1994) 1, 25-dihydroxyvitamin D3 modulates phosphorylation of serine 205 in the human vitamin D receptor: site-directed mutagenesis of this residue promotes alternative phosphorylation. Biochemistry 33:4300–4311

67. Hsieh JC et al (1991) Human vitamin D receptor is selectively phosphorylated by protein kinase C on serine 51, a residue crucial to its trans-activation function. Proc Natl Acad Sci USA 88:9315–9319

68. Macoritto M et al (2008) Phosphorylation of the human retinoid X receptor alpha at serine 260 impairs coactivator(s) recruitment and induces hormone resistance to multiple ligands. J Biol Chem 283:4943–4956

69. Arriagada G et al (2007) Phosphorylation at serine 208 of the 1[alpha], 25-dihydroxy vitamin D3 receptor modulates the interaction with transcriptional coactivators. J Steroid Biochem Mol Biol 103:425–429

70. Jurutka PW et al (1996) Human vitamin D receptor phosphorylation by casein kinase II at Ser-208 potentiates transcriptional activation. Proc Natl Acad Sci USA 93:3519–3524

71. Barletta F, Freedman LP, Christakos S (2002) Enhancement of VDR-mediated transcription by phosphorylation: correlation with increased interaction between the VDR and DRIP205, a subunit of the VDR-interacting protein coactivator complex. Mol Endocrinol 16:301–314

72. Carlberg C, Seuter S (2007) The vitamin D receptor. Dermatol Clin 25:515–523, viii

73. Thompson PD et al (2002) Liganded VDR induces CYP3A4 in small intestinal and colon cancer cells via DR3 and ER6 vitamin D responsive elements. Biochem Biophys Res Commun 299:730–738

74. Song CS et al (2006) An essential role of the CAAT/enhancer binding protein-alpha in the vitamin D-induced expression of the human steroid/bile acid-sulfotransferase (SULT2A1). Mol Endocrinol 20:795–808

75. Rosenfeld MG, Lunyak VV, Glass CK (2006) Sensors and signals: a coactivator/corepressor/epigenetic code for integrating signal-dependent programs of transcriptional response. Genes Dev 20:1405–1428

76. Chen CD et al (2004) Molecular determinants of resistance to antiandrogen therapy. Nat Med 10:33–39

2 The Molecular Cancer Biology of the VDR

77. Murayama A, Kim MS, Yanagisawa J, Takeyama K, Kato S (2004) Transrepression by a liganded nuclear receptor via a bHLH activator through co-regulator switching. EMBO J 23:1598–1608
78. Fujiki R et al (2005) Ligand-induced transrepression by VDR through association of WSTF with acetylated histones. EMBO J 24:3881–3894
79. Kim MS et al (2007) 1Alpha, 25(OH)2D3-induced transrepression by vitamin D receptor through E-box-type elements in the human parathyroid hormone gene promoter. Mol Endocrinol 21:334–342
80. Turunen MM, Dunlop TW, Carlberg C, Vaisanen S (2007) Selective use of multiple vitamin D response elements underlies the 1 alpha, 25-dihydroxyvitamin D3-mediated negative regulation of the human CYP27B1 gene. Nucleic Acids Res 35:2734–2747
81. Song CS et al (2005) An essential role of the CAAT/enhancer binding protein-{alpha} in the vitamin D induced expression of the human steroid/bile acid-sulfotransferase (SULT2A1). Mol Endocrinol 20(4):795–808
82. Eeckhoute J, Carroll JS, Geistlinger TR, Torres-Arzayus MI, Brown M (2006) A cell-type-specific transcriptional network required for estrogen regulation of cyclin D1 and cell cycle progression in breast cancer. Genes Dev 20:2513–2526
83. Turner BM (1998) Histone acetylation as an epigenetic determinant of long-term transcriptional competence. Cell Mol Life Sci 54:21–31
84. Jenuwein T, Allis CD (2001) Translating the histone code. Science 293:1074–1080
85. Turner BM (2002) Cellular memory and the histone code. Cell 111:285–291
86. Hartman HB, Yu J, Alenghat T, Ishizuka T, Lazar MA (2005) The histone-binding code of nuclear receptor co-repressors matches the substrate specificity of histone deacetylase 3. EMBO Rep 6:445–451
87. Strahl BD et al (2001) Methylation of histone H4 at arginine 3 occurs in vivo and is mediated by the nuclear receptor coactivator PRMT1. Curr Biol 11:996–1000
88. Shogren-Knaak M et al (2006) Histone H4-K16 acetylation controls chromatin structure and protein interactions. Science 311:844–847
89. Shi X et al (2006) ING2 PHD domain links histone H3 lysine 4 methylation to active gene repression. Nature 442:96–99
90. Varambally S et al (2002) The polycomb group protein EZH2 is involved in progression of prostate cancer. Nature 419:624–629
91. Yu J, Li Y, Ishizuka T, Guenther MG, Lazar MA (2003) A SANT motif in the SMRT corepressor interprets the histone code and promotes histone deacetylation. EMBO J 22:3403–3410
92. Roux C et al (2008) New insights into the role of vitamin D and calcium in osteoporosis management: an expert roundtable discussion. Curr Med Res Opin 24:1363–1370
93. Dawson-Hughes B et al (2005) Estimates of optimal vitamin D status. Osteoporos Int 16:713–716
94. Yoshizawa T et al (1997) Mice lacking the vitamin D receptor exhibit impaired bone formation, uterine hypoplasia and growth retardation after weaning. Nat Genet 16:391–396
95. Li YC et al (1997) Targeted ablation of the vitamin D receptor: an animal model of vitamin D-dependent rickets type II with alopecia. Proc Natl Acad Sci USA 94:9831–9835
96. Van Cromphaut SJ et al (2001) Duodenal calcium absorption in vitamin D receptor-knockout mice: functional and molecular aspects. Proc Natl Acad Sci USA 98:13324–13329
97. Dontu G, Al Hajj M, Abdallah WM, Clarke MF, Wicha MS (2003) Stem cells in normal breast development and breast cancer. Cell Prolif 36(Suppl 1):59–72
98. Reya T, Clevers H (2005) Wnt signalling in stem cells and cancer. Nature 434:843–850
99. Al Hajj M, Clarke MF (2004) Self-renewal and solid tumor stem cells. Oncogene 23:7274–7282
100. Al Hajj M, Becker MW, Wicha M, Weissman I, Clarke MF (2004) Therapeutic implications of cancer stem cells. Curr Opin Genet Dev 14:43–47
101. De Marzo AM, Nelson WG, Meeker AK, Coffey DS (1998) Stem cell features of benign and malignant prostate epithelial cells. J Urol 160:2381–2392

102. Huss WJ, Gray DR, Werdin ES, Funkhouser WK Jr, Smith GJ (2004) Evidence of pluripotent human prostate stem cells in a human prostate primary xenograft model. Prostate 60:77–90

103. Bikle DD, Gee E, Pillai S (1993) Regulation of keratinocyte growth, differentiation, and vitamin D metabolism by analogs of 1, 25-dihydroxyvitamin D. J Invest Dermatol 101:713–718

104. Reichrath J et al (1994) Hair follicle expression of 1, 25-dihydroxyvitamin D3 receptors during the murine hair cycle. Br J Dermatol 131:477–482

105. Sakai Y, Kishimoto J, Demay MB (2001) Metabolic and cellular analysis of alopecia in vitamin D receptor knockout mice. J Clin Invest 107:961–966

106. Palmer HG, Martinez D, Carmeliet G, Watt FM (2008) The vitamin D receptor is required for mouse hair cycle progression but not for maintenance of the epidermal stem cell compartment. J Invest Dermatol 128:2113–2117

107. Ellison TI, Eckert RL, MacDonald PN (2007) Evidence for 1, 25-dihydroxyvitamin D3-independent transactivation by the vitamin D receptor: uncoupling the receptor and ligand in keratinocytes. J Biol Chem 282:10953–10962

108. Nam Y et al (2006) A novel missense mutation in the mouse hairless gene causes irreversible hair loss: genetic and molecular analyses of Hr m1Enu. Genomics 87:520–526

109. Bikle DD, Elalieh H, Chang S, Xie Z, Sundberg JP (2006) Development and progression of alopecia in the vitamin D receptor null mouse. J Cell Physiol 207:340–353

110. Beaudoin GM 3rd, Sisk JM, Coulombe PA, Thompson CC (2005) Hairless triggers reactivation of hair growth by promoting Wnt signaling. Proc Natl Acad Sci USA 102:14653–14658

111. Thompson CC, Sisk JM, Beaudoin GM 3rd (2006) Hairless and Wnt signaling: allies in epithelial stem cell differentiation. Cell Cycle 5:1913–1917

112. Palmer HG, Anjos-Afonso F, Carmeliet G, Takeda H, Watt FM (2008) The vitamin D receptor is a Wnt effector that controls hair follicle differentiation and specifies tumor type in adult epidermis. PLoS ONE 3:e1483

113. Zinser G, Packman K, Welsh J (2002) Vitamin D(3) receptor ablation alters mammary gland morphogenesis. Development 129:3067–3076

114. Zinser GM, Welsh J (2004) Accelerated mammary gland development during pregnancy and delayed postlactational involution in vitamin D3 receptor null mice. Mol Endocrinol 18:2208–2223

115. Colston K, Colston MJ, Feldman D (1981) 1, 25-dihydroxyvitamin D3 and malignant melanoma: the presence of receptors and inhibition of cell growth in culture. Endocrinology 108:1083–1086

116. Miyaura C et al (1981) 1 alpha, 25-Dihydroxyvitamin D3 induces differentiation of human myeloid leukemia cells. Biochem Biophys Res Commun 102:937–943

117. Abe E et al (1981) Differentiation of mouse myeloid leukemia cells induced by 1 alpha, 25-dihydroxyvitamin D3. Proc Natl Acad Sci USA 78:4990–4994

118. Palmer HG et al (2003) Genetic signatures of differentiation induced by 1alpha, 25-dihydroxyvitamin D3 in human colon cancer cells. Cancer Res 63:7799–7806

119. Koike M et al (1997) 19-nor-hexafluoride analogue of vitamin D3: a novel class of potent inhibitors of proliferation of human breast cell lines. Cancer Res 57:4545–4550

120. Campbell MJ, Elstner E, Holden S, Uskokovic M, Koeffler HP (1997) Inhibition of proliferation of prostate cancer cells by a 19-nor-hexafluoride vitamin D3 analogue involves the induction of p21waf1, p27kip1 and E-cadherin. J Mol Endocrinol 19:15–27

121. Elstner E et al (1999) Novel 20-epi-vitamin D3 analog combined with 9-cis-retinoic acid markedly inhibits colony growth of prostate cancer cells. Prostate 40:141–149

122. Peehl DM et al (1994) Antiproliferative effects of 1, 25-dihydroxyvitamin D3 on primary cultures of human prostatic cells. Cancer Res 54:805–810

123. Welsh J et al (2002) Impact of the Vitamin D3 receptor on growth-regulatory pathways in mammary gland and breast cancer. J Steroid Biochem Mol Biol 83:85–92

124. Colston KW, Berger U, Coombes RC (1989) Possible role for vitamin D in controlling breast cancer cell proliferation. Lancet 1:188–191

2 The Molecular Cancer Biology of the VDR   49

125. Colston K, Colston MJ, Fieldsteel AH, Feldman D (1982) 1, 25-dihydroxyvitamin D3 receptors in human epithelial cancer cell lines. Cancer Res 42:856–859
126. Eelen G et al (2004) Microarray analysis of 1alpha, 25-dihydroxyvitamin D3-treated MC3T3–E1 cells. J Steroid Biochem Mol Biol 89–90:405–407
127. Akutsu N et al (2001) Regulation of gene expression by 1alpha, 25-dihydroxyvitamin D3 and its analog EB1089 under growth-inhibitory conditions in squamous carcinoma cells. Mol Endocrinol 15:1127–1139
128. Wang TT et al (2005) Large-scale in silico and microarray-based identification of direct 1, 25-dihydroxyvitamin D3 target genes. Mol Endocrinol 19(11):2685–95
129. Liu M, Lee MH, Cohen M, Bommakanti M, Freedman LP (1996) Transcriptional activation of the Cdk inhibitor p21 by vitamin D3 leads to the induced differentiation of the myelomonocytic cell line U937. Genes Dev 10:142–153
130. Wang QM, Jones JB, Studzinski GP (1996) Cyclin-dependent kinase inhibitor p27 as a mediator of the G1-S phase block induced by 1, 25-dihydroxyvitamin D3 in HL60 cells. Cancer Res 56:264–267
131. Li P et al (2004) p27(Kip1) stabilization and G(1) arrest by 1, 25-dihydroxyvitamin D(3) in ovarian cancer cells mediated through down-regulation of cyclin E/cyclin-dependent kinase 2 and Skp1-Cullin-F-box protein/Skp2 ubiquitin ligase. J Biol Chem 279:25260–25267
132. Huang YC, Chen JY, Hung WC (2004) Vitamin D(3) receptor/Sp1 complex is required for the induction of p27(Kip1) expression by vitamin D(3). Oncogene 23(28):4856–61
133. Hengst L, Reed S (1996) Translational control of p27Kip1 accumulation during the cell cycle. Science 271:1861–1864
134. Jiang F, Li P, Fornace AJ Jr, Nicosia SV, Bai W (2003) G2/M arrest by 1, 25-dihydroxyvitamin D3 in ovarian cancer cells mediated through the induction of GADD45 via an exonic enhancer. J Biol Chem 278:48030–48040
135. Khanim FL et al (2004) Altered SMRT levels disrupt vitamin D3 receptor signalling in prostate cancer cells. Oncogene 23:6712–6725
136. Lubbert M et al (1991) Stable methylation patterns of MYC and other genes regulated during terminal myeloid differentiation. Leukemia 5:533–539
137. Koeffler HP (1983) Induction of differentiation of human acute myelogenous leukemia cells: therapeutic implications. Blood 62:709–721
138. Studzinski GP, Bhandal AK, Brelvi ZS (1986) Potentiation by 1-alpha, 25-dihydroxyvitamin D3 of cytotoxicity to HL-60 cells produced by cytarabine and hydroxyurea. J Natl Cancer Inst 76:641–648
139. Lin RJ et al (1998) Role of the histone deacetylase complex in acute promyelocytic leukaemia. Nature 391:811–814
140. Song G, Wang L (2008) Transcriptional mechanism for the paired miR-433 and miR-127 genes by nuclear receptors SHP and ERRgamma. Nucleic Acids Res 36:5727–5735
141. Shah YM et al (2007) Peroxisome proliferator-activated receptor alpha regulates a microRNA-mediated signaling cascade responsible for hepatocellular proliferation. Mol Cell Biol 27:4238–4247
142. Barbieri CE, Pietenpol JA (2006) p63 and epithelial biology. Exp Cell Res 312:695–706
143. Blackburn AC, Jerry DJ (2002) Knockout and transgenic mice of Trp53: what have we learned about p53 in breast cancer? Breast Cancer Res 4:101–111
144. Kommagani R, Caserta TM, Kadakia MP (2006) Identification of vitamin D receptor as a target of p63. Oncogene 25:3745–3751
145. Maruyama R et al (2006) Comparative genome analysis identifies the vitamin D receptor gene as a direct target of p53-mediated transcriptional activation. Cancer Res 66:4574–4583
146. Lu J et al (2005) Transcriptional profiling of keratinocytes reveals a vitamin D-regulated epidermal differentiation network. J Invest Dermatol 124:778–785
147. Yang L, Yang J, Venkateswarlu S, Ko T, Brattain MG (2001) Autocrine TGFbeta signaling mediates vitamin D3 analog-induced growth inhibition in breast cells. J Cell Physiol 188:383–393

148. Wu Y, Craig TA, Lutz WH, Kumar R (1999) Identification of 1 alpha, 25-dihydroxyvitamin D3 response elements in the human transforming growth factor beta 2 gene. Biochemistry 38:2654–2660

149. Matilainen M, Malinen M, Saavalainen K, Carlberg C (2005) Regulation of multiple insulin-like growth factor binding protein genes by 1alpha, 25-dihydroxyvitamin D3. Nucleic Acids Res 33:5521–5532

150. Vousden KH, Lu X (2002) Live or let die: the cell's response to p53. Nat Rev Cancer 2:594–604

151. Lee D et al (2002) SWI/SNF complex interacts with tumor suppressor p53 and is necessary for the activation of p53-mediated transcription. J Biol Chem 277:22330–22337

152. Wilkinson DS et al (2005) A direct intersection between p53 and transforming growth factor beta pathways targets chromatin modification and transcription repression of the alpha-fetoprotein gene. Mol Cell Biol 25:1200–1212

153. Szeto FL et al (2007) Involvement of the vitamin D receptor in the regulation of NF-[kappa] B activity in fibroblasts. J Steroid Biochem Mol Biol 103:563–566

154. Froicu M, Cantorna M (2007) Vitamin D and the vitamin D receptor are critical for control of the innate immune response to colonic injury. BMC Immunol 8:5

155. Gombart AF, Borregaard N, Koeffler HP (2005) Human cathelicidin antimicrobial peptide (CAMP) gene is a direct target of the vitamin D receptor and is strongly up-regulated in myeloid cells by 1, 25-dihydroxyvitamin D3. FASEB J 19:1067–1077

156. Wan T-T et al (2004) Cutting edge: 1, 25-dihydroxyvitamin D3 is a direct inducer of antimicrobial peptide gene expression. J Immunol 173:2909–2912

157. Mallbris LED, Sundblad L, Granath F, Stahle M (2005) UVB Upregulates the antimicrobial protein hCAP18 mRNA in human skin. J Investig Dermatol 125:1072–1074

158. Zasloff M (2005) Sunlight, vitamin D, and the innate immune defenses of the human skin. J Investig Dermatol 125:xvi–xvii

159. Quigley DA et al (2009) Genetic architecture of mouse skin inflammation and tumour susceptibility. Nature 458(7237):505–8

160. Blutt SE, McDonnell TJ, Polek TC, Weigel NL (2000) Calcitriol-induced apoptosis in LNCaP cells is blocked by overexpression of Bcl-2. Endocrinology 141:10–17

161. Mathiasen I, Lademann U, Jaattela M (1999) Apoptosis induced by vitamin D compounds in breast cancer cells is inhibited by Bcl-2 but does not involve known caspases or p53. Cancer Res 59:4848–4856

162. Song H et al (2003) Vitamin D(3) up-regulating protein 1 (VDUP1) antisense DNA regulates tumorigenicity and melanogenesis of murine melanoma cells via regulating the expression of fas ligand and reactive oxygen species. Immunol Lett 86:235–247

163. Han SH et al (2003) VDUP1 upregulated by TGF-beta1 and 1, 25-dihydorxyvitamin D3 inhibits tumor cell growth by blocking cell-cycle progression. Oncogene 22:4035–4046

164. Albertson DG et al (2000) Quantitative mapping of amplicon structure by array CGH identifies CYP24 as a candidate oncogene. Nat Genet 25:144–146

165. Townsend K et al (2005) Autocrine metabolism of vitamin D in normal and malignant breast tissue. Clin Cancer Res 11:3579–3586

166. Abedin SA et al (2009) Elevated NCOR1 disrupts a network of dietary-sensing nuclear receptors in bladder cancer cells. Carcinogenesis 30(3):449–56

167. Menezes RJ et al (2008) Vitamin D receptor expression in normal, premalignant, and malignant human lung tissue. Cancer Epidemiol Biomarkers Prev 17:1104–1110

168. Pendas-Franco N et al (2008) DICKKOPF-4 is induced by TCF/beta-catenin and upregulated in human colon cancer, promotes tumour cell invasion and angiogenesis and is repressed by 1alpha, 25-dihydroxyvitamin D3. Oncogene 27:4467–4477

169. Palmer HG et al (2004) The transcription factor SNAIL represses vitamin D receptor expression and responsiveness in human colon cancer. Nat Med 10:917–919

170. Palmer HG et al (2001) Vitamin D(3) promotes the differentiation of colon carcinoma cells by the induction of E-cadherin and the inhibition of beta-catenin signaling. J Cell Biol 154:369–387

2 The Molecular Cancer Biology of the VDR

171. Shah S et al (2006) The molecular basis of vitamin D receptor and beta-catenin crossregulation. Mol Cell 21:799–809
172. Miller CW, Morosetti R, Campbell MJ, Mendoza S, Koeffler HP (1997) Integrity of the 1, 25-dihydroxyvitamin D3 receptor in bone, lung, and other cancers. Mol Carcinog 19:254–257
173. Guy M, Lowe LC, Bretherton-Watt D, Mansi JL, Colston KW (2003) Approaches to evaluating the association of vitamin D receptor gene polymorphisms with breast cancer risk. Recent Results Cancer Res 164:43–54
174. John EM, Schwartz GG, Koo J, Van Den BD, Ingles SA (2005) Sun exposure, vitamin D receptor gene polymorphisms, and risk of advanced prostate cancer. Cancer Res 65:5470–5479
175. Ingles SA et al (1998) Association of prostate cancer with vitamin D receptor haplotypes in African-Americans. Cancer Res 58:1620–1623
176. Ma J et al (1998) Vitamin D receptor polymorphisms, circulating vitamin D metabolites, and risk of prostate cancer in United States physicians. Cancer Epidemiol Biomarkers Prev 7:385–390
177. Gorlach M, Burd CG, Dreyfuss G (1994) The mRNA poly(A)-binding protein: localization, abundance, and RNA-binding specificity. Exp Cell Res 211:400–407
178. Kim JG et al (2003) Association between vitamin D receptor gene haplotypes and bone mass in postmenopausal Korean women. Am J Obstet Gynecol 189:1234–1240
179. Kuraishi T, Sun Y, Aoki F, Imakawa K, Sakai S (2000) The poly(A) tail length of casein mRNA in the lactating mammary gland changes depending upon the accumulation and removal of milk. Biochem J 347:579–583
180. Schondorf T et al (2003) Association of the vitamin D receptor genotype with bone metastases in breast cancer patients. Oncology 64:154–159
181. Lundin AC, Soderkvist P, Eriksson B, Bergman-Jungestrom M, Wingren S (1999) Association of breast cancer progression with a vitamin D receptor gene polymorphism. South-East Sweden Breast Cancer Group. Cancer Res 59:2332–2334
182. Guy M et al (2004) Vitamin d receptor gene polymorphisms and breast cancer risk. Clin Cancer Res 10:5472–5481
183. Ntais C, Polycarpou A, Ioannidis JP (2003) Vitamin D receptor gene polymorphisms and risk of prostate cancer: a meta-analysis. Cancer Epidemiol Biomarkers Prev 12:1395–1402
184. Rashid SF et al (2001) Synergistic growth inhibition of prostate cancer cells by 1 alpha, 25 Dihydroxyvitamin D(3) and its 19-nor-hexafluoride analogs in combination with either sodium butyrate or trichostatin A. Oncogene 20:1860–1872
185. Campbell MJ, Gombart AF, Kwok SH, Park S, Koeffler HP (2000) The anti-proliferative effects of 1alpha, 25(OH)2D3 on breast and prostate cancer cells are associated with induction of BRCA1 gene expression. Oncogene 19:5091–5097
186. Miller GJ, Stapleton GE, Hedlund TE, Moffat KA (1995) Vitamin D receptor expression, 24-hydroxylase activity, and inhibition of growth by 1alpha, 25-dihydroxyvitamin D3 in seven human prostatic carcinoma cell lines. Clin Cancer Res 1:997–1003
187. Rashid SF, Mountford JC, Gombart AF, Campbell MJ (2001) 1alpha, 25-dihydroxyvitamin D(3) displays divergent growth effects in both normal and malignant cells. Steroids 66:433–440
188. Campbell MJ, Adorini L (2006) The vitamin D receptor as a therapeutic target. Expert Opin Ther Targets 10:735–748
189. Battaglia S, Maguire O, Thorne JL, Hornung LB, Doig CL, Liu S, Sucheston LE, Bianchi A, Khanim F, Gommersall LM, Coulter HS, Rakha S, Giddings I, O'Neill LP, Cooper CS, McCabe CJ, Bunce CM, Campbell MJ (2010) Elevated NCOR1 disrupts PPAR{alpha}/{gamma} signaling in prostate cancer and forms a targetable epigenetic lesion. Carcinogenesis 31(9):1650–60
190. Kim JY, Son YL, Lee YC (2008) Involvement of SMRT Corepressor in Transcriptional Repression by the Vitamin D Receptor. Mol Endocrinol 23(2):251–64
191. Ting HJ, Bao BY, Reeder JE, Messing EM, Lee YF (2007) Increased expression of corepressors in aggressive androgen-independent prostate cancer cells results in loss of 1alpha, 25-dihydroxyvitamin D3 responsiveness. Mol Cancer Res 5:967–980

192. Banwell CM et al (2006) Altered nuclear receptor corepressor expression attenuates vitamin D receptor signaling in breast cancer cells. Clin Cancer Res 12:2004–2013
193. Makishima M et al (2002) Vitamin D receptor as an intestinal bile acid sensor. Science 296:1313–1316
194. Banwell CM, O'Neill LP, Uskokovic MR, Campbell MJ (2004) Targeting 1alpha, 25-dihydroxyvitamin D3 antiproliferative insensitivity in breast cancer cells by co-treatment with histone deacetylation inhibitors. J Steroid Biochem Mol Biol 89–90:245–249
195. Tavera-Mendoza LE et al (2008) Incorporation of histone deacetylase inhibition into the structure of a nuclear receptor agonist. Proc Natl Acad Sci USA 105:8250–8255
196. Costa EM, Feldman D (1987) Modulation of 1, 25-dihydroxyvitamin D3 receptor binding and action by sodium butyrate in cultured pig kidney cells (LLC-PK1). J Bone Miner Res 2:151–159
197. Gaschott T, Stein J (2003) Short-chain fatty acids and colon cancer cells: the vitamin D receptor–butyrate connection. Recent Results Cancer Res 164:247–257
198. Daniel C, Schroder O, Zahn N, Gaschott T, Stein J (2004) p38 MAPK signaling pathway is involved in butyrate-induced vitamin D receptor expression. Biochem Biophys Res Commun 324:1220–1226
199. Chen JS, Faller DV, Spanjaard RA (2003) Short-chain fatty acid inhibitors of histone deacetylases: promising anticancer therapeutics? Curr Cancer Drug Targets 3:219–236
200. Gaschott T, Werz O, Steinmeyer A, Steinhilber D, Stein J (2001) Butyrate-induced differentiation of Caco-2 cells is mediated by vitamin D receptor. Biochem Biophys Res Commun 288:690–696
201. Tanaka Y, Bush KK, Klauck TM, Higgins PJ (1989) Enhancement of butyrate-induced differentiation of HT-29 human colon carcinoma cells by 1, 25-dihydroxyvitamin D3. Biochem Pharmacol 38:3859–3865
202. Krasowski MD, Yasuda K, Hagey LR, Schuetz EG (2005) Evolutionary selection across the nuclear hormone receptor superfamily with a focus on the NR1I subfamily (vitamin D, pregnane X, and constitutive androstane receptors). Nucl Recept 3:2
203. Bookout AL et al (2006) Anatomical profiling of nuclear receptor expression reveals a hierarchical transcriptional network. Cell 126:789–799
204. Carlberg C, Dunlop TW (2006) An integrated biological approach to nuclear receptor signaling in physiological control and disease. Crit Rev Eukaryot Gene Expr 16:1–22
205. Adorini L, Daniel KC, Penna G (2006) Vitamin D receptor agonists, cancer and the immune system: an intricate relationship. Curr Top Med Chem 6:1297–1301
206. Evans RM (2005) The nuclear receptor superfamily: a rosetta stone for physiology. Mol Endocrinol 19:1429–1438
207. Westerhoff HV, Palsson BO (2004) The evolution of molecular biology into systems biology. Nat Biotechnol 22:1249–1252
208. Muller M, Kersten S (2003) Nutrigenomics: goals and strategies. Nat Rev Genet 4:315–322

Chapter 3
Anti-inflammatory Activity of Calcitriol in Cancer

Aruna V. Krishnan and David Feldman

Abstract Calcitriol exerts antiproliferative and pro-differentiating actions in many malignant cells and in animal models of cancer and its use as an anticancer agent in patients is currently being evaluated. Several molecular pathways are involved in the growth inhibitory effects of calcitriol, resulting in cell cycle arrest, induction of apoptosis, and the inhibition of invasion, metastasis, and angiogenesis. This chapter describes recent research revealing that anti-inflammatory effects are an additional anticancer pathway of calcitriol action and some of the molecular pathways underlying these effects are discussed. In normal and malignant prostate epithelial cells, calcitriol inhibits the synthesis and biological actions of pro-inflammatory prostaglandins (PGs) by three actions: (1) the inhibition of the expression of cyclooxygenase-2 (COX-2), the enzyme that synthesizes PGs; (2) the upregulation of the expression of 15-prostaglandin dehydrogenase (15-PGDH), the enzyme that inactivates PGs; and (3) decreasing the expression of EP and FP PG receptors that are essential for PG signaling. The combination of calcitriol and non-steroidal anti-inflammatory drugs (NSAIDs) results in a synergistic inhibition of the growth of prostate cancer (PCa) cells and offers a potential therapeutic strategy for PCa. Calcitriol also increases the expression of mitogen-activated protein kinase phosphatase 5 (MKP5) in prostate cells resulting in the subsequent inhibition of p38 stress kinase signaling and the attenuation of the production of pro-inflammatory cytokines. There is also considerable evidence for an anti-inflammatory role for calcitriol through the inhibition of nuclear factor kappa B (NFκB) signaling in several cancer cells. The discovery of these novel calcitriol-regulated molecular pathways reveals that calcitriol has anti-inflammatory actions, which in addition to its other anticancer effects may play an important role in cancer prevention and treatment.

D. Feldman (✉)
Department of Medicine, Division of Endocrinology,
Stanford University School of Medicine,
300 Pasteur Drive, Stanford, CA 94305–5103, USA
e-mail: dfeldman@stanford.edu

D.L. Trump and C.S. Johnson (eds.), *Vitamin D and Cancer*,
DOI 10.1007/978-1-4419-7188-3_3, © Springer Science+Business Media, LLC 2011

Keywords Calcitriol • Prostate cancer • Breast cancer • Colorectal cancer • Anti-inflammatory actions • Prostaglandins • IGFBP-3 • MKP5 • Inflammatory cytokines • NFκB • Chemoprevention and treatment

3.1 Introduction

Calcitriol (1,25-dihydroxyvitamin D_3), the biologically most active form of vitamin D, exerts antiproliferative and pro-differentiating effects in a number of malignant cells raising the possibility of its use as an anticancer agent as described in many chapters of this volume. In vivo studies have also demonstrated an anticancer effect of calcitriol to retard the development and growth of tumors in animal models. Many molecular pathways mediate the anticancer effects of calcitriol [1]. Recent research, including observations from our laboratory, suggests that calcitriol exhibits anti-inflammatory actions that may contribute to its beneficial effects in several cancers, in addition to the other actions described in this book. Inflammation has been suggested to contribute to the development and progression of many cancers [2] including prostate [3], breast [4], colon [5], lung [6], ovarian [7], liver [8], and skin [9] cancers. Inflammatory mediators enhance tumorigenesis through the activation of multiple signaling pathways. Our observations in prostate cancer (PCa) cells reveal that calcitriol exerts important regulatory effects on some of the key molecular pathways involved in inflammation. In this chapter, we will discuss the role of the anti-inflammatory actions of calcitriol and its potential chemopreventive and therapeutic utility in cancer.

3.2 Inflammation and Cancer

Chronic inflammation has been recognized as a risk factor for cancer development [10, 11]. Inflammation can be triggered by a variety of stimuli such as injury or infection, autoimmune disease, the development of benign or malignant tumors, or other pathologies. The responses of the immune system in fighting the development of tumors may also fuel the process of tumorigenesis. Cancer-related inflammation is characterized by the presence of inflammatory cells at the tumor sites and the overexpression of inflammatory mediators such as cytokines, chemokines, prostaglandins (PGs), and reactive oxygen and nitrogen species in tumor tissue [10–13]. Many of these pro-inflammatory mediators activate angiogenic switches usually under the control of vascular endothelial growth factor (VGEF) and thereby promote tumor angiogenesis, metastasis, and invasion [2, 14]. Epidemiological studies show a decrease in the risk of developing several cancers associated with the intake of antioxidants and non-steroidal anti-inflammatory drugs (NSAIDs) [14–16]. Current research has begun to unravel several molecular pathways that link inflammation and cancer. Our observations in PCa as well as those of others in several cancers

3   Anti-inflammatory Activity of Calcitriol in Cancer   55

have shown that calcitriol exerts regulatory effects on some of these inflammatory networks, revealing important anti-inflammatory actions of calcitriol.

3.3   Anti-inflammatory Effects of Calcitriol

Calcitriol exerts antiproliferative and pro-differentiating effects in many malignant cells and retards tumor growth in animal models of cancer [1, 17–29]. Several important mechanisms have been implicated in the anticancer effect of calcitriol including the induction of cell cycle arrest, stimulation of apoptosis, and inhibition of metastasis and angiogenesis [1, 20–32]. We used cDNA microarrays as a means to achieve our research goal of gaining a more complete understanding of the molecular pathways through which calcitriol mediates its antiproliferative and pro-differentiation effects in PCa cells [33, 34]. Our results have revealed that calcitriol regulates the expression of genes involved in PG metabolism and signaling, thereby reducing the levels and biological activity of PGs [35]. PGs are pro-inflammatory molecules that promote tumorigenesis and cancer growth [4, 36–39]. We have also shown that calcitriol up-regulates the expression of mitogen-activated protein kinase phosphatase-5 (MKP5; also known as dual specificity phosphatase-10 [DUSP10]) and thereby promotes down-stream anti-inflammatory effects, including a reduction in the level of expression of pro-inflammatory cytokines [40]. Recent research also indicates that calcitriol interferes with the activation and signaling of nuclear factor-kappaB (NFκB), a transcription factor that regulates the expression of numerous genes involved in inflammatory and immune responses and cellular proliferation [41] and thought to play a key role in the process leading from inflammation to carcinogenesis [42]. In the following sections, we will discuss the importance of these molecular pathways of inflammation in the development and progression of PCa, breast cancer (BCa), and colorectal cancer (CRC) and the therapeutic significance of the inhibition of these of pro-inflammatory signals by calcitriol.

3.3.1   Regulation of Prostaglandin Metabolism and Signaling

PGs have been shown to play a role in the development and progression of many cancers and extensive data support the idea that cyclooxygenase-2 (COX-2), the enzyme responsible for PG synthesis, is an important molecular target in cancer therapy [4, 36–39]. PGs promote carcinogenesis by stimulating cellular proliferation, inhibiting apoptosis, promoting angiogenesis, and by activating carcinogens [43, 44]. We have recently discovered that calcitriol regulates the expression of several key genes involved in the PG pathway causing a decrease in PG synthesis, an increase in PG catabolism, and the inhibition of PG signaling through their receptors in PCa cells [35].

3.3.1.1 COX-2

Cyclooxygenase (COX)/prostaglandin endoperoxidase synthase is the rate-limiting enzyme that catalyzes the conversion of arachidonic acid to PGs and related eicosanoids. COX exists as two isoforms, COX-1, which is constitutively expressed in many tissues and cell types and COX-2, which is inducible by a variety of stimuli. COX-2 is regarded as an immediate-early response gene whose expression is rapidly induced by mitogens, cytokines, tumor promoters, and growth factors [37]. Genetic and clinical studies indicate that increased COX-2 expression is one of the key steps in carcinogenesis [45]. Long-term use of NSAIDs or aspirin has been shown to be associated with a decrease in death rate from several cancers such as colorectal, stomach, breast, lung, prostate, bladder, and ovarian cancers [15, 16, 46, 47].

Several studies suggest a causative and/or stimulatory role for COX-2 in prostate tumorigenesis and demonstrate its overexpression in prostate adenocarcinoma [48, 49]. However, not all PCa are associated with elevated COX-2 expression [50, 51]. Zha et al. [51] did not find consistent overexpression of COX-2 in established PCa. However, they detected appreciable COX-2 expression in areas of proliferative inflammatory atrophy (PIA), lesions that have been implicated in prostate carcinogenesis. Silencing of COX-2 in metastatic PCa cells induces cell growth arrest and causes morphological changes associated with enhanced differentiation, highlighting the role of COX-2 in prostate carcinogenesis [52]. COX-2 protein expression in prostate biopsy cores and PCa surgical specimens is inversely correlated with disease-free survival [53]. A recent analysis of archival radical prostatectomy specimens concluded that COX-2 expression was an independent predictor of recurrence [54]. Elevated COX-2 protein levels have been reported in ~40% of invasive breast carcinomas [4]. NSAIDs inhibit the development of BCa in a variety of animal models (reviewed in [4]). Interestingly, PG signaling stimulates the transcription of the aromatase gene [55] and a positive correlation between COX-2 and aromatase expression in human breast carcinomas reflects this causal link [56, 57]. COX-2 overexpression in BCa correlates with features of aggressive breast disease including larger tumor size, high-grade, increased proliferation, negative hormone receptor status, and overexpression of the Her-2/neu oncogene [58–61]. An inverse relationship between COX-2 protein levels and disease-free survival in BCa patients has also been shown [59, 62]. Epidemiological observations show a significant reduction in the incidence of CRC among chronic users of NSAIDs (reviewed in [63]). A critical link between COX-2 and colorectal tumorigenesis was demonstrated when *Apc* delta716 mutant mice were mated to COX-2 knockout mice and a dramatic reduction in the number of intestinal polyps was seen in the doubly null progeny compared to COX-2 wild-type mice [64]. COX-2 protein is significantly overexpressed in CRC [38, 39, 63] and increased COX-2 expression correlates with a larger polyp size and progression to invasive carcinoma [65, 66].

Local production of PGs at the tumor sites by infiltrating inflammatory cells also increases the risk of carcinogenesis and/or cancer progression [3, 39, 51, 67, 68]. In colon cancer, COX-2 expression has been found in the carcinoma cells as well

3 Anti-inflammatory Activity of Calcitriol in Cancer   57

as infiltrating macrophages within the tumors [69, 70]. In other cancers, COX-2 expression has been demonstrated in vascular endothelial cells, fibroblasts, and smooth muscle cells around the cancer [71, 72]. PGs generated by COX-2 act in an autocrine and paracrine manner to stimulate cell growth. At the cellular level both arachidonic acid, the substrate for COX, and the product prostaglandin E_2 (PGE_2) stimulate proliferation by regulating the expression of genes that are involved in growth regulation including c-fos [73]. Studies in experimental models of cancer have shown that COX-2 enhances tumor development and progression by promoting resistance to apoptosis and stimulating angiogenesis and tumor invasion, and it is therefore regarded as an oncogene [14, 39].

3.3.1.2 15-PGDH

15-PGDH is the enzyme that catalyzes the conversion of PGs to their corresponding 15-keto derivatives, which exhibit greatly reduced biological activity. Therefore, 15-PGDH can be regarded as a physiological antagonist of COX-2. 15-PGDH has been described as an oncogene antagonist in colon cancer by Yan et al. [74]. Their studies show that 15-PGDH is universally expressed in normal colon but is routinely absent or severely reduced in cancer specimens. Most importantly, the stable transfection of a 15-PGDH expression vector into colon cancer cells greatly reduces the ability of the cells to form tumors and/or slows tumor growth in nude mice demonstrating that 15-PGDH functions as a tumor suppressor [74]. Another study in mice also demonstrates that 15-PGDH acts in vivo as a highly potent suppressor of colon neoplasia development [75]. Low expression of 15-PGDH and methylation of the 15-PGDH promoter in 30–40% of primary breast tumors has been reported by Wolf et al. [76]. Their studies in BCa cells also demonstrated a suppression of cell proliferation in vitro and decreased tumorigenicity in vivo following the overexpression of 15-PGDH, thus supporting a tumor suppressor role for 15-PGDH in BCa [76].

3.3.1.3 PG Receptors

PGE and PGF are the major PGs stimulating the proliferation of PCa cells and they act by binding to G-protein coupled membrane receptors (prostanoid receptors). There are eight members in the prostanoid receptor subfamily and they are distinguished by their ligand-binding profile and the signal transduction pathways that they activate accounting for some of the diverse and often opposing effects of PGs [77]. PGE acts through four different receptor subtypes (EP1-EP4), while PGF acts through the FP receptor. PCa cells express both EP and FP receptors [35, 73]. PG receptors are also expressed in most endothelial cells, macrophages, and stromal cells found in the tumor microenvironment. PG interaction with its receptors can send positive feedback signals to increase COX-2 mRNA levels [73, 78, 79]. Therefore, irrespective of the initial trigger of COX-2 expression, PGs could

mediate a wave of COX-2 expression at the tumor sites not only in the cancer cells themselves but also in the surrounding stromal cells and infiltrating macrophages as well as endothelial cells, thereby promoting tumor progression.

3.3.1.4 Calcitriol Effects on the PG Pathway in Prostate Cells

Our studies demonstrate that calcitriol regulates the expression of PG pathway genes in multiple PCa cell lines as well as primary prostatic epithelial cells established from surgically removed prostate tissue from PCa patients [35]. We found measurable amounts of COX-2 mRNA and protein in various PCa cell lines as well as primary prostatic epithelial cells derived from normal and cancerous prostate tissue, which were significantly decreased by calcitriol treatment. We also found that calcitriol significantly increased the expression of 15-PGDH mRNA and protein in various PCa cells. We further showed that by inhibiting COX-2 and stimulating 15-PGDH expression, calcitriol decreased the levels of biologically active PGs in PCa cells, thereby reducing the growth stimulation due to PGs. Our data also revealed that calcitriol decreased the expression of EP and FP PG receptors. The calcitriol-induced decrease in PG receptor levels resulted in the attenuation of PG-mediated functional responses even when exogenous PGs were added to the cell cultures. Calcitriol suppressed the induction of the immediate-early gene *c-fos* and the growth stimulation seen following the addition of exogenous PGs or the PG precursor arachidonic acid to PCa cell cultures [35]. We postulate that the down-regulation of PG receptors by calcitriol would inhibit the positive feedback exerted by PGs on COX-2, thereby limiting the wave of COX-2 expression at the tumor sites and slowing down tumor progression. Thus, calcitriol inhibits the PG pathway in PCa cells by three separate mechanisms: decreasing COX-2 expression, increasing 15-PGDH expression, and reducing PG receptor levels. We believe that these actions contribute to the suppression of the proliferative and angiogeneic stimuli provided by PGs in PCa cells. The regulation of PG metabolism and biological actions constitutes an important novel pathway of calcitriol action mediating its anti-inflammatory effects.

3.3.1.5 Combination of Calcitriol and NSAIDs as a Therapeutic Approach in PCa

NSAIDs are a class of drugs that decrease PG synthesis by inhibiting COX-1 and COX-2 enzymatic activities. Several NSAIDs nonselectively inhibit both the constitutively expressed COX-1 and the inducible COX-2, while others have been shown to be more selective in preferentially inhibiting COX-2 enzymatic activity. We tested the effect of combinations of calcitriol and various NSAIDs on PCa cell proliferation [35]. These studies were based on our hypothesis that the action of calcitriol at the genomic level to reduce COX-2 expression, leading to decreased COX-2 protein levels, will allow the use of lower concentrations of NSAIDs to

3   Anti-inflammatory Activity of Calcitriol in Cancer   59

inhibit COX-2 enzyme activity. Further, an increase in the expression of 15-PGDH and a decrease in PG receptor levels due to calcitriol actions will lower the concentrations and biological activity of PGs, thereby enhancing the NSAID effect. Therefore, we hypothesized that the combination of calcitriol and NSAIDs would exhibit an additive/synergistic activity to inhibit PCa cell growth. In cell culture studies, we examined the growth inhibitory effects of the combinations of calcitriol with the COX-2-selective NSAIDs NS398 and SC-58125 and the nonselective NSAIDs, naproxen and ibuprofen. The combinations caused a synergistic enhancement of the inhibition of PCa cell proliferation, compared to the individual agents [35]. These results led us to further hypothesize that the combination of calcitriol and NSAIDs may have clinical utility in PCa therapy [35].

Preclinical [80] and clinical studies [81] on colon and other cancers have successfully used the strategy of combining low doses of two active drugs to achieve a more effective chemoprevention and therapeutic outcome than those using the individual agents [82]. The combination approach would also minimize the toxicities of the individual drugs by allowing them to be used at lower doses while achieving a significant therapeutic effect. Based on our preclinical observations, we proposed that a combination of calcitriol with a NSAID would be a beneficial approach in PCa therapy. The combination strategy allows the use of lower concentrations of NSAIDs, thereby minimizing their undesirable side effects. It has become clear that the long-term use of COX-2-selective inhibitors such as rofecoxib (Vioxx) causes an increase in cardiovascular complications in patients [83–86]. Very recently, even the use of nonselective NSAIDs has been shown to increase cardiovascular risk in patients with heart disease [87]. However, in comparison to COX-2-selective inhibitors, nonselective NSAIDs such as naproxen may be associated with fewer cardiovascular adverse effects [87, 88]. Our preclinical data showed that the combinations of calcitriol with nonselective or selective NSAIDs were equally effective in inducing synergistic growth inhibition [35]. We therefore proposed that the combination of calcitriol with a nonselective NSAID would be a useful therapeutic approach in PCa that would allow both drugs to be used at reduced dosages leading to increased cardiovascular safety [89].

Calcitriol, in fact, is already being used in combination therapy with other agents that may enhance its antiproliferative activity while reducing its tendency to cause hypercalcemia [90]. The results of the ASCENT I clinical trial in advanced PCa patients who failed other therapies demonstrated that the administration of a very high dose (45 μg) of calcitriol (DN101, Novacea, South San Francisco, CA) once weekly along with the regimen of the chemotherapy drug docetaxel (taxotere) in use at the time of that trial (once weekly) caused a very significant improvement in overall survival and time to progression, providing evidence indicating that calcitriol could enhance the efficacy of active drugs in cancer treatment [91]. The ASCENT I trial did not meet its primary endpoint, i.e., a lowering of serum PSA. However, on the basis of promising survival results (16.4 months in the docetaxel arm vs. 24.5 months in the docetaxel plus calcitriol arm), a larger, phase III trial (ASCENT II) with survival as an endpoint was initiated. A new, improved docetaxel regimen (every 3 week dosing) was used in the control arm of the ASCENT

II trial, which was compared to DN101 plus the older docetaxel dosing regimen (once a week), resulting in an asymmetric study design. Unfortunately the improved survival due to the combination demonstrated in the ASCENT I trial could not be confirmed in the ASCENT II trial [92 http://novacea.com/ #85 2008]. In fact, the trial was prematurely stopped by the data safety monitoring committee after 900 patients were enrolled, when an excess number of deaths was noted in the study arm (DN101 plus old docetaxel regimen) versus the control arm (new docetaxel regimen). Since the trial was stopped, further analysis [93 http://novacea.com/ #129 2008] suggests that the increased deaths in the treatment arm compared to the control arm were not due to calcitriol toxicity but due to better survival in the control arm that received the new and improved docetaxel regimen.

Based on our preclinical observations in PCa cells, we recently carried out a single-arm, open-label phase II study evaluating the combination of the nonselective NSAID naproxen and calcitriol in patients with early recurrent PCa [94]. Patients in our study had no evidence of metastases. All the patients received 45 micrograms of calcitriol (DN101) orally once a week and 375 mg naproxen twice a day for 1 year. The trial was prematurely stopped after 21 patients had been enrolled when the FDA put a temporary hold on DN101 based on the data from the ASCENT II trial described above. The therapy was well-tolerated by most patients. Only four patients showed evidence of progression and were removed from the study. We monitored serum PSA levels every 8 weeks. Bone scans were done every 3 months along with ultrasound of the kidney to assess asymptomatic renal stones. Serum testosterone levels were not affected by the therapy and there were no sexual side effects. There was mild gastro-intestinal toxicity in three patients presumably from the naproxen and one patient had to be removed from the study. One patient developed a small asymptomatic renal stone and was removed from the study. He required no intervention for his renal stone. Changes in PSA doubling time (PSA-DT) postintervention were compared to baseline PSA-DT values. A prolongation of the PSA-DT was achieved in 75% of the patients suggesting a beneficial effect of the combination therapy [89, 94].

3.3.2 Induction of MKP5 and Inhibition of Stress-Activated Kinase Signaling

Our cDNA microarray analysis in normal human prostate epithelial cells [34] revealed another novel calcitriol responsive gene, MKP5, also known as DUSP10. Calcitriol upregulates MKP5 expression leading to downstream anti-inflammatory responses in cells derived from normal prostatic epithelium and primary, localized adenocarcinoma, supporting a role for calcitriol in the prevention and early treatment of PCa [40]. In primary cultures of normal prostatic epithelial cells from the peripheral zone, calcitriol increased MKP5 transcription [40]. We identified a putative positive vitamin D response element (VDRE) in the MKP5 promoter mediating this calcitriol effect [40]. Interestingly, calcitriol upregulation of MKP5

3   Anti-inflammatory Activity of Calcitriol in Cancer   61

was seen only in primary cells derived from normal prostatic epithelium and
primary, localized adenocarcinoma but not in the established PCa cell lines derived
from PCa metastasis such as LNCaP, PC-3, or DU145. MKP5 is a member of the
dual specificity MKP family of enzymes that dephosphorylate, and thereby inactivate, mitogen activated protein kinases (MAPKs). MKP5 specifically dephos-
phorylates p38 MAPK and the stress-activated protein kinase Jun-N-terminal
kinase (JNK), leading to their inactivation. Calcitriol inhibited the phosphorylation and activation of p38 in normal primary prostate cells in a MKP-5-dependent
manner as MKP5 siRNA completely abolished p38 inactivation by calcitriol [40].
A consequence of p38 stress-induced kinase activation is an increase in the production of pro-inflammatory cytokines that sustain and amplify the inflammatory
response [95]. As interleukin-6 (IL-6) is a p38-regulated pro-inflammatory
cytokine implicated in PCa progression [96], we investigated the effect of calcitriol on IL-6 production. Stimulation of primary prostate cells with the pro-inflammatory factor, tumor necrosis factor α (TNFα), increased IL-6 mRNA stability
and concentrations of IL-6 in the conditioned media. Pretreatment of the cells with
calcitriol significantly attenuated the increase in IL-6 production following TNFα
treatment [40].

IL-6 is a major pro-inflammatory cytokine that participates in inflammation-
associated carcinogenesis [97] and has been implicated in the pathogenesis of
several cancers [96, 98, 99]. Serum IL-6 levels were significantly elevated and positively correlated to tumor burden in colon cancer patients [100]. Serum IL-6 levels
were also significantly elevated in BCa patients [101, 102] and in PCa patients,
where in addition a positive correlation between IL-6 levels and the number of bone
metastases was also seen [102]. IL-6 is known to be associated with PCa progression [96]. Our data demonstrate the ability of calcitriol to reduce the production of
pro-inflammatory cytokines such as IL-6 by inhibiting p38 activation via MKP5
upregulation as well as to interfere with the signaling of pleitropic inflammatory
cytokines such as TNFα [40]. These observations provide evidence of significant
anti-inflammatory effects of calcitriol in cancer cells. Interestingly, established
metastasis-derived PCa cell lines exhibited low levels of MKP5 and were unable to
induce MKP5 in response to calcitriol. We therefore speculate that a loss of MKP5
might occur during PCa progression, as a result of a selective pressure to eliminate
the tumor suppressor activity of MKP5 and/or calcitriol.

3.3.3   Inhibition of NFκB Activation and Signaling

NFκB comprises a family of inducible transcription factors ubiquitously present in
all cells. NFκB transcription factors are important regulators of innate immune
responses and inflammation [103]. In the basal state, most NFκB dimers are bound
to specific inhibitory proteins called IκB and pro-inflammatory signals activate
NFκB mainly through IκB kinase (IKK)-dependent phoshorylation and degradation of the inhibitory IκB proteins [42]. Free NFκB then translocates to the nucleus

and activates the transcription of pro-inflammatory cytokines, chemokines, and anti-apoptotic factors [104]. In contrast to normal cells many cancer cells have elevated levels of active NFκB [105, 106]. Constitutive activation of NFκB has been observed in androgen-independent PCa [107–109]. The NFκB protein RelB is uniquely expressed at high levels in PCa with high Gleason scores [110]. NFκB plays a major role in the control of immune responses and inflammation and promotes malignant behavior by increasing the transcription of the anti-apoptotic gene Bcl2 [111], cell cycle progression factors such as c-myc and cyclin D1, proteolytic enzymes such as matrix metalloproteinase 9 (MMP-9), urokinase-type plasminogen activator (uPA), and angiogenic factors such as VEGF and interleukin-8 (IL-8) [109, 112]. IL-8, an angiogenic factor and a downstream target of NFκB, is also a potent chemotactic factor for neutrophils and is associated with the initiation of the inflammatory response [113].

Calcitriol is known to directly modulate basal and cytokine-induced NFκB activity in many cells including human lymphocytes [114], fibroblasts [115], and peripheral blood monocytes [116]. A reduction in the levels of the NFκB inhibitory protein IκBα has been reported in mice lacking the VDR [117]. IKKβ-mediated activation of NFκB contributes to the development of colitis-associated cancer through the activation of anti-apoptotic genes and the production IL-6 [42]. Addition of a VDR antagonist to colon cancer cells upregulates NFκB activity by decreasing the levels of IκBα, suggesting that vitamin D ligands exert a suppressive effect on NFκB activation [118]. Calcitriol and its analogs have been shown to block NFκB activation by increasing the expression of IκB in macrophages and peripheral blood mononuclear cells [116, 119, 120]. There is considerable evidence for the inhibition of NFκB signaling by calcitriol in PCa cells. Calcitriol decreases the levels of the angiogenic and pro-inflammatory cytokine IL-8 in immortalized normal human prostate epithelial cell lines (HPr-1 and RWPE-1) and established PCa cell lines (LNCaP, PC-3 and DU145) [121]. The suppression of IL-8 by calcitriol appears to be due to the inhibition of NFκB signaling. Calcitriol reduces the nuclear translocation of the NFκB subunit p65 thereby inhibiting the NFκB complex from binding to its DNA response element and consequently suppressing the NFκB stimulation of transcription of downstream targets such as IL-8 [121]. Thus calcitriol could delay the progression of PCa by suppressing the expression of angiogenic and pro-inflammatory factors such as VEGF and IL-8. In addition, calcitriol also indirectly inhibits NFκB signaling by up-regulating the expression of IGFBP-3, which has been shown to interfere with NFκB signaling in PCa cells by suppressing p65 NFκB protein levels and the phosphorylation of IκBα [122]. NFκB also provides an adaptive response to PCa cells against cytotoxicity induced by redox-active therapeutic agents and is implicated in radiation resistance of cancers [123, 124]. A recent study shows that calcitriol significantly enhances the sensitivity of PCa cells to ionizing radiation by selectively suppressing radiation-mediated RelB activation [125]. Thus calcitriol may serve as an effective agent for sensitizing PCa cells to radiation therapy via suppression of the NFκB pathway.

3.4 The Role of Anti-inflammatory Effects of Calcitriol in Cancer Prevention and Treatment

As already discussed, current perspectives in cancer biology suggest that inflammation plays a role in the development of cancer [67, 126, 127]. De Marzo et al. [128] have proposed that the PIA lesions in the prostate, which are associated with acute or chronic inflammation, are precursors of prostatic intraepithelial neoplasia (PIN) and PCa. The epithelial cells in PIA lesions have been shown to exhibit many molecular signs of stress including elevated expression of COX-2 [51, 126]. Inflammatory bowel disease is associated with the development of CRC [129–131]. Based on the recent research demonstrating anti-inflammatory effects of calcitriol (as discussed in the preceding sections) in the malignant cells as well as the infiltrating cells at the tumor sites, we postulate that calcitriol may play a role in delaying or preventing cancer development and/or progression.

3.4.1 Calcitriol and Prostate Cancer Chemoprevention

PCa generally progresses very slowly, likely for decades, before symptoms become obvious and diagnosis is made [132]. Recently, inflammation in the prostate has been proposed to be an etiological factor in the development of PCa [67]. The observed latency in PCa provides a long window of opportunity for intervention by chemopreventive agents. Dietary supplementation of COX-2 selective NSAIDs such as celecoxib has been shown to suppress prostate carcinogenesis in the Transgenic Adenocarcinoma of the Mouse Prostate (TRAMP) model [133]. Our studies on the inhibitory effects of calcitriol on COX-2 expression and the PG pathway and MKP5 induction with the resultant stress kinase inactivation and inhibition of pro-inflammatory cytokine production as well as published observations of calcitriol actions to inhibit NFκB signaling suggest that calcitriol exhibits significant anti-inflammatory effects in vitro. Therefore, we hypothesize that calcitriol has the potential to be useful as a chemopreventive agent in PCa. Recently, Foster and coworkers have demonstrated that administration of high dose calcitriol (20 µg/kg), intermittently 3 days/week for up to 14–30 weeks, suppresses prostate tumor development in TRAMP mice [134, 135]. The efficacy of calcitriol as a chemopreventive agent has also been examined in Nkx3.1; Pten mutant mice, which recapitulate stages of prostate carcinogenesis from PIN lesions to adenocarcinoma [136]. The data reveal that calcitriol significantly reduces the progression of PIN from a lower to a higher grade. Calcitriol is more effective when administered before, rather than subsequent to, the initial occurrence of PIN. These animal studies as well as our in vitro observations suggest that clinical trials in PCa patients with PIN or early disease evaluating calcitriol and its analogs as agents that prevent and/or delay progression, are warranted.

3.5 Summary and Conclusions

Our recent research has identified several new calcitriol target genes revealing novel molecular pathways of calcitriol action in prostate cells. The data suggest that calcitriol has anti-inflammatory actions that contribute to its therapeutic and cancer-preventive effects in PCa. Calcitriol reduces both PG production (by suppressing COX-2 and increasing 15-PGDH expression) and PG biological actions (by PG receptor down-regulation). We propose that calcitriol inhibition of the PG pathway contributes significantly to its anti-inflammatory actions. Combinations of calcitriol with NSAIDs exhibit synergistic enhancement of growth inhibition in PCa cell cultures, suggesting that they may have therapeutic utility in PCa. The results of our recent clinical trial in patients with early recurrent PCa indicate that the combination of a weekly high dose calcitriol with the nonselective NSAID naproxen has activity to slow the rate of rise of PSA in most patients. Another novel molecular pathway of calcitriol action in prostate cells involves the induction of MKP5 expression and the subsequent inhibition of p38 stress kinase signaling, resulting in the attenuation of the production of pro-inflammatory cytokines. There is also considerable evidence for an anti-inflammatory role for calcitriol in several cancers through the inhibition of NFκB signaling in many cancer cells as well as the infiltrating cells present at the tumor sites. The discovery of these novel calcitriol-regulated pathways suggest that calcitriol has anti-inflammatory actions, which in addition to its other anticancer effects, may play an important role in the prevention and/or treatment of cancer. We conclude that calcitriol may have utility as a cancer chemopreventive agent. Calcitriol and its analogs may also have therapeutic utility, particularly in PCa and should therefore be evaluated in clinical trials in PCa patients with early or precancerous disease.

References

1. Deeb KK, Trump DL, Johnson CS (2007) Vitamin D signalling pathways in cancer: potential for anticancer therapeutics. Nat Rev Cancer 7(9):684–700
2. Angelo LS, Kurzrock R (2007) Vascular endothelial growth factor and its relationship to inflammatory mediators. Clin Cancer Res 13(10):2825–2830
3. De Marzo AM, Platz EA, Sutcliffe S, Xu J, Gronberg H, Drake CG, Nakai Y, Isaacs WB, Nelson WG (2007) Inflammation in prostate carcinogenesis. Nat Rev Cancer 7(4):256–269
4. Howe LR (2007) Inflammation and breast cancer. Cyclooxygenase/prostaglandin signaling and breast cancer. Breast Cancer Res 9(4):210
5. Fantini MC, Pallone F (2008) Cytokines: from gut inflammation to colorectal cancer. Curr Drug Targets 9(5):375–380
6. Ardies CM (2003) Inflammation as cause for scar cancers of the lung. Integr Cancer Ther 2(3):238–246
7. Altinoz MA, Korkmaz R (2004) NF-kappaB, macrophage migration inhibitory factor and cyclooxygenase-inhibitions as likely mechanisms behind the acetaminophen- and NSAID-prevention of the ovarian cancer. Neoplasma 51(4):239–247

3 Anti-inflammatory Activity of Calcitriol in Cancer

8. Bartsch H, Nair J (2004) Oxidative stress and lipid peroxidation-derived DNA-lesions in inflammation driven carcinogenesis. Cancer Detect Prev 28(6):385–391

9. Hussein MR, Ahmed RA (2005) Analysis of the mononuclear inflammatory cell infiltrate in the non-tumorigenic, pre-tumorigenic and tumorigenic keratinocytic hyperproliferative lesions of the skin. Cancer Biol Ther 4(8):819–821

10. Allavena P, Garlanda C, Borrello MG, Sica A, Mantovani A (2008) Pathways connecting inflammation and cancer. Curr Opin Genet Dev 18(1):3–10

11. Mantovani A, Allavena P, Sica A, Balkwill F (2008) Cancer-related inflammation. Nature 454(7203):436–444

12. Lucia MS, Torkko KC (2004) Inflammation as a target for prostate cancer chemoprevention: pathological and laboratory rationale. J Urol 171(2 Pt 2):S30–S34, discussion S5

13. Mantovani A, Pierotti MA (2008) Cancer and inflammation: a complex relationship. Cancer Lett 267(2):180–181

14. Kundu JK, Surh YJ (2008) Inflammation: gearing the journey to cancer. Mutat Res 659(1–2):15–30

15. Moran EM (2002) Epidemiological and clinical aspects of nonsteroidal anti-inflammatory drugs and cancer risks. J Environ Pathol Toxicol Oncol 21(2):193–201

16. Thun MJ, Henley SJ, Patrono C (2002) Nonsteroidal anti-inflammatory drugs as anticancer agents: mechanistic, pharmacologic, and clinical issues. J Natl Cancer Inst 94(4):252–266

17. Colston KW, Welsh J (2005) Vitamin D and breast cancer. In: Feldman D, Pike JW, Glorieux FH (eds) Vitamin D. Elsevier Academic Press, San Diego, pp 1663–1677

18. Cross HS (2005) Vitamin D and colon cancer. In: Feldman D, Pike JW, Glorieux FH (eds) Vitamin D. Elsevier Academic Press, San Diego, pp 1709–1725

19. Gombart AF, Luong QT, Koeffler HP (2006) Vitamin D compounds: activity against microbes and cancer. Anticancer Res 26(4A):2531–2542

20. Gonzalez-Sancho JM, Larriba MJ, Ordonez-Moran P, Palmer HG, Munoz A (2006) Effects of 1alpha,25-dihydroxyvitamin D3 in human colon cancer cells. Anticancer Res 26(4A):2669–2681

21. Gross MD (2005) Vitamin D and calcium in the prevention of prostate and colon cancer: new approaches for the identification of needs. J Nutr 135(2):326–331

22. Harris DM, Go VL (2004) Vitamin D and colon carcinogenesis. J Nutr 134(12 Suppl):3463S–3471S

23. Krishnan AV, Peehl DM, Feldman D (2003) Inhibition of prostate cancer growth by vitamin D: Regulation of target gene expression. J Cell Biochem 88(2):363–371

24. Krishnan AV, Peehl DM, Feldman D (2005) Vitamin D and prostate cancer. In: Feldman D, Pike JW, Glorieux FH (eds) Vitamin D, 2nd edn. Elsevier Academic Press, San Diego, pp 1679–1707

25. Luong QT, Koeffler HP (2005) Vitamin D compounds in leukemia. J Steroid Biochem Mol Biol 97(1–2):195–202

26. Mordan-McCombs S, Valrance M, Zinser G, Tenniswood M, Welsh J (2007) Calcium, vitamin D and the vitamin D receptor: impact on prostate and breast cancer in preclinical models. Nutr Rev 65(8 Pt 2):S131–S133

27. Schwartz GG, Skinner HG (2007) Vitamin D status and cancer: new insights. Curr Opin Clin Nutr Metab Care 10(1):6–11

28. Stewart LV, Weigel NL (2004) Vitamin D and prostate cancer. Exp Biol Med (Maywood) 229(4):277–284

29. Welsh J (2007) Targets of vitamin D receptor signaling in the mammary gland. J Bone Miner Res 22(Suppl 2):V86–V90

30. Fleet JC (2008) Molecular actions of vitamin D contributing to cancer prevention. Mol Aspects Med 29(6):388–396

31. Krishnan AV, Moreno J, Nonn L, Malloy P, Swami S, Peng L, Peehl DM, Feldman D (2007) Novel pathways that contribute to the anti-proliferative and chemopreventive activities of calcitriol in prostate cancer. J Steroid Biochem Mol Biol 103(3–5):694–702

32. Krishnan AV, Moreno J, Nonn L, Swami S, Peehl DM, Feldman D (2007) Calcitriol as a chemopreventive and therapeutic agent in prostate cancer: role of anti-inflammatory activity. J Bone Miner Res 22(Suppl 2):V74–V80

33. Krishnan AV, Shinghal R, Raghavachari N, Brooks JD, Peehl DM, Feldman D (2004) Analysis of vitamin D-regulated gene expression in LNCaP human prostate cancer cells using cDNA microarrays. Prostate 59(3):243–251

34. Peehl DM, Shinghal R, Nonn L, Seto E, Krishnan AV, Brooks JD, Feldman D (2004) Molecular activity of 1, 25-dihydroxyvitamin D3 in primary cultures of human prostatic epithelial cells revealed by cDNA microarray analysis. J Steroid Biochem Mol Biol 92(3):131–141

35. Moreno J, Krishnan AV, Swami S, Nonn L, Peehl DM, Feldman D (2005) Regulation of prostaglandin metabolism by calcitriol attenuates growth stimulation in prostate cancer cells. Cancer Res 65(17):7917–7925

36. Badawi AF (2000) The role of prostaglandin synthesis in prostate cancer. BJU Int 85(4): 451–462

37. Hussain T, Gupta S, Mukhtar H (2003) Cyclooxygenase-2 and prostate carcinogenesis. Cancer Lett 191(2):125–135

38. Sinicrope FA (2006) Targeting cyclooxygenase-2 for prevention and therapy of colorectal cancer. Mol Carcinog 45(6):447–454

39. Zha S, Yegnasubramanian V, Nelson WG, Isaacs WB, De Marzo AM (2004) Cyclooxygenases in cancer: progress and perspective. Cancer Lett 215(1):1–20

40. Nonn L, Peng L, Feldman D, Peehl DM (2006) Inhibition of p38 by vitamin D reduces interleukin-6 production in normal prostate cells via mitogen-activated protein kinase phosphatase 5: implications for prostate cancer prevention by vitamin D. Cancer Res 66(8):4516–4524

41. McCarty MF (2004) Targeting multiple signaling pathways as a strategy for managing prostate cancer: multifocal signal modulation therapy. Integr Cancer Ther 3(4):349–380

42. Maeda S, Omata M (2008) Inflammation and cancer: role of nuclear factor-kappaB activation. Cancer Sci 99(5):836–842

43. Dubois RN, Abramson SB, Crofford L, Gupta RA, Simon LS, Van De Putte LB, Lipsky PE (1998) Cyclooxygenase in biology and disease. Faseb J 12(12):1063–1073

44. Hawk ET, Viner JL, Dannenberg A, DuBois RN (2002) COX-2 in cancer–a player that's defining the rules. J Natl Cancer Inst 94(8):545–546

45. Markowitz SD (2007) Aspirin and colon cancer–targeting prevention? N Engl J Med 356(21):2195–2198

46. Garcia Rodriguez LA, Huerta-Alvarez C (2000) Reduced incidence of colorectal adenoma among long-term users of nonsteroidal antiinflammatory drugs: a pooled analysis of published studies and a new population-based study. Epidemiology 11(4):376–381

47. Nelson JE, Harris RE (2000) Inverse association of prostate cancer and non-steroidal anti-inflammatory drugs (NSAIDs): results of a case-control study. Oncol Rep 7(1):169–170

48. Gupta S, Srivastava M, Ahmad N, Bostwick DG, Mukhtar H (2000) Over-expression of cyclooxygenase-2 in human prostate adenocarcinoma. Prostate 42(1):73–78

49. Yoshimura R, Sano H, Masuda C, Kawamura M, Tsubouchi Y, Chargui J, Yoshimura N, Hla T, Wada S (2000) Expression of cyclooxygenase-2 in prostate carcinoma. Cancer 89(3): 589–596

50. Wagner M, Loos J, Weksler N, Gantner M, Corless CL, Barry JM, Beer TM, Garzotto M (2005) Resistance of prostate cancer cell lines to COX-2 inhibitor treatment. Biochem Biophys Res Commun 332(3):800–807

51. Zha S, Gage WR, Sauvageot J, Saria EA, Putzi MJ, Ewing CM, Faith DA, Nelson WG, De Marzo AM, Isaacs WB (2001) Cyclooxygenase-2 is up-regulated in proliferative inflammatory atrophy of the prostate, but not in prostate carcinoma. Cancer Res 61(24):8617–8623

52. Narayanan BA, Narayanan NK, Davis L, Nargi D (2006) RNA interference-mediated cyclooxygenase-2 inhibition prevents prostate cancer cell growth and induces differentiation: modulation of neuronal protein synaptophysin, cyclin D1, and androgen receptor. Mol Cancer Ther 5(5):1117–1125

3 Anti-inflammatory Activity of Calcitriol in Cancer 67

53. Rubio J, Ramos D, Lopez-Guerrero JA, Iborra I, Collado A, Solsona E, Almenar S, Llombart-Bosch A (2005) Immunohistochemical Expression of Ki-67 Antigen, Cox-2 and Bax/Bcl-2 in Prostate Cancer; Prognostic Value in Biopsies and Radical Prostatectomy Specimens. Eur Urol 31:31

54. Cohen BL, Gomez P, Omori Y, Duncan RC, Civantos F, Soloway MS, Lokeshwar VB, Lokeshwar BL (2006) Cyclooxygenase-2 (cox-2) expression is an independent predictor of prostate cancer recurrence. Int J Cancer 119(5):1082–1087

55. Simpson ER, Clyne C, Rubin G, Boon WC, Robertson K, Britt K, Speed C, Jones M (2002) Aromatase–a brief overview. Annu Rev Physiol 64:93–127

56. Brodie AM, Lu Q, Long BJ, Fulton A, Chen T, Macpherson N, DeJong PC, Blankenstein MA, Nortier JW, Slee PH, van de Ven J, van Gorp JM, Elbers JR, Schipper ME, Blijham GH, Thijssen JH (2001) Aromatase and COX-2 expression in human breast cancers. J Steroid Biochem Mol Biol 79(1–5):41–47

57. Brueggemeier RW, Quinn AL, Parrett ML, Joarder FS, Harris RE, Robertson FM (1999) Correlation of aromatase and cyclooxygenase gene expression in human breast cancer specimens. Cancer Lett 140(1–2):27–35

58. Boland GP, Butt IS, Prasad R, Knox WF, Bundred NJ (2004) COX-2 expression is associated with an aggressive phenotype in ductal carcinoma in situ. Br J Cancer 90(2):423–429

59. Ristimaki A, Sivula A, Lundin J, Lundin M, Salminen T, Haglund C, Joensuu H, Isola J (2002) Prognostic significance of elevated cyclooxygenase-2 expression in breast cancer. Cancer Res 62(3):632–635

60. Subbaramaiah K, Norton L, Gerald W, Dannenberg AJ (2002) Cyclooxygenase-2 is overexpressed in HER-2/neu-positive breast cancer: evidence for involvement of AP-1 and PEA3. J Biol Chem 277(21):18649–18657

61. Wulfing P, Diallo R, Muller C, Wulfing C, Poremba C, Heinecke A, Rody A, Greb RR, Bocker W, Kiesel L (2003) Analysis of cyclooxygenase-2 expression in human breast cancer: high throughput tissue microarray analysis. J Cancer Res Clin Oncol 129(7):375–382

62. Denkert C, Winzer KJ, Muller BM, Weichert W, Pest S, Kobel M, Kristiansen G, Reles A, Siegert A, Guski H, Hauptmann S (2003) Elevated expression of cyclooxygenase-2 is a negative prognostic factor for disease free survival and overall survival in patients with breast carcinoma. Cancer 97(12):2978–2987

63. Sinicrope FA, Gill S (2004) Role of cyclooxygenase-2 in colorectal cancer. Cancer Metastasis Rev 23(1–2):63–75

64. Oshima M, Dinchuk JE, Kargman SL, Oshima H, Hancock B, Kwong E, Trzaskos JM, Evans JF, Taketo MM (1996) Suppression of intestinal polyposis in Apc delta716 knockout mice by inhibition of cyclooxygenase 2 (COX-2). Cell 87(5):803–809

65. Humar B, Giovanoli O, Wolf A, Attenhofer M, Bendik I, Meier R, Muller H, Dobbie Z (2000) Germline alterations in the cyclooxygenase-2 gene are not associated with the development of extracolonic manifestations in a large swiss familial adenomatous polyposis kindred. Int J Cancer 87(6):812–817

66. Khan KN, Masferrer JL, Woerner BM, Soslow R, Koki AT (2001) Enhanced cyclooxygenase-2 expression in sporadic and familial adenomatous polyposis of the human colon. Scand J Gastroenterol 36(8):865–869

67. Nelson WG, De Marzo AM, DeWeese TL, Isaacs WB (2004) The role of inflammation in the pathogenesis of prostate cancer. J Urol 172(5 Pt 2):S6–S11, discussion S-2

68. Wang W, Bergh A, Damber JE (2005) Cyclooxygenase-2 expression correlates with local chronic inflammation and tumor neovascularization in human prostate cancer. Clin Cancer Res 11(9):3250–3256

69. Bamba H, Ota S, Kato A, Adachi A, Itoyama S, Matsuzaki F (1999) High expression of cyclooxygenase-2 in macrophages of human colonic adenoma. Int J Cancer 83(4):470–475

70. Chapple KS, Cartwright EJ, Hawcroft G, Tisbury A, Bonifer C, Scott N, Windsor AC, Guillou PJ, Markham AF, Coletta PL, Hull MA (2000) Localization of cyclooxygenase-2 in human sporadic colorectal adenomas. Am J Pathol 156(2):545–553

71. Goluboff ET, Shabsigh A, Saidi JA, Weinstein IB, Mitra N, Heitjan D, Piazza GA, Pamukcu R, Buttyan R, Olsson CA (1999) Exisulind (sulindac sulfone) suppresses growth of human prostate cancer in a nude mouse xenograft model by increasing apoptosis. Urology 53(2):440–445

72. Mifflin RC, Saada JI, Di Mari JF, Adegboyega PA, Valentich JD, Powell DW (2002) Regulation of COX-2 expression in human intestinal myofibroblasts: mechanisms of IL-1-mediated induction. Am J Physiol Cell Physiol 282(4):C824–C834

73. Chen Y, Hughes-Fulford M (2000) Prostaglandin E2 and the protein kinase A pathway mediate arachidonic acid induction of c-fos in human prostate cancer cell. Br J Cancer 82(12):2000–2006

74. Yan M, Rerko RM, Platzer P, Dawson D, Willis J, Tong M, Lawrence E, Lutterbaugh J, Lu S, Willson JK, Luo G, Hensold J, Tai HH, Wilson K, Markowitz SD (2004) 15-Hydroxyprostaglandin dehydrogenase, a COX-2 oncogene antagonist, is a TGF-beta-induced suppressor of human gastrointestinal cancers. Proc Natl Acad Sci USA 101(50):17468–17473

75. Myung SJ, Rerko RM, Yan M, Platzer P, Guda K, Dotson A, Lawrence E, Dannenberg AJ, Lovgren AK, Luo G, Pretlow TP, Newman RA, Willis J, Dawson D, Markowitz SD (2006) 15-Hydroxyprostaglandin dehydrogenase is an in vivo suppressor of colon tumorigenesis. Proc Natl Acad Sci USA 103(32):12098–12102

76. Wolf I, O'Kelly J, Rubinek T, Tong M, Nguyen A, Lin BT, Tai HH, Karlan BY, Koeffler HP (2006) 15-hydroxyprostaglandin dehydrogenase is a tumor suppressor of human breast cancer. Cancer Res 66(15):7818–7823

77. Breyer RM, Bagdassarian CK, Myers SA, Breyer MD (2001) Prostanoid receptors: subtypes and signaling. Annu Rev Pharmacol Toxicol 41:661–690

78. Tjandrawinata RR, Dahiya R, Hughes-Fulford M (1997) Induction of cyclo-oxygenase-2 mRNA by prostaglandin E2 in human prostatic carcinoma cells. Br J Cancer 75(8):1111–1118

79. Tjandrawinata RR, Hughes-Fulford M (1997) Up-regulation of cyclooxygenase-2 by product-prostaglandin E2. Adv Exp Med Biol 407:163–170

80. Torrance CJ, Jackson PE, Montgomery E, Kinzler KW, Vogelstein B, Wissner A, Nunes M, Frost P, Discafani CM (2000) Combinatorial chemoprevention of intestinal neoplasia. Nat Med 6(9):1024–1028

81. Meyskens FL, McLaren CE, Pelot D, Fujikawa-Brooks S, Carpenter PM, Hawk ET, Kelloff G, Lawson MJ, Kidao J, McCracken J, Albers G, Ahnen DJ, Turgeon K, Goldschmid S, Lance P, Hagedorn CH, Gillen DL, Gerner EW (2008) Difluoromethylornithine plus sulindac for the prevention of sporadic colorectal adenomas: a randomized placebo-controlled, double-blind trial. Cancer Prev Res 1:32–38

82. Sporn MB, Hong WK (2008) Clinical prevention of recurrence of colorectal adenomas by the combination of difluoromethylornithine and sulindac: an important milestone. Cancer Prev Res 1:9–11

83. Antman EM, DeMets D, Loscalzo J (2005) Cyclooxygenase inhibition and cardiovascular risk. Circulation 112(5):759–770

84. Graham DJ, Campen D, Hui R, Spence M, Cheetham C, Levy G, Shoor S, Ray WA (2005) Risk of acute myocardial infarction and sudden cardiac death in patients treated with cyclo-oxygenase 2 selective and non-selective non-steroidal anti-inflammatory drugs: nested case-control study. Lancet 365(9458):475–481

85. Ray WA, Stein CM, Daugherty JR, Hall K, Arbogast PG, Griffin MR (2002) COX-2 selective non-steroidal anti-inflammatory drugs and risk of serious coronary heart disease. Lancet 360(9339):1071–1073

86. Solomon SD, Pfeffer MA, McMurray JJ, Fowler R, Finn P, Levin B, Eagle C, Hawk E, Lechuga M, Zauber AG, Bertagnolli MM, Arber N, Wittes J (2006) Effect of celecoxib on cardiovascular events and blood pressure in two trials for the prevention of colorectal adenomas. Circulation 114(10):1028–1035

87. Gislason GH, Rasmussen JN, Abildstrom SZ, Schramm TK, Hansen ML, Fosbol EL, Sorensen R, Folke F, Buch P, Gadsboll N, Rasmussen S, Poulsen HE, Kober L, Madsen M, Torp-Pedersen C (2009) Increased mortality and cardiovascular morbidity associated with

use of nonsteroidal anti-inflammatory drugs in chronic heart failure. Arch Intern Med 169(2):141–149

88. Bombardier C, Laine L, Reicin A, Shapiro D, Burgos-Vargas R, Davis B, Day R, Ferraz MB, Hawkey CJ, Hochberg MC, Kvien TK, Schnitzer TJ (2000) Comparison of upper gastrointestinal toxicity of rofecoxib and naproxen in patients with rheumatoid arthritis. VIGOR Study Group. N Engl J Med 343(21):1520–1528

89. Krishnan AV, Srinivas S, Feldman D (2009) Inhibition of prostaglandin synthesis and actions contributes to the beneficial effects of calcitriol in prostate cancer. Dermatoendocrinol 1(1):7–11

90. Johnson CS, Hershberger PA, Trump DL (2002) Vitamin D-related therapies in prostate cancer. Cancer Metastasis Rev 21(2):147–158

91. Beer TM, Ryan CW, Venner PM, Petrylak DP, Chatta GS, Ruether JD, Redfern CH, Fehrenbacher L, Saleh MN, Waterhouse DM, Carducci MA, Vicario D, Dreicer R, Higano CS, Ahmann FR, Chi KN, Henner WD, Arroyo A, Clow FW (2007) Double-blinded randomized study of high-dose calcitriol plus docetaxel compared with placebo plus docetaxel in androgen-independent prostate cancer: a report from the ASCENT Investigators. J Clin Oncol 25(6):669–674

92. Release NP. Novacea announces preliminary findings from data analysis of Ascent-2 Phase 3 trial. June 04, 2008

93. Release NP. Novacea update on Asentar(TM). September 11, 2008

94. Srinivas S, Feldman D (2008) A phase II trial of calcitriol and naproxen in recurrent prostate cancer. Anticancer Res 29:3605–3610

95. Park JI, Lee MG, Cho K, Park BJ, Chae KS, Byun DS, Ryu BK, Park YK, Chi SG (2003) Transforming growth factor-beta1 activates interleukin-6 expression in prostate cancer cells through the synergistic collaboration of the Smad2, p38-NF-kappaB, JNK, and Ras signaling pathways. Oncogene 22(28):4314–4332

96. Culig Z, Steiner H, Bartsch G, Hobisch A (2005) Interleukin-6 regulation of prostate cancer cell growth. J Cell Biochem 95(3):497–505

97. Rose-John S, Schooltink H (2007) Cytokines are a therapeutic target for the prevention of inflammation-induced cancers. Recent Results Cancer Res 174:57–66

98. Kai H, Kitadai Y, Kodama M, Cho S, Kuroda T, Ito M, Tanaka S, Ohmoto Y, Chayama K (2005) Involvement of proinflammatory cytokines IL-1beta and IL-6 in progression of human gastric carcinoma. Anticancer Res 25(2A):709–713

99. Schneider MR, Hoeflich A, Fischer JR, Wolf E, Sordat B, Lahm H (2000) Interleukin-6 stimulates clonogenic growth of primary and metastatic human colon carcinoma cells. Cancer Lett 151(1):31–38

100. Chung YC, Chang YF (2003) Serum interleukin-6 levels reflect the disease status of colorectal cancer. J Surg Oncol 83(4):222–226

101. Lyon DE, McCain NL, Walter J, Schubert C (2008) Cytokine comparisons between women with breast cancer and women with a negative breast biopsy. Nurs Res 57(1):51–58

102. Tumminello FM, Badalamenti G, Incorvaia L, Fulfaro F, D'Amico C, Leto G (2009) Serum interleukin-6 in patients with metastatic bone disease: correlation with cystatin C. Med Oncol 26(1):10–15

103. Karin M, Lin A (2002) NF-kappaB at the crossroads of life and death. Nat Immunol 3(3):221–227

104. Ghosh S, Karin M (2002) Missing pieces in the NF-kappaB puzzle. Cell 109(Suppl):S81–S96

105. Palayoor ST, Youmell MY, Calderwood SK, Coleman CN, Price BD (1999) Constitutive activation of IkappaB kinase alpha and NF-kappaB in prostate cancer cells is inhibited by ibuprofen. Oncogene 18(51):7389–7394

106. Sovak MA, Bellas RE, Kim DW, Zanieski GJ, Rogers AE, Traish AM, Sonenshein GE (1997) Aberrant nuclear factor-kappaB/Rel expression and the pathogenesis of breast cancer. J Clin Invest 100(12):2952–2960

107. Gasparian AV, Yao YJ, Kowalczyk D, Lyakh LA, Karseladze A, Slaga TJ, Budunova IV (2002) The role of IKK in constitutive activation of NF-kappaB transcription factor in prostate carcinoma cells. J Cell Sci 115(Pt 1):141–151
108. Ismail HA, Lessard L, Mes-Masson AM, Saad F (2004) Expression of NF-kappaB in prostate cancer lymph node metastases. Prostate 58(3):308–313
109. Suh J, Rabson AB (2004) NF-kappaB activation in human prostate cancer: important mediator or epiphenomenon? J Cell Biochem 91(1):100–117
110. Lessard L, Begin LR, Gleave ME, Mes-Masson AM, Saad F (2005) Nuclear localisation of nuclear factor-kappaB transcription factors in prostate cancer: an immunohistochemical study. Br J Cancer 93(9):1019–1023
111. Catz SD, Johnson JL (2001) Transcriptional regulation of bcl-2 by nuclear factor kappa B and its significance in prostate cancer. Oncogene 20(50):7342–7351
112. Huang S, Pettaway CA, Uehara H, Bucana CD, Fidler IJ (2001) Blockade of NF-kappaB activity in human prostate cancer cells is associated with suppression of angiogenesis, invasion, and metastasis. Oncogene 20(31):4188–4197
113. Ferrer FA, Miller LJ, Andrawis RI, Kurtzman SH, Albertsen PC, Laudone VP, Kreutzer DL (1998) Angiogenesis and prostate cancer: in vivo and in vitro expression of angiogenesis factors by prostate cancer cells. Urology 51(1):161–167
114. Yu XP, Bellido T, Manolagas SC (1995) Down-regulation of NF-kappa B protein levels in activated human lymphocytes by 1, 25-dihydroxyvitamin D3. Proc Natl Acad Sci USA 92(24):10990–10994
115. Harant H, Wolff B, Lindley IJ (1998) 1Alpha, 25-dihydroxyvitamin D3 decreases DNA binding of nuclear factor-kappaB in human fibroblasts. FEBS Lett 436(3):329–334
116. Stio M, Martinesi M, Bruni S, Treves C, Mathieu C, Verstuyf A, d'Albasio G, Bagnoli S, Bonanomi AG (2007) The Vitamin D analogue TX 527 blocks NF-kappaB activation in peripheral blood mononuclear cells of patients with Crohn's disease. J Steroid Biochem Mol Biol 103(1):51–60
117. Sun J, Kong J, Duan Y, Szeto FL, Liao A, Madara JL, Li YC (2006) Increased NF-kappaB activity in fibroblasts lacking the vitamin D receptor. Am J Physiol Endocrinol Metab 291(2):E315–E322
118. Schwab M, Reynders V, Loitsch S, Steinhilber D, Stein J, Schroder O (2007) Involvement of different nuclear hormone receptors in butyrate-mediated inhibition of inducible NF kappa B signalling. Mol Immunol 44(15):3625–3632
119. Cohen-Lahav M, Douvdevani A, Chaimovitz C, Shany S (2007) The anti-inflammatory activity of 1, 25-dihydroxyvitamin D3 in macrophages. J Steroid Biochem Mol Biol 103(3–5):558–562
120. Cohen-Lahav M, Shany S, Tobvin D, Chaimovitz C, Douvdevani A (2006) Vitamin D decreases NFkappaB activity by increasing IkappaBalpha levels. Nephrol Dial Transplant 21(4):889–897
121. Bao BY, Yao J, Lee YF (2006) 1alpha, 25-dihydroxyvitamin D3 suppresses interleukin-8-mediated prostate cancer cell angiogenesis. Carcinogenesis 27(9):1883–1893
122. Han J, Jogie-Brahim S, Oh Y (2007). New paradigm foe antitumor action of IGF binding protein-3 (IGFBP-3): novel NF-kB inhibitory effect of IGFBP-3 in prostate cancer. In: Proceedings from the Endocrine Society Meeting, Toronto, Canada; pp P3–370
123. Criswell T, Leskov K, Miyamoto S, Luo G, Boothman DA (2003) Transcription factors activated in mammalian cells after clinically relevant doses of ionizing radiation. Oncogene 22(37):5813–5827
124. Kimura K, Bowen C, Spiegel S, Gelmann EP (1999) Tumor necrosis factor-alpha sensitizes prostate cancer cells to gamma-irradiation-induced apoptosis. Cancer Res 59(7):1606–1614
125. Xu Y, Fang F, St Clair DK, Josson S, Sompol P, Spasojevic I, St Clair WH (2007) Suppression of RelB-mediated manganese superoxide dismutase expression reveals a primary mechanism for radiosensitization effect of 1alpha, 25-dihydroxyvitamin D(3) in prostate cancer cells. Mol Cancer Ther 6(7):2048–2056

126. De Marzo AM, Marchi VL, Epstein JI, Nelson WG (1999) Proliferative inflammatory atrophy of the prostate: implications for prostatic carcinogenesis. Am J Pathol 155(6):1985–1992
127. Nelson WG, De Marzo AM, Isaacs WB (2003) Prostate cancer. N Engl J Med 349(4):366–381
128. De Marzo AM, DeWeese TL, Platz EA, Meeker AK, Nakayama M, Epstein JI, Isaacs WB, Nelson WG (2004) Pathological and molecular mechanisms of prostate carcinogenesis: implications for diagnosis, detection, prevention, and treatment. J Cell Biochem 91(3):459–477
129. Herszenyi L, Miheller P, Tulassay Z (2007) Carcinogenesis in inflammatory bowel disease. Dig Dis 25(3):267–269
130. Itzkowitz SH, Yio X (2004) Inflammation and cancer IV. Colorectal cancer in inflammatory bowel disease: the role of inflammation. Am J Physiol Gastrointest Liver Physiol 287(1):G7–G17
131. Seril DN, Liao J, Yang GY, Yang CS (2003) Oxidative stress and ulcerative colitis-associated carcinogenesis: studies in humans and animal models. Carcinogenesis 24(3):353–362
132. Tsai CJ, Cohn BA, Cirillo PM, Feldman D, Stanczyk FZ, Whittemore AS (2006) Sex steroid hormones in young manhood and the risk of subsequent prostate cancer: a longitudinal study in African-Americans and Caucasians (United States). Cancer Causes Control 17(10):1237–1244
133. Gupta S, Adhami VM, Subbarayan M, MacLennan GT, Lewin JS, Hafeli UO, Fu P, Mukhtar H (2004) Suppression of prostate carcinogenesis by dietary supplementation of celecoxib in transgenic adenocarcinoma of the mouse prostate model. Cancer Res 64(9):3334–3343
134. Alagbala A, Moser MT, Johnson CS, Trump DL, Posner GH, Foster BA (2005) Prevention of prostate cancer progression with vitamin D compounds in the transgenic adenocarcinoma of the mouse prostate (TRAMP) model. In: Fourth annual AACR International Conference on Frontiers in Cancer Prevention Research, 2005, Baltimore, MD, pp 107
135. Alagbala AA, Moser MT, Johnson CS, Trump DL, Posner GH, Foster BA (2006) 1α, 25-dihydroxyvitamin D3 and its analog (QW-1624F2–2) prevent prostate cancer progression. In: 13th Workshop on Vitamin D, 2006, Victoria, BC, Canada, pp 95
136. Banach-Petrosky W, Ouyang X, Gao H, Nader K, Ji Y, Suh N, DiPaola RS, Abate-Shen C (2006) Vitamin D inhibits the formation of prostatic intraepithelial neoplasia in Nkx3.1;Pten mutant mice. Clin Cancer Res 12(19):5895–5901

Chapter 4
The Epidemiology of Vitamin D and Cancer Risk

Edward Giovannucci

Abstract Vitamin D status and cancer risk has been investigated in a number of epidemiologic studies. The methods to estimate vitamin D status have included direct measures of circulating 25(OH)vitamin D (25(OH)D) levels, surrogates or determinants of 25(OH)D, including region of residence, intake, and sun exposure estimates. For colorectal cancer, the evidence for an inverse association between vitamin D status and risk is quite consistent. Evidence for breast cancer is intriguing, but prospective studies of 25(OH)D are sparse and conflicting. For prostate cancer, the data on circulating 25(OH)D have suggested no association or a weak inverse association, but studies of sun exposure on prostate cancer risk are more suggestive. It is plausible that for prostate cancer, vitamin D level, much longer before the time of diagnosis, is the most relevant exposure. Most of the epidemiologic studies to date have examined vitamin D status in relation to risk of cancer, but emerging evidence suggests that vitamin D may also be an important factor for cancer progression and mortality. Further study is needed to establish when in the life span and on what stages of carcinogenesis vitamin D is relevant, the precise intakes and levels required for benefit, and which cancer sites are most affected.

Keywords Epidemiology • Cancer risks • Vitamin D level • Vitamin D intake • Colorectal cancer • Prostate cancer • Breast cancer • Pancreatic cancer • Ovarian cancer • Esophageal cancer • Gastric cancer • Non-Hodgkin lymphoma • 25(OH)-vitamin D • UV radiation

---

E. Giovannucci (✉)
Department of Nutrition, 2–371, Harvard School of Public Health,
665 Huntington Avenue, Boston, MA 02115, USA
and
Department of Epidemiology, Harvard School of Public Health,
Boston, MA 02115, USA
and
Channing Laboratory, Department of Medicine,
Brigham and Women's Hospital and Harvard Medical School,
181 Longwood Avenue, Boston, MA 02115, USA
e-mail: egiovann@hsph.harvard.edu

D.L. Trump and C.S. Johnson (eds.), *Vitamin D and Cancer*,
DOI 10.1007/978-1-4419-7188-3_4, © Springer Science+Business Media, LLC 2011

Abbreviations

Ca	Calcium
CI	Confidence interval
D2	Ergocalciferol
D3	Cholecalciferol
1,25(OH)$_2$D	1,25-Dihydroxyvitamin D
25(OH)D	25-Hydroxyvitamin D
IU	International unit
nmol/L	Nanomoles per liter
ng/mL	Nanograms per milliliter
NHL	Non-Hodgkin lymphoma
RCT	Randomized controlled trial
RR	Relative risk
UV-B	Ultraviolet B light

4.1 Introduction

The hypothesis that vitamin D confers protection against some cancers was first based on some epidemiologic observations. As early as 1937, Peller and Stephenson hypothesized that sunlight exposure, by inducing skin cancer, could induce some degree of immunity against some internal cancers [1]. Then in 1941, Apperly demonstrated an association between latitude and cancer mortality, leading him to hypothesize a direct benefit of sunlight on cancer mortality independent of any effect on skin cancer [2]. These observations and hypotheses went largely ignored until the early 1980s when Garland and Garland hypothesized that inadequate vitamin D status resulting from lower solar UV-B radiation exposure accounted for the association between higher latitudes and increased mortality of colon cancer [3], breast cancer [4], and ovarian cancer [5]. Thereafter, this proposed anticarcinogenic effect of vitamin D was extended to prostate cancer [6, 7] and to other malignancies [8].

These initial observations formed the basis of the vitamin D cancer hypothesis. In the past several decades, laboratory studies have discovered numerous anticarcinogenic properties of vitamin D, including inducing differentiation and inhibiting proliferation, invasiveness, angiogenesis, and metastatic potential. Over this time, a variety of epidemiologic study designs have been utilized to assess exposure to vitamin D at the individual level, and then to examine the estimated vitamin D level to the risk of a specific cancer or to total cancer. This chapter will review the epidemiologic evidence from cohort and case–control studies of the association between vitamin D status and cancer risk, including studies directly measuring circulating levels of 25(OH)vitamin D (25(OH)D), the presumed relevant metabolite of vitamin D status, and surrogates or determinants of 25(OH)D

4 The Epidemiology of Vitamin D and Cancer Risk

level. Before the specific studies are reviewed, the major strengths and limitations of the various approaches to assess vitamin D status that have been used will be summarized.

4.2 Overview of Study Designs

4.2.1 Prospective Studies of Circulating 25(OH) Vitamin D and Cancer Risk

Some studies have examined plasma or serum 25(OH) level in relation to cancer risk, especially for colorectal cancer and for prostate cancer. There are a few other studies for other endpoints, including breast, ovarian, and pancreatic cancers. The studies based on circulating 25(OH)D level are arguably the "gold standard" among observational studies for testing the vitamin D cancer hypothesis because 25(OH) D accounts not only for skin exposure to UV-B radiation, but also for factors that determine vitamin D status, such as total vitamin D intake and skin pigmentation. In addition, 25(OH)D has a relatively long half-life $(t_{1/2})$ in the circulation of about 2–3 weeks, and thus can provide a fairly good indicator of long-term vitamin D status. For example, in one study of middle-aged to elderly men, the correlation of two 25(OH)D measures approximately 3 years apart was 0.7 [9]. However, it is not clear how the consistency of 25(OH)D over time would be across other populations.

In epidemiologic studies, circulating 25(OH)D has typically been based on a measure in archived blood samples using a nested case–control study design. Because the sample is taken before the diagnosis of cancer, in some cases over a decade before, it is unlikely that any association observed is due to reverse causation, that is, spuriously due to the cancer influencing the blood level. One complexity in studies of 25(OH)D is that typically only one measurement is made, and levels fluctuate seasonally throughout the year due to variances in sun exposure. Several studies have been based on the measurement of 25(OH)D in individuals already diagnosed with cancer; these studies need to be interpreted very cautiously because of the potential for the phenomenon of reverse causation. For example, during treatment period for cancer, exposure to sunlight is likely to be very skewed due to hospitalizations, disability, change in habit, etc. Thus, these types of studies are not summarized in detail here.

4.2.2 Studies of Vitamin D Intake

Vitamin D intakes are relatively low in general because of the scarcity of vitamin D in natural foods and fortification of this vitamin is limited. For example, a glass

of fortified milk (in the USA) contains only 100 IU vitamin D, whereas being exposed to enough UV-B radiation to cause a slight pinkness to the skin with most of the skin uncovered (one minimal erythemal dose) produces vitamin D equivalent to an oral dose of 20,000 IU vitamin D [10, 11]. In most populations, with some exceptions such as in Iceland, much more vitamin D is made from sun exposure than is ingested. Nonetheless, vitamin D intake is an important contributor to 25(OH)D levels, especially in winter months in regions at high latitudes, when it may be the sole contributor. Yet, even with added vitamin D from supplementation and fortification, vitamin D intake at typical levels currently do not raise 25(OH)D levels substantially, and most variability in populations comes from sun exposure. One important consideration of studies of vitamin D intake is that, depending on the specific population, intake of vitamin D may be predominantly from one or a few sources, such as fatty fish, fortified milk, or supplements. Thus, there will tend to be high correlations with other dietary factors (e.g., omega-3 fatty acids in fish, calcium in milk, and other vitamins and minerals in supplements) increasing the possibility of confounding. One important issue is that ergocalciferol (D2) is often used in supplements, and ergocalciferol has been estimated to be only one-fourth as potent as cholecalciferol (D3) in raising 25(OH)D) [12].

4.2.3 Studies of Predicted 25(OH)D Level

A study can use known predictors of 25(OH)D level based on data on the individual level to formulate a predicted 25(OH)D score. For example, based on individuals' reported vitamin D intake, region of residence (surrogate of UV-B exposure), outdoor activity level, skin color, and body mass index, a quantitative estimate of the expected vitamin D level can be made. The predicted 25(OH)D approach may have some advantages and disadvantages compared to the use of a single measurement of circulating 25(OH)D in epidemiologic studies. The measurement of 25(OH)D is more direct, intuitive, and encompasses some of the sources of variability of 25(OH)D not taken into account by the score. The most important of these is actual sun exposure behaviors, such as type of clothing and use of sunscreen. However, in some aspects, the predicted 25(OH)D measure may provide a comparable or superior estimate of long-term vitamin D status over a single measurement of circulating 25(OH)D. Most importantly, some factors accounted by the predicted 25(OH)D score are immutable (e.g., skin color) or relatively stable (region of residence, body mass index). In contrast, circulating 25(OH)D level has a half-life of 2–3 weeks, and thus a substantial proportion of variability picked up by a single blood measure would likely be due to relatively recent exposures, which may not be representative of long-term exposure. The predicted 25(OH)D approach has been rarely used.

4 The Epidemiology of Vitamin D and Cancer Risk

4.2.4 Case–Control and Cohort Studies of Sun Exposure

Self-reported sun exposure or surrogates such as region of residence and number of sunburns can be used in epidemiologic studies. A number of ecologic studies have examined the vitamin D and cancer hypotheses at the population level, but some case–control and cohort studies, which assess exposure and outcome at the individual level are now available. In principle, confounding may be better controlled because typically more detailed information can be assessed on other covariates in analytic studies. In addition, the study population may be relatively homogeneous, which may reduce the potential for residual or uncontrolled confounding. An additional strength of such studies is that exposure is actually assessed for the individual, whereas in ecologic studies, exposure is inferred – for example, presumably living in sunnier regions may allow for greater opportunity for sun exposure, but actual exposure will depend on individuals' behaviors.

The sun exposure studies have some strengths and some limitations. They do not directly assess vitamin D exposure, and some surrogates that have been used (such as sunburns) may represent acute short-term exposures to sun rather than chronic exposures, which may be more relevant for vitamin D synthesis. There also may be measurement error and perhaps recall bias in case–control studies in assessing past exposures. Some objective methods to assess sun exposure, such as the use of reflectometry, may be useful. One important advantage of these studies is that most blood-based and dietary cohorts are in middle-aged individuals, and the assessment of past sun exposures allows the possibility of estimating vitamin D status at points earlier in life. For some cancers, it is plausible that these earlier time periods may be most relevant.

4.2.5 Randomized Trials

A double-blinded, placebo-controlled, randomized intervention is the "gold standard" in establishing a causal association because in theory, confounding can be largely eliminated as an explanation of a positive result. Because of their expense, these studies have been rarely done in the context of vitamin D and cancer. In practice, these studies have practical limitations, including selection of the effective dose, varying baseline levels of the exposure of interest, poor compliance, contamination by the placebo group adopting the change, and the unknown but presumably long induction period for cancer. Thus, when these studies show a null association, caution must be given not to overinterpret the results. Besides the absence of a true association, one or more of the limitations mentioned above could produce a null association. If a significant association is found, such studies are the strongest evidence of a causal association.

4.3 Colorectal Cancer

4.3.1 25(OH)D Level

Colorectal cancer has been the most frequently studied cancer in relation to vitamin D status. Prospective studies that have examined circulating 25(OH)D levels in relation to colorectal cancer risk have tended to support a lower risk of colorectal cancer among those with higher circulating 25(OH)D levels [13–21]. This finding was demonstrated in a recent meta-analysis of studies of 25(OH)D level and colorectal cancer risk, which was based on 535 colorectal cases in total. [22]. In the meta-analysis, individuals with serum 25(OH) level ≥ 82 nmol/L had a 50% lower incidence of colorectal cancer ($p < 0.01$) when compared to those with levels less than 30 nmol/L. The two largest studies included in the meta-analysis were the Nurses' Health Study and the Women's Health Initiative. In the Nurses' Health Study [15], based on 193 cases of colorectal cancer, the relative risk (RR) decreased in a monotonic fashion across increasing quintiles of plasma 25(OH)D level. The RR was 0.53 (95% confidence interval (CI)=0.27–1.04) comparing the top to bottom quintiles after adjusting for age, body mass index, physical activity, smoking, family history, use of hormone replacement therapy, aspirin use, and dietary intakes. The observational component of the Women's Health Initiative (which was also a randomized trial (RCT) of calcium and vitamin D), based on 322 total cases of colorectal cancer, showed a similar inverse association between baseline 25(OH)D level and colorectal cancer risk; however, detailed analyses on potential confounders were not shown. [21].

Since this meta-analysis was reported, three additional studies on 25(OH)D and colorectal cancer risk have been published. In the Health Professionals Follow-Up Study [23], a nonstatistically significant inverse association between higher plasma 25(OH)D concentration and risk of colorectal cancer was observed, and a statistically significant inverse association for colon cancer (highest versus lowest quintile: multivariate RR=0.46, 95% CI=0.24 to 0.89; P(trend) =.005). In the Japan Public Health Center-based Prospective Study [24], a nested case–control study of 375 incident cases of colorectal cancer from 38,373 study subjects during 11.5 years of follow-up after blood collection, plasma 25(OH)D was not significantly associated with colorectal cancer. However, the lowest category of plasma 25(OH)D was associated with an elevated risk of rectal cancer in both men (RR=4.6; 95% CI= 1.0–20) and women (RR, 2.7, 95% CI, 0.94–7.6), compared with the combined category of the other quartiles. This analysis adjusted for multiple factors, including sex, age, study area, date of blood draw, and fasting time, smoking, alcohol consumption, body mass index, physical exercise, vitamin supplement use, and family history of colorectal cancer. Finally, 25(OH)D levels was examined in relation to colorectal cancer mortality risk in the Third National Health and Nutrition Examination Survey [25]. That analysis examined 16,818 participants, who were followed from 1988–1994 through 2000, over which 66 cases of fatal colorectal cancer were identified. The risk of colorectal cancer mortality was inversely related

4   The Epidemiology of Vitamin D and Cancer Risk   79

to baseline serum 25(OH)D level, with levels 80 nmol/L or higher associated with a 72% risk reduction (95% CI = 32% to 89%) compared with levels <50 nmol/L, P(trend) = 0.02.

4.3.2   Predicted 25(OH)D Level

Predicted 25(OH)D was examined in relation to risk of colorectal cancer in the Health Professionals Follow-Up Study [26]. This approach required two steps. First, plasma 25(OH)D levels were measured in a sample of 1,095 men of this cohort. Then, factors hypothesized to influence circulating 25(OH)D levels, including geographical region, skin pigmentation, dietary intake, supplement intake, body mass index, and leisure-time physical activity (a surrogate of potential exposure to sunlight UV-B) were used as the independent variables in multiple linear regression model to develop a predicted 25(OH)D score, the dependent variable [26]. Secondly, the score, after being validated in an additional sample of men with 25(OH)D measured, was calculated for each of 47,800 cohort members and examined in relation to subsequent risk of cancer using Cox proportional hazards regression. There were 691 cases of colorectal cancer diagnosed from 1986 to 2000 in this cohort. The analysis showed that a 25-nmol/L (10 ng/mL) increment in predicted 25(OH)D was associated with a reduced risk of colorectal cancer (multivariate RR = 0.63; 95% CI 0.48–0.83), an association which persisted after controlling for body mass index or physical activity, which are contributors to the 25(OH)D score, and known risk factors for colorectal cancer.

4.3.3   Dietary Intake

As discussed above, dietary and supplementary intake of vitamin D are relatively moderate predictors of 25(OH)D status, but may be relatively more important in winter months in high latitude climates, when sunlight UV-B exposure is low. Dietary or supplementary vitamin D has been investigated in relation to colorectal cancer risk in cohort studies of men [27, 28] and women [29–31] or both sexes [32, 33], as well as in case–control studies [34–41]. The majority of these studies suggested inverse associations for colon or rectal cancer, or both endpoints combined [27–30, 33, 35, 37, 39, 40, 42]. The studies that took into account supplementary vitamin D may be more informative as dietary vitamin D intake alone tends to be low in most populations. For studies that also assessed supplementary vitamin D, the average intake of the top category was approximately 700–800 IU/day, whereas in populations where vitamin D in supplements are rarely consumed, the highest intake category averaged around 200–300 IU/day. An association with vitamin D, if one exists, is more likely to be observed in the higher intake populations with supplements assessed. In fact, in these studies, a risk reduction in the top versus bottom

category was generally seen (risk reduction of 34% [28], 46% [29], 58% [30], 24% [31], 30% [40], 29% male, 0% female [33], 50% males, 40% females [41], and 28% male, 11% female [42]). In the other studies, weaker reductions or no reductions were seen. These studies tend to support a role of vitamin D, though the high intake groups tend to be enriched with multivitamin users and consumers of (fortified) milk and fatty fish, which could have an anticancer effect unrelated to vitamin D.

4.3.4 Sun Exposure

Besides ecological studies that examine sun exposure (estimated by region) studies can examine sun exposure at the individual level. One such study was a death certificate-based case–control study, which examined mortality from female breast, ovarian, colon, and prostate cancers in relation to residential and occupational exposure to sunlight [43]. In this study, the cases consisted of all deaths from these cancers between 1984 and 1995 in 24 states of the USA, allowing for a very large number of 153,511 deaths from colorectal cancer. The controls in this study were age-frequency-matched to a series of cases, and deaths from cancer and certain neurological diseases were excluded because of possible relationships with sun exposure. Non-melanoma skin cancer served as a positive "control" group, and an expected positive association was found between individuals with presumably higher opportunity to sun exposure and skin cancer risk. The authors used multivariate analyses, which controlled for age, sex, race, and mutual adjustment for residence, occupation (outdoor versus indoor), occupational physical activity levels and socioeconomic status. For colon cancer, individuals with a high compared to low exposure to sun based on residence were at decreased risk (RR=0.73, 95% CI, 0.71–0.74), and individuals with outdoor occupations (RR = 0.90; 95% CI, 0.86–0.94) and occupations that required more physical activity (RR = 0.89; 95% CI, 0.86–0.92) were at lower risk. The inverse association with outdoor occupation was strongest among those living in the highest sunlight region (RR = 0.81; 95% CI, 0.74–0.90), suggesting that sunlight was a key factor associated with outdoor occupation that reduced the risk.

4.3.5 Vitamin D and Colorectal Adenoma

Adenomas are precursors to the majority of colorectal cancers. Because adenomas can be detected decades prior to development of cancer, they can serve as a predictive indicator for cancer [44, 45]. The malignancy transformation rate for adenomas ranges from 5% for small adenomas to 50% for villous adenomas over 2 cm in diameter [46, 47]. Some studies have examined circulating 25(OH)D or vitamin D intake and risk colorectal adenomas. A recent meta-analysis of colorectal adenoma, comprised of seven studies on 25(OH)D and 12 on vitamin D intake published before December 2007, was performed [48]. The meta-analysis found that circulating

4   The Epidemiology of Vitamin D and Cancer Risk   81

25(OH)D was inversely associated with risk of total colorectal adenoma (RR = 0.70 (95% CI: 0.56–0.87, for high versus low circulating 25(OH)D) and advanced adenoma (RR = 0.64, 95% CI: 0.45–0.90). In addition, the highest quintile of vitamin D intake was associated with a decreased risk of colorectal adenoma compared to low vitamin D intake (RR = 0.89; 95% CI: 0.78–1.02), recurrent adenoma (RR = 0.88; 95% CI: 0.72–1.07), and advanced adenoma (RR = 0.75, 95% CI: 0.57–0.99). The overall results of this meta-analysis indicate that vitamin D status, assessed through intake and circulating 25(OH)D, is associated with a decreased risk of colorectal adenoma, especially advanced adenoma.

4.3.6   Randomized Controlled Trial

The Women's Health Initiative was a randomized placebo-controlled trial that examined 400 IU vitamin D plus 1,000 mg/day of elemental calcium in 36,282 postmenopausal women in relation to risk of colorectal cancer (n = 322 cases) and other endpoints over 7 years [21]. This study found no suggestion of a benefit of the intervention on incidence of colorectal cancer, but this trial had some important limitations, which preclude a definitive answer. First, and most importantly, the dose of 400 IU/day of vitamin D was likely insufficient to yield a meaningful contrast of 25(OH)D between the treated and the control groups. The anticipated increase of circulating 25(OH)D level following an increment of 400 IU/day would be approximately 7.5 nmol/L, and was likely even less given the suboptimal compliance in this study. In the epidemiologic studies of 25(OH)D, the contrast between the high and low quintiles was generally at least 50 nmol/L (20 ng/mL) [22]. Second, epidemiologic data, although limited, suggest that any influence of vitamin D (and calcium) intakes may require at least 10 years to demonstrate a risk reduction for colorectal cancer [30], so possibly the time duration of the trial may not have been sufficiently long. Third, the Women's Health Initiative study had a factorial design with hormonal replacement use, and a post hoc analysis suggested an interaction whereby women who had not taken hormones may have benefited from the vitamin D and calcium intervention, but those on hormones did not [49]. If so, the effect of vitamin D may have been diluted in the overall study population.

4.4   Prostate Cancer

4.4.1   25(OH) Vitamin D

Most of the studies of circulating 25(OH)D level and prostate cancer risk have not found clear risk reductions for prostate cancer associated with higher 25(OH)D levels, although some of the studies suggested weak inverse associations [20, 50–54]. The only two studies [55, 56] that support an inverse association were conducted in

Nordic countries, where the vitamin D levels may be particularly low due to low solar UV-B exposure at higher latitudes. However, even these findings were equivocal, because one of these studies also found an increased risk in men with the highest 25(OH)D values, which suggested a U-shaped relationship between vitamin D and prostate cancer risk [56]. Several studies found supportive [50] or suggestive [51] inverse associations for circulating 1,25(OH)$_2$D levels and prostate cancer risk, especially for aggressive prostate cancer. In the Physicians' Health Study, the participants with both low 25(OH)D and 1,25(OH)$_2$D were at about a twofold higher risk of aggressive prostate cancer [57]. In the Health Professionals Follow-up Study, both lower 25(OH)D and 1,25(OH)$_2$D levels were associated with lower prostate cancer risk [53], but these were mostly organ-confined prostate cancers detected through PSA testing. In fact, although numbers of advanced cases were limited (n = 60), there was a suggestive inverse association between 25(OH)D levels and risk of advanced prostate cancer [53]. Finally, in the Prostate, Lung, Colorectal, and Ovarian Cancer Screening Trial, an analysis based on 749 cases and 781 controls found no association, and, in fact, even a suggestively increased risk of aggressive prostate cancer among men with higher circulating 25(OH)D levels [58]. Clearly, studies of circulating 25(OH)D have tended not to support an association for prostate cancer, or at best, have yielded equivocal results.

4.4.2 Predicted 25(OH)D Level

Predicted 25(OH)D was examined in relation to advanced stage prostate cancer in the Health Professionals Follow-Up Study. The method for this analysis was summarized above (section 4.3.2) [26]. Over follow-up from 1986 to 2002, 461 cases of advanced prostate cancer were documented. In the multivariate model, a 25 nmol/L increment in predicted 25(OH)D level was associated with a modest nonsignificant 20% reduction in risk, providing modest support of an association.

4.4.3 Vitamin D Intake

Only four studies were identified in the literature that examined vitamin D intake and prostate cancer risk. None of these studies supported an association between vitamin D intake and prostate cancer incidence [59–62]. Two of these studies [59, 62] assessed supplemental vitamin D in addition to diet.

4.4.4 Sun Exposure

A death-certificate-based case–control study of cancer mortality described previously for colon cancer also examined prostate cancer mortality based on

4 The Epidemiology of Vitamin D and Cancer Risk

97,873 prostate cancer deaths. In this study, residential exposure to sunlight had an inverse association with prostate cancer mortality, though this association was rather modest in magnitude (RR=0.90; 95% CI, 0.86–0.91) [43]. Further, occupation exposure to sunlight was found not to be associated with fatal prostate cancer risk (RR=1.00; 95% CI, 0.96–1.05). Thus, the evidence for a link between sun exposure and prostate cancer mortality was relatively weak, and, of note, the association was weaker than that observed for other cancer sites, including colon cancer, breast cancer, ovarian cancer, and non-Hodgkin's lymphoma using the same study design.

In several case–control and cohort studies, surrogates of sun exposure were examined in relation to prostate cancer risk. One case–control study of advanced prostate cancer is of special interest because it was based on use of a reflectometer to measure overall sun exposure [63]. In this method, the difference between facultative skin pigmentation on the forehead (a sun-exposed site) and constitutive pigmentation on the upper underarm (a sun-protected site) is used to estimate sun exposure. Sun exposure estimated by reflectometry was inversely associated with risk of advanced prostate cancer (RR=0.51; 95% CI, 0.33–0.80). Further, this study found that high occupational outdoor activity level was associated with a suggestively reduced risk of advanced prostate cancer relative to low exposure (RR = 0.73; 95% CI, 0.48–1.11).

A cohort study was based on 5,811 non-Hispanic white men using National Health and Nutrition Examination Survey I data; of these men, 151 (102 nonfatal, 59 fatal) were diagnosed with prostate cancer over follow-up from 1971 to 1992. Several measures of presumed sun exposure were associated with significantly lower risk of prostate cancer; these were longest residence in regions with high solar radiation (RR=0.66; 95% CI, 0.47–0.93), and high solar radiation in the state of birth (RR = 0.49; 95% CI, 0.27–0.90) [64]. The associations were stronger for fatal prostate cancer. Frequent recreational sun exposure in adulthood was associated with a lower risk of fatal prostate cancer only (RR = 0.47; 95% CI, 0.23–0.99). On the basis of these findings, the authors hypothesized that both early-life and adult exposure to sun are critical for prostate carcinogenesis, although the study did not have adequate power to simultaneously adjust for adult and early-life residences.

Studies in the UK are of especial interest given the low sun exposure in that region. Several case–control studies in the UK reported on factors such as childhood sunburns, holidays in a hot climate, and skin type in relation to prostate cancer risk. Rather striking findings were found in subgroups characterized by childhood sunburns, holidays in a hot climate, and skin type; specifically, a significant 13-fold higher risk of prostate cancer was observed in men with combinations of high sun exposure/light skin compared to low sun exposure/darker skin type [65, 66]. Furthermore, self-reported UV exposure parameters and skin type in 553 men with prostate cancer were studied in association with stage, Gleason score, and survival after starting hormone manipulation therapy [67]. UV exposures 10, 20, and 30 years before diagnosis were inversely associated with stage, and the RR for UV exposure 10 years before diagnosis was lowest (RR=0.69, 95% CI=0.56–0.86). RRs were lower in men with (lighter)

skin types I/II than III/IV. Also, men with skin types I/II experienced longer survival after beginning hormone therapy (RR=0.62, 95% CI=0.40–0.95). These findings also support that vitamin D may influence prostate cancer mortality.

4.5 Breast Cancer

4.5.1 25(OH)D Level

Two large prospective studies have examined circulating 25(OH)D levels in relation to breast cancer risk. The first of these was the Nurses' Health Study, which was based on 701 breast cancer cases and 724 controls [68]. The results suggested a moderate association; women in the highest quintile of 25(OH)D had an RR of 0.73, 95% CI=0.49–1.07 (P trend=0.06) when compared with women in the lowest quintile of 25(OH)D. In a subgroup analysis, this inverse association was primarily in women of ages 60 years and older, suggesting that vitamin D may be more important for postmenopausal than for premenopausal breast cancer. Another large prospective study of 25(OH)D level and breast cancer risk was based on the Prostate, Lung, Colorectal, and Ovarian Cancer Screening Trial study, over which 1,005 incident cases of breast cancer were followed from 1993 to 2005, with a mean time between blood draw and diagnosis of 3.9 years [69]. In this cohort, women with 25(OH)D levels in the highest quintile were not at lower risk for breast cancer when compared to women with values in the low quintile (RR=1.04; 95%CI=0.75–1.45) nor was any trend observed p(trend)=0.81). Unlike in the Nurses' Health Study, risk of breast cancer was not reduced even in the stratum of older women. The range of 25(OH)D was comparable to that in the Nurses' Health Study.

Two other small studies are noteworthy. A small nested case–control study, based on only 28 cases, reported a nonsignificant inverse association for breast cancer risk [25]. Also, a nested case–control study based on 96 breast cancer cases found no association between prediagnostic 1,25(OH)$_2$D concentration and risk of breast cancer, but circulating 25(OH)D was not examined in this study.

4.5.2 Vitamin D Intake

A number of studies have examined vitamin D intake in relation to breast cancer risk. A meta-analysis for studies identified six such studies conducted up to June 2007 [70]. In the meta-analysis, vitamin D intake was not associated with risk of breast cancer (summary RR=0.98; 95%CI=0.93–1.03). However, significant heterogeneity (p<0.01) appeared to be due to the level of vitamin D intake. When the studies were stratified into those with vitamin D intakes higher than 400 IU or lower than this amount, a modest association was observed only in those three studies where intakes

were ≥ 400 IU (summary RR=0.92, 95%CI=0.87–0.97; p(heterogeneity)=0.14). One of the studies with high intakes, the Nurses' Health Study, is of interest because vitamin D intake was updated every 2–4 years, which allowed for an improved estimate of long-term intake [71]. That study, which was based on 3,482 cases of breast cancer, found that total vitamin D intake (dietary plus supplementary intake) was inversely associated with the risk of incident breast cancer (multivariate RR=0.72; 95%CI=0.55–0.94) for >500 versus ≤ 150 IU/day of vitamin D. Notably, similar inverse associations were observed with other components of dairy foods, including lactose and calcium, indicating the difficulty of teasing out the independent effects. Nonetheless, total vitamin D intake had a stronger inverse association than did either dietary or supplemental vitamin D intake individually, which suggested that vitamin D was indeed the relevant causal factor.

4.5.3 Sun Exposure

The death certificate-based case–control study of cancer mortality described above found that greater residential exposure to sunlight (RR=0.74; 95% CI, 0.72–0.76) and occupational exposure to sunlight (RR = 0.82, 95% CI, 0.70–0.97) were associated with reduced mortality from female breast cancer (n = 130,261 cases) [43]. The study also found that the magnitude of the association between outdoor employment and reduced breast cancer mortality was strongest in regions of greatest residential sunlight (OR=0.75, 95% CI, 0.55–1.03), suggesting that sun light exposure was the primary reason underlying the reduced risk with outdoor employment.

A population-based case–control study of 972 cases and 1,135 controls conducted in Canada, examined self-reported sun exposure behaviors at different age periods in relation to risk of breast cancer [72]. The study found a significantly reduced risk of breast cancer associated with increasing estimated sun exposure from ages 10 to 19 (RR=0.65; 95% CI, 0.50–0.85 for the highest quartile of outdoor activities versus the lowest; P for trend=0.0006). Notably, the associations from ages 20 to 29 years were weaker, and no evidence was observed for exposures for ages 45–54 years. These results suggest that the relevant time for vitamin D exposure and reduced breast cancer risk occurs primarily or solely during adolescence.

A population-based case–control study was conducted based on 1,788 incident cases of advanced breast cancer and 2,129 controls over the years 1995–2003 among Hispanic, African-American, and non-Hispanic White women from California [73]. In this study, among women with light constitutive skin pigmentation, those with high sun exposure index based on reflectometry had a reduced risk of advanced breast cancer (RR=0.53, 95% CI: 0.31, 0.91). However, among women with medium or dark pigmentation, high sun exposure index was not associated with risk. To explain these discordant findings, the investigators posited that these measures based on reflectometry may reflect vitamin D status better in more lightly pigmented women than in darker skinned women. Finally, in a relatively small

cohort of 5,009 women, among whom 190 women developed incident breast cancer, several measures of sunlight exposure and dietary vitamin D intake showed a moderate inverse association with risk of breast cancer [74].

4.6 Pancreatic Cancer

4.6.1 25(OH)D Level

Only one report of circulating 25(OH)D in relation to pancreatic cancer was found in the literature. This study was based on the Alpha-Tocopherol, Beta-Carotene Cancer Prevention Cohort of male Finnish smokers [75]. The analysis was based on 200 cases of pancreatic cancer and 400 matched controls. In this study, men with higher vitamin D concentrations were at significantly increased risk for pancreatic cancer (highest versus lowest quintile, >65.5 versus <32.0 nmol/L: multivariate RR, 2.92; 95% CI, 1.56–5.48, P(trend)=0.001). This finding was unanticipated and persisted in detailed multivariate analysis and in a number of sensitivity analyses.

4.6.2 Predicted 25(OH)D

Only one analysis, based on the Health Professionals Follow-up Study, was based on predicted 25(OH)D to examine risk of pancreatic cancer (n=170) [26]. In this study, a 25 nmol/L increment in predicted 25(OH)D was associated with a significant reduction in pancreatic cancer risk, even after detailed multivariate adjustment (multivariate RR=0.49; 95% CI=0.28–0.86). These results were confirmed in the Nurses' Health Study [76]. Why this result differs markedly from those based on circulating 25(OH)D in the Alpha-Tocopherol, Beta-Carotene Cancer Prevention Cohort is unclear, but some differences include that in the Health Professionals Study very few men were current smokers (<10%), the method of assessing vitamin D status was different, the range of vitamin D was much lower in Finland due to lower sun exposure, and the men from the Health Professionals study generally had a healthier lifestyle.

4.6.3 Vitamin D Intake

Only one report examining vitamin D intake in relation to pancreatic cancer risk was identified. This was a prospective study, which combined data from the Nurses' Health Study and the Health Professionals Follow-Up Study, and assessed total vitamin D intake from diet and supplements [77]. The analysis was based on 365 incident cases of pancreatic cancer over 16 years of follow-up with repeated dietary measures generally every 4 years. The analysis showed a significant reduction in risk of pancreatic cancer when comparing vitamin D intakes of ≥600 IU/day to total

4 The Epidemiology of Vitamin D and Cancer Risk

vitamin D intake <150 IU/day (multivariate RR=0.59; 95% CI, 0.40–0.88; p(trend)=0.01). Controlling for a number of other dietary and lifestyle factors did not alter this inverse association.

4.7 Ovarian Cancer

4.7.1 25(OH)D Level

Only one report of plasma 25(OH)D in relation to risk of epithelial ovarian cancer was identified in the literature. This study was conducted using data from three prospective cohorts: the Nurses' Health Study, the Nurses' Health Study II, and the Women's Health Study [78]. The analysis was based on 224 cases and 603 controls from the combined cohorts. The findings showed no significant association between 25(OH)D and ovarian cancer risk (top versus bottom quartile: RR=0.83; 95% CI, 0.49–1.39; P(trend)=0.57). However, after the first 2 years of follow-up were excluded, an inverse association was suggested (RR=0.67, 95%CI, 0.43–1.05). This finding is noteworthy because ovarian cancer is often diagnosed at advanced stages, so reverse causation may obscure the results from the early follow-up period. Another finding was that a significant inverse association with 25(OH)D levels was observed among overweight and obese women (RR = 0.39; 95% CI, 0.16–0.93; P(trend)=0.04). Finally, women with adequate versus inadequate 25(OH)D levels had a modestly decreased risk of the subgroup of serous ovarian cancer (RR, 0.64; 95% CI, 0.39–1.05). Though these subgroup findings are noteworthy, they require replication.

4.7.2 Sun Exposure

In the death certificate-based case–control study of ovarian cancer mortality (n=39,002 cases) in association with residential and occupational exposure to sunlight described above (see section 4.3.4) [43], residential (RR=0.84; 95% CI, 0.81–0.88) but not occupational exposure to sunlight was inversely associated with ovarian cancer mortality. Thus, this evidence is suggestive of a role of sunlight on ovarian cancer risk, but of a magnitude weaker than that for colon and breast cancer.

4.8 Esophageal and Gastric Cancers

4.8.1 25(OH)D Level

Cancers of the esophagus and stomach are relatively rare in developed countries, such as the USA, but are extremely common in some areas, particularly in Linxian, China. One study of vitamin D, nested in a randomized trial of micronutrients [79],

was conducted in Linxian, China. The analysis included 545 squamous cell carcinomas of the esophagus, 353 adenocarcinomas of the gastric cardia, and 81 gastric noncardia adenocarcinomas diagnosed over 5.25 years of follow-up. For squamous cell carcinomas of the esophagus, when comparing men in the fourth quartile of serum 25(OH)D concentrations to those in the first, a positive association was found (RR = 1.77; 95%CI, 1.16–2.70, P trend = 0.0033). In contrast, no association was found in women (RR = 1.06 (95% CI = 0.71–1.59), P trend = 0.70), or for gastric cardia or noncardia adenocarcinoma. The cut-point for the top quartile was only 48.7 nmol/L.

The other study, from Linxian, China, was a cross-sectional analysis of 720 subjects who underwent endoscopy and biopsy, and were categorized by the presence or absence of histologic esophageal squamous dysplasia [80]. The mean level of 25(OH)D in this population was only 35 nmol/L. In this high-risk area, 230 of 720 subjects were diagnosed with squamous dysplasia. In multivariate analyses, the subjects in the highest compared with the lowest quartile of 25(OH)D were at a significantly increased risk of squamous dysplasia (RR = 1.86; 95% CI, 1.35–2.62). This association was observed both in men (RR = 1.74; 95% CI, 1.08–2.93) and women (RR = 1.96; 95% CI, 1.28–3.18).

4.9 Non-Hodgkin Lymphoma

4.9.1 Sun Exposure

The relationship between sun exposure and non-Hodgkin Lymphoma (NHL) is of special interest because some studies suggest a positive association between NHL and skin cancer, suggesting that sunlight may increase risk of NHL. Partly based on this relationship, a number of case–control studies have examined sun exposure and risk for NHL. The International Lymphoma Epidemiology Consortium (InterLymph) recently presented results summarizing the association between sun exposure and NHL risk in a pooled analysis of 10 case–control studies [81]. The studies comprised 8,243 cases and 9,697 controls of European origin and were conducted in the USA, Europe, and Australia. Four measures of self-reported personal sun exposure were assessed at interview; these included time (1) outdoors and not in the shade in warmer months or summer, (2) in the sun in leisure activities, (3) in sun light, and (4) sun bathing in summer. The risk of NHL fell significantly with the composite measure of increasing recreational sun exposure; the multivariate pooled RR (adjusting for smoking and alcohol) = 0.76 (95% CI 0.63–0.91) for the highest exposure category, and the trend was significant (p for trend 0.005). For increasing total sun exposure, a nonsignificant inverse trend was observed with NHL risk (RR = 0.87; 95% CI 0.71–1.05; $P = 0.08$). Of note, the inverse association between recreational sun exposure and NHL risk was statistically significant at 18–40 years of age and in the 10 years before diagnosis, and statistically significant

4   The Epidemiology of Vitamin D and Cancer Risk   89

for B cell lymphomas, but not for T cell lymphomas. However, the numbers for the T cell lymphomas were small and thus the results were inconclusive.

A case–control study based on death certificates of residential and occupational sun exposure and NHL mortality was conducted, as described above (Sect. 4.3.4) [82]. The study, conducted in 24 states in the USA, and based on over 33,000 fatal cases of NHL, found a 17% reduction in risk of NHL mortality that the RR for those residing in states with the highest sunlight exposure (multivariate RR=0.83 (95%CI=0.81 to 0.86). Intriguingly, the risk reduction was remarkably high for those under 45 years of age (RR=0.44 (95%CI=0.28–0.67). The risk of NHL mortality was also reduced with higher occupational sunlight exposure (RR=0.88; 95% CI=0.81–0.96). Besides its effects on vitamin D levels, chronic UV exposure has effects on the immune system [83], and hence sun light exposure could potentially influence neoplasms of the immune system through mechanisms besides vitamin D.

4.10   Total Cancer

4.10.1   Circulating 25(OH)D

Three relatively small studies examined circulating 25(OH)D in relation to risk of total cancer. One analysis was conducted in the Third National Health and Nutrition Examination Survey [25]. In this analysis, there were 16,818 participants who were followed from 1988 to 1994 through 2000. Over this follow-up, 536 cancer deaths were identified. Baseline vitamin D status was not significantly associated with total cancer mortality, although a nonsignificant inverse trend ($P=0.12$) was observed in women only. There were generally too few specific cancer sites to be examined, but colorectal cancer mortality was inversely related to serum 25(OH)D level (discussed above), and a nonsignificant inverse association was observed for breast cancer.

Two small studies were conducted in specialized populations. In the Ludwigshafen Risk and Cardiovascular Health study, 25(OH)D was measured in 3,299 patients who provided a blood sample in the morning before coronary angiography [84]. These subjects were followed for a median period of about 8 years, over which 95 cancer deaths were recorded. The multivariate analysis adjusted for age, sex, body mass index, smoking, retinol, exercise, alcohol, and diabetes history. Higher 25(OH)D level at baseline appeared to be associated with a lower risk of total cancer (multivariate RR=0.45; 95%CI=0.22–0.93) for the fourth quartile versus the first quartile of 25(OH)D. The risk decrease was monotonic, and the RR per increase of 25 nmol/L in serum 25(OH)D concentrations was 0.66 (95%CI=0.49–0.89).

The other study examined pre-transplant 25(OH)D levels in 363 renal transplant recipients at Saint-Jacques University Hospital at Besancon, France [85]. Mean

25(OH)D was low at pre-transplant (17.6 ng/mL) and further reduced after the transplant (post-transplant patients are advised to avoid sun exposure). Thirty-two cancers were diagnosed over 5 years of follow-up. 25(OH)D levels were lower in patients who developed cancer after transplantation (13.7 ± 6 vs 18.3 ± 17.8 ng/mL, $P = 0.022$). The risk of total cancer increased by 12% for each 1 ng/mL (2.5 nmol/L) decrement in 25(OH)D (RR = 1.12; 95% CI = 1.04–1.23; $P = 0.021$).

4.10.2 Predicted 25(OH)D

In the Health Professionals Follow-up Study cohort, predicted 25(OH)D levels were examined in relation to risk of total cancer in men. The methods for this analysis were discussed above (Sect. 4.3.2). From 1986 through January 31, 2000, 4,286 incident cancers (excluding organ-confined prostate cancer and non-melanoma skin cancer) and 2,025 cancer deaths from cancer were identified. An increment of 25 nmol/L in predicted 25(OH)D level was associated with a 17% reduction in total cancer incidence (multivariate RR = 0.83, 95%CI = 0.74–0.92) and a 29% reduction in total cancer mortality (multivariate RR = 0.71, 95% CI = 0.60–0.83). The reduction was largely confined to cancers of the digestive tract system, including esophagus, stomach, pancreas, colon, and rectum; as a group, there was a 45% reduction in mortality associated with a 25 nmol/L increment in 25(OH)D (multivariate RR = 0.55, 95% CI = 0.41–0.74).

4.10.3 Randomized Trials (RCT)

Two RCTs of vitamin D supplementation and total cancer risk were identified. The first was an RCT of 2,037 men and 649 women aged 65–85 years living in the general community in the UK. The subjects took either 100,000 IU oral vitamin D (cholecalciferol) supplementation or placebo every 4 months over 5 years [86]. After treatment, the 25(OH)D level was 74.3 nmol/L in the vitamin D group and 53.4 nmol/L in the placebo group. There were 188 incident cancer cases in the vitamin D group and 173 in the placebo group, and no overall reduction was observed for cancer risk (RR = 1.09, 95%CI = 86–1.36), although a slight, nonsignificant reduction in risk of cancer mortality was suggested (RR = 0.86; 95%CI = 0.61–1.20).

The other RCT was a 4-year, community-based, double-blind, placebo-controlled RCT of vitamin D and calcium of 1,179 US women aged >55 years living in Nebraska; the primary outcome was fracture incidence and the principal secondary outcome was cancer incidence [87]. The subjects were randomly assigned to receive daily 1,400–1,500 mg supplemental calcium/d alone (Ca-only), supplemental calcium plus 1,100 IU vitamin D (Ca+D), or placebo. The achieved 25(OH)D level after treatment was 96 nmol/L in the vitamin D group and 71 in the non-vitamin D

4   The Epidemiology of Vitamin D and Cancer Risk   91

groups. A limitation of the study was the relatively small number of total cancers of 50 in total and 37 after the first year. Nonetheless, total cancer incidence was lower in the Ca + D women than in the placebo-control subjects ($P<0.03$), and the RR of incident cancer was 0.40 ($P=0.01$) in the Ca+D group and 0.53 ($P=0.06$) in the Ca-only group. In a sub-analysis confined to cancers diagnosed after the first year, the RR for the Ca+D group was 0.23 (95% CI=0.09–0.60; $P<0.005$); no significant risk reduction was observed for the Ca-only group. In multivariate models, both vitamin D treatment and higher 25(OH)D levels were significant, independent predictors of reduced cancer incidence.

4.11   Summary

Ecologic studies that compared cancer mortality rates in different regions within the USA initiated the hypothesis that high vitamin D levels may lower risk of various cancers, a hypothesis that was subsequently supported by biologic evidence. Colorectal cancer was the first cancer type hypothesized to be related to vitamin D status [3]. Subsequently, although regional UV-B was shown to be associated with a number of cancers, the magnitude of the association appeared to be strongest for colorectal cancer [8]. This finding for colorectal cancer was confirmed in epidemiologic studies of circulating 25(OH)D and colorectal cancer risk, in which individuals in the high quartile or quintile of 25(OH)D had a 40–50% risk reduction relative to those in the lowest group. Inverse associations have also been observed for predicted vitamin D, sun light exposure and dietary intake, and for the colorectal cancer precursor, the adenoma. The consistency of this association in diverse circumstances indicates that an uncontrolled or unaccounted confounding factor is unlikely to account for these associations.

   For breast cancer, for which an inverse correlation has been observed between regionally estimated UV-B in ecologic studies, the epidemiologic data are sparser and less consistent. The evidence from analytic epidemiologic studies of vitamin D and breast cancer are somewhat conflicting. There have been only two adequately powered prospective studies of circulating 25(OH)D levels, and these have yielded inconsistent results. The studies of vitamin D intake are modestly supportive but limited by the generally low intakes of vitamin D. One case–control study provided intriguing findings: more sun exposure primarily during ages 10–19, but not at other ages, than controls, was inversely associated with risk of breast cancer. Because recall bias is a possible explanation, replicating these results in prospective settings is important. Interestingly, adolescent exposures have often been found to be critical in determining subsequent breast cancer risk, probably because the breast tissue are rapidly developing over this time period.

   Ecologic studies of regional UV-B and cancer mortality find an inverse association with prostate cancer mortality. However, this association appears not as strong as that for colorectal or breast cancer [8], and in one study, was limited to counties north of 40°N latitude, in the USA [88]. The studies of circulating 25(OH)D have

found no or relatively weak nonsignificant associations, and vitamin D intake studies, while sparse, are not supportive of any protection for prostate cancer. In contrast to these findings, studies generally support that more sun exposure is associated with a lower risk of prostate cancer. Two factors are important to consider for prostate cancer. First, most of the evidence to date has focused on incident cancer, while for prostate cancer the association with vitamin D may be stronger for progression and mortality. Second, it has been observed that prostate cancer cells lose 1-alpha-hydroxylase activity early in the carcinogenesis process [89, 90]. This fact may suggest that prostate cancers are insensitive to the effects of circulating 25(OH)D or are only sensitive to it at very early stage decades before the diagnosis. Thus, future studies should focus on studying vitamin D level early in life and on risk of fatal prostate cancer.

For other cancer sites, the data are generally too sparse to support strong conclusions. Some noteworthy findings bear acknowledgement. A study of predicted 25(OH)D in men found associations, particularly for mortality, largely with cancers along the gastro-intestinal tract. This result is interesting especially given that a similar pattern has been observed from some ecologic studies based on region of residence. Gastrointestinal cancers account from one-quarter to one-third of all cancer deaths across different countries, so confirming or refuting this finding is important. In contrast, in some special high-risk populations, circulating 25(OH)D was associated with an increased risk of pancreatic, gastric, and esophageal cancers. This puzzling finding could relate to different etiologies of cancer across populations. In particular, the study of esophageal and gastric cancers was conducted in a very high-risk region in China; no studies have been conducted in regions with traditional risk factors for esophageal cancer. Other cancers that deserve further study in particular are ovarian cancer and NHL.

Few studies have examined the potential influence of vitamin D on cancer mortality or survival from cancer. Some preliminary evidence has suggested that vitamin D status (estimated by season of diagnosis [91] or by blood sample directly [92, 93] around the time of diagnosis) may influence survival from cancer. Also noteworthy is that vitamin D status has been sometimes found to be more strongly related to cancer mortality than cancer incidence. These findings suggest that vitamin D may affect progression of cancer, or prognosis, in addition to incidence. Intervention studies could relatively feasible test the hypothesis that administering vitamin D after diagnosis improves survival.

References

1. Peller S, Stephenson CS (1937) Skin irritation and cancer in the United States Navy. Am J Med Sci 194:326–333
2. Apperly FL (1941) The relation of solar radiation to cancer mortality in North American. Cancer Res 1:191–195
3. Garland CF, Garland FC (1980) Do sunlight and vitamin D reduce the likelihood of colon cancer? Int J Epidemiol 9:227–231

4 The Epidemiology of Vitamin D and Cancer Risk

4. Garland FC, Garland CF, Gorham ED, Young JF (1990) Geographic variation in breast cancer mortality in the United States: a hypothesis involving exposure to solar radiation. Prev Med 19:614–622
5. Lefkowitz ES, Garland CF (1994) Sunlight, vitamin D, and ovarian cancer mortality rates in US women. Int J Epidemiol 23:1133–1136
6. Schwartz GG, Hulka BS (1990) Is vitamin D deficiency a risk factor for prostate cancer? (Hypothesis). Anticancer Res 10:1307–1311
7. Hanchette CL, Schwartz GG (1992) Geographic patterns of prostate cancer mortality. Cancer 70:2861–2869
8. Grant WB (2002) An estimate of premature cancer mortality in the US due to inadequate doses of solar ultraviolet-B radiation. Cancer 94:1867–1875
9. Platz EA, Rimm EB, Willett WC, Kantoff PW, Giovannucci E (2000) Racial variation in prostate cancer incidence and in hormonal system markers among male health professionals. J Natl Cancer Inst 92:2009–2017
10. Holick MF (2004) Vitamin D: importance in the prevention of cancers, type 1 diabetes, heart disease, and osteoporosis. Am J Clin Nutr 79:362–371
11. Hollis BW (2005) Circulating 25-hydroxyvitamin D levels indicative of vitamin D sufficiency: implications for establishing a new effective dietary intake recommendation for vitamin D. J Nutr 135:317–322
12. Vieth R (2005) The pharmacology of vitamin D, including fortification strategies. In: Feldman D et al (eds) Vitamin D, 2nd edn. Elsevier Academic Press, Amsterdam, pp 995–1015
13. Garland CF, Comstock GW, Garland FC, Helsing KJ, Shaw EK, Gorham ED (1989) Serum 25-hydroxyvitamin D and colon cancer: eight-year prospective study. Lancet 2: 1176–1178
14. Tangrea J, Helzlsouer K, Pietinen P et al (1997) Serum levels of vitamin D metabolites and the subsequent risk of colon and rectal cancer in Finnish men. Cancer Causes Control 8:615–625
15. Feskanich D, Ma J, Fuchs CS et al (2004) Plasma vitamin D metabolites and risk of colorectal cancer in women. Cancer Epidemiol Biomarkers Prev 13:1502–1508
16. Levine AJ, Harper JM, Ervin CM et al (2001) Serum 25-hydroxyvitamin D, dietary calcium intake, and distal colorectal adenoma risk. Nutr Cancer 39:35–41
17. Peters U, McGlynn KA, Chatterjee N et al (2001) Vitamin D, calcium, and vitamin D receptor polymorphism in colorectal adenomas. Cancer Epidemiol Biomarkers Prev 10:1267–1274
18. Platz EA, Hankinson SE, Hollis BW et al (2000) Plasma 1, 25-dihydroxy-and 25-hydroxyvitamin D and adenomatous polyps of the distal colorectum. Cancer Epidemiol Biomarkers Prev 9:1059–1065
19. Grau MV, Baron JA, Sandler RS et al (2003) Vitamin D, calcium supplementation, and colorectal adenomas: results of a randomized trial. J Natl Cancer Inst 95:1765–1771
20. Braun MM, Helzlsouer KJ, Hollis BW, Comstock GW (1995) Prostate cancer and prediagnostic levels of serum vitamin D metabolites (Maryland, United States). Cancer Causes Control 6:235–239
21. Wactawski-Wende J, Kotchen JM, Anderson GL et al (2006) Calcium plus vitamin D supplementation and the risk of colorectal cancer. N Engl J Med 354:684–696
22. Gorham ED, Garland CF, Garland FC et al (2007) Optimal vitamin D status for colorectal cancer prevention A quantitative meta analysis. Am J Prev Med 32:210–216
23. Wu K, Feskanich D, Fuchs CS, Willett WC, Hollis BW, Giovannucci EL (2007) A nested case–control study of plasma 25-hydroxyvitamin D concentrations and risk of colorectal cancer. J Natl Cancer Inst 99:1120–1129
24. Otani T, Iwasaki M, Sasazuki S, Inoue M, Tsugane S (2007) Plasma vitamin D and risk of colorectal cancer: the Japan Public Health Center-Based Prospective Study. Br J Cancer 97:446–451
25. Freedman DM, Looker AC, Chang SC, Graubard BI (2007) Prospective study of serum vitamin D and cancer mortality in the United States. J Natl Cancer Inst 99:1594–1602

26. Giovannucci E, Liu Y, Rimm EB et al (2006) Prospective study of predictors of vitamin D status and cancer incidence and mortality in men. J Natl Cancer Inst 98:451–459

27. Garland C, Shekelle RB, Barrett-Conner E, Criqui MH, Rossof AH, Paul O (1985) Dietary vitamin D and calcium and risk of colorectal cancer: A 19-year prospective study in men. Lancet 1:307–309

28. Kearney J, Giovannucci E, Rimm EB et al (1996) Calcium, vitamin D and dairy foods and the occurrence of colon cancer in men. Am J Epidemiol 143:907–917

29. Bostick RM, Potter JD, Sellers TA, McKenszie DR, Kushi II, Folsom AR (1993) Relation of calcium, vitamin D, and dairy food intake to incidence of colon cancer in older women. Am J Epidemiol 137:1302–1317

30. Martinez ME, Giovannucci EL, Colditz GA et al (1996) Calcium, vitamin D, and the occurrence of colorectal cancer among women. J Natl Cancer Inst 88:1375–1382

31. Zheng W, Anderson KE, Kushi LH et al (1998) A prospective cohort study of intake of calcium, vitamin D, and other micronutrients in relation to incidence of rectal cancer among postmenopausal women. Cancer Epidemiol Biomarkers Prev 7:221–225

32. Jarvinen R, Knekt P, Hakulinen T, Aromaa A (2001) Prospective study on milk products, calcium and cancers of the colon and rectum. Eur J Clin Nutr 55:1000–1007

33. McCullough ML, Robertson AS, Rodriguez C et al (2003) Calcium, vitamin D, dairy products, and risk of colorectal cancer in the cancer prevention study II nutrition cohort (United States). Cancer Causes Control 14:1–12

34. Heilbrun LK, Nomura A, Hankin JH, Stemmermann GN (1985) Dietary vitamin D and calcium and risk of colorectal cancer (letter). Lancet 1:925

35. Benito E, Stiggelbout A, Bosch FX et al (1991) Nutritional factors in colorectal cancer risk: a case–control study in Majorca. Int J Cancer 49:161–167

36. Peters RK, Pike MC, Garabrandt D, Mack TM (1992) Diet and colon cancer in Los Angeles County, California. Cancer Causes Control 3:457–473

37. Ferraroni M, La Vecchia C, D'Avanzo B, Negri E, Franceschi S, Decarli A (1994) Selected micronutrient intake and the risk of colorectal cancer. Br J Cancer 70:1150–1155

38. Boutron MC, Faivre J, Marteau P, Couillault C, Senesse P, Quipourt V (1996) Calcium, phosphorus, vitamin D, dairy products and colorectal carcinogenesis: a French case-control study. Br J Cancer 74:145–151

39. Pritchard RS, BJ A, Gerhardsson de Verdier M (1996) Dietary calcium, vitamin D, and the risk of colorectal cancer in Stockholm, Sweden. Cancer Epidemiol Biomarkers Prev 5: 897–900

40. Marcus PM, Newcomb PA (1998) The association of calcium and vitamin D, and colon and rectal cancer in Wisconsin women. Int J Epidemiol 27:788–793

41. Kampman E, Slattery ML, Caan B, Potter JD (2000) Calcium, vitamin D, sunshine exposures, dairy products and colon cancer risk (United States). Cancer Causes Control 11:459–466

42. Park SY, Murphy SP, Wilkens LR, Stram DO, Henderson BE, Kolonel LN (2007) Calcium, vitamin D, and dairy product intake and prostate cancer risk: the Multiethnic Cohort Study. Am J Epidemiol 166:1259–1269

43. Freedman DM, Dosemeci M, McGlynn K (2002) Sunlight and mortality from breast, ovarian, colon, prostate, and non-melanoma skin cancer: a composite death certificate based case–control study. Occup Environ Med 59:257–262

44. Lane N, Fenoglio CM (1976) I. Observations on the adenoma as precursor to ordinary large bowel carcinoma. Gastrointest Radiol 1:111–119

45. Giovannucci E (2001) An updated review of the epidemiological evidence that cigarette smoking increases risk of colorectal cancer. Cancer Epidemiol Biomarkers Prev 10:725–731

46. Morson B (1976) Polyps and cancer of the large bowel. West J Med 125:93–99

47. Eide TJ (1986) Risk of colorectal cancer in adenoma-bearing individuals within a defined population. Int J Cancer 38:173–176

48. Wei MY, Garland CF, Gorham ED, Mohr SB, Giovannucci E (2008) Vitamin D and prevention of colorectal adenoma: a meta-analysis. Cancer Epidemiol Biomarkers Prev 17:2958–2969

4 The Epidemiology of Vitamin D and Cancer Risk   95

49. Ding EL, Mehta S, Fawzi WW, Giovannucci EL (2008) Interaction of estrogen therapy with calcium and vitamin D supplementation on colorectal cancer risk: Reanalysis of Women's Health Initiative randomized trial. Int J Cancer 122:1690–1694
50. Corder EH, Guess HA, Hulka BS et al (1993) Vitamin D and prostate cancer: a prediagnostic study with stored sera. Cancer Epidemiol Biomarkers Prev 2:467–472
51. Gann PH, Ma J, Hennekens CH, Hollis BW, Haddad JG, Stampfer MJ (1996) Circulating vitamin D metabolites in relation to subsequent development of prostate cancer. Cancer Epidemiol Biomarkers Prev 5:121–126
52. Nomura AM, Stemmermann GN, Lee J et al (1998) Serum vitamin D metabolite levels and the subsequent development of prostate cancer. Cancer Causes Control 9:425–432
53. Platz EA, Leitzmann MF, Hollis BW, Willett WC, Giovannucci E (2004) Plasma 1, 25-dihydroxy- and 25-hydroxyvitamin D and subsequent risk of prostate cancer. Cancer Causes Control 15:255–265
54. Jacobs ET, Giuliano AR, Martinez ME, Hollis BW, Reid ME, Marshall JR (2004) Plasma levels of 25-hydroxyvitamin D, 1, 25-dihydroxyvitamin D and the risk of prostate cancer. J Steroid Biochem Mol Biol 89–90:533–537
55. Ahonen MH, Tenkanen L, Teppo L, Hakama M, Tuohimaa P (2000) Prostate cancer risk and prediagnostic serum 25-hydroxyvitamin D levels (Finland). Cancer Causes Control 11:847–852
56. Tuohimaa P, Tenkanen L, Ahonen M et al (2004) Both high and low levels of blood vitamin D are associated with a higher prostate cancer risk: a longitudinal, nested case-control study in the Nordic countries. Int J Cancer 108:104–108
57. Li H, Stampfer MJ, Hollis BW et al (2007) A prospective study of plasma vitamin D metabolites, vitamin D receptor polymorphisms, and prostate cancer. PLoS Med 4:e103
58. Ahn J, Peters U, Albanes D et al (2008) Serum vitamin D concentration and prostate cancer risk: a nested case-control study. J Natl Cancer Inst 100:796–804
59. Giovannucci E, Rimm EB, Wolk A et al (1998) Calcium and fructose intake in relation to risk of prostate cancer. Cancer Res 58:442–447
60. Chan JM, Giovannucci E, Andersson SO, Yuen J, Adami HO, Wolk A (1998) Dairy products, calcium, phosphorous, vitamin D, and risk of prostate cancer. Cancer Causes Control 9:559–566
61. Chan JM, Pietinen P, Virtanen M, Malila N, Tangrea J (2000) Diet and prostate cancer risk in a cohort of smokers, with a specific focus on calcium and phosphorus (Finland). Cancer Causes Control 11:859–867
62. Kristal AR, Cohen JH, Qu P, Stanford JL (2002) Associations of energy, fat, calcium, and vitamin D with prostate cancer risk. Cancer Epidemiol Biomarkers Prev 11:719–725
63. John EM, Schwartz GG, Koo J, Van Den Berg D, Ingles SA (2005) Sun exposure, vitamin D receptor gene polymorphisms, and risk of advanced prostate cancer. Cancer Res 65:5470–5479
64. John EM, Koo J, Schwartz GG (2007) Sun exposure and prostate cancer risk: evidence for a protective effect of early-life exposure. Cancer Epidemiol Biomarkers Prev 16:1283–1286
65. Luscombe CJ, Fryer AA, French ME et al (2001) Exposure to ultraviolet radiation: association with susceptibility and age at presentation with prostate cancer. Lancet 358:641–642
66. Bodiwala D, Luscombe CJ, French ME et al (2003) Associations between prostate cancer susceptibility and parameters of exposure to ultraviolet radiation. Cancer Lett 200:141–148
67. Rukin N, Blagojevic M, Luscombe CJ et al (2007) Associations between timing of exposure to ultraviolet radiation, T-stage and survival in prostate cancer. Cancer Detect Prev 31:443–449
68. Bertone-Johnson E, Chen WY, Holick MF et al (2005) Plasma 25-hydroxyvitamin D and 1, 25-dihydroxyvitamin D and risk of breast cancer. Cancer Epidemiol Biomarkers Prev 14: 1991–1997
69. Freedman DM, Chang SC, Falk RT et al (2008) Serum levels of vitamin D metabolites and breast cancer risk in the prostate, lung, colorectal, and ovarian cancer screening trial. Cancer Epidemiol Biomarkers Prev 17:889–894

70. Gissel T, Rejnmark L, Mosekilde L, Vestergaard P (2008) Intake of vitamin D and risk of breast cancer: a meta-analysis. J Steroid Biochem Mol Biol 111:195–199
71. Shin MH, Holmes MD, Hankinson SE, Wu K, Colditz GA, Willett WC (2002) Intake of dairy products, calcium, and vitamin D and risk of breast cancer. J Natl Cancer Inst 94:1301–1310
72. Knight JA, Lesosky M, Barnett H, Raboud JM, Vieth R (2007) Vitamin D and reduced risk of breast cancer: a population-based case-control study. Cancer Epidemiol Biomarkers Prev 16:422–429
73. John EM, Schwartz GG, Koo J, Wang W, Ingles SA (2007) Sun exposure, vitamin D receptor gene polymorphisms, and breast cancer risk in a multiethnic population. Am J Epidemiol 166:1409–1419
74. John EM, Schwartz GG, Dreon DM, Koo J (1999) Vitamin D and breast cancer risk: the NHANES I Epidemiologic follow-up study, 1971–1975 to 1992. National Health and Nutrition Examination Survey. Cancer Epidemiol Biomarkers Prev 8:399–406
75. Stolzenberg-Solomon RZ, Vieth R, Azad A et al (2006) A prospective nested case–control study of vitamin D status and pancreatic cancer risk in male smokers. Cancer Res 66:10213–10219
76. Bao Y, Ng K, Wolpin BM, Michaud DS, Giovannucci E, Fuchs CS (2010) Predicted vitamin D status and pancreatic cancer risk in two prospective cohort studies. Br J Cancer 102:1422–1427
77. Skinner HG, Michaud DS, Giovannucci E, Willett WC, Colditz GA, Fuchs CS (2006) Vitamin D intake and the risk for pancreatic cancer in two cohort studies. Cancer Epidemiol Biomarkers Prev 15:1688–1695
78. Tworoger SS, Lee IM, Buring JE, Rosner B, Hollis BW, Hankinson SE (2007) Plasma 25-hydroxyvitamin D and 1, 25-dihydroxyvitamin D and risk of incident ovarian cancer. Cancer Epidemiol Biomarkers Prev 16:783–788
79. Chen W, Dawsey SM, Qiao YL et al (2007) Prospective study of serum 25(OH)-vitamin D concentration and risk of oesophageal and gastric cancers. Br J Cancer 97:123–128
80. Abnet CC, Chen W, Dawsey SM et al (2007) Serum 25(OH)-vitamin D concentration and risk of esophageal squamous dysplasia. Cancer Epidemiol Biomarkers Prev 16:1889–1893
81. Kricker A, Armstrong BK, Hughes AM et al (2008) Personal sun exposure and risk of non Hodgkin lymphoma: a pooled analysis from the Interlymph Consortium. Int J Cancer 122:144–154
82. Freedman DM, Zahm SH, Dosemeci M (1997) Residential and occupational exposure to sunlight and mortality from non-Hodgkin's lymphoma: composite (threefold) case-control study. BMJ 314:1451–1455
83. Norval M, McLoone P, Lesiak A, Narbutt J (2008) The effect of chronic ultraviolet radiation on the human immune system. Photochem Photobiol 84:19–28
84. Pilz S, Dobnig H, Winklhofer-Roob B et al (2008) Low serum levels of 25-hydroxyvitamin D predict fatal cancer in patients referred to coronary angiography. Cancer Epidemiol Biomarkers Prev 17:1228–1233
85. Ducloux D, Courivaud C, Bamoulid J, Kazory A, Dumoulin G, Chalopin JM (2008) Pretransplant serum vitamin D levels and risk of cancer after renal transplantation. Transplantation 85:1755–1759
86. Trivedi DP, Doll R, Khaw KT (2003) Effect of four monthly oral vitamin D3 (cholecalciferol) supplementation on fractures and mortality in men and women living in the community: randomized double blind controlled trial. BMJ 326:469–475
87. Lappe JM, Travers-Gustafson D, Davies KM, Recker RR, Heaney RP (2007) Vitamin D and calcium supplementation reduces cancer risk: results of a randomized trial. Am J Clin Nutr 85:1586–1591
88. Schwartz GG, Hanchette CL (2006) UV, latitude, and spatial trends in prostate cancer mortality: all sunlight is not the same (United States). Cancer Causes Control 17:1091–1101

89. Whitlatch LW, Young MV, Schwartz GG et al (2002) 25-Hydroxyvitamin D-1alpha-hydroxylase activity is diminished in human prostate cancer cells and is enhanced by gene transfer. J Steroid Biochem Mol Biol 81:135–140
90. Chen TC, Wang L, Whitlatch LW, Flanagan JN, Holick MF (2003) Prostatic 25-hydroxyvitamin D-1alpha-hydroxylase and its implication in prostate cancer. J Cell Biochem 88:315–322
91. Porojnicu A, Robsahm TE, Berg JP, Moan J (2007) Season of diagnosis is a predictor of cancer survival Sun-induced vitamin D may be involved: a possible role of sun-induced Vitamin D. J Steroid Biochem Mol Biol 103:675–678
92. Ng K, Meyerhardt JA, Wu K et al (2008) Circulating 25-hydroxyvitamin d levels and survival in patients with colorectal cancer. J Clin Oncol 26:2984–1991
93. Zhou W, Heist RS, Liu G et al (2007) Circulating 25-hydroxyvitamin d levels predict survival in early-stage non-small-cell lung cancer patients. J Clin Oncol 25:479–485

Chapter 5
Vitamin D and Angiogenesis

Yingyu Ma, Candace S. Johnson, and Donald L. Trump

Abstract Angiogenesis is a physiological process involving the formation of new blood vessels from existing vessels. It is essential for the growth of primary tumor and local tumor invasion and metastasis. This chapter reviews the general angiogenesis process, the endogenous factors that regulate angiogenesis, and therapeutic angiogenesis inhibitors. It also reviews the effect of vitamin D on angiogenesis. Vitamin D receptor is detected on endothelial cells and vascular smooth muscle cells (VSMCs). $1,25D_3$ has anti-proliferative effects on tumor-derived endothelial cells through the induction of cell cycle arrest and apoptosis. Increasing evidence supports an anti-angiogenic role of $1,25D_3$ in a number of in vivo tumor model systems. However, vitamin D promotes angiogenesis in more physiological settings. Besides endothelial cells, vitamin D affects the physiological functions and pathology of VSMCs, including cell growth, contractility, motility, and the evolution of vascular calcification, which are involved in cardiovascular diseases. In summary, vitamin D plays important roles in vasculature and angiogenesis. Preclinical studies support the anti-angiogenic effect and the use of $1,25D_3$ in cancer therapy.

Keywords $1,25D_3$ (calcitriol) • Angiogenesis • Vasculature • Endothelial cells • Vascular smooth muscle cells (VSMCs)

C.S. Johnson (✉)
Chair, Pharmacology & Therapeutics,
Roswell Park Cancer Institute,
Elm & Carlton Streets, Buffalo, NY 14263, USA
e-mail: candace.johnson@roswellpark.org

D.L. Trump and C.S. Johnson (eds.), *Vitamin D and Cancer*,
DOI 10.1007/978-1-4419-7188-3_5, © Springer Science+Business Media, LLC 2011

5.1 Overview of Angiogenesis

5.1.1 Angiogenesis Process

Angiogenesis generally refers to the formation of new capillaries from existing vessels [1]. Angiogenesis is an essential and complex process involved in development, reproduction, and wound healing. Pathological angiogenesis can be found in many disorders such as cancer, atherosclerosis, autoimmune diseases, and age-related macular degeneration [1]. Although quiescent in adulthood, endothelial cells proliferate rapidly in response to a stimulus such as hypoxia [2]. Folkman first proposed the hypothesis that tumor growth is dependent on angiogenesis in 1971 [3]. This is based on the observation that solid tumors cannot grow beyond a size of approximately 2 mm diameter without having their own blood supply to provide oxygen and nutrients. In addition to the growth of primary tumor, angiogenesis is also essential for local tumor invasion and metastasis.

Angiogenesis occurs in several differentiated steps, including initiation, endothelial cell proliferation and migration, lumen formation, maturation, and remodeling [4]. The angiogenesis process begins with vasodilation and increased vascular permeability of existing vasculature, which subsequently leads to the extravasation of plasma proteins that form scaffold to support the migrating endothelial cells. Angiopoietin-2, which inhibits Tie2 signaling, promotes the loosening of the support cells [5]. It is followed by the degradation of the basement membrane and extracellular matrix (ECM) by proteases including matrix metalloproteinase (MMP), plasminogen activator, chymase, and heparinase secreted by activated endothelial cells [4]. Once the path is cleared, endothelial cells migrate through the degraded ECM. A variety of growth factors are released from the ECM and stimulate the proliferation of endothelial cells, which results in the formation of solid endothelial cell sprouts into the stromal space of previously avascular tissue. Adhesion molecules involved in cell–cell and cell–matrix interactions, such as integrin $\alpha v \beta 3$, vascular endothelial cadherin, intercellular adhesion molecule-1 (ICAM-1), vascular adhesion molecule-1 (VCAM-1), P-selectin, and E-selectin, also contribute to the processes of endothelial cell migration, spreading, invasion, and proliferation [6]. Adhesion molecules also determine the polarity of the endothelial cells, a necessary step for lumen formation in the solid sprouts [6]. Then, capillary loops are formed and tubes developed with the formation of tight junctions and deposition of new basement membrane. The newly formed capillaries are stabilized by the recruitment of pericytes and smooth muscle cells, which is regulated by platelet-derived growth factor (PDGF). Finally, vessel maturation involves remodeling by which the initial capillary network is modified by pruning and vessel enlargement.

Besides this sprouting angiogenesis, several other mechanisms for neovascularization in tumors have been discovered, including intussusceptive angiogenesis, endothelial progenitor cells recruitment, vasculogenic mimicry, and lymph angiogenesis [7]. Intussusceptive angiogenesis, also known as splitting angiogenesis, is a non-sprouting vessel formation which results in the expansion of capillary plexus.

5 Vitamin D and Angiogenesis

In this process, the capillary wall protrudes into the lumen to split a single vessel in two [8]. It is a fast and energy-efficient process since the proliferation of endothelial cells is not required. Endothelial cells are rearranged and remodeled instead. Both intussusceptive and sprouting angiogenesis occur in the leading edge of the tumor, while in the stabilized tumor regions, intussusception mostly leads to network remodeling and occlusion of vascular segments [9]. New vessels can also grow by the recruitment of circulating endothelial progenitor cells. The contribution of endothelial progenitor cells to tumor angiogenesis is controversial. Some studies support that the recruitment of endothelial progenitor cells is sufficient for tumor angiogenesis [10–12], while others show minimal involvement of endothelial progenitor cells [13, 14]. Transplantation of wild-type bone marrow or vascular endothelial growth factor (VEGF)-mobilized stem cells is able to restore tumor angiogenesis in the angiogenic-defective, tumor-resistant Id-mutant mice [10]. Low levels (4.9%) of endothelial progenitor cells are found in tumor endothelium in patients who developed tumors after receiving bone marrow transplantation [12]. A study using genetically tagged endothelial cells fails to detect bone-marrow-derived cells in newly formed tumor endothelium [14]. Vasculogenic mimicry is a phenomenon when highly aggressive tumor cells, such as melanoma, form patterned vascular channels in the absence of endothelial cells, which provides tumors with a secondary circulation mechanism [15].

Lymph angiogenesis, the formation of new sprouts on existing lymphatic vessels, is another mechanism for tumor cells to receive better circulation. Tumor cells and inflammatory cells produce a variety of lymph angiogenic factors, such as VEGF-C, PDGF-BB, and Angiopoietin-2, to stimulate the formation of new lymphatic vessels [16].

5.1.2 Endogenous Activators and Inhibitors

Angiogenesis is regulated by a delicate balance of activators and inhibitors. This balance is disrupted in favor of angiogenic events during tumor development, which is described as the angiogenic switch is turned on. The endogenous angiogenic factors are released by the tumor cells and degraded extracellular matrix in the tumor microenvironment. Angiogenic activators include hypoxia which activates hypoxia inducible factor α (HIFα) [17], growth factors such as A VEGFA (also known as vascular permeability factor, VPF), basic fibroblast growth factor (bFGF) [18], PDGF [19], pleiotrophin (PTN) [20], granulocyte colony-stimulating factor (G-CSF) [21], hepatocyte growth factor (HGF)/scatter factor (SF) [22], placental growth factor [23], transforming growth factor-α (TGF-α) [24], and TGF-β [25]. VEGFA is the most important molecule that stimulates angiogenesis [26]. It not only promotes endothelial cell proliferation and mobility, but also induces vasodilatation of the existing blood vessels and enhances vessel wall permeability. VEGF facilitates the degradation of ECM by upregulating the expression of MMPs and plasminogen activators. In addition to growth factors, other molecules also stimulate

angiogenesis, which include cytokines and chemokines such as tumor necrosis factor-α (TNF- α) [27], interleukin-8 (IL-8) [28]; oncogenes such as Ras [29]; as well as angiogenin [30], angiopoietin-1 [31], prostaglandins E1 and E2 [32, 33].

In 1980, interferon α was reported as the first angiogenesis inhibitor [34–36]. Since then, many more endogenous angiogenesis inhibitors have been described, which can be divided into two categories: matrix-derived which are fragments of naturally occurring basement membrane and ECM proteins and nonmatrix-derived. Matrix-derived inhibitors including endostatin [37], a fragment of collagen XVIII; arresten [38], a fragment of the noncollagenous (NC1) domain of the α1 chain of type IV collagen; canstatin [39], a fragment of the NC1 domain of the α2 chain of type IV collagen; endorepellin [40], a peptide derived from the carboxy terminus of perlecan; fibulins [41], fragments released by elastases and cathepsins-mediated digestion of basement membrane; thrombospondin-1 [42], an ECM adhesive glycoprotein; and tumstatin [43, 44], a peptide derived from the α3 chain of type IV collagen NC1 domain. Non-matrix-derived inhibitors including angiostatin [45], which is an internal fragment of plasminogen; truncated antithrombin III [46]; interferons [36]; interleukin-12 [47]; 2-methoxyestradiol [48]; pigment epithelial-derived factor (PEDF) [49, 50]; platelet factor 4 [51]; prolactin fragment [52]; tissue inhibitors of matrix metalloproteinase-2 (TIMP-2) [53]; troponin I (Tn 1) [54]; and vasostatin [55].

These inhibitors suppress angiogenesis by inhibiting endothelial cell proliferation, adhesion, migration, and tube formation and promoting apoptosis and cell cycle arrest in endothelial cells through common and distinct signaling mechanisms. How they function together to inhibit angiogenesis is not fully understood.

5.1.3 Therapeutic Angiogenesis Inhibitors

Several angiogenesis inhibitors have been approved for the use in treating cancer and many others are currently in clinical trials. Bevacizumab (Avastin), a monoclonal antibody against VEGF, is the first angiogenesis inhibitor approved by FDA [56]. It is currently used to treat various cancers, including metastatic colorectal, nonsmall-cell lung, and breast cancer. In addition to VEGF, other targets of angiogenesis inhibitors include VEGF receptor (VEGFR), epidermal growth factor receptor (EGFR), mammalian target of rapamycin (mTOR), and MMPs. Cetuximab (Erbitux) is a chimeric monoclonal antibody directed against EGFR and inhibits EGFR signaling, thereby inhibiting angiogenesis and cell proliferation [57]. There are several receptor tyrosine kinase inhibitors developed against angiogenesis, including sorafenib (Nexavar), a dual-function tyrosine kinase inhibitor of VEGFR and Raf kinase that exhibits antiproliferative and anti-angiogenic activities [58]; sunitinib (Sutent), an inhibitor of VEGFR and PDGFR [59]; and erlotinib (Tarveca), an inhibitor of EGFR [60]. Other inhibitors include temsirolimus (Torisel), a small molecule inhibitor of mTOR [61]; bortezomib (Velcade), a proteasome inhibitor that inhibits cancer cell survival and angiogenesis [62]; thalidomide (Thalomid),

5 Vitamin D and Angiogenesis

a synthetic glutamic acid derivative that inhibits the expression of VEGF and beta fibroblast growth factor and thus suppressing angiogenesis [63].

The standard maximum tolerated dose (MTD) chemotherapy requires long drug-free intervals for bone marrow recovery. In contrast, angiogenesis inhibitors are administered with a low-dose metronomic regimen without breaks [1]. Chemotherapy usually targets dividing cells and does not differentiate tumor cells and normal cells, thereby causing more severe side effects such as bone marrow suppression, severe vomiting, and diarrhea. Compared with classic chemotherapeutic drugs, angiogenesis inhibitors have several advantages. They target high levels of angiogenesis as found in tumors, and the stable vasculature of the host is spared. Therefore, their side effects are usually mild and include thrombotic complications, intratumoral bleeding, hypertension, and peripheral neuropathy [1]. They do interfere with fetal development and wound healing since these processes also depend on angiogenesis. Tumor resistance to angiogenesis inhibitors is not as common as with chemotherapy. Angiogenesis inhibitors have been reported to enhance the antitumor activity of some standard chemotherapy agents [56, 64, 65]. Notably, every class of chemotherapeutic drugs has been shown to have anti-angiogenic effects in either in vitro or in vivo angiogenesis assays [66].

5.2 Vitamin D Effects on Angiogenesis

5.2.1 VDR Expression in Cells of Vasculature

Both endothelial cells [67–69] and vascular smooth muscle cells (VSMCs) [70–74] have been demonstrated to express functional vitamin D receptor (VDR). High-affinity VDR is detected in cultured bovine aortic endothelial cells using receptor-binding assays [67]. Immunoblot analysis shows that VDR protein is expressed endogenously and readily induced by $1,25D_3$, the active metabolite of vitamin D, in endothelial cells isolated from Matrigel plugs or murine squamous cell carcinoma (SCC) [69]. Receptor-binding assay and immunoblot analysis reveal the expression of VDR in VSMCs [70, 73]. 1α-Hydroxylase (1α-OHase), the enzyme that leads to local production of $1,25D_3$ from its precursor $25(OH)D_3$, is expressed in endothelial cells isolated from human renal arteries, postcapillary venules from lymphoid tissue, and human umbilical vein endothelial cells (HUVEC) [75]. The 1α-OHase expressed in endothelial cells is enzymatically active since treatment with $1,25D_3$ or $25(OH)D_3$ suppresses HUVEC proliferation [75].

5.2.2 Effect on Endothelial Cells

$1,25D_3$ suppresses VEGF-induced proliferation of bovine aortic endothelial cells. It also reduces VEGF-induced endothelial cell sprouting and elongation in vitro

and induces apoptosis in sprouting endothelial cells [76]. $1,25D_3$ prevents retinal endothelial cells from forming capillary networks in Matrigel, while cell proliferation or migration is not affected at similar concentrations of $1,25D_3$ [77]. $1,25D_3$ and vitamin D analogs 7553, 6760, and EB1089 exert anti-proliferative effects on tumor-derived endothelial cells (TDEC) [68]. The TDECs are isolated by enzyme digestion from SCC VII/SF tumors in C3H/HeJ mice, and sorted by flow cytometry using antibodies against endothelial cells markers [68]. $1,25D_3$ differentially regulates cell growth of Matrigel-derived endothelial cells (MDEC) and TDEC isolated from SCC tumors [69]. VDR protein is expressed and its signaling axis intact in both MDEC and TDEC. $1,25D_3$ induces G0/G1 cell cycle arrest and apoptosis in TDEC, which is accompanied by decreased p21 expression, increased p27 expression, and reduced phosphorylation of Akt and ERK1/2. Increased cleavage of pro-caspase 3 and poly (ADP-ribose) polymerase is observed in TDEC following $1,25D_3$ treatment. In contrast, these effects are not observed in MDEC treated with $1,25D_3$ [69]. The difference in methylation status of the 24-hydroxylase (CYP24) promoter may be one of the mechanisms for these observations. $1,25D_3$ induces CYP24 mRNA and protein expression and enzymatic activity in MDEC but not in TDEC [78]. VDR is recruited to the CYP24 promoter in MDEC but not TDEC. Further studies show hypermethylation in two CpG islands located at the 5′ end in TDEC but not in MDEC, indicating methylation silencing of CYP24 [78]. Knocking down CYP24 by siRNA sensitizes MDEC to $1,25D_3$-mediated growth inhibitory effect. On the other hand, when TDEC is treated with DNA methyltransferase inhibitor 5-aza-2′-deoxycytidine, $1,25D_3$ induces CYP24 expression and TDEC loses its sensitivity to $1,25D_3$ [78]. These results indicate that the methylation-mediated silencing of CYP24 in TDEC contributes to the differential growth inhibitory effects of $1,25D_3$ on endothelial cells isolated from different microenvironments.

5.2.3 Effect on Angiogenesis Models

The effect of vitamin D on angiogenesis was first reported in 1990, when $1,25D_3$ and a synthetic analog 22-oxa-$1,25D_3$ were found to inhibit embryonic angiogenesis in chorioallantoic membranes in a dose-dependent manner [79].

Growing evidence supports an anti-angiogenic role of vitamin D in vivo in various model systems. $1,25D_3$ inhibits the proliferation of TDEC from VDR wild-type mice but not from VDR knockout mice [80]. Tumors from VDR knockout mice show enlarged blood vessels, increased vascular volume, less pericyte coverage on vessels, and higher vascular leakage compared to those from wild-type mice. In addition, HIF-1α, VEGF, Ang1, and PDGF-BB expressions are higher in tumors from VDR knockout mice [80]. $1,25D_3$ reduces the mean vessel counts in retinoblastoma in a transgenic murine retinoblastoma model system [81]. In an MCF-7 tumor xenografts model which overexpress VEGF, treatment with $1,25D_3$ resulted in less vascularized tumors compared with vehicle-control-treated tumors [76].

1,25D$_3$ or 1(OH)D$_3$ inhibits tumor growth and prolongs the survival time in a murine renal cell carcinoma model [82]. Angiogenesis is also inhibited by either agent as assessed by the blood volumes in the tumors. The number and size of pulmonary and hepatic metastatic foci are reduced by either agent [82]. 1,25D$_3$ and 1(OH)D$_3$ have also been shown to inhibit the development and angiogenesis in azoxymethane-induced colon cancer model in Wistar rats, which is associated with reduced VEGF expression in tumors [83]. 1,25D$_3$ or 22-oxa-1,25D$_3$ inhibits angiogenesis in a mouse dorsal air sac model and an in vivo chamber angiogenesis model, using Lewis lung carcinoma (LLC) tumor cells and bFGF as angiogenesis activators, respectively [84]. 1,25D$_3$ also shows an anti-angiogenic effect in a suture-induced cornea inflammation mouse model [85] and a mouse oxygen-induced ischemic retinopathy model [77]. 1,25D$_3$ and retinoids (all-trans retinoic acid, 13-cis retinoic, and 9-*cis* retinoic acid) synergistically inhibit tumor cell-induced angiogenesis in vivo in mouse xenograft models [86, 87]. The same group also reported that 1,25D$_3$ potentiates the anti-angiogenic effects of IL-12 in the tumor cell-induced angiogenesis model, which may partially contribute to the anti-tumor activity of 1,25D$_3$ and IL-12 [88].

Interestingly, vitamin D promotes angiogenesis in more physiological settings. Vascular invasion of the chondro–osseous junction of growth plate, in which VEGF plays an important role, is essential in endochondral bone formation. Mice treated with 1,25D$_3$ show enhanced vascularization of growth plate cartilage compared with vehicle-control-treated mice [89]. 1,25D$_3$ enhances VEGF isoforms expression in CFK2 chondrogenic cell line in vitro and in growth plate chondrocytes and osteoblasts in the tibia and juvenile of mice in vivo [89]. 1,25D$_3$ also stimulates osteoclasts in tibias to express MMP-9, which activates VEGF stored in the cartilage matrix [89]. Vitamin D analog ED-71 promotes blood vessel formation in bone marrow cavity following bone marrow ablation in mice, which is associated with enhanced VEGF120 expression in bone marrow cells [90]. It is beneficial that 1,25D$_3$ differentially regulate angiogenesis in normal and tumor microenvironments.

5.2.4 Potential Mechanisms for the Effects of Vitamin D on Angiogenesis

The mechanisms for anti-angiogenic effects of vitamin D remain unclear. 1,25D$_3$ or 22-oxa-1,25D$_3$ suppresses the expression of MMP-2, MMP-9 and VEGF in Lewis lung carcinoma (LLC) cells, which may partially contribute to the anti-angiogenic activity of the two agents in LLC-induced angiogenesis in vivo [84]. One study indicates a role of interleukin-8 (IL-8) in the anti-angiogenic effect of vitamin D. 1,25D$_3$ suppresses the secretion of IL-8 and TNF-induced IL-8, which is overexpressed in prostate cancer development, in prostate cancer cell lines [91]. 1,25D$_3$ also reduces the activation of nuclear factor-κB (NF-κB), which is a main regulator of IL-8. 1,25D$_3$ or IL-8-neutralizaing antibody inhibits prostate cancer cell conditioned media-induced HUVEC tube formation, migration, and MMP-9 expression [91].

These results indicate that $1,25D_3$-mediated interruption of IL-8 signaling may prevent the progression of prostate cancer. Hypoxia is a pathophysiologic condition that promotes angiogenesis, and is mediated by transcription factor HIF-1. HIF-1 is overexpressed in many cancer cells and is positively related to disease progression [92]. $1,25D_3$ suppresses the expression of HIF-1α and VEGF in human prostate and colorectal cancer cells [93]. $1,25D_3$ also inhibits HIF-1 transcriptional activity and reduces the transcript levels of HIF-1 target genes including VEGF, ET-1, Glut-1. $1,25D_3$-mediated suppression of hypoxia-induced VEGF expression is HIF-pathway-dependent as studied in HIF-1α knockout colon cancer cells [93]. HIF-1 pathway may also be involved in $1,25D_3$ anti-angiogenic effects.

5.2.5 Effect on VSMC

Endothelial cells alone cannot complete angiogenesis nor maintain the newly formed vessels. Peri endothelial cells, including smooth muscle cells and pericytes, play an essential role in vessel maturation and stabilization [4].

Vitamin D has a variety of effects on the function and pathology of VSMCs, including cell growth, contractility, migration, and the evolution of vascular calcifications [72, 94–100]. VSMCs have been shown to express an enzymatically active 1α-hydroxylase, which can be increased by parathyroid hormone (PTH) and native and synthetic phytoestrogens [101]. $1,25D_3$ inhibits the DNA synthesis and thus proliferation of VSMC, but increases metabolic turnover as assessed by creatine kinase activity, suggesting a potential role of the $1,25D_3$ synthesized intracellularly in these cells [101, 102]. $1,25D_3$ also inhibits epidermal growth factor (EGF)-induced VSMC proliferation [102]. In contrast, other studies support a role of $1,25D_3$ in promoting VSMC proliferation [71, 73] by upregulating VEGF expression [73]. VEGF receptor antagonist or VEGF-neutralizing antibody abrogates the effect of $1,25D_3$ on VSMC proliferation [73]. In fact, $1,25D_3$ may inhibit or promote the growth of VSMC, depending on the underling culture conditions [72]. In nonquiescent cells, $1,25D_3$ inhibits thrombin or PDGF-induced VSMC growth, as well as thrombin-mediated induction of c-myc RNA. While in quiescent cells, $1,25D_3$ promotes the cell growth and the induction of c-myc RNA by thrombin [72].

The role of vitamin D in vascular calcification and cardiovascular disease is controversial. Epidemiological data show that there is correlation between ischemic heart disease mortality rate and geographic latitude for several European and Western countries [103]. High latitude has been shown to associate with low serum vitamin D levels [104]. In addition, an inverse association between coronary heart disease mortality rate in males and altitude was observed [105]. It is a fact that the intensity of UVB radiation increases exponentially at higher altitude. The mortality rate of coronary heart disease also displays seasonal variation. Multiple studies show that ischemic heart disease death rate is low in summer and high in winter [106–108]. Serum levels of $1,25D_3$ show the opposite pattern with a peak in summer and a nadir in winter [109–111]. These studies suggest that lower level of serum vitamin D is associated with higher ischemic heart disease mortality rate.

5 Vitamin D and Angiogenesis

Some studies suggest that vitamin D insufficiency may contribute to the pathogenesis of cardiovascular disease. Vascular calcification is a risk factor for cardiovascular mortality. Almost all significant atherosclerotic lesions observed by angiography are calcified [112]. In a study with two populations (173 subjects) at high and moderate risk for coronary heart disease, serum levels of $1,25D_3$ are inversely correlated with the extent of vascular calcification [113]. In contrast, the extent of calcification is not correlated with the levels of osteocalcin or parathyroid hormone [113]. Another study showed that serum levels of calcitriol are an independent negative indicator of coronary calcium mass measured by electron-beam computed tomography [114]. The protective role of $1,25D_3$ in vascular calcification and atherosclerosis may be due to following mechanisms (reviewed in [104]). $1,25D_3$ promotes the synthesis of the matrix Gla protein which inhibits vascular calcification. Low serum level of calcitriol leads to increased level of PTH, which may promote cardiovascular disease. $1,25D_3$ has been shown to inhibit the proliferation of VSMCs, which express VDR. $1,25D_3$ also inhibits the production of the pro-inflammatory cytokines TNF and IL-6 in monocytes. In a short-term supplementation study, vitamin D_3 and calcium result in increased serum $25(OH)D_3$ level, and reduce systolic blood pressure, heart rate, and PTH levels in elderly women [115]. Vitamin D_3 supplementation reduces TNF serum levels while increases the levels of anti-inflammatory cytokine interleukin 10 in a study on 93 chronic heart failure patients [116].

However, other studies have found $1,25D_3$ may contribute to the pathogenesis of atherosclerotic lesions. $1,25D_3$ stimulates calcium influx (94) and VEGF expression in VSMCs [117]. $1,25D_3$ enhances vascular calcification in a dose-dependent manner through increasing the expression of bone matrix protein osteopontin and inhibiting the expression and secretion of PTH-related peptide (PTHrP) by VSMCs [118]. Additionally, $1,25D_3$ induces VSMC migration in a dose-dependent manner [100]. This effect is independent of gene transcription and involves non-genomic activation of PI3K pathway [100].

In summary, there is growing evidence that vitamin D has an impact on vasculature and angiogenesis. $1,25D_3$ has growth inhibitory effects in endothelial cells through the induction of cell cycle arrest and apoptosis. $1,25D_3$ exerts anti-angiogenic effects in a variety of tumor model systems in vivo. These observations provide additional preclinical rationale for the use of $1,25D_3$ in cancer therapy. $1,25D_3$ also regulates vascular calcification and plays important roles in cardiovascular diseases. Further investigations into the mechanisms of vitamin D anti-angiogenic effects will be needed to enhance our understanding on its role in vasculature.

References

1. Folkman J (2007) Angiogenesis: an organizing principle for drug discovery? Nat Rev Drug Discov 6:273–286
2. Carmeliet P (2003) Angiogenesis in health and disease. Nat Med 9:653–660
3. Folkman J (1971) Tumor angiogenesis: therapeutic implications. N Engl J Med 285:1182–1186
4. Carmeliet P (2000) Mechanisms of angiogenesis and arteriogenesis. Nat Med 6:389–395

5. Maisonpierre PC, Suri C, Jones PF, Bartunkova S, Wiegand SJ, Radziejewski C, Compton D, McClain J, Aldrich TH, Papadopoulos N, Daly TJ, Davis S, Sato TN, Yancopoulos GD (1997) Angiopoietin-2, a natural antagonist for Tie2 that disrupts in vivo angiogenesis. Science 277:55–60

6. Bischoff J (1997) Cell adhesion and angiogenesis. J Clin Invest 100:S37–39

7. Hillen F, Griffioen AW (2007) Tumour vascularization: sprouting angiogenesis and beyond. Cancer Metastasis Rev 26:489–502

8. Djonov V, Baum O, Burri PH (2003) Vascular remodeling by intussusceptive angiogenesis. Cell Tissue Res 314:107–117

9. Patan S, Munn LL, Jain RK (1996) Intussusceptive microvascular growth in a human colon adenocarcinoma xenograft: a novel mechanism of tumor angiogenesis. Microvasc Res 51:260–272

10. Lyden D, Hattori K, Dias S, Costa C, Blaikie P, Butros L, Chadburn A, Heissig B, Marks W, Witte L, Wu Y, Hicklin D, Zhu Z, Hackett NR, Crystal RG, Moore MA, Hajjar KA, Manova K, Benezra R, Rafii S (2001) Impaired recruitment of bone-marrow-derived endothelial and hematopoietic precursor cells blocks tumor angiogenesis and growth. Nat Med 7:1194–1201

11. Ruzinova MB, Schoer RA, Gerald W, Egan JE, Pandolfi PP, Rafii S, Manova K, Mittal V, Benezra R (2003) Effect of angiogenesis inhibition by Id loss and the contribution of bone-marrow-derived endothelial cells in spontaneous murine tumors. Cancer Cell 4:277–289

12. Peters BA, Diaz LA, Polyak K, Meszler L, Romans K, Guinan EC, Antin JH, Myerson D, Hamilton SR, Vogelstein B, Kinzler KW, Lengauer C (2005) Contribution of bone marrow-derived endothelial cells to human tumor vasculature. Nat Med 11:261–262

13. Machein MR, Renninger S, de Lima-Hahn E, Plate KH (2003) Minor contribution of bone marrow-derived endothelial progenitors to the vascularization of murine gliomas. Brain Pathol 13:582–597

14. Gothert JR, Gustin SE, van Eekelen JA, Schmidt U, Hall MA, Jane SM, Green AR, Gottgens B, Izon DJ, Begley CG (2004) Genetically tagging endothelial cells in vivo: bone marrow-derived cells do not contribute to tumor endothelium. Blood 104:1769–1777

15. Maniotis AJ, Folberg R, Hess A, Seftor EA, Gardner LM, Pe'er J, Trent JM, Meltzer PS, Hendrix MJ (1999) Vascular channel formation by human melanoma cells in vivo and in vitro: vasculogenic mimicry. Am J Pathol 155:739–752

16. Cao Y (2005) Opinion: emerging mechanisms of tumour lymphangiogenesis and lymphatic metastasis. Nat Rev Cancer 5:735–743

17. Carmeliet P, Dor Y, Herbert JM, Fukumura D, Brusselmans K, Dewerchin M, Neeman M, Bono F, Abramovitch R, Maxwell P, Koch CJ, Ratcliffe P, Moons L, Jain RK, Collen D, Keshert E (1998) Role of HIF-1alpha in hypoxia-mediated apoptosis, cell proliferation and tumour angiogenesis. Nature 394:485–490

18. Gospodarowicz D (1976) Humoral control of cell proliferation: the role of fibroblast growth factor in regeneration, angiogenesis, wound healing, and neoplastic growth. Prog Clin Biol Res 9:1–19

19. Mustoe TA, Purdy J, Gramates P, Deuel TF, Thomason A, Pierce GF (1989) Reversal of impaired wound healing in irradiated rats by platelet-derived growth factor-BB. Am J Surg 158:345–350

20. Fang W, Hartmann N, Chow DT, Riegel AT, Wellstein A (1992) Pleiotrophin stimulates fibroblasts and endothelial and epithelial cells and is expressed in human cancer. J Biol Chem 267:25889–25897

21. Takahashi T, Kalka C, Masuda H, Chen D, Silver M, Kearney M, Magner M, Isner JM, Asahara T (1999) Ischemia- and cytokine-induced mobilization of bone marrow-derived endothelial progenitor cells for neovascularization. Nat Med 5:434–438

22. Grant DS, Kleinman HK, Goldberg ID, Bhargava MM, Nickoloff BJ, Kinsella JL, Polverini P, Rosen EM (1993) Scatter factor induces blood vessel formation in vivo. Proc Natl Acad Sci USA 90:1937–1941

5 Vitamin D and Angiogenesis

23. Maglione D, Guerriero V, Viglietto G, Delli-Bovi P, Persico MG (1991) Isolation of a human placenta cDNA coding for a protein related to the vascular permeability factor. Proc Natl Acad Sci USA 88:9267–9271
24. Schreiber AB, Winkler ME, Derynck R (1986) Transforming growth factor-alpha: a more potent angiogenic mediator than epidermal growth factor. Science 232:1250–1253
25. Pepper MS, Belin D, Montesano R, Orci L, Vassalli JD (1990) Transforming growth factor-beta 1 modulates basic fibroblast growth factor-induced proteolytic and angiogenic properties of endothelial cells in vitro. J Cell Biol 111:743–755
26. Adams RH, Alitalo K (2007) Molecular regulation of angiogenesis and lymphangiogenesis. Nat Rev Mol Cell Biol 8:464–478
27. Leibovich SJ, Polverini PJ, Shepard HM, Wiseman DM, Shively V, Nuseir N (1987) Macrophage-induced angiogenesis is mediated by tumour necrosis factor-alpha. Nature 329:630–632
28. Strieter RM, Kunkel SL, Elner VM, Martonyi CL, Koch AE, Polverini PJ, Elner SG (1992) Interleukin-8. A corneal factor that induces neovascularization. Am J Pathol 141:1279–1284
29. Thompson TC, Southgate J, Kitchener G, Land H (1989) Multistage carcinogenesis induced by ras and myc oncogenes in a reconstituted organ. Cell 56:917–930
30. Fett JW, Strydom DJ, Lobb RR, Alderman EM, Bethune JL, Riordan JF, Vallee BL (1985) Isolation and characterization of angiogenin, an angiogenic protein from human carcinoma cells. Biochemistry 24:5480–5486
31. Davis S, Aldrich TH, Jones PF, Acheson A, Compton DL, Jain V, Ryan TE, Bruno J, Radziejewski C, Maisonpierre PC, Yancopoulos GD (1996) Isolation of angiopoietin-1, a ligand for the TIE2 receptor, by secretion-trap expression cloning. Cell 87:1161–1169
32. Ziche M, Jones J, Gullino PM (1982) Role of prostaglandin E1 and copper in angiogenesis. J Natl Cancer Inst 69:475–482
33. Harada S, Nagy JA, Sullivan KA, Thomas KA, Endo N, Rodan GA, Rodan SB (1994) Induction of vascular endothelial growth factor expression by prostaglandin E2 and E1 in osteoblasts. J Clin Invest 93:2490–2496
34. Brouty-Boye D, Zetter BR (1980) Inhibition of cell motility by interferon. Science 208:516–518
35. Dvorak HF, Gresser I (1989) Microvascular injury in pathogenesis of interferon-induced necrosis of subcutaneous tumors in mice. J Natl Cancer Inst 81:497–502
36. Sidky YA, Borden EC (1987) Inhibition of angiogenesis by interferons: effects on tumor- and lymphocyte-induced vascular responses. Cancer Res 47:5155–5161
37. ÓReilly MS, Boehm T, Shing Y, Fukai N, Vasios G, Lane WS, Flynn E, Birkhead JR, Olsen BR, Folkman J (1997) Endostatin: an endogenous inhibitor of angiogenesis and tumor growth. Cell 88:277–285
38. Colorado PC, Torre A, Kamphaus G, Maeshima Y, Hopfer H, Takahashi K, Volk R, Zamborsky ED, Herman S, Sarkar PK, Ericksen MB, Dhanabal M, Simons M, Post M, Kufe DW, Weichselbaum RR, Sukhatme VP, Kalluri R (2000) Anti-angiogenic cues from vascular basement membrane collagen. Cancer Res 60:2520–2526
39. Kamphaus GD, Colorado PC, Panka DJ, Hopfer H, Ramchandran R, Torre A, Maeshima Y, Mier JW, Sukhatme VP, Kalluri R (2000) Canstatin, a novel matrix-derived inhibitor of angiogenesis and tumor growth. J Biol Chem 275:1209–1215
40. Mongiat M, Sweeney SM, San Antonio JD, Fu J, Iozzo RV (2003) Endorepellin, a novel inhibitor of angiogenesis derived from the C terminus of perlecan. J Biol Chem 278:4238–4249
41. Kalluri R (2002) Discovery of type IV collagen non-collagenous domains as novel integrin ligands and endogenous inhibitors of angiogenesis. Cold Spring Harb Symp Quant Biol 67:255–266
42. Good DJ, Polverini PJ, Rastinejad F, Le Beau MM, Lemons RS, Frazier WA, Bouck NP (1990) A tumor suppressor-dependent inhibitor of angiogenesis is immunologically and functionally indistinguishable from a fragment of thrombospondin. Proc Natl Acad Sci USA 87:6624–6628

43. Maeshima Y, Colorado PC, Kalluri R (2000) Two RGD-independent alpha vbeta 3 integrin binding sites on tumstatin regulate distinct anti-tumor properties. J Biol Chem 275: 23745–23750
44. Maeshima Y, Colorado PC, Torre A, Holthaus KA, Grunkemeyer JA, Ericksen MB, Hopfer H, Xiao Y, Stillman IE, Kalluri R (2000) Distinct antitumor properties of a type IV collagen domain derived from basement membrane. J Biol Chem 275:21340–21348
45. ÓReilly MS, Holmgren L, Shing Y, Chen C, Rosenthal RA, Moses M, Lane WS, Cao Y, Sage EH, Folkman J (1994) Angiostatin: a novel angiogenesis inhibitor that mediates the suppression of metastases by a Lewis lung carcinoma. Cell 79:315–328
46. ÓReilly MS, Pirie-Shepherd S, Lane WS, Folkman J (1999) Antiangiogenic activity of the cleaved conformation of the serpin antithrombin. Science 285:1926–1928
47. Voest EE, Kenyon BM, ÓReilly MS, Truitt G, D'Amato RJ, Folkman J (1995) Inhibition of angiogenesis in vivo by interleukin 12. J Natl Cancer Inst 87:581–586
48. Yue TL, Wang X, Louden CS, Gupta S, Pillarisetti K, Gu JL, Hart TK, Lysko PG, Feuerstein GZ (1997) 2-Methoxyestradiol, an endogenous estrogen metabolite, induces apoptosis in endothelial cells and inhibits angiogenesis: possible role for stress-activated protein kinase signaling pathway and Fas expression. Mol Pharmacol 51:951–962
49. Abe R, Shimizu T, Yamagishi S, Shibaki A, Amano S, Inagaki Y, Watanabe H, Sugawara H, Nakamura H, Takeuchi M, Imaizumi T, Shimizu H (2004) Overexpression of pigment epithelium-derived factor decreases angiogenesis and inhibits the growth of human malignant melanoma cells in vivo. Am J Pathol 164:1225–1232
50. Doll JA, Stellmach VM, Bouck NP, Bergh AR, Lee C, Abramson LP, Cornwell ML, Pins MR, Borensztajn J, Crawford SE (2003) Pigment epithelium-derived factor regulates the vasculature and mass of the prostate and pancreas. Nat Med 9:774–780
51. Maione TE, Gray GS, Petro J, Hunt AJ, Donner AL, Bauer SI, Carson HF, Sharpe RJ (1990) Inhibition of angiogenesis by recombinant human platelet factor-4 and related peptides. Science 247:77–79
52. Corbacho AM, Martinez De La Escalera G, Clapp C (2002) Roles of prolactin and related members of the prolactin/growth hormone/placental lactogen family in angiogenesis. J Endocrinol 173:219–238
53. Seo DW, Li H, Guedez L, Wingfield PT, Diaz T, Salloum R, Wei BY, Stetler-Stevenson WG (2003) TIMP-2 mediated inhibition of angiogenesis: an MMP-independent mechanism. Cell 114:171–180
54. Moses MA, Wiederschain D, Wu I, Fernandez CA, Ghazizadeh V, Lane WS, Flynn E, Sytkowski A, Tao T, Langer R (1999) Troponin I is present in human cartilage and inhibits angiogenesis. Proc Natl Acad Sci USA 96:2645–2650
55. Pike SE, Yao L, Jones KD, Cherney B, Appella E, Sakaguchi K, Nakhasi H, Teruya-Feldstein J, Wirth P, Gupta G, Tosato G (1998) Vasostatin, a calreticulin fragment, inhibits angiogenesis and suppresses tumor growth. J Exp Med 188:2349–2356
56. Hurwitz H, Fehrenbacher L, Novotny W, Cartwright T, Hainsworth J, Heim W, Berlin J, Baron A, Griffing S, Holmgren E, Ferrara N, Fyfe G, Rogers B, Ross R, Kabbinavar F (2004) Bevacizumab plus irinotecan, fluorouracil, and leucovorin for metastatic colorectal cancer. N Engl J Med 350:2335–2342
57. Huang SM, Li J, Harari PM (2002) Molecular inhibition of angiogenesis and metastatic potential in human squamous cell carcinomas after epidermal growth factor receptor blockade. Mol Cancer Ther 1:507–514
58. Strumberg D, Richly H, Hilger RA, Schleucher N, Korfee S, Tewes M, Faghih M, Brendel E, Voliotis D, Haase CG, Schwartz B, Awada A, Voigtmann R, Scheulen ME, Seeber S (2005) Phase I clinical and pharmacokinetic study of the Novel Raf kinase and vascular endothelial growth factor receptor inhibitor BAY 43–9006 in patients with advanced refractory solid tumors. J Clin Oncol 23:965–972
59. Marzola P, Degrassi A, Calderan L, Farace P, Nicolato E, Crescimanno C, Sandri M, Giusti A, Pesenti E, Terron A, Sbarbati A, Osculati F (2005) Early antiangiogenic activity of

5 Vitamin D and Angiogenesis

SU11248 evaluated in vivo by dynamic contrast-enhanced magnetic resonance imaging in an experimental model of colon carcinoma. Clin Cancer Res 11:5827–5832

60. Tabernero J (2007) The role of VEGF and EGFR inhibition: implications for combining anti-VEGF and anti-EGFR agents. Mol Cancer Res 5:203–220

61. Frost P, Moatamed F, Hoang B, Shi Y, Gera J, Yan H, Frost P, Gibbons J, Lichtenstein A (2004) In vivo antitumor effects of the mTOR inhibitor CCI-779 against human multiple myeloma cells in a xenograft model. Blood 104:4181–4187

62. Sunwoo JB, Chen Z, Dong G, Yeh N, Crowl Bancroft C, Sausville E, Adams J, Elliott P, Van Waes C (2001) Novel proteasome inhibitor PS-341 inhibits activation of nuclear factor-kappa B, cell survival, tumor growth, and angiogenesis in squamous cell carcinoma. Clin Cancer Res 7:1419–1428

63. Melchert M, List A (2007) The thalidomide saga. Int J Biochem Cell Biol 39:1489–1499

64. Teicher BA, Sotomayor EA, Huang ZD (1992) Antiangiogenic agents potentiate cytotoxic cancer therapies against primary and metastatic disease. Cancer Res 52:6702–6704

65. Kakeji Y, Teicher BA (1997) Preclinical studies of the combination of angiogenic inhibitors with cytotoxic agents. Invest New Drugs 15:39–48

66. Miller KD, Sweeney CJ, Sledge GW Jr (2001) Redefining the target: chemotherapeutics as antiangiogenics. J Clin Oncol 19:1195–1206

67. Merke J, Milde P, Lewicka S, Hugel U, Klaus G, Mangelsdorf DJ, Haussler MR, Rauterberg EW, Ritz E (1989) Identification and regulation of 1, 25-dihydroxyvitamin D3 receptor activity and biosynthesis of 1, 25-dihydroxyvitamin D3. Studies in cultured bovine aortic endothelial cells and human dermal capillaries. J Clin Invest 83:1903–1915

68. Bernardi RJ, Johnson CS, Modzelewski RA, Trump DL (2002) Antiproliferative effects of 1alpha, 25-dihydroxyvitamin D(3) and vitamin D analogs on tumor-derived endothelial cells. Endocrinology 143:2508–2514

69. Chung I, Wong MK, Flynn G, Yu WD, Johnson CS, Trump DL (2006) Differential antiproliferative effects of calcitriol on tumor-derived and matrigel-derived endothelial cells. Cancer Res 66:8565–8573

70. Merke J, Hofmann W, Goldschmidt D, Ritz E (1987) Demonstration of 1, 25(OH)2 vitamin D3 receptors and actions in vascular smooth muscle cells in vitro. Calcif Tissue Int 41:112–114

71. Koh E, Morimoto S, Fukuo K, Itoh K, Hironaka T, Shiraishi T, Onishi T, Kumahara Y (1988) 1, 25-Dihydroxyvitamin D3 binds specifically to rat vascular smooth muscle cells and stimulates their proliferation in vitro. Life Sci 42:215–223

72. Mitsuhashi T, Morris RC Jr, Ives HE (1991) 1, 25-dihydroxyvitamin D3 modulates growth of vascular smooth muscle cells. J Clin Invest 87:1889–1895

73. Cardus A, Parisi E, Gallego C, Aldea M, Fernandez E, Valdivielso JM (2006) 1, 25-Dihydroxyvitamin D3 stimulates vascular smooth muscle cell proliferation through a VEGF-mediated pathway. Kidney Int 69:1377–1384

74. Rajasree S, Umashankar PR, Lal AV, Sarma PS, Kartha CC (2002) 1, 25-dihydroxyvitamin D3 receptor is upregulated in aortic smooth muscle cells during hypervitaminosis D. Life Sci 70:1777–1788

75. Zehnder D, Bland R, Chana RS, Wheeler DC, Howie AJ, Williams MC, Stewart PM, Hewison M (2002) Synthesis of 1, 25-dihydroxyvitamin D(3) by human endothelial cells is regulated by inflammatory cytokines: a novel autocrine determinant of vascular cell adhesion. J Am Soc Nephrol 13:621–629

76. Mantell DJ, Owen PE, Bundred NJ, Mawer EB, Canfield AE (2000) 1 alpha, 25-dihydroxyvitamin D(3) inhibits angiogenesis in vitro and in vivo. Circ Res 87:214–220

77. Albert DM, Scheef EA, Wang S, Mehraein F, Darjatmoko SR, Sorenson CM, Sheibani N (2007) Calcitriol is a potent inhibitor of retinal neovascularization. Invest Ophthalmol Vis Sci 48:2327–2334

78. Chung I, Karpf AR, Muindi JR, Conroy JM, Nowak NJ, Johnson CS, Trump DL (2007) Epigenetic silencing of CYP24 in tumor-derived endothelial cells contributes to selective growth inhibition by calcitriol. J Biol Chem 282:8704–8714

79. Oikawa T, Hirotani K, Ogasawara H, Katayama T, Nakamura O, Iwaguchi T, Hiragun A (1990) Inhibition of angiogenesis by vitamin D3 analogues. Eur J Pharmacol 178:247–250

80. Chung I, Han G, Seshadri M, Gillard BM, Yu WD, Foster BA, Trump DL, Johnson CS (2009) Role of vitamin D receptor in the antiproliferative effects of calcitriol in tumor-derived endothelial cells and tumor angiogenesis in vivo. Cancer Res 69:967–975

81. Shokravi MT, Marcus DM, Alroy J, Egan K, Saornil MA, Albert DM (1995) Vitamin D inhibits angiogenesis in transgenic murine retinoblastoma. Invest Ophthalmol Vis Sci 36:83–87

82. Fujioka T, Hasegawa M, Ishikura K, Matsushita Y, Sato M, Tanji S (1998) Inhibition of tumor growth and angiogenesis by vitamin D3 agents in murine renal cell carcinoma. J Urol 160:247–251

83. Iseki K, Tatsuta M, Uehara H, Iishi H, Yano H, Sakai N, Ishiguro S (1999) Inhibition of angiogenesis as a mechanism for inhibition by 1alpha-hydroxyvitamin D3 and 1, 25-dihydroxyvitamin D3 of colon carcinogenesis induced by azoxymethane in Wistar rats. Int J Cancer 81:730–733

84. Nakagawa K, Sasaki Y, Kato S, Kubodera N, Okano T (2005) 22-Oxa-1alpha, 25-dihydroxyvitamin D3 inhibits metastasis and angiogenesis in lung cancer. Carcinogenesis 26:1044–1054

85. Suzuki T, Sano Y, Kinoshita S (2000) Effects of 1alpha, 25-dihydroxyvitamin D3 on Langerhans cell migration and corneal neovascularization in mice. Invest Ophthalmol Vis Sci 41:154–158

86. Majewski S, Szmurlo A, Marczak M, Jablonska S, Bollag W (1993) Inhibition of tumor cell-induced angiogenesis by retinoids, 1, 25-dihydroxyvitamin D3 and their combination. Cancer Lett 75:35–39

87. Majewski S, Skopinska M, Marczak M, Szmurlo A, Bollag W, Jablonska S (1996) Vitamin D3 is a potent inhibitor of tumor cell-induced angiogenesis. J Investig Dermatol Symp Proc 1:97–101

88. Nowicka D, Zagozdzon R, Majewski S, Marczak M, Jablonska S, Bollag W (1998) Calcitriol enhances antineoplastic and antiangiogenic effects of interleukin-12. Arch Dermatol Res 290:696–700

89. Lin R, Amizuka N, Sasaki T, Aarts MM, Ozawa GD, Henderson JE, White JH (2002) 1Alpha, 25-dihydroxyvitamin D3 promotes vascularization of the chondro-osseous junction by stimulating expression of vascular endothelial growth factor and matrix metalloproteinase 9. J Bone Miner Res 17:1604–1612

90. Okuda N, Takeda S, Shinomiya K, Muneta T, Itoh S, Noda M, Asou Y (2007) ED-71, a novel vitamin D analog, promotes bone formation and angiogenesis and inhibits bone resorption after bone marrow ablation. Bone 40:281–292

91. Bao BY, Yao J, Lee YF (2006) 1alpha, 25-dihydroxyvitamin D3 suppresses interleukin-8-mediated prostate cancer cell angiogenesis. Carcinogenesis 27:1883–1893

92. Melillo G (2006) Inhibiting hypoxia-inducible factor 1 for cancer therapy. Mol Cancer Res 4:601–605

93. Ben-Shoshan M, Amir S, Dang DT, Dang LH, Weisman Y, Mabjeesh NJ (2007) 1alpha, 25-dihydroxyvitamin D3 (Calcitriol) inhibits hypoxia-inducible factor-1/vascular endothelial growth factor pathway in human cancer cells. Mol Cancer Ther 6:1433–1439

94. Shan JJ, Li B, Taniguchi N, Pang PK (1996) Inhibition of membrane L-type calcium channel activity and intracellular calcium concentration by 24R, 25-dihydroxyvitamin D3 in vascular smooth muscle. Steroids 61:657–663

95. Shan J, Resnick LM, Lewanczuk RZ, Karpinski E, Li B, Pang PK (1993) 1, 25-Dihydroxyvitamin D as a cardiovascular hormone. Effects on calcium current and cytosolic free calcium in vascular smooth muscle cells. Am J Hypertens 6:983–988

96. Kawashima H (1988) 1, 25-Dihydroxyvitamin D3 stimulates Ca-ATPase in a vascular smooth muscle cell line. Biochem Biophys Res Commun 150:1138–1143

5 Vitamin D and Angiogenesis

97. Inoue T, Kawashima H (1988) 1, 25-Dihydroxyvitamin D3 stimulates 45Ca2+-uptake by cultured vascular smooth muscle cells derived from rat aorta. Biochem Biophys Res Commun 152:1388–1394

98. Wakasugi M, Noguchi T, Inoue M, Kazama Y, Tawata M, Kanemaru Y, Onaya T (1991) Vitamin D3 stimulates the production of prostacyclin by vascular smooth muscle cells. Prostaglandins 42:127–136

99. Somjen D, Kohen F, Amir-Zaltsman Y, Knoll E, Stern N (2000) Vitamin D analogs modulate the action of gonadal steroids in human vascular cells in vitro. Am J Hypertens 13:396–403

100. Rebsamen MC, Sun J, Norman AW, Liao JK (2002) 1Alpha, 25-dihydroxyvitamin D3 induces vascular smooth muscle cell migration via activation of phosphatidylinositol 3-kinase. Circ Res 91:17–24

101. Somjen D, Weisman Y, Kohen F, Gayer B, Limor R, Sharon O, Jaccard N, Knoll E, Stern N (2005) 25-Hydroxyvitamin D3-1alpha-hydroxylase is expressed in human vascular smooth muscle cells and is upregulated by parathyroid hormone and estrogenic compounds. Circulation 111:1666–1671

102. Carthy EP, Yamashita W, Hsu A, Ooi BS (1989) 1, 25-Dihydroxyvitamin D3 and rat vascular smooth muscle cell growth. Hypertension 13:954–959

103. Fleck A (1989) Latitude and ischaemic heart disease. Lancet 1:613

104. Zittermann A, Schleithoff SS, Koerfer R (2005) Putting cardiovascular disease and vitamin D insufficiency into perspective. Br J Nutr 94:483–492

105. Mortimer EA Jr, Monson RR, MacMahon B (1977) Reduction in mortality from coronary heart disease in men residing at high altitude. N Engl J Med 296:581–585

106. Douglas AS, Allan TM, Rawles JM (1991) Composition of seasonality of disease. Scott Med J 36:76–82

107. Douglas AS, Dunnigan MG, Allan TM, Rawles JM (1995) Seasonal variation in coronary heart disease in Scotland. J Epidemiol Commun Health 49:575–582

108. Grimes DS, Hindle E, Dyer T (1996) Sunlight, cholesterol and coronary heart disease. QJM 89:579–589

109. Docio S, Riancho JA, Perez A, Olmos JM, Amado JA, Gonzalez-Macias J (1998) Seasonal deficiency of vitamin D in children: a potential target for osteoporosis-preventing strategies? J Bone Miner Res 13:544–548

110. Zittermann A, Scheld K, Stehle P (1998) Seasonal variations in vitamin D status and calcium absorption do not influence bone turnover in young women. Eur J Clin Nutr 52:501–506

111. Vecino-Vecino C, Gratton M, Kremer R, Rodriguez-Manas L, Duque G (2006) Seasonal variance in serum levels of vitamin d determines a compensatory response by parathyroid hormone: study in an ambulatory elderly population in Quebec. Gerontology 52:33–39

112. Honye J, Mahon DJ, Jain A, White CJ, Ramee SR, Wallis JB, al-Zarka A, Tobis JM (1992) Morphological effects of coronary balloon angioplasty in vivo assessed by intravascular ultrasound imaging. Circulation 85:1012–1025

113. Watson KE, Abrolat ML, Malone LL, Hoeg JM, Doherty T, Detrano R, Demer LL (1997) Active serum vitamin D levels are inversely correlated with coronary calcification. Circulation 96:1755–1760

114. Doherty TM, Tang W, Dascalos S, Watson KE, Demer LL, Shavelle RM, Detrano RC (1997) Ethnic origin and serum levels of 1alpha, 25-dihydroxyvitamin D3 are independent predictors of coronary calcium mass measured by electron-beam computed tomography. Circulation 96:1477–1481

115. Pfeifer M, Begerow B, Minne HW, Nachtigall D, Hansen C (2001) Effects of a short-term vitamin D(3) and calcium supplementation on blood pressure and parathyroid hormone levels in elderly women. J Clin Endocrinol Metab 86:1633–1637

116. Schleithoff SS, Zittermann A, Tenderich G, Berthold HK, Stehle P, Koerfer R (2006) Vitamin D supplementation improves cytokine profiles in patients with congestive heart failure: a double-blind, randomized, placebo-controlled trial. Am J Clin Nutr 83:754–759

117. Yamamoto T, Kozawa O, Tanabe K, Akamatsu S, Matsuno H, Dohi S, Hirose H, Uematsu T (2002) 1, 25-Dihydroxyvitamin D3 stimulates vascular endothelial growth factor release in aortic smooth muscle cells: role of p38 mitogen-activated protein kinase. Arch Biochem Biophys 398:1–6
118. Jono S, Nishizawa Y, Shioi A, Morii H (1998) 1, 25-Dihydroxyvitamin D3 increases in vitro vascular calcification by modulating secretion of endogenous parathyroid hormone-related peptide. Circulation 98:1302–1306

Chapter 6
Vitamin D: Cardiovascular Function and Disease

Robert Scragg

Abstract Opinions about the effect of vitamin D on risk of cardiovascular disease have changed substantially over the last half century. During the 1950s and 1970s, the dominant view was that vitamin D was a cause of cardiovascular disease. During the 1980s and 1990s, an increasing number of studies showed benefits from vitamin D, challenging earlier opinions that vitamin D was harmful. During the first decade of this century, the weight of scientific opinion has shifted 180° from that of 50 years ago, and the prevailing focus of research is on identifying the potential beneficial effects of vitamin D against cardiovascular disease. Since 2003, large epidemiological studies of hemodialysis patients and general population samples have shown inverse associations between vitamin D and cardiovascular disease. A growing body of laboratory and clinical research has identified several possible mechanisms to explain this association. These include adverse effects of vitamin D deficiency on immune and inflammatory processes, endothelial function, matrix-metalloproteinases and insulin resistance, which result in cardiac hypertrophy, thickened arteries, increased plaque formation, and rupture and thrombosis. Large randomized trials are required to determine with certainty whether vitamin D protects against cardiovascular disease.

Keywords 25-hydroxyvitamin D • Cardiovascular disease • Hypertension • Vitamin D

Abbreviations

CI	Confidence interval
CRP	C-reactive protein
CV	Cardiovascular
$1,25(OH)_2D$	1,25-Dihydroxyvitamin D
25OHD	25-Hydroxyvitamin D
IL	Interleukin
MMP	Matrix-metallo-protease

R. Scragg (✉)
School of Population Health, University of Auckland, Private Bag 92019,
Auckland 1142, New Zealand
e-mail: r.scragg@auckland.ac.nz

D.L. Trump and C.S. Johnson (eds.), *Vitamin D and Cancer*,
DOI 10.1007/978-1-4419-7188-3_6, © Springer Science+Business Media, LLC 2011

NHANES	National Health and Nutrition Examination Survey
PTH	Parathyroid hormone
TNF	Tumor necrosis factor
UV	Ultraviolet

6.1 Introduction

Opinions about the effect of vitamin D on risk of cardiovascular (CV) disease have changed substantially over the last half century. From the 1950s to the end of the 1970s, the dominant viewpoint was that vitamin D was a cause of CV disease. During the 1980s and 1990s, an increasing number of studies were published showing benefits from vitamin D, challenging the earlier opinions that vitamin D was harmful and resulting in a period of flux where researchers increasingly were open to the possibility that vitamin D could protect against CV disease. This coincided with a substantial increase in research on vitamin D and cancer, which along with the identification of vitamin D receptors in many body tissues, resulted in an increased acceptance by vitamin D researchers that the effects of vitamin D were not restricted to bone disease, but could affect the health of many organs and body systems. During the first decade of this century, the weight of scientific opinion has shifted 180° from that of 50 years ago, and the prevailing focus of research is on identifying the potential beneficial effects of vitamin D against CV disease. This latter period has coincided with a rapid increase in the number of publications on vitamin D and CV disease (Fig. 6.1). There are lessons to be learnt from this story, and the current generation of researchers needs to be mindful of the possibility that opinions may change again in the future.

The purpose of this review is to describe the key developments in research on vitamin D and CV disease over the last 50 years, to summarize the findings from recent large epidemiological studies which strongly support a beneficial effect from vitamin D against CV disease, and to give an overview of the possible mechanisms by which vitamin D may protect against CV disease.

6.2 1950s to 1970s: Adverse Vascular Effects from Vitamin D

6.2.1 Vascular Lesions from Vitamin D Intoxication

An epidemic of cases of infantile hypercalcemia occurred in Great Britain during 1953–1955 which was attributed to vitamin D fortification of commercial milk powders and infant cereals, and vitamin D supplements [1]. In response, the British government reduced the amount of vitamin D in fortified foods so that by 1957–1958 daily intake of vitamin D by infants had halved. However, the number of

6   Vitamin D: Cardiovascular Function and Disease   117

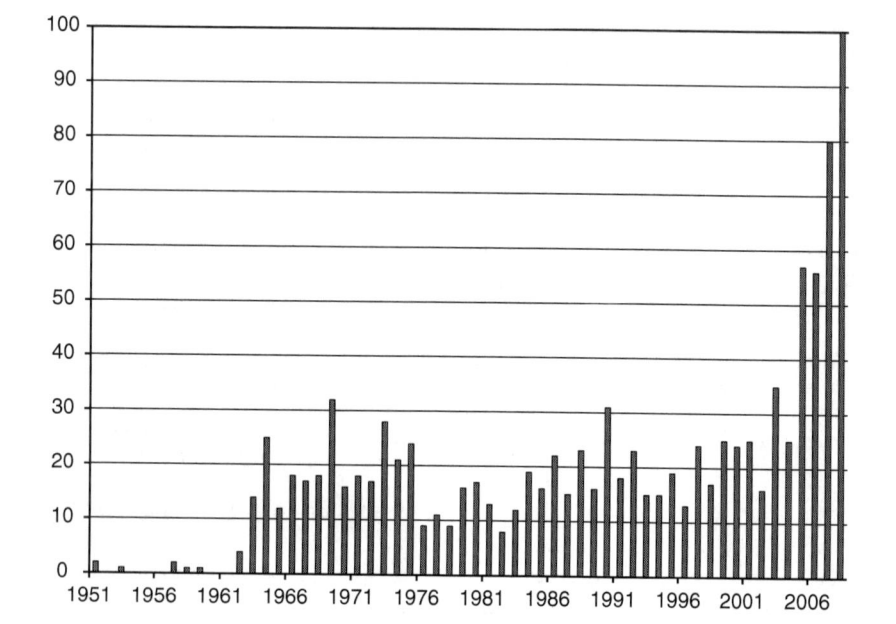

Fig. 6.1 Number of PubMed publications by year (1950–2008): search terms "Vitamin D" and "Cardiovascular Disease"

hypercalcemia cases in 1959 remained at the same level as before the reduction in vitamin D and did not decrease until 1960–1961 [1]. Thus, a report by the American Academy of Pediatrics in 1967 concluded that the hypothesis that vitamin D caused infantile hypercalcemia was unproven [1].

Despite this official report, many researchers still held the opinion that vitamin D was a cause of infantile hypercalcemia, and went further by linking the condition to a rare congenital abnormality in infants characterized by supravalvular aortic stenosis, an elfin facies, and severe mental retardation [2–5]. The basis for linking the two conditions was the similarity in the vascular lesions between those observed in supravalvular aortic stenosis and those produced by vitamin D intoxication [6]. Evidence thought confirmatory at the time came from animal studies in which pregnant rabbits were given intramuscular vitamin D doses of up to 4.5 million IU/day, and their offspring 250 IU/day, with the latter at autopsy found to have medial degeneration, calcification, and necrosis of the aorta that was similar to the pathology of the congenital anomaly in children [2, 7]; while high doses of vitamin D (up to 770 IU/g over 10 days) were found to cause both aortic and cardiac lesions in young rabbits [8]. In contrast, several case reports of infants with arterial calcification concluded it was not caused by vitamin D [9–12], and that high doses of vitamin D taken during pregnancy did not result in infantile hypercalcemia or arterial lesions such as aortic stenosis [13, 14].

However, the prevailing opinion remained that vitamin D was a risk factor for vascular damage and CV disease [3, 15, 16]. This was supported by the development

of animal models of arteriosclerosis caused by hypervitaminosis D, using mega-doses of 5,000–10,000 IU/kg/day [17, 18], equivalent to daily doses of 350,000–700,000 IU for a 70 kg adult human; and case reports of arterial calcification and hypertension in patients taking up to 170,000 IU of vitamin D/day [19–21]. Despite dissenting opinion [22] and isolated reports that vitamin D could prevent myocardial calcification in rats [23] and assist with the treatment of vascular calcification, heart failure, and cardiac arrthymias in humans [24–27], reviews in the 1970s were convinced about the causal role of vitamin D in atherosclerosis and coronary heart disease [15, 16, 28]. The authors of one of these reviews, although writing nearly 4 decades ago, could be addressing contemporary concerns by stating:

> The tragic "operation over-kill" of adding vitamin D to almost everything excepting cigars may well be one of the most important pathogenic factors in human atherosclerosis. People in the USA may well be the victims of Madison Avenue advertising tycoons, food manufacturers, unsuspecting dietitians, and indifferent physicians who have probably all played a role in adding excessive amounts of vitamin D to many foods [15].

Looking back at these reports raises an obvious question: Why were their opinions so different to the results from recent cohort studies (see Sect. 6.4.2.1) which have consistently reported inverse associations between vitamin D status and risk of CV disease? Two points can be made about the approach to research in the 1960s and 1970s. Firstly, there was an overreliance on case reports in determining causation. These case reports were limited because they often had very small numbers of selected patients, who may not have been representative of all patients with a particular condition, and did not include a control group to help decide whether cases had a higher-than-expected intake of vitamin D. Secondly, there was a lack of appreciation that the doses of vitamin D given in the animal models of arteriosclerosis were orders of magnitude higher than those normally ingested by the general human population. A recent review of the evidence on vitamin D and vascular calcification has concluded that vitamin D exerts a biphasic effect on vascular calcification with adverse effects occurring when body vitamin D levels are very low or very high [29].

6.2.2 Early Epidemiological Studies

6.2.2.1 Dietary Vitamin D and Cardiovascular Disease

Epidemiological studies carried out in the 1970s were influenced by the prevailing opinion of the time that vitamin D was a cause of CV disease. Positive correlations between vitamin D intake measured in national surveys and standardized mortality ratios for ischemic heart disease ($r=0.58$) and cerebrovascular disease ($r=0.49$) during 1964–1969 were reported in an ecological study of eight regions within England and Wales [30]. A population-based myocardial infarction case–control study in Tromso (Norway) reported significantly higher mean daily intake of vitamin D in cases (males 31.28 µg, females 34.05 µg) compared to age- and sex-matched controls (males 22.68 µg, females 20.68 µg) [31]. The limitations of

6   Vitamin D: Cardiovascular Function and Disease   119

this study include measurement of dietary intake in cases several years after their heart attack since these individuals are likely to have changed their dietary patterns after such a major medical event, and reliance on dietary vitamin D intake to determine vitamin D status.

6.2.2.2   25-Hydroxyvitamin D and Myocardial Infarction

The small contribution of dietary vitamin D to overall body vitamin D levels was revealed by the development in the 1970s of competitive protein-binding assays for 25-hydroxyvitamin D (25OHD), the main marker of vitamin D status [32]. Diet was shown to contribute less than 20% of vitamin D stored in the body, with the major component (more than 80%) coming from vitamin D synthesized in the skin through sun exposure.[33, 34]

The first epidemiological study of cardiovascular disease to report results with this new method of measuring vitamin D status was from Heidelberg, Germany [35]. The study recruited only 15 myocardial infarction cases, an unstated time after their heart attack, and found that their mean serum 25OHD level (32 nmol/L) was within the normal range for other controls at that time of year. The authors concluded, somewhat surprisingly at the time, that "nutritional vitamin D status or exposure to sunlight cannot account for the development of myocardial infarction."

The next report came from a case control study in Copenhagen, Denmark [36]. The authors of this report, concerned about the possible effect of the acute-phase reaction from a myocardial infarction on serum 25OHD levels, first showed in a pilot study of 12 patients that 25OHD did not fluctuate in the first 4 days after onset of symptoms. They then recruited 128 consecutive patients admitted with chest pain (53 who had a myocardial infarction and 75 with angina) and compared them with 409 controls, although no details are provided on how the latter were selected. Mean serum 25OHD was slightly lower in cases (myocardial infarction 24.0 ng/mL, angina 23.5 ng/mL) than in controls (28.8 ng/mL), with case–control differences being statistically significant during May–August ($p<0.05$). The authors concurred with the conclusion of the earlier German report by stating that the "present results do not support the theory that patients with ischaemic heart disease have a higher vitamin D intake than the rest of the population."

The third report came from the Tromso Heart Study, Norway, which had previously reported higher vitamin D intakes in cases [31]. This was a nested case–control study which avoided possible bias, from a systematic error caused by the effect of the disease on measures of vitamin D status, by measuring 25OHD levels in blood samples collected at baseline interviews when participants were enrolled into the study [37]. Mean serum 25OHD in 23 patients free of disease at baseline who had myocardial infarctions during the 4-year follow-up period was again slightly lower than in 46 controls matched for age and time of year (59.0 vs 63.4 nmol/L); with the case–control difference being significant after correcting for vitamin D binding protein ($p=0.024$), indicating that cases had a lower concentration of free-25OHD.

120   R. Scragg

In summary, the results from these three studies clearly called for a reevaluation of the hypothesis that vitamin D was a cause of coronary heart disease. Their overall conclusion was that vitamin D levels in patients with coronary heart disease were either the same as, or lower than, in healthy controls. The major limitation of these studies is their small sample sizes, which is a likely reason for their insufficient statistical power to observe consistent significant reductions in serum 25OHD levels among heart disease cases, as reported in subsequent larger epidemiological studies (see below).

6.2.2.3   25-Hydroxyvitamin D and Serum Cholesterol

These three early case–control studies of serum 25OHD and coronary heart disease also provided important information about the association between vitamin D status and serum cholesterol. Animal studies in the 1950s and 1960s had shown previously that the combination of high dietary intake of vitamin D and cholesterol could produce raised blood cholesterol levels and atherosclerotic lesions [38, 39]. An experimental study in humans found that daily vitamin D doses of 50,000 or 1,000 IU for 21 days significantly increased serum cholesterol levels, although the study can be criticized because of the lack of a control group [40]. Analyses of baseline cross-sectional data from the Tromso Heart Study reported a significant positive association ($p = 0.0013$) between dietary vitamin D intake and serum cholesterol in men aged 20–50 years [41].

However, after the advent of assays for 25OHD, the Danish and Norwegian studies found no association between serum 25OHD and serum cholesterol [36, 37]. This result has been confirmed by subsequent epidemiological studies [42–45]. Thus, the overall evidence to date suggests that any association between vitamin D and CV disease does not involve serum cholesterol.

6.3   1980s to 1990s: Vitamin D May Protect Against Cardiovascular Disease

6.3.1   Hypothesis

The early studies showing that more than 80% of vitamin D comes from sun exposure [33, 34] emphasized the importance of solar ultraviolet (UV) radiation in determining vitamin D status, and provided a possible link between vitamin D and some of the descriptive epidemiological variations in CV disease rates. UV-B irradiation (wavelengths 280–320 nm), acting on the skin, converts the precursor 7-dehydrocholesterol into vitamin D_3, which comprises most of the vitamin D in humans [46]. The intensity of UV radiation on the surface of the earth varies with season being highest in summer and lowest in winter, decreases with increasing latitude from the equator, and increases with altitude by up to 18% per 1,000 m [47].

6 Vitamin D: Cardiovascular Function and Disease

The descriptive epidemiology of CV disease shows that rates are highest in winter in both the northern and southern hemispheres, increase with increasing latitude, and decrease with increasing altitude. Drawing this evidence together, the author published a hypothesis in 1981 that sunlight and vitamin D may protect against CV disease [48]. This hypothesis was also consistent with the increased CV disease rates in population groups with lower vitamin D levels due to decreased skin synthesis, such as older people and those with increased skin pigmentation (e.g., African-Americans) [49, 50]. A more detailed review of the evidence in support of the hypothesis was subsequently published [51].

Recent ecological studies of CV disease have continued to provide support for the hypothesis. For example, an inverse association between UV insolation and coronary heart disease mortality in men has recently been reported for the countries of Western Europe [52]. Seasonal variations in vitamin D status, with low 25OHD levels in winter, have been shown in both the northern and southern hemispheres [53, 54]. Winter excesses in mortality and incidence have been reported for the full spectrum of CV disease, including coronary heart disease [55–57], stroke [57–59], heart failure [60, 61], ventricular arrhythmias [62], endocarditis [63], and pulmonary embolism [64].

Importantly, the winter excess in CV disease has been observed in warm climates, such as Los Angeles [65], and in Hawaii despite a small seasonal variation in temperature between 22.8°C and 27.8°C [66]. The winter excess in cardiovascular disease is attributed frequently to the cold temperatures of winter [67], but it does not seem plausible that the mild temperatures experienced by people in the above two locations during the winter months is a major factor in their raised CV disease rates at that time of year.

The hypothesis that vitamin D protected against CV disease was tested in a population-based case–control study of myocardial infarction carried out in New Zealand by the author and colleagues in the 1980s and published in 1990 [68]. The sample was restricted to incident cases from a register which provided blood samples within 12 h of onset of symptoms since a pilot study showed that plasma 25OHD was unchanged during this period [69]. The unit of measurement for 25OHD in this study was actually nanograms per milliliter (ng/mL), rather than nanomoles per liter (nmol/L) as reported. Mean plasma 25OHD was significantly lower in cases ($n = 179$) than controls selected from the electoral roll who were individually matched by age, sex, and date of blood collection (32.0 vs 35.0 ng/mL; $p = 0.017$). An inverse association between plasma 25OHD and risk of myocardial infarction, with the odds ratio for those in the highest 25OHD quartile being 0.30 (95% confidence interval [CI]: 0.15, 0.61) compared with the lowest quartile [68].

6.3.2 Animal Studies

Independently of the above epidemiological studies, research from animal models in the 1980s was beginning to better define the effect of vitamin D on CV function

when given in physiological doses. This was stimulated by the identification of a receptor to 1,25-dihydroxyvitamin D (1,25(OH)$_2$D), firstly in cultured rat heart identified in 1983 [70, 71], and subsequently confirmed by others [72], which was found to be located in the nucleus [73]. Together with the additional finding of a vitamin-D-dependent calcium-binding protein in myocardial tissue in 1982 [74], these studies supported a role for 1,25(OH)$_2$D in regulating CV function.

Further studies were carried out with the aim of elucidating the possible CV mechanisms involved with vitamin D. When rats reared deficient in vitamin D were compared to those given 30 IU of vitamin D$_3$/day (equivalent to about 8,500 IU/day for 70 kg human adult), the vitamin-D-deficient rats had increased cardiac contraction [75], and myocardial hypertrophy due to myocardial collagen deposition and myocyte hyperplasia and hypertrophy [76–79]. These effects were independent of changes in serum calcium, suggesting a direct effect of vitamin D, since myocardial accumulation of calcium after very high vitamin D doses could be blocked by calcium channel blockers [80, 81]; and in contrast with the earlier studies showing adverse effects from excessive vitamin D which were secondary to increases in serum calcium (Sect. 6.2.1). However, the health implications of these studies were unclear as the increased cardiac contractility in vitamin D deficiency could be interpreted as beneficial, while the myocardial hypertrophy could be detrimental.

Evidence was also accumulating of a role for vitamin D in regulating blood pressure. A receptor to 1,25(OH)$_2$D was described in smooth muscle tissue [82], and also in endothelial cells with early evidence of autocrine synthesis of 1,25(OH)$_2$D that was a function of 25OHD substrate concentration [83]. Alterations in vitamin D metabolism were observed in spontaneously hypertensive rats which were shown to have decreased plasma levels of 1,25(OH)$_2$D [84]; while injection of the same metabolite in normotensive rats resulted in a delayed increase in blood pressure consistent with a genomic mechanism [85].

6.3.3 Human Studies

6.3.3.1 Blood Pressure

A key stimulus for research on vitamin D and hypertension were the studies in the early 1980s showing elevated parathyroid hormone (PTH) levels in hypertension cases [86, 87], which was speculated as being a possible response to increased urinary calcium loss, along with research showing inverse associations between blood levels of both 1,25(OH)$_2$D and PTH with renin in hypertension patients [88]. Given the well-documented inverse association between PTH and vitamin D status, these studies suggested that low vitamin D levels might be a risk factor for hypertension. A US cross-sectional study reported an inverse association between dietary vitamin and systolic blood pressure [89]. However, results from studies of the association between blood levels of 25OHD and blood pressure were inconclusive. An early Polish study found that plasma levels of 25OHD were lower in hypertension

6 Vitamin D: Cardiovascular Function and Disease

cases compared with controls, which was attributed by the authors to a vitamin D lowering effect from thiazide diuretics [90]. A cross-sectional study from New Zealand found a weak inverse association between plasma 25OHD and diastolic blood pressure which was not significant after adjusting for age and season [42]. Small case–control studies (with <30 cases) reported either increased [91] or similar [92] serum 25OHD levels in cases compared with controls. A further nested case–control study from New Zealand with a much larger sample (186 cases), reported similar 25OHD levels in cases and controls matched by age, sex, ethnicity, and season [93].

Inconsistent findings have also been reported in studies of blood levels of the active metabolite 1,25(OH)$_2$D and blood pressure. A cross-sectional study of 373 women from Iowa reported a significant positive association between serum 1,25(OH)$_2$D and blood pressure after adjusting for age, BMI, and current thiazide use [94]. This finding was confirmed in small case–control studies which reported· significantly higher levels of both 1,25(OH)$_2$D and PTH in cases [91, 92]. However, other studies have reported inverse associations between 1,25-dihydroxyvitamin D and blood pressure [95–97].

A small number of experimental studies were also carried out during this period. The first two in Sweden showed that active vitamin D (alphacalcidol) lowered blood pressure in patients with intermittent hypercalcemia or impaired glucose tolerance [98, 99]. Although both of these studies were double-blind with controls, in one of them there is a reported high dropout rate, from 86 participants at baseline to 25 remaining at follow-up after 6 months treatment, raising the possibility of a withdrawal bias [98]; while in the other, the reduction in blood pressure was limited to those with hypertension (blood pressure ≥ 150/90) [99]. A further study by the same research group reported a reduction in blood pressure in 14 men with impaired glucose tolerance given alphacalcidol over 18 months, but the lack of control group negates the findings from this study [100]. In contrast, two studies of participants sampled from the community did not show an effect of vitamin D$_3$ on blood pressure. One was a US study from Oregon ($n=65$) which found that 1,000 IU vitamin D$_3$/day (with calcium) for 3 years did not show any effect on blood pressure [101], despite this dose increasing 25OHD levels by about 30 nmol/L [102]. The other study was carried out in the UK ($n=189$) and found that a single 100,000 IU dose of vitamin D$_3$ had no effect on blood pressure after 6 weeks, when compared with controls, although the difference in 25OHD at 6 weeks between the groups was only 8.6 ug/mL (21.5 nmol/L) [44].

6.3.3.2 Cardiac Function

At this time, isolated case reports started to appear of congestive heart failure with vitamin D deficiency and hypocalcemia, in both adults and children, being successfully treated by vitamin D (in combination with calcium) [103–105]; while children with severe rickets without clinical signs of heart failure were found pretreatment to have thickened interventricular septa which returned to normal

after treatment [106]. Consistent with these case reports, vitamin D supplementation (with 1-α-hydroxyvitamin D) of hemodialysis patients was found to improve left ventricular cardiac function, as measured with echocardiography, by increasing fractional fiber shortening [107] and decreasing end-systolic and end-diastolic diameter [108]; although the results from the latter two studies are not entirely consistent with each other, perhaps because of their limited statistical power due to very small samples (12 and 5, respectively). Benefits in cardiac function have also been reported for $1,25(OH)_2D$. This metabolite reduced end-systolic diameter and increased fractional shortening, but only in hemodialysis patients ($n=5$) with very high PTH levels in a Finnish report [109]; and reduced measures of cardiac size (intraventricular wall thickness and left ventricle mass), without any change in blood pressure or cardiac output, in 15 hemodialysis patients compared with 10 control patients from Korea [110]. In a US case series of 101 patients with severe congestive heart failure undergoing evaluation for cardiac transplantation, patients with more severe disease had significantly lower 25OHD levels, although this could have been a consequence from less outdoor sun exposure due to feeling unwell from their disease [111].

6.3.3.3 Calcification

Research using very high doses of vitamin D to produce vascular and cardiac lesions from calcification continued throughout this period with animal models [112–117]. However, human studies reported either inverse associations [118, 119], or no association [120], between blood levels of $1,25(OH)_2D$ and coronary calcification. Since blood levels of $1,25(OH)_2D$ can be influenced by a number of variables, including dietary calcium and vitamin D status [121], the significance of these findings was unclear in the absence of studies of the relationship of 25OHD and calcification.

6.3.4 Summary

Although animal studies of vitamin D toxicity and arteriosclerosis continued during the 1980s and 1990s, this period was characterized by a shift in emphasis from studies of adverse effects toward those looking at potential beneficial effects of vitamin D on CV function. The identification of vitamin D receptors in cardiac and smooth muscle was compelling evidence for a role by vitamin D in regulating CV function. However, the number of epidemiological studies, which are essential for determining etiology, was still limited, with the majority of reports being either animal studies or human studies of patients. The latter often had very small numbers which limited their statistical power for evaluating vitamin D, or selected groups of patients and controls who may have not been representative of the wider populations from which they were sampled. These deficiencies in design are

6   Vitamin D: Cardiovascular Function and Disease

a possible explanation for the inconsistent results reported during this period. Thus, by the end of this period, it was still not possible to conclude whether vitamin D in physiological doses was beneficial, harmful or irrelevant to CV health.

6.4   2000s: Increasing Evidence of a Beneficial Cardiovascular Effect

The number of publications on vitamin D and CV disease has rapidly increased in the first decade of the new millennium (Fig. 6.1). This new research has been influenced by reports from large epidemiological studies showing inverse associations between vitamin D and CV disease, initially from cohorts of hemodialysis patients, but in 2008 from general population cohorts. Coinciding with these new epidemiological findings has been the publications from patient and animal studies providing new insights into possible mechanisms linking vitamin D and CV disease.

6.4.1   Studies in Hemodialysis Patients

CV disease is the main cause of death in developed countries. Interest in the beneficial effects from vitamin D against CV disease was stimulated by a landmark publication by US researchers showing that a cohort of hemodialysis patients on paricalcitol had a 16% reduction in all-cause mortality compared with those on calcitriol [122]. The authors of this report restricted their comparisons to those on either form of activated vitamin D by excluding patients not on any form of vitamin D to avoid confounding by indication. Thus, the possibility remained that the reduced mortality in those taking paricalcitol was an artifact caused by increased mortality in those taking calcitriol. However, the latter possibility was dispelled by a Japanese cohort study showing decreased CV mortality in dialysis patients on alfacalcidol compared to no vitamin D, the adjusted hazard ratio being 0.38 (95% CI: 0.25, 0.58) of CV mortality over 5 years [123].

This finding was confirmed by a further cohort study of 51,000 US hemodialysis patients, which found a CV disease incidence rate of 7.6 per 100 person-years in the vitamin D-treated group (mainly calcitriol or paricalcitol) compared with 14.6 per 100 person-years in the nonvitamin D group ($p < 0.001$), with the relative reduction in all-cause mortality being 20% [124]. Of interest in relation to the possible protective mechanisms associated with vitamin D (see Sect. 6.4.4), this study also reported a significant reduction in mortality from an infectious disease among the vitamin D-treated group compared with the untreated (1.1 vs 2.8 deaths per 100 person-years, $p < 0.0001$). Similar findings were observed in a recent cohort study of hemodialysis patients from six Latin American countries, with patients given oral active vitamin D having reduced mortality (of about 50%) from all-causes,

CV disease and infectious disease, compared to those who did not receive vitamin D [125]. Other publications from cohort studies of patients with chronic kidney disease have reported relative reductions in all-cause mortality of about 20% for those who received activated vitamin D, regardless of whether patients are on dialysis [126–129] or not [130].

Vitamin D supplementation can remove the association between vitamin D status and mortality in dialysis patients. A cohort study of incident hemodialysis patients, using the nested case–control design, observed increased CV mortality after 90 days follow-up in those with low baseline 25OHD levels in patients not on vitamin D therapy; while no association with baseline 25OHD was observed in those on vitamin D [131]. Recently, activated vitamin D has been associated with racial differences in survival in US hemodialysis patients, with all-cause mortality being 16% lower in treated-black versus treated-white patients, and 35% higher in untreated-black versus untreated-white patients [132]. The consistent findings from cohort studies of vitamin D treatment and mortality are compelling, but we need results from randomized trials before we can be certain that activated vitamin D improves survival in hemodialysis patients [133, 134].

6.4.2 Studies in Healthy Populations

6.4.2.1 Cardiovascular Disease

In 2008, a tipping point was reached with the publication of results from four large cohort studies showing that low baseline blood levels of 25OHD predict subsequent increased risk of CV disease and all-cause mortality. The first study was from the Framingham Study Offspring cohort ($n = 1,739$) which found that participants with baseline serum 25OHD levels <10 ng/mL (25 nmol/L) had an adjusted hazard ratio of 1.80 (95% CI: 1.05, 3.08) for CV disease during the 5-year follow-up period, compared with those >15 ng/mL (37.4 nmol/L) [135]. The effect was evident in participants with hypertension (blood pressure $\geq 140/90$ mmHg), but not in those with normal blood pressure, suggesting that hypertension could magnify the beneficial effects of vitamin D on the CV system. The second report was from the US Health Professionals Follow-up Study ($n = 18,225$) which found in a nested case–control comparison that men with baseline plasma 25OHD levels ≤ 15 ng/mL (37.4 nmol/L) had a relative risk of 2.09 (95% CI: 1.24, 3.54) for myocardial infarction (fatal plus nonfatal) over 10-year follow-up compared to those with 25OHD < 30 ng/mL (74.9 nmol/L) adjusting for covariates [136]. The third was from the follow-up cohort ($n = 13,331$) of the Third National Health and Nutrition examination Survey (NHANES III), a representative sample of the US population surveyed during 1988–1994, which found that participants in the lowest quartile of baseline serum 25OHD < 17.8 ng/mL (44.4 nmol/L) had a 26% (95% CI: 8, 46) increased risk of all-cause mortality during a median 8.7-year follow-up, compared with those in the highest 25OHD quartile [137].

6   Vitamin D: Cardiovascular Function and Disease   127

The fourth study, from Germany on patients ($n=3,258$) referred for coronary angiography and followed for a median period of 7.7 years, is described here because it had a similar design as the above studies [138]. Patients with baseline serum 25OHD levels in the bottom quartile had a significantly increased relative risk of all-cause mortality (hazard ratio=2.08; 95% CI: 1.60, 2.70) and CV mortality (hazard ratio = 2.22; 95% CI: 1.57, 3.13) compared with those in the highest baseline 25OHD quartile, after adjusting for the full range of covariates, including baseline serum $1,25(OH)_2D$ which also was independently and inversely associated with follow-up risk of all-cause and CV mortality.

The study designs used in the first three of these studies provide the best-quality evidence to date on the association between vitamin D status and risk of CV disease in the general population [135–137], aside from the Women's Health Initiative randomized control trial of vitamin D supplementation which has methodological weaknesses (see Sect. 6.4.3) [139].

Consistent with the above cohort studies, a 2006 case–control study from Cambridge (UK) found that mean Z score of 25OHD for incident stroke cases, measured within 30 days of disease onset, was significantly below that expected for a sample of healthy controls (−1.4, 95% CI: −1.7, −1.1; $p<0.0001$) [140]. In contrast, a hospital-based case–control study of coronary artery disease from India reported in 2001 that a significantly ($p<0.001$) higher proportion of cases (59.4%) than controls (22.1%) had serum 25OHD levels above 222.5 nmol/L [141]. A limitation of this study is that it recruited prevalent cases of coronary artery disease, an unknown time after their heart attacks, when their vitamin D status may not have reflected that at the time of disease onset.

Further studies have been published on vitamin D and congestive heart failure. A case control study from Germany found significantly lower serum 25OHD and $1,25(OH)_2D$ levels in cases and controls [142], while low serum $1,25(OH)_2D$ (but not 25OHD) predicted increased risk of death or need for heart transplant in patients with end-stage congestive heart failure [143]. In contrast, a recent German randomized controlled trial of 93 patients with congestive heart failure failed to show an effect of vitamin D supplementation on measures of cardiac function with echocardiography [144]. Information has recently been reported on vitamin D and arterial disease. Analyses of NHANES data (for 2001–2004) found that the prevalence ratio of peripheral arterial disease increased by 1.35 (95% CI: 1.15, 1.59) for each 10 ng/mL (25 nmol/L) decrease in serum 25OHD [145].

6.4.2.2   Blood Pressure

This decade has also seen the publication of large epidemiological studies of vitamin D and blood pressure. A large cross-sectional study from Norway ($n=15,596$) found that dietary vitamin D was unrelated to blood pressure [146]. Results from three large US health professional cohorts (total $n=209,313$) did not show an association between dietary vitamin D and incident hypertension [147]; although a recent US study of female health professionals ($n=28,886$) reported that risk of

incident hypertension had a weak inverse association with vitamin D but not vitamin D supplements [148]. A possible explanation for the failure of most of these studies to find an association is that dietary sources of vitamin D contribute only a small proportion of the total vitamin D entering the body each day, which is mainly derived from sun exposure [34]. Interestingly, when two of these cohort studies were reanalyzed using plasma 25OHD, which measures vitamin D from all sources, both measured 25OHD and estimated 25OHD were inversely associated with risk of incident hypertension in both men and women [149]. For example, participants in the lowest baseline quartile of plasma 25OHD (<15 ng/mL) had a 3.18 (95% CI: 1.39, 7.29) increased risk of developing hypertension over 4 years than those in the highest 25OHD quartile (≥30 ng/mL). This finding is supported by a recent publication from the cross-sectional NHANES III study ($n = 12,644$) which found that serum 25OHD was inversely associated with both systolic blood pressure and pulse pressure [53]. However, another cross-sectional study from the Netherlands ($n = 1,205$) did not observe any association between serum 25OHD and blood pressure, possibly because the elderly sample had relatively high vitamin D levels, although there was a significant positive association between serum parathyroid hormone and blood pressure [150].

Further intervention studies, both from Germany, have also been carried out. A randomized clinical trial in elderly women found that 800 IU of vitamin D3/day (with 1,200 mg of calcium) after 8 weeks significantly decreased systolic blood pressure by 5 mmHg, but not diastolic, compared with placebo [151]. Another randomized trial of patients with hypertension found that exposure to UV-B radiation over 6 weeks, which increases vitamin D, lowered blood pressure by 6 mmHg compared with the UV-A control group ($p < 0.05$) [152].

6.4.3 Studies of Vitamin D Supplementation

The Women's Health Initiative trial is the only randomized trial to date which has examined the effect of vitamin D on CV disease in the general population [139]. Postmenopausal women aged 50–79 years ($n = 36,282$) at 40 clinical sites in the USA were randomized to take calcium carbonate 500 mg with vitamin D 200 IU twice daily or placebo. Both fatal and nonfatal disease events were recorded. After 7 years of follow-up, the adjusted hazard ratios in the treated group versus control were 1.04 (95% CI: 0.92, 1.18) for coronary heart disease and 0.95 (95% CI: 0.82, 1.10) for stroke. Thus, this study did not detect any effect of vitamin D and calcium supplementation on CV disease.

However, this study has some major design limitations which prevent it from being a proper test of the hypothesis that vitamin D protects against CV disease [139, 153, 154]. Firstly, the dose of vitamin D was only 400 IU/day, which would have raised serum 25OHD levels only by about 10 nmol/L [155], way below the daily dose of 1,700 IU required to raise 25OHD levels above 80 nmol/L that is currently considered optimum [156]. The actual vitamin D ingested would have

6   Vitamin D: Cardiovascular Function and Disease

been further reduced by poor compliance as only 59% of participants took ≥ 80% of the study medication. Lastly, the control group was able to continue taking vitamin D supplements, resulting in contamination.

Further evidence from randomized trials suggesting a beneficial effect of vitamin D against CV disease comes from a recent meta-analysis of vitamin D supplementation and all-cause mortality [157]. The results of this meta-analysis are relevant since CV disease is the main cause of mortality in developed countries. It summarized 18 randomized clinical trials published from 1992 to 2006, which included data from the Women's Health Initiative trial [158], 15 studies in Europe, and two studies from Australia and New Zealand. The meta-analysis found that vitamin D supplementation produces a 7% relative reduction in all-cause mortality [157]. Most of the prevented deaths in the treated group are likely to have been from CV and infectious diseases, since the weighted mean follow-up period was 5.7 years, too short to detect any benefit in preventing cancer deaths [159]. These findings are consistent with the cohort studies of dialysis patients (described above) which have reported lower all-cause mortality in patients prescribed active vitamin D [122, 124–129, 131, 132].

A 7% relative reduction in all-cause mortality may seem small. However, the weighted vitamin D dose of 528 IU/day for all studies combined is likely to have only increased blood 25OHD levels by 10–15 nmol/L [155]. As mentioned above, this daily vitamin D dose is much lower than that currently recommended to maintain serum 25OHD at optimum levels [156]. Thus, the potential beneficial effect of vitamin D supplementation on all-cause mortality may be higher than 7% if larger vitamin D doses (>2,000 IU/day) are given which increase blood 25OHD levels up to 100 nmol/L [160].

6.4.4   Cardiovascular Pathophysiology

Since the start of the millennium, numerous publications from research on animal models and from patients with CV disease have greatly increased understanding of the mechanisms involved in the possible protective effect of vitamin D against CV disease. These mechanisms, reviewed below, involve beneficial changes in inflammatory processes, endothelial function, matrix metalloproteinases (MMPs), and the renin–angiotensin system (Fig. 6.2).

6.4.4.1   Inflammatory Factors

Until the 1990s, the dominant view held that the major risk factors of CV disease were cigarette smoking, hypercholesterolemia, and hypertension (the latter two caused by dietary saturated fats and physical inactivity), which exerted their effects over many years of exposure [161]. There was no place in this chronic disease model for inflammation, despite evidence from many countries showing winter

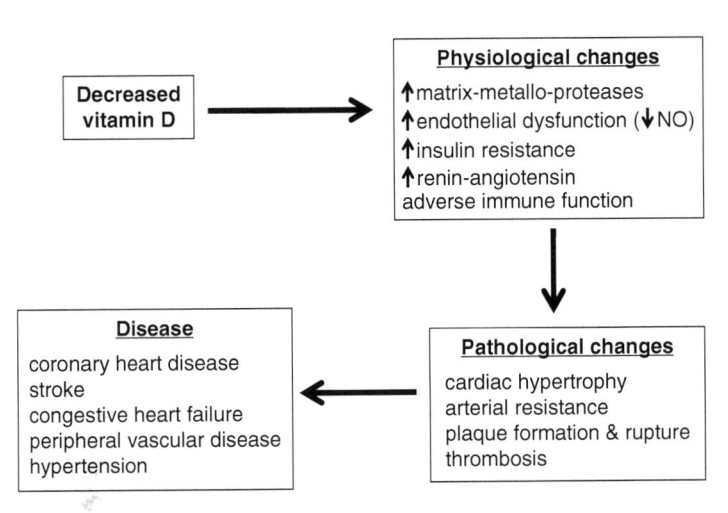

Fig. 6.2 Mechanisms by which low vitamin D status may increase the risk of cardiovascular disease

excesses in CV disease that coincided with winter respiratory infections (see Sect. 6.3.1). Opinion has changed substantially over the last 10 years, and it is now well established that subclinical inflammatory factors mediate the traditional chronic risk factors (such as smoking) and are centrally involved in the process of atherosclerosis and plaque rupture [162–164]. Blood levels of inflammatory markers, such as C-reactive protein (CRP) and the cytokine interleukin-6 (IL-6), predict subsequent risk of cardiovascular disease [164, 165]. Inflammatory cytokines also influence endothelial function [165, 166], which is an independent predictor of CV disease [167], and synthesis of MMPs [168] which also have a role in CV disease [169, 170]; while positive associations have been reported between IL-6 and insulin resistance [171–173] which are consistent with a role for pro-inflammatory factors in the etiology of type 2 diabetes [174, 175].

In a recent landmark paper, vitamin D was shown to have an important role in the innate immune system by stimulating the synthesis of the antimicrobial peptide cathelicidin [176]. This new finding provides a biological explanation for the historical link between sun exposure, vitamin D, and tuberculosis [177], as well as the association between rickets and infection which has been known since the 1960s [178]. Further, subclinical vitamin D deficiency has been reported in newborns and young adults without rickets, suffering from acute respiratory infections [179, 180], while women receiving vitamin D supplements in a clinical trial reported fewer respiratory symptoms than controls [181].

Laboratory in vitro studies have shown that $1,25(OH)_2D_3$ decreases production of pro-inflammatory cytokines such as IL-6 and tumor necrosis factor α (TNFα) by macrophages and lymphocytes[182–184] and up-regulates synthesis of anti-inflammatory IL-10 [185]. However, human studies of vitamin D supplementation have produced conflicting results perhaps due to varying doses of vitamin D. Vitamin D supplementation (2,000 IU/day for 9 months) decreased TNFα and

6   Vitamin D: Cardiovascular Function and Disease   131

increased IL-10, with no effect on CRP, in German patients with congestive heart failure [144]. Calcitriol supplementation decreased blood levels of IL-1 and IL-6 in hemodialysis patients [171]. Serum levels of 25OHD were inversely associated with CRP and IL-6 in German coronary angiography patients [138]. In contrast, a study which gave lower doses of vitamin D (≤800 IU/day) did not find any effect from it on IL-6 or [186].

6.4.4.2   Cardiovascular Function

Evidence has continued to emerge from animal models that vitamin D deficiency results in cardiac hypertrophy and fibrosis [187], possibly involving activation of the renin–angiotensin system [188]. MMPs may be involved in this cardiac hypertrophy since vitamin D supplementation lowers blood MMP-9 and MMP-2 [189]; and raised plasma levels of MMP-9 have been reported in men who had increased left-ventricular end-diastolic dimensions and wall thickness from the Framingham study [190].

Evidence has continued to emerge indicating that vitamin D deficiency may influence arterial function. As mentioned above, vitamin D suppresses MMPs which may prevent MMP-induced intimal thickening of blood vessels [168], and thereby reduce arterial stiffness. This possibility is supported by an earlier study of hypertension patients, which found that serum 25OHD levels, after 3 min of arterial occlusion of the calf, were associated positively with blood flow ($r=0.72$) and negatively with vascular resistance ($r=-0.78$) [191]. Serum 25OHD also was correlated positively with brachial artery distensibility and flow-mediated dilatation, after adjustment for age and blood pressure, in patients with end-stage renal disease [192]. Carotid artery intimal medial thickening is associated inversely with serum $25OHD_3$ in type 2 diabetes patients [193]; while vitamin D supplementation increases flow-mediated brachial artery dilatation in type 2 diabetes patients who have 25OHD levels below 50 nmol/L [194]. The above studies provide an explanation for an inverse association between serum 25OHD and microvascular complications observed in Japanese patients with type 2 diabetes [195]. The inverse associations between serum 25OHD and flow-mediated dilatation suggest vitamin D may improve impaired endothelial function arising from reduced nitric oxide synthesis by the endothelium [196] and thereby reduce risk of coronary heart disease [197, 198].

These changes in endothelial function may also reduce blood pressure since serum 25OHD levels are inversely associated both with pulse pressure, a marker of vascular resistance, and with systolic blood pressure [53]. Alternatively, vitamin D may lower blood pressure by downregulating the renin–angiotensin system [88, 199].

6.4.5   Summary

The past decade has seen a dramatic increase in the number of publications on vitamin D and CV disease. The critical development has been the publication in

2008 of results from three cohort studies showing that blood levels of 25OHD predict risk of CV disease and total mortality in the general population (see Sect. 6.4.2). These findings are strengthened further by a meta-analysis showing that vitamin D reduces total mortality [157] and results from cohort studies of dialysis patients showing that active vitamin D reduces total mortality and CV disease (see Sect. 6.4.1). This rush of publications stands in stark contrast with the dearth of large-scale epidemiological studies published during the previous half-century.

When this recent evidence is looked at in its totality, it meets many of the criteria for causation proposed originally for epidemiological studies by Bradford-Hill [200]. These include the temporality requirement of evidence from cohort studies that exposure (low vitamin D status) precedes the onset of the disease, evidence of reversibility from clinical trials that vitamin D supplementation reduces total mortality (most probably through preventing CV disease), consistency of evidence as shown by the agreement in the findings from cohorts studies in the general population and dialysis patients, and evidence of a moderately strong association with a doubling in the risk of CV disease between highest and lowest quantiles of 25OHD. Lastly, evidence of biological plausibility has come from recent animal and clinical studies identifying a number of mechanisms to explain the possible link between vitamin D and CV disease. These include effects of vitamin D on immune and inflammatory processes, endothelial function, the rennin–angiotensin system, MMPs and insulin resistance.

The evidence from recent cohort studies is now so compelling that large-scale clinical trials are required to determine, once and for all, whether vitamin D supplementation prevents CV disease, both in the general population and in patient populations [133–135, 137, 157]. As the great British epidemiologist Sir Richard Doll, who changed his view from opposing to supporting the beneficial effects of vitamin D and sun exposure before he died, said "This isn't difficult science. We should have answers" [201]. Neither is it expensive science since vitamin D is very cheap compared with most other treatments for CV disease. If clinical trials were to confirm that vitamin D prevents CV disease, the potential benefits would be substantial.

References

1. Fraser D (1967) The relation between infantile hypercalcemia and vitamin D–public health implications in North America. Pediatrics 40:1050–1061
2. Friedman WF (1967) Vitamin D as a cause of the supravalvular aortic stenosis syndrome. Am Heart J 73:718–720
3. Taussig HB (1966) Possible injury to the cardiovascular system from vitamin D. Ann Intern Med 65:1195–1200
4. Stare FJ et al (1966) Congenital supravalvular aortic stenosis, idiopathic hypercalcemia, and vitamin D. Nutr Rev 24:311–313
5. Seelig MS (1969) Vitamin D and cardiovascular, renal, and brain damage in infancy and childhood. Ann N Y Acad Sci 147:539–582

6 Vitamin D: Cardiovascular Function and Disease 133

6. Garcia RE, Friedman WF, Kaback MM, Rowe RD (1964) Idiopathic hypercalcemia and supravalvular aortic stenosis. Documentation of a new syndrome. N Engl J Med 271:117–120

7. Friedman WF, Roberts WC (1966) Vitamin D and the supravalvar aortic stenosis syndrome. The transplacental effects of vitamin D on the aorta of the rabbit. Circulation 34:77–86

8. Coleman EN (1965) Infantile hypercalcaemia and cardiovascular lesions. Evidence, hypothesis, and speculation. Arch Dis Child 40:535–540

9. Antia AU, Wiltse HE, Rowe RD, Pitt EL, Levin S, Ottesen OE, Cooke RE (1967) Pathogenesis of the supravalvular aortic stenosis syndrome. J Pediatr 71:431–441

10. Barold SS, Linhart JW, Samet P (1968) Coarctation of the aorta with unusual facies and mental retardation. Ann Intern Med 69:103–106

11. Beuren AJ, Schulz R, Sinapius D, Stoermer J (1969) Calcinosis of the arteries with coronary calcification in infancy. Am Heart J 78:87–93

12. Bird T (1974) Idiopathic arterial calcification in infancy. Arch Dis Child 49:82–89

13. Forbes GB, Cafarelli C, Manning J (1968) Vitamin D and infantile hypercalcemia. Pediatrics 42:203–204

14. Goodenday LS, Gordon GS (1971) No risk from vitamin D in pregnancy. Ann Intern Med 75:807–808

15. Taylor CB, Hass GM, Ho KJ, Liu LB (1972) Risk factors in the pathogenesis of atherosclerotic heart disease and generalized atherosclerosis. Ann Clin Lab Sci 2:239–243

16. Kummerow FA (1979) Nutrition imbalance and angiotoxins as dietary risk factors in coronary heart disease. Am J Clin Nutr 32:58–83

17. Eisenstein R, Zeruolis L (1964) Vitamin D-induced aortic calcification. Arch Pathol 77:27–35

18. Schenk EA, Penn I, Schwartz S (1965) Experimental atherosclerosis in the dog: a morphologic evaluation. Arch Pathol 80:102–109

19. Mallick NP, Berlyne GM (1968) Arterial calcification after vitamin-D therapy in hyperphosphataemic renal failure. Lancet 2:1316–1320

20. Way WG, Morgan DL, Sutton LE Jr (1958) Hypertension and hypercalcemic nephropathy due to vitamin D intoxication; a case report. Pediatrics 21:59–69

21. Blum M, Kirsten M, Worth MH Jr (1977) Reversible hypertension. Caused by the hypercalcemia of hyperparathyroidism, vitamin D toxicity, and calcium infusion. Jama 237:262–263

22. Price JD, Sookochoff MM (1969) Arterial calcification after vitamin D. Lancet 1:416

23. Bajusz E, Jasmin G (1963) Action of parathyroidectomy and dihydrotachysterol on myocardial calcification due to coronary vein obstruction. Proc Soc Exp Biol Med 112:752–755

24. Verberckmoes R, Bouillon R, Krempien B (1975) Disappearance of vascular calcifications during treatment of renal osteodystrophy. Two patients treated with high doses of vitamin D and aluminum hydroxide. Ann Intern Med 82:529–533

25. Aryanpur I, Farhoudi A, Zangeneh F (1974) Congestive heart failure secondary to idiopathic hypoparathyroidism. Am J Dis Child 127:738–739

26. Falko JM, Bush CA, Tzagournis M, Thomas FB (1976) Case report. Congestive heart failure complicating the hungry bone syndrome. Am J Med Sci 271:85–89

27. Johnson JD, Jennings R (1968) Hypocalcemia and cardiac arrhythmias. Am J Dis Child 115:373–376

28. Yogamundi Moon J (1972) Factors affecting arterial calcification associated with atherosclerosis. A review. Atherosclerosis 16:119–126

29. Zittermann A, Schleithoff SS, Koerfer R (2007) Vitamin D and vascular calcification. Curr Opin Lipidol 18:41–46

30. Knox EG (1973) Ischaemic-heart-disease mortality and dietary intake of calcium. Lancet 1:1465–1467

31. Linden V (1974) Vitamin D and myocardial infarction. Br Med J 3:647–650

32. Hollis BW, Horst RL (2007) The assessment of circulating 25(OH)D and 1, 25(OH)2D: where we are and where we are going. J Steroid Biochem Mol Biol 103:473–476

33. Haddad JG Jr, Hahn TJ (1973) Natural and synthetic sources of circulating 25-hydroxyvitamin D in man. Nature 244:515–517
34. Poskitt EM, Cole TJ, Lawson DE (1979) Diet, sunlight, and 25-hydroxy vitamin D in healthy children and adults. Br Med J 1:221–223
35. Schmidt-Gayk H, Goossen J, Lendle F, Seidel D (1977) Serum 25-hydroxycalciferol in myocardial infarction. Atherosclerosis 26:55–58
36. Lund B, Badskjaer J, Lund B, Soerensen OH (1978) Vitamin D and ischaemic heart disease. Horm Metab Res 10:553–556
37. Vik B, Try K, Thelle DS, Forde OH (1979) Tromso Heart Study: vitamin D metabolism and myocardial infarction. Br Med J 2:176
38. De Langen CD, Donath WF (1956) Vitamin D sclerosis of the arteries and the danger of feeding extra vitamin D to older people, with a view on the development of different forms of arteriosclerosis. Acta Med Scand 156:317–323
39. Penn I, Schenk E, Rob C, Deweese J, Schwartz SI (1965) Evaluation of the development of athero-arteriosclerosis in autogenous venous grafts inserted into the peripheral arterial system. Circulation 31(SUPPL 1):192–196
40. Fleischman AI, Bierenbaum ML, Raichelson R, Hayton T, Watson P (1970) Vitamin D and hypercholesterolemia in adult humans: Atherosclerosis, proceedings of the second international symposium. Springer-Verlag, New York, pp 468–472
41. Linden V, Linden V (1975) Vitamin D and serum cholesterol. Scand J Soc Med 3:83–85
42. Scragg R, Holdaway I, Jackson R, Lim T (1992) Plasma 25-hydroxyvitamin D3 and its relation to physical activity and other heart disease risk factors in the general population. Ann Epidemiol 2:697–703
43. Scragg R, Holdaway I, Singh V, Metcalf P, Baker J, Dryson E (1995) Serum 25-hydroxyvitamin D3 is related to physical activity and ethnicity but not obesity in a multicultural workforce. Aust N Z J Med 25:218–223
44. Scragg R, Khaw KT, Murphy S (1995) Effect of winter oral vitamin D3 supplementation on cardiovascular risk factors in elderly adults. Eur J Clin Nutr 49:640–646
45. Martins D, Wolf M, Pan D, Zadshir A, Tareen N, Thadhani R, Felsenfeld A, Levine B, Mehrotra R, Norris K (2007) Prevalence of cardiovascular risk factors and the serum levels of 25-hydroxyvitamin D in the United States: data from the Third National Health and Nutrition Examination Survey. Arch Intern Med 167:1159–1165
46. Holick MF, MacLaughlin JA, Doppelt SH (1981) Regulation of cutaneous previtamin D3 photosynthesis in man: skin pigment is not an essential regulator. Science 211:590–593
47. Kimlin MG (2008) Geographic location and vitamin D synthesis. Mol Aspects Med 29:453–461
48. Scragg R (1981) Seasonality of cardiovascular disease mortality and the possible protective effect of ultra-violet radiation. Int J Epidemiol 10:337–341
49. MacLaughlin J, Holick MF (1985) Aging decreases the capacity of human skin to produce vitamin D3. J Clin Invest 76:1536–1538
50. Clemens TL, Adams JS, Henderson SL, Holick MF (1982) Increased skin pigment reduces the capacity of skin to synthesise vitamin D3. Lancet 1:74–76
51. Scragg R (1995) Sunlight, vitamin D and cardiovascular disease. In: Crass MF, Avioloi LV (eds) Calcium-regulating hormones and cardiovascular function. CRC Press, Boca Raton, pp 213–237
52. Wong A (2008) Incident solar radiation and coronary heart disease mortality rates in Europe. Eur J Epidemiol 23:609–614
53. Scragg R, Sowers M, Bell C (2007) Serum 25-hydroxyvitamin D, ethnicity, and blood pressure in the Third National Health and Nutrition Examination Survey. Am J Hypertens 20:713–719
54. Rockell JE, Skeaff CM, Williams SM, Green TJ (2006) Serum 25-hydroxyvitamin D concentrations of New Zealanders aged 15 years and older. Osteoporos Int 17:1382–1389
55. Ornato JP, Peberdy MA, Chandra NC, Bush DE (1996) Seasonal pattern of acute myocardial infarction in the National Registry of Myocardial Infarction. J Am Coll Cardiol 28:1684–1688

6 Vitamin D: Cardiovascular Function and Disease

56. Spencer FA, Goldberg RJ, Becker RC, Gore JM (1998) Seasonal distribution of acute myocardial infarction in the second National Registry of Myocardial Infarction. J Am Coll Cardiol 31:1226–1233
57. Sheth T, Nair C, Muller J, Yusuf S (1999) Increased winter mortality from acute myocardial infarction and stroke: the effect of age. J Am Coll Cardiol 33:1916–1919
58. Oberg AL, Ferguson JA, McIntyre LM, Horner RD (2000) Incidence of stroke and season of the year: evidence of an association. Am J Epidemiol 152:558–564
59. Vinall PE, Maislin G, Michele JJ, Deitch C, Simeone FA (1994) Seasonal and latitudinal occurrence of cerebral vasospasm and subarachnoid hemorrhage in the northern hemisphere. Epidemiology 5:302–308
60. Allegra JR, Cochrane DG, Biglow R (2001) Monthly, weekly, and daily patterns in the incidence of congestive heart failure. Acad Emerg Med 8:682–685
61. Stewart S, McIntyre K, Capewell S, McMurray JJ (2002) Heart failure in a cold climate. Seasonal variation in heart failure-related morbidity and mortality. J Am Coll Cardiol 39:760–766
62. Page RL, Zipes DP, Powell JL, Luceri RM, Gold MR, Peters R, Russo AM, Bigger JT Jr, Sung RJ, McBurnie MA (2004) Seasonal variation of mortality in the Antiarrhythmics Versus Implantable Defibrillators (AVID) study registry. Heart Rhythm 1:435–440
63. Finkelhor RS, Cater G, Qureshi A, Einstadter D, Hecker MT, Bosich G (2005) Seasonal diagnosis of echocardiographically demonstrated endocarditis. Chest 128:2588–2592
64. Stein PD, Kayali F, Beemath A, Skaf E, Alnas M, Alesh I, Olson RE (2005) Mortality from acute pulmonary embolism according to season. Chest 128:3156–3158
65. Kloner RA, Poole WK, Perritt RL (1999) When throughout the year is coronary death most likely to occur? A 12-year population-based analysis of more than 220 000 cases. Circulation 100:1630–1634
66. Seto TB, Mittleman MA, Davis RB, Taira DA, Kawachi I (1998) Seasonal variation in coronary artery disease mortality in Hawaii: observational study. Bmj 316:1946–1947
67. Kvaloy JT, Skogvoll E (2007) Modelling seasonal and weather dependency of cardiac arrests using the covariate order method. Stat Med 26:3315–3329
68. Scragg R, Jackson R, Holdaway IM, Lim T, Beaglehole R (1990) Myocardial infarction is inversely associated with plasma 25-hydroxyvitamin D3 levels: a community-based study. Int J Epidemiol 19:559–563
69. Scragg R, Jackson R, Holdaway I, Woollard G, Woollard D (1989) Changes in plasma vitamin levels in the first 48 hours after onset of acute myocardial infarction. Am J Cardiol 64:971–974
70. Simpson RU (1983) Evidence for a specific 1, 25-dihydroxyvitamin D3 receptor in rat heart (abstract). Circulation 68:239
71. Simpson RU, Thomas GA, Arnold AJ (1985) Identification of 1, 25-dihydroxyvitamin D3 receptors and activities in muscle. J Biol Chem 260:8882–8891
72. Walters MR, Wicker DC, Riggle PC (1986) 1, 25-Dihydroxyvitamin D3 receptors identified in the rat heart. J Mol Cell Cardiol 18:67–72
73. Stumpf WES, Madhabananda S, deLuca H (1985) Sites of action of 1,25(OH)2 vitamin D3 identified by thaw-mount autoradiography. In: AW Norman (ed) Proceedings of Sixth Workshop on Vitamin D. Walter DeGruyter, Berlin, pp 416–418
74. Thomasset M, Parkes CO, Cuisinier-Gleizes P (1982) Rat calcium-binding proteins: distribution, development, and vitamin D dependence. Am J Physiol 243:E483–488
75. Weishaar RE, Simpson RU (1987) Vitamin D3 and cardiovascular function in rats. J Clin Invest 79:1706–1712
76. Weishaar RE, Simpson RU (1987) Involvement of vitamin D3 with cardiovascular function. II. Direct and indirect effects. Am J Physiol 253:E675–683
77. Weishaar RE, Kim SN, Saunders DE, Simpson RU (1990) Involvement of vitamin D3 with cardiovascular function. III. Effects on physical and morphological properties. Am J Physiol 258:E134–142
78. ÓConnell TD, Simpson RU (1995) 1, 25-Dihydroxyvitamin D3 regulation of myocardial growth and c-myc levels in the rat heart. Biochem Biophys Res Commun 213:59–65

79. ÓConnell TD, Berry JE, Jarvis AK, Somerman MJ, Simpson RU (1997) 1,25-Dihydroxyvitamin D3 regulation of cardiac myocyte proliferation and hypertrophy. Am J Physiol 272:H1751–1758

80. Simpson RU, Weishaar RE (1988) Involvement of 1, 25-dihydroxyvitamin D3 in regulating myocardial calcium metabolism: physiological and pathological actions. Cell Calcium 9:285–292

81. Weishaar RE, Burrows SD, Kim SN, Kobylarz-Singer DC, Andrews LK, Quade MM, Overhiser R, Kaplan HR (1987) Protection of the failing heart: Comparative effects of chronic administration of digitalis and diltiazem on myocardial metabolism in the cardiomyopathic hamster. J Appl Cardiol 2:339–360

82. Kawashima H (1987) Receptor for 1, 25-dihydroxyvitamin D in a vascular smooth muscle cell line derived from rat aorta. Biochem Biophys Res Commun 146:1–6

83. Merke J, Milde P, Lewicka S, Hugel U, Klaus G, Mangelsdorf DJ, Haussler MR, Rauterberg EW, Ritz E (1989) Identification and regulation of 1, 25-dihydroxyvitamin D3 receptor activity and biosynthesis of 1, 25-dihydroxyvitamin D3. Studies in cultured bovine aortic endothelial cells and human dermal capillaries. J Clin Invest 83:1903–1915

84. Kurtz TW, Portale AA, Morris RC Jr (1986) Evidence for a difference in vitamin D metabolism between spontaneously hypertensive and Wistar-Kyoto rats. Hypertension 8:1015–1020

85. Bukoski RD, Xue H (1993) On the vascular inotropic action of 1, 25-(OH)2 vitamin D3. Am J Hypertens 6:388–396

86. McCarron DA, Pingree PA, Rubin RJ, Gaucher SM, Molitch M, Krutzik S (1980) Enhanced parathyroid function in essential hypertension: a homeostatic response to a urinary calcium leak. Hypertension 2:162–168

87. Zachariah PK, Schwartz GL, Strong CG, Ritter SG (1988) Parathyroid hormone and calcium. A relationship in hypertension. Am J Hypertens 1:79S–82S

88. Resnick LM, Muller FB, Laragh JH (1986) Calcium-regulating hormones in essential hypertension. Relation to plasma renin activity and sodium metabolism. Ann Intern Med 105:649–654

89. Sowers MR, Wallace RB, Lemke JH (1985) The association of intakes of vitamin D and calcium with blood pressure among women. Am J Clin Nutr 42:135–142

90. Kokot F, Pietrek J, Srokowska S, Wartenberg W, Kuska J, Jedrychowska M, Duda G, Zielinska K, Wartenberg Z, Kuzmiak M (1981) 25-Hydroxyvitamin D in patients with essential hypertension. Clin Nephrol 16:188–192

91. Brickman AS, Nyby MD, von Hungen K, Eggena P, Tuck ML (1990) Calcitropic hormones, platelet calcium, and blood pressure in essential hypertension. Hypertension 16:515–522

92. Morimoto S, Imaoka M, Kitano S, Imanaka S, Fukuo K, Miyashita Y, Koh E, Ogihara T (1991) Exaggerated natri-calci-uresis and increased circulating levels of parathyroid hormone and 1, 25-dihydroxyvitamin D in patients with senile hypertension. Contrib Nephrol 90:94–98

93. Scragg R, Holdaway I, Singh V, Metcalf P, Baker J, Dryson E (1995) Serum 25-hydroxycholecalciferol concentration in newly detected hypertension. Am J Hypertens 8:429–432

94. Sowers MF, Wallace RB, Hollis BW, Lemke JH (1988) Relationship between 1, 25-dihydroxyvitamin D and blood pressure in a geographically defined population. Am J Clin Nutr 48:1053–1056

95. Young EW, McCarron DA, Morris CD (1990) Calcium regulating hormones in essential hypertension. Importance of gender. Am J Hypertens 3:161S–166S

96. Lind L, Hanni A, Lithell H, Hvarfner A, Sorensen OH, Ljunghall S (1995) Vitamin D is related to blood pressure and other cardiovascular risk factors in middle-aged men. Am J Hypertens 8:894–901

97. Kristal-Boneh E, Froom P, Harari G, Ribak J (1997) Association of calcitriol and blood pressure in normotensive men. Hypertension 30:1289–1294

6   Vitamin D: Cardiovascular Function and Disease   137

98. Lind L, Wengle B, Ljunghall S (1987) Blood pressure is lowered by vitamin D (alphacalcidol) during long-term treatment of patients with intermittent hypercalcaemia. A double-blind, placebo-controlled study. Acta Med Scand 222:423–427

99. Lind L, Lithell H, Skarfors E, Wide L, Ljunghall S (1988) Reduction of blood pressure by treatment with alphacalcidol. A double-blind, placebo-controlled study in subjects with impaired glucose tolerance. Acta Med Scand 223:211–217

100. Lind L, Pollare T, Hvarfner A, Lithell H, Sorensen OH, Ljunghall S (1989) Long-term' treatment with active vitamin D (alphacalcidol) in middle-aged men with impaired glucose tolerance. Effects on insulin secretion and sensitivity, glucose tolerance and blood pressure. Diabetes Res 11:141–147

101. Orwoll ES, Oviatt S (1990) Relationship of mineral metabolism and long-term calcium and cholecalciferol supplementation to blood pressure in normotensive men. Am J Clin Nutr 52:717–721

102. Orwoll ES, Weigel RM, Oviatt SK, McClung MR, Deftos LJ (1988) Calcium and cholecalciferol: effects of small supplements in normal men. Am J Clin Nutr 48:127–130

103. Gillor A, Groneck P, Kaiser J, Schmitz-Stolbrink A (1989) Congestive heart failure in rickets caused by vitamin D deficiency. Monatsschr Kinderheilkd 137:108–110

104. Connor TB, Rosen BL, Blaustein MP, Applefeld MM, Doyle LA (1982) Hypocalcemia precipitating congestive heart failure. N Engl J Med 307:869–872

105. Brunvand L, Haga P, Tangsrud SE, Haug E (1995) Congestive heart failure caused by vitamin D deficiency? Acta Paediatr 84:106–108

106. Uysal S, Kalayci AG, Baysal K (1999) Cardiac functions in children with vitamin D deficiency rickets. Pediatr Cardiol 20:283–286

107. McGonigle RJ, Timmis AD, Keenan J, Jewitt DE, Weston MJ, Parsons V (1981) The influence of 1 alpha-hydroxycholecalciferol on left ventricular function in end-stage renal failure. Proc Eur Dial Transplant Assoc 18:579–585

108. Coratelli P, Petrarulo F, Buongiorno E, Giannattasio M, Antonelli G, Amerio A (1984) Improvement in left ventricular function during treatment of hemodialysis patients with 25-OHD3. Contrib Nephrol 41:433–437

109. Lemmila S, Saha H, Virtanen V, Ala-Houhala I, Pasternack A (1998) Effect of intravenous calcitriol on cardiac systolic and diastolic function in patients on hemodialysis. Am J Nephrol 18:404–410

110. Park CW, Oh YS, Shin YS, Kim CM, Kim YS, Kim SY, Choi EJ, Chang YS, Bang BK (1999) Intravenous calcitriol regresses myocardial hypertrophy in hemodialysis patients with secondary hyperparathyroidism. Am J Kidney Dis 33:73–81

111. Shane E, Mancini D, Aaronson K, Silverberg SJ, Seibel MJ, Addesso V, McMahon DJ (1997) Bone mass, vitamin D deficiency, and hyperparathyroidism in congestive heart failure. Am J Med 103:197–207

112. Kunitomo M, Kinoshita K, Bando Y (1981) Experimental atherosclerosis in rats fed a vitamin D, cholesterol-rich diet. J Pharmacobiodyn 4:718–723

113. Takeo S, Schraven E, Keil M, Nitz RE (1982) Vitamin D-induced myocardial lesions and the protection by carbocromen. Arzneimittelforschung 32:1412–1417

114. Toda T, Leszczynski DE, Kummerow FA (1983) The role of 25-hydroxy-vitamin D3 in the induction of atherosclerosis in swine and rabbit by hypervitaminosis D. Acta Pathol Jpn 33:37–44

115. Atkinson J (1992) Vascular calcium overload. Physiological and pharmacological consequences. Drugs 44(Suppl 1):111–118

116. Fleckenstein-Grun G, Thimm F, Frey M, Matyas S (1995) Progression and regression by verapamil of vitamin D3-induced calcific medial degeneration in coronary arteries of rats. J Cardiovasc Pharmacol 26:207–213

117. Bennani-Kabchi N, Kehel L, el Bouayadi F, Fdhil H, Amarti A, Saidi A, Marquie G (1999) New model of atherosclerosis in sand rats subjected to a high cholesterol diet and vitamin D2. Therapie 54:559–565

118. Watson KE, Abrolat ML, Malone LL, Hoeg JM, Doherty T, Detrano R, Demer LL (1997) Active serum vitamin D levels are inversely correlated with coronary calcification. Circulation 96:1755–1760

119. Doherty TM, Tang W, Dascalos S, Watson KE, Demer LL, Shavelle RM, Detrano RC (1997) Ethnic origin and serum levels of 1alpha, 25-dihydroxyvitamin D3 are independent predictors of coronary calcium mass measured by electron-beam computed tomography. Circulation 96:1477–1481

120. Arad Y, Spadaro LA, Roth M, Scordo J, Goodman K, Sherman S, Lerner G, Newstein D, Guerci AD (1998) Serum concentration of calcium, 1, 25 vitamin D and parathyroid hormone are not correlated with coronary calcifications. An electron beam computed tomography study. Coron Artery Dis 9:513–518

121. Bonjour JP, Chevalley T, Fardellone P (2007) Calcium intake and vitamin D metabolism and action, in healthy conditions and in prostate cancer. Br J Nutr 97:611–616

122. Teng M, Wolf M, Lowrie E, Ofsthun N, Lazarus JM, Thadhani R (2003) Survival of patients undergoing hemodialysis with paricalcitol or calcitriol therapy. N Engl J Med 349:446–456

123. Shoji T, Shinohara K, Kimoto E, Emoto M, Tahara H, Koyama H, Inaba M, Fukumoto S, Ishimura E, Miki T et al (2004) Lower risk for cardiovascular mortality in oral 1alpha-hydroxy vitamin D3 users in a haemodialysis population. Nephrol Dial Transplant 19:179–184

124. Teng M, Wolf M, Ofsthun MN, Lazarus JM, Hernan MA, Camargo CA Jr, Thadhani R (2005) Activated injectable vitamin D and hemodialysis survival: a historical cohort study. J Am Soc Nephrol 16:1115–1125

125. Naves-Diaz M, Alvarez-Hernandez D, Passlick-Deetjen J, Guinsburg A, Marelli C, Rodriguez-Puyol D, Cannata-Andia JB (2008) Oral active vitamin D is associated with improved survival in hemodialysis patients. Kidney Int 74:1070–1078

126. Melamed ML, Eustace JA, Plantinga L, Jaar BG, Fink NE, Coresh J, Klag MJ, Powe NR (2006) Changes in serum calcium, phosphate, and PTH and the risk of death in incident dialysis patients: a longitudinal study. Kidney Int 70:351–357

127. Kalantar-Zadeh K, Kuwae N, Regidor DL, Kovesdy CP, Kilpatrick RD, Shinaberger CS, McAllister CJ, Budoff MJ, Salusky IB, Kopple JD (2006) Survival predictability of time-varying indicators of bone disease in maintenance hemodialysis patients. Kidney Int 70:771–780

128. Tentori F, Hunt WC, Stidley CA, Rohrscheib MR, Bedrick EJ, Meyer KB, Johnson HK, Zager PG (2006) Mortality risk among hemodialysis patients receiving different vitamin D analogs. Kidney Int 70:1858–1865

129. Kovesdy CP, Ahmadzadeh S, Anderson JE, Kalantar-Zadeh K (2008) Association of activated vitamin D treatment and mortality in chronic kidney disease. Arch Intern Med 168:397–403

130. Shoben AB, Rudser KD, de Boer IH, Young B, Kestenbaum B (2008) Association of oral calcitriol with improved survival in nondialyzed CKD. J Am Soc Nephrol 19:1613–1619

131. Wolf M, Shah A, Gutierrez O, Ankers E, Monroy M, Tamez H, Steele D, Chang Y, Camargo CA Jr, Tonelli M et al (2007) Vitamin D levels and early mortality among incident hemodialysis patients. Kidney Int 72:1004–1013

132. Wolf M, Betancourt J, Chang Y, Shah A, Teng M, Tamez H, Gutierrez O, Camargo CA Jr, Melamed M, Norris K et al (2008) Impact of activated vitamin D and race on survival among hemodialysis patients. J Am Soc Nephrol 19(7):1379–1388

133. Al-Aly Z (2007) Vitamin D as a novel nontraditional risk factor for mortality in hemodialysis patients: the need for randomized trials. Kidney Int 72:909–911

134. de Boer IH, Kestenbaum B (2008) Vitamin D in chronic kidney disease: is the jury in? Kidney Int 74:985–987

135. Wang TJ, Pencina MJ, Booth SL, Jacques PF, Ingelsson E, Lanier K, Benjamin EJ, D'Agostino RB, Wolf M, Vasan RS (2008) Vitamin D deficiency and risk of cardiovascular disease. Circulation 117:503–511

136. Giovannucci E, Liu Y, Hollis BW, Rimm EB (2008) 25-hydroxyvitamin D and risk of myocardial infarction in men: a prospective study. Arch Intern Med 168:1174–1180

137. Melamed ML, Michos ED, Post W, Astor B (2008) 25-hydroxyvitamin D levels and the risk of mortality in the general population. Arch Intern Med 168:1629–1637
138. Dobnig H, Pilz S, Scharnagl H, Renner W, Seelhorst U, Wellnitz B, Kinkeldei J, Boehm BO, Weihrauch G, Maerz W (2008) Independent association of low serum 25-hydroxyvitamin d and 1, 25-dihydroxyvitamin d levels with all-cause and cardiovascular mortality. Arch Intern Med 168:1340–1349
139. Hsia J, Heiss G, Ren H, Allison M, Dolan NC, Greenland P, Heckbert SR, Johnson KC, Manson JE, Sidney S et al (2007) Calcium/vitamin D supplementation and cardiovascular events. Circulation 115:846–854
140. Poole KE, Loveridge N, Barker PJ, Halsall DJ, Rose C, Reeve J, Warburton EA (2006) Reduced vitamin D in acute stroke. Stroke 37:243–245
141. Rajasree S, Rajpal K, Kartha CC, Sarma PS, Kutty VR, Iyer CS, Girija G (2001) Serum 25-hydroxyvitamin D3 levels are elevated in South Indian patients with ischemic heart disease. Eur J Epidemiol 17:567–571
142. Zittermann A, Schleithoff SS, Tenderich G, Berthold HK, Korfer R, Stehle P (2003) Low vitamin D status: a contributing factor in the pathogenesis of congestive heart failure? J Am Coll Cardiol 41:105–112
143. Zittermann A, Schleithoff SS, Gotting C, Dronow O, Fuchs U, Kuhn J, Kleesiek K, Tenderich G, Koerfer R (2008) Poor outcome in end-stage heart failure patients with low circulating calcitriol levels. Eur J Heart Fail 10:321–327
144. Schleithoff SS, Zittermann A, Tenderich G, Berthold HK, Stehle P, Koerfer R (2006) Vitamin D supplementation improves cytokine profiles in patients with congestive heart failure: a double-blind, randomized, placebo-controlled trial. Am J Clin Nutr 83:754–759
145. Melamed ML, Muntner P, Michos ED, Uribarri J, Weber C, Sharma J, Raggi P (2008) Serum 25-hydroxyvitamin D levels and the prevalence of peripheral arterial disease. Results from NHANES 2001 to 2004. Arterioscler Thromb Vasc Biol 28(6):1179–1185
146. Jorde R, Bonaa KH (2000) Calcium from dairy products, vitamin D intake, and blood pressure: the Tromso Study. Am J Clin Nutr 71:1530–1535
147. Forman JP, Bischoff-Ferrari HA, Willett WC, Stampfer MJ, Curhan GC (2005) Vitamin D intake and risk of incident hypertension: results from three large prospective cohort studies. Hypertension 46:676–682
148. Wang L, Manson JE, Buring JE, Lee IM, Sesso HD (2008) Dietary intake of dairy products, calcium, and vitamin D and the risk of hypertension in middle-aged and older women. Hypertension 51:1073–1079
149. Forman JP, Giovannucci E, Holmes MD, Bischoff-Ferrari HA, Tworoger SS, Willett WC, Curhan GC (2007) Plasma 25-hydroxyvitamin D levels and risk of incident hypertension. Hypertension 49:1063–1069
150. Snijder MB, Lips P, Seidell JC, Visser M, Deeg DJ, Dekker JM, van Dam RM (2007) Vitamin D status and parathyroid hormone levels in relation to blood pressure: a population-based study in older men and women. J Intern Med 261:558–565
151. Pfeifer M, Begerow B, Minne HW, Nachtigall D, Hansen C (2001) Effects of a short-term vitamin D(3) and calcium supplementation on blood pressure and parathyroid hormone levels in elderly women. J Clin Endocrinol Metab 86:1633–1637
152. Krause R, Buhring M, Hopfenmuller W, Holick MF, Sharma AM (1998) Ultraviolet B and blood pressure. Lancet 352:709–710
153. Newmark HL, Heaney RP (2006) Calcium, vitamin D, and risk reduction of colorectal cancer. Nutr Cancer 56:1–2
154. Michos ED, Blumenthal RS (2007) Vitamin D supplementation and cardiovascular disease risk. Circulation 115:827–828
155. Heaney RP (2008) Vitamin D in health and disease. Clin J Am Soc Nephrol 3:1535–1541
156. Vieth R, Bischoff-Ferrari H, Boucher BJ, Dawson-Hughes B, Garland CF, Heaney RP, Holick MF, Hollis BW, Lamberg-Allardt C, McGrath JJ et al (2007) The urgent need to recommend an intake of vitamin D that is effective. Am J Clin Nutr 85:649–650

157. Autier P, Gandini S (2007) Vitamin D supplementation and total mortality: a meta-analysis of randomized controlled trials. Arch Intern Med 167:1730–1737
158. Jackson RD, LaCroix AZ, Gass M, Wallace RB, Robbins J, Lewis CE, Bassford T, Beresford SA, Black HR, Blanchette P et al (2006) Calcium plus vitamin D supplementation and the risk of fractures. N Engl J Med 354:669–683
159. Giovannucci E (2007) Can vitamin D reduce total mortality? Arch Intern Med 167:1709–1710
160. Vieth R, Chan PC, MacFarlane GD (2001) Efficacy and safety of vitamin D3 intake exceeding the lowest observed adverse effect level. Am J Clin Nutr 73:288–294
161. Beaglehole R, Magnus P (2002) The search for new risk factors for coronary heart disease: occupational therapy for epidemiologists? Int J Epidemiol 31:1117–1122, author reply 1134–1115
162. Ross R (1999) Atherosclerosis – an inflammatory disease. N Engl J Med 340:115–126
163. Hansson GK, Libby P, Schonbeck U, Yan ZQ (2002) Innate and adaptive immunity in the pathogenesis of atherosclerosis. Circ Res 91:281–291
164. Rao M, Jaber BL, Balakrishnan VS (2006) Inflammatory biomarkers and cardiovascular risk: association or cause and effect? Semin Dial 19:129–135
165. Tousoulis D, Antoniades C, Koumallos N, Stefanadis C (2006) Pro-inflammatory cytokines in acute coronary syndromes: from bench to bedside. Cytokine Growth Factor Rev 17:225–233
166. Cardaropoli S, Silvagno F, Morra E, Pescarmona GP, Todros T (2003) Infectious and inflammatory stimuli decrease endothelial nitric oxide synthase activity in vitro. J Hypertens 21:2103–2110
167. Khoshdel AR, Carney SL, Nair BR, Gillies A (2007) Better management of cardiovascular diseases by pulse wave velocity: combining clinical practice with clinical research using evidence-based medicine. Clin Med Res 5:45–52
168. Newby AC (2005) Dual role of matrix metalloproteinases (matrixins) in intimal thickening and atherosclerotic plaque rupture. Physiol Rev 85:1–31
169. Loftus IM, Thompson MM (2002) The role of matrix metalloproteinases in vascular disease. Vasc Med 7:117–133
170. Perlstein TS, Lee RT (2006) Smoking, metalloproteinases, and vascular disease. Arterioscler Thromb Vasc Biol 26:250–256
171. Turk S, Akbulut M, Yildiz A, Gurbilek M, Gonen S, Tombul Z, Yeksan M (2002) Comparative effect of oral pulse and intravenous calcitriol treatment in hemodialysis patients: the effect on serum IL-1 and IL-6 levels and bone mineral density. Nephron 90:188–194
172. Mohamed-Ali V, Goodrick S, Rawesh A, Katz DR, Miles JM, Yudkin JS, Klein S, Coppack SW (1997) Subcutaneous adipose tissue releases interleukin-6, but not tumor necrosis factor-alpha, in vivo. J Clin Endocrinol Metab 82:4196–4200
173. Bastard JP, Jardel C, Bruckert E, Blondy P, Capeau J, Laville M, Vidal H, Hainque B (2000) Elevated levels of interleukin 6 are reduced in serum and subcutaneous adipose tissue of obese women after weight loss. J Clin Endocrinol Metab 85:3338–3342
174. Pickup JC (2004) Inflammation and activated innate immunity in the pathogenesis of type 2 diabetes. Diabetes Care 27:813–823
175. Giulietti A, van Etten E, Overbergh L, Stoffels K, Bouillon R, Mathieu C (2007) Monocytes from type 2 diabetic patients have a pro-inflammatory profile. 1, 25-Dihydroxyvitamin D(3) works as anti-inflammatory. Diabetes Res Clin Pract 77:47–57
176. Liu PT, Stenger S, Li H, Wenzel L, Tan BH, Krutzik SR, Ochoa MT, Schauber J, Wu K, Meinken C et al (2006) Toll-like receptor triggering of a vitamin D-mediated human antimicrobial response. Science 311:1770–1773
177. Martineau AR, Honecker FU, Wilkinson RJ, Griffiths CJ (2007) Vitamin D in the treatment of pulmonary tuberculosis. J Steroid Biochem Mol Biol 103:793–798
178. Stroder J, Kasal P (1970) Phagocytosis in vitamin D deficient rickets. Klin Wochenschr 48:383–384

6 Vitamin D: Cardiovascular Function and Disease

179. Karatekin G, Kaya A, Salihoglu O, Balci H, Nuhoglu A (2007) Association of subclinical vitamin D deficiency in newborns with acute lower respiratory infection and their mothers. Eur J Clin Nutr 117:803–811
180. Laaksi I, Ruohola JP, Tuohimaa P, Auvinen A, Haataja R, Pihlajamaki H, Ylikomi T (2007) An association of serum vitamin D concentrations <40 nmol/L with acute respiratory tract infection in young Finnish men. Am J Clin Nutr 86:714–717
181. Aloia JF, Li-Ng M (2007) Re: epidemic influenza and vitamin D. Epidemiol Infect 135:1095–1096, author reply 1097–1098
182. Muller K, Diamant M, Bendtzen K (1991) Inhibition of production and function of interleukin-6 by 1, 25-dihydroxyvitamin D3. Immunol Lett 28:115–120
183. Willheim M, Thien R, Schrattbauer K, Bajna E, Holub M, Gruber R, Baier K, Pietschmann P, Reinisch W, Scheiner O et al (1999) Regulatory effects of 1alpha, 25-dihydroxyvitamin D3 on the cytokine production of human peripheral blood lymphocytes. J Clin Endocrinol Metab 84:3739–3744
184. Zhu Y, Mahon BD, Froicu M, Cantorna MT (2005) Calcium and 1 alpha, 25-dihydroxyvitamin D3 target the TNF-alpha pathway to suppress experimental inflammatory bowel disease. Eur J Immunol 35:217–224
185. Canning MO, Grotenhuis K, de Wit H, Ruwhof C, Drexhage HA (2001) 1-alpha, 25-Dihydroxyvitamin D3 (1, 25(OH)(2)D(3)) hampers the maturation of fully active immature dendritic cells from monocytes. Eur J Endocrinol 145:351–357
186. Pittas AG, Harris SS, Stark PC, Dawson-Hughes B (2007) The effects of calcium and vitamin D supplementation on blood glucose and markers of inflammation in nondiabetic adults. Diabetes Care 30:980–986
187. Simpson RU, Hershey SH, Nibbelink KA (2007) Characterization of heart size and blood pressure in the vitamin D receptor knockout mouse. J Steroid Biochem Mol Biol 103:521–524
188. Xiang W, Kong J, Chen S, Cao LP, Qiao G, Zheng W, Liu W, Li X, Gardner DG, Li YC (2005) Cardiac hypertrophy in vitamin D receptor knockout mice: role of the systemic and cardiac renin-angiotensin systems. Am J Physiol Endocrinol Metab 288:E125–132
189. Timms PM, Mannan N, Hitman GA, Noonan K, Mills PG, Syndercombe-Court D, Aganna E, Price CP, Boucher BJ (2002) Circulating MMP9, vitamin D and variation in the TIMP-1 response with VDR genotype: mechanisms for inflammatory damage in chronic disorders? Qjm 95:787–796
190. Sundstrom J, Evans JC, Benjamin EJ, Levy D, Larson MG, Sawyer DB, Siwik DA, Colucci WS, Sutherland P, Wilson PW et al (2004) Relations of plasma matrix metalloproteinase-9 to clinical cardiovascular risk factors and echocardiographic left ventricular measures: the Framingham Heart Study. Circulation 109:2850–2856
191. Duprez D, De Buyzere M, De Backer T, Clement D (1993) Relationship between vitamin D and the regional blood flow and vascular resistance in moderate arterial hypertension. J Hypertens Suppl 11:S304–305
192. London GM, Guerin AP, Verbeke FH, Pannier B, Boutouyrie P, Marchais SJ, Metivier F (2007) Mineral metabolism and arterial functions in end-stage renal disease: potential role of 25-hydroxyvitamin D deficiency. J Am Soc Nephrol 18:613–620
193. Targher G, Bertolini L, Padovani R, Zenari L, Scala L, Cigolini M, Arcaro G (2006) Serum 25-hydroxyvitamin D3 concentrations and carotid artery intima-media thickness among type 2 diabetic patients. Clin Endocrinol (Oxf) 65:593–597
194. Sugden JA, Davies JI, Witham MD, Morris AD, Struthers AD (2008) Vitamin D improves endothelial function in patients with Type 2 diabetes mellitus and low vitamin D levels. Diabet Med 25:320–325
195. Suzuki A, Kotake M, Ono Y, Kato T, Oda N, Hayakawa N, Hashimoto S, Itoh M (2006) Hypovitaminosis D in type 2 diabetes mellitus: Association with microvascular complications and type of treatment. Endocr J 53:503–510
196. Zieman SJ, Melenovsky V, Kass DA (2005) Mechanisms, pathophysiology, and therapy of arterial stiffness. Arterioscler Thromb Vasc Biol 25:932–943

197. Abrams J (1997) Role of endothelial dysfunction in coronary artery disease. Am J Cardiol 79:2–9
198. Monnink SH, Tio RA, van Boven AJ, van Gilst WH, van Veldhuisen DJ (2004) The role of coronary endothelial function testing in patients suspected for angina pectoris. Int J Cardiol 96:123–129
199. Li YC, Kong J, Wei M, Chen ZF, Liu SQ, Cao LP (2002) 1, 25-Dihydroxyvitamin D(3) is a negative endocrine regulator of the renin-angiotensin system. J Clin Invest 110:229–238
200. Hill AB (1965) The environment and disease: association or causation? Proc R Soc Med 58:293–300
201. Gillie O (2006) Sunlight, Vitamin D and Health. In: Gillie O (ed) Occasional Reports. Health Research Forum, London

Chapter 7
Induction of Differentiation in Cancer Cells by Vitamin D: Recognition and Mechanisms*

Elzbieta Gocek and George P. Studzinski

Abstract Current understanding of the vitamin D-induced differentiation of neoplastic cells, which results in the generation of cells that acquire near-normal, mature phenotype is summarized here. The criteria by which differentiation is recognized in each cell type are provided, and only those effects of $1\alpha,25$-dihydroxyvitamin D_3 (1,25D) on cell proliferation and survival which are associated with the differentiation process are emphasized. The existing knowledge of the signaling pathways that lead to vitamin-D-induced differentiation of colon, breast, prostate, squamous cell carcinoma (SCC), osteosarcoma, and myeloid leukemia cancer cells is outlined. Where known, the distinctions between the different mechanisms of 1,25D-induced differentiation which are cell-type-specific and cell-context-specific are pointed out. A considerable body of evidence suggests that several types of human cancer cells can be suitable candidates for chemoprevention or differentiation therapy with vitamin D. However, further studies of the underlying mechanisms are needed to gain further insights on how to improve the therapeutic approaches that incorporate vitamin D derivatives.

Abbreviations

A	Androgen
AKT	Serine/threonine-specific protein kinase B
Alk Pase	Alkaline phosphatase
AML	Acute myeloid leukemia
AP-1	Activating protein 1
APC	Adenomatous polyposis coli
APL	Acute promyelocytic leukemia

---

*The substance of this chapter has been reported as an Invited Review on the same topic in "Crit Rev Clin Lab Sci."

G.P. Studzinski (✉)
Department of Pathology and Laboratory Medicine, UMDNJ-New Jersey Medical School,
185 So. Orange Ave., Room 543, Newark, NJ 07101–1709, USA
e-mail: studzins@umdnj.edu

AR	Androgen receptor
ATRA	All-trans retinoic acid
BMP	Bone morphogenetic protein
CaR	Calcium-sensing surface receptor
Cdk5	Cyclin-dependent kinase 5
C/EBP	CCAAT/enhancer binding protein
CoA	Coactivator
1,25D	$1\alpha,25$-Dihydroxyvitamin D_3
E_2	Estrogen
EGFR	Epidermal growth factor receptor
EGR-1	Early growth response protein 1
EP	Early progenitor
ER	Estrogen receptor
ERK	Extracellular-signal regulated kinase
FC	Flow Cytometry
GF	Growth factor
GFR	Growth factor receptor
hOC	Human osteocalcin
hOC	Human osteopontin
IBP-5	IGF binding protein-5
IGFBP-3	Insulin-like growth factor binding protein-3
IP_3	Inositol triphosphate
JNK	Jun N-terminal kinase
KLF-4	Kruppel-like factor 4
KSR-1	Kinase suppressor of Ras-1
LPS	Lipopolysaccharides
MALDI-TOFMS	Matrix-assisted laser desorption/ionization-time-of-flight mass spectrometry
MAPK	Mitogen activated protein kinase
MSE	Monocyte-specific esterase ("non-specific" esterase)
NBT	Nitroblue tetrazolium
Nck5a	"Cyclin-like" neuronal Cdk5 activator
NR	Nuclear receptor
NSE	Nonspecific esterase
24OHase	24-Hydroxylase
P	Progenitor
p90RSK	Ribosomal s6 kinase (MAPK-activated protein kinase-1)
PI3K	Phosphatidylinositol 3-kinase
PIP_3	Phosphatidylinositol 3, 4, 5-triphosphate
PKC	Protein kinase C
PLC-γ1	Phospholipase C gamma-1
Rb	Retinoblastoma protein
PSA	Prostate specific antigen
RAR	Retinoic acid receptor
ROS	Reactive oxygen species
RXRα	Retinoid X receptor alpha

SCC	Squamous cell carcinoma
Sp-1	Specificity protein 1
TCF4	T-cell transcription factor 4
Wnt	Wingless-related MMTV integration site
VDR	Vitamin D receptor
VDRE	Vitamin D$_3$ response element

7.1 Introduction

In general, differentiation is a term that signifies the structural and functional changes that lead to maturation of cells during development of various lineages. Cancer cells are unable, in varying degrees, to achieve such maturation, and thus malignant neoplastic cells show a lack of, or only partial, evidence of differentiation, known as anaplasia. Since the basic underlying cause for the failure to differentiate can be attributed to structural changes in the cell's DNA, i.e., mutations, which are essentially irreversible, it is remarkable that some compounds can induce several types of malignant cells to undergo differentiation toward the more mature phenotypes. The physiological form of vitamin D, 1α,25-dihydroxyvitamin D$_3$ (1,25D), is one such compound, and the importance of this finding is that it offers the potential to be an alternative to, or to provide an adjunctive intervention to the therapy, as well as the prevention of neoplastic diseases.

The feasibility of differentiation therapy of cancer is supported by the early observations that some cases of neuroblastoma, a childhood malignancy, can spontaneously differentiate into tumors that are composed of normal-appearing neuronal cells, and the child's life is spared [1, 2]. The reasons for this conversion have not been elucidated, but it seems reasonable to assume that as the child matures, the endocrine and the immune systems become more efficient, and one or more of such factors are able to induce differentiation of neural precursor cells to the more mature, noninvasive forms.

An example of an already successful interventional approach to differentiation therapy of a neoplastic disease is the use of all-trans retinoic acid (ATRA) for the treatment of acute promyelocytic leukemia (APL) and perhaps other leukemias [3–5]. Additionally, a synthetic analog of ATRA, Fenretinide, can potentially serve as an agent which can prevent breast cancer in women [6], illustrating the fact that a demonstration of a clear clinical therapeutic effect of a differentiation agent opens up the possibility that it may also serve as a cancer chemopreventive compound.

While the role of 1,25D in cancer chemotherapy and cancer chemoprevention is only beginning to be established, there are several reasons to believe that its promise will be fulfilled. These reasons include the fact that 1,25D is a naturally occurring physiological substance, and thus unlikely to cause the adverse reactions which occur when xenobiotics are administered to patients, unless given in very high concentrations. Second, the issue of hypercalcemia, which occurs when the concentrations of 1,25D greatly exceed the physiological range, and has previously limited its clinical applications [7, 8], can be addressed by the dual strategy of developing analogs of 1,25D

with reduced calcium-mobilizing properties [9–12], and combining these with other compounds which enhance the differentiation-inducing actions of 1,25D or its analogs [13–15]. Also, progress is being made in understanding the mechanisms responsible for 1,25D-induced differentiation, summarized later in this review, and although this understanding is by no means complete, it is likely that insights will be obtained that can be translated to clinical applications.

Differentiation of neoplastic cells induced by 1,25D and other agents rarely, if ever, results in the generation of completely normal, functioning cells. Indeed, the appearance of cells resulting from induced progenitors has been aptly described as resembling "caricatures" rather than normal cells. Such cells may exhibit, and are recognized by, some features of the normal, mature cells of the particular developmental lineage, but seldom function like the mature normal cells. However, this is not the major objective of differentiation therapy of neoplastic diseases; the real benefits are due to the cessation of the proliferation of these cells, which is a consequence of cell cycle arrest associated with differentiation [16–19], and in some cases to the reduced survival of the differentiated cells. For instance, 1,25D-induced monocytic differentiation of myeloid leukemia cells can result in the G1 phase cell cycle block, resulting in cessation of cell proliferation [19], while 1,25D treatment of breast or prostate cancer cells can induce cell death by apoptosis as well as differentiation [20–22].

An important consideration in the area of 1,25D-induced differentiation is cell-type and cell-context specificity. For instance, in contrast to breast and prostate cancer cells which are induced to undergo apoptosis, in myeloid leukemia cells 1,25D-induced differentiation is accompanied by increased cell survival [23, 24]. The pathways which are known to signal 1,25D-induced differentiation and the associated cell cycle and survival effects also differ, though they may overlap, in different cell types. This may further be complicated by the type of mutations that are responsible for the block of differentiation, and the resulting uncontrolled proliferation of the neoplastic cells. We therefore discuss separately the principal cancer cell types known to be candidates for differentiation therapy or chemoprevention by 1,25D.

7.2 Solid Tumors

7.2.1 Colon Cancer

It is well established that colon cancer cells in culture can undergo differentiation to a more mature phenotype, and the inducing agents include the short-chain fatty acid butyrate and 1,25D. The evidence for differentiation has traditionally been the expression of the hydrolytic enzyme alkaline phosphatase (Alk Pase), which can be demonstrated on the microvilli and tubulovacuolar system of the surface "principal cells" of the colon mucosa [25, 26], but is poorly expressed in proliferating colon cancer cells [27]. More recently, other markers of colonic epithelial cell differentiation

7   Induction of Differentiation in Cancer Cells by Vitamin D   147

have been identified, and these include changes in "transepithelial electrical resistance" and ubiquitin, as based on matrix-assisted laser desorption/ionization time-of-flight mass spectrometry (MALDI-TOFMS). The latter procedure generates specific mass spectral fingerprints characteristics of cell differentiation, and it was suggested that ubiquitin can be a marker of differentiation of the T84 human colon carcinoma cell line [28]. In another colon cancer cell line, SW80, 1,25D was shown to induce easily recognizable morphological changes indicative of differentiated epithelial-like phenotype [29]. These morphological changes include consequences of the adherence to the culture substratum, which make the cells look flat and polygonal, and it was demonstrated that these cells have reduced tumorigenicity when implanted into athymic mice. Thus, the epidemiological data which indicate that 1,25D has a negative effect on the incidence of human colorectal cancer [30, 31] are well supported by the in vitro studies of 1,25D-induced differentiation of colon carcinoma cell lines.

How 1,25D signals differentiation of colon cancer cells is not entirely clear, but several groups of key molecules have been identified that appear to govern this process, and an outline of their postulated interactions is integrated in Fig. 7.1.

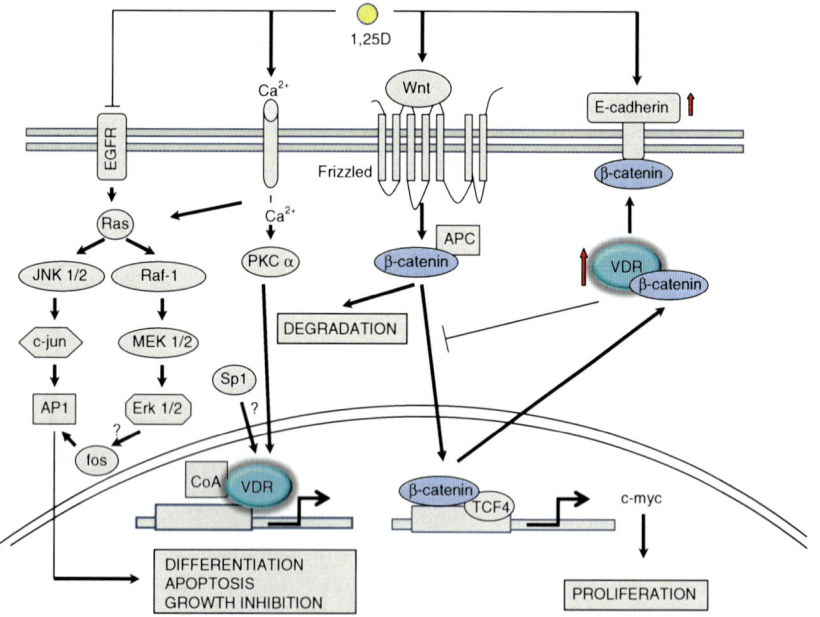

Fig. 7.1 The suggested pathways of 1,25D-induced differentiation in colon cancer. In proliferating colon epithelial cells, the β-catenin complexed with TCF-4 drives the expression of growth promoting genes such as c-myc. This is under the control of Wnt and its surface receptor Frizzled, which inactivate GSK-3β (not shown) and allow the accumulation of β-catenin and thus growth promotion. Binding of β-catenin by VDR, or by other proteins including E-cadherin, the expression of which is induced by 1,25D (formula shown) leads to the loss of β-catenin from the transcriptional complex in the nucleus and, as a consequence, decreased cell proliferation. Also shown is the activation of PKCα by 1,25D-induced influx of calcium (Ca²⁺) which can activate by phosphorylation the transcriptional activity of VDR and repression of EGFR by 1,25D in colon-derived cells

One mechanism that can explain the reduced cell proliferation which accompanies differentiation is the marked inhibitory effect of 1,25D on the expression of epidermal growth factor receptor (EGFR), apparent at both mRNA and protein levels in CaCo-2 cells [32]. The accumulated data also suggest that the central role in 1,25D-induced differentiation is played by the vitamin D receptor (VDR). An early study demonstrated that 1,25D has a protective effect on chemically induced rat colon carcinogenesis [33], and others showed that VDR can be a marker for colon cancer cell differentiation [34, 35]. This was followed up by Cross and colleagues in a series of experiments which showed that VDR levels increased in early stages of carcinogenesis, or in human colonic mucosa during early tumor development, but that VDR levels were low in poorly differentiated late-stage carcinomas [36, 37]. This suggested that VDR levels have a restraining effect on the growth of colon cells. A mechanism that can explain the increased levels of VDR in differentiated colon cells was provided by the Brasitus group, indicating that in CaCo-2 cells 1,25D causes an increased activity of the AP-1 transcription factor [27], which is downstream from the mitogen-activated protein kinases (MAPK) pathways and can transactivate VDR gene expression [38]. The consequent up-regulation of VDR may further be increased in the presence of 1,25D by stabilization of the VDR protein [39], but the nature of the initial activation of MAPK pathways in colon cancer cells is not entirely clear. The suggested calcium-induced activation of protein kinase C alpha (PKC α) as an upstream event in MAPK activation [27, 40] appears to be feasible, as an influx of calcium into the cells is known to occur after 1,25D exposure of many types of cells including colon cancer [41], but this pathway remains to be further investigated. Nonetheless, the importance of VDR in colon cancer cell differentiation is further underscored by the suggestion that butyrate-induced differentiation of CaCo-2 cells is mediated by VDR [42], and by the recent report that decreased recruitment of VDR to the vitamin D response elements (VDRE) contributes to the reduced transcriptional responsiveness of proliferating CaCo-2 cells to 1,25D [43].

An emerging role for VDR, other than its function as a transcription factor that binds to VDRE in the promoter regions of 1,25D-responsive genes, is exemplified by the finding that VDR can interact with β-catenin, and thereby repress in colon cells the oncogenic gene-regulatory activity of β-catenin [29]. The transrepression of β-catenin signaling is not limited to an interaction with VDR, as such interactions can take place with other nuclear receptors, such as the retinoic acid receptor (RAR) and the androgen receptor (AR) [29, 44]. This interaction has been shown to involve also the coactivator p300, a histone acetyl transferase [45]. The recently reported repression of the VDR gene by the transcription factor SNAIL [46], and the repression by 1,25D of the Wingless-related MMTV integration site (Wnt) antagonist DICKOPF-4 [47] may also be important for the inhibition of Wnt/β-catenin signaling by 1,25D, and for its induction of differentiation in colon cancer cells.

Signaling by β-catenin can also be repressed by the 1,25D-induced up-regulation of the expression of E-cadherin [29], a transmembrane protein that plays a major role in the maintenance of the adhesive and polarized phenotype of epithelial cells [48]. The presence of E- cadherin can promote nuclear export of β-catenin, and this

7   Induction of Differentiation in Cancer Cells by Vitamin D   149

may be augmented by direct VDR/β-catenin interaction [48]. Since β-catenin/T-cell transcription factor 4 (TCF-4) complex is the nuclear effector of the Wnt growth-signaling pathway, responsible for the expression of c-myc and other growth pro-moting genes [49], the repressive effects of 1,25D on the growth of colon cancer cells may be explained by the ability of 1,25D to regulate the expression of VDR, E-cadherin, and the activity of the β-catenin/TCF pathway, as illustrated in Fig. 7.1.

In addition to protein–protein complex formation with β-catenin, VDR has also been reported to interact with the transcription factor – Specificity protein 1 (Sp1) in SW 620 human cancer cells, and thus induce the expression of p27/Kip1 inhibitor of the cell cycle [50]. However, it is not clear precisely how this is achieved, given the ubiquitous nature of Sp1 binding sites in gene promoters. Nonetheless, the direct binding of VDR to other proteins, which may be ligand-independent, is an area that deserves further study, and has been reported to occur in cell types other than colon carcinoma, such as osteoblastic cells and myeloid leukemia, as discussed later.

7.2.2   Breast Cancer

The induction of differentiation of breast cancer cell lines by 1,25D and the role of 1,25D in normal development of rodent mammary tissue are well established. For instance, studies of VDR knockout mice in the Welsh laboratory have shown that 1,25D participates in the growth inhibition of the normal mammary gland [51]. Further, the disruption of 1,25D/VDR signaling leads to distorted morphology of murine mammary gland with duct abnormalities and increased numbers of preneo-plastic lesions, suggesting that 1,25D-liganded VDR serves to maintain differentiation of normal mammary epithelium [52].

Induction of differentiation of breast cancer cells by 1,25D can be demonstrated by β-casein production [53], or by a change in overall cell size and shape, associated with changed cytoarchitecture of actin filaments and microtubules in MDA-MB-453 cells [54]. Treatment of these cells with 1,25D resulted in accumulation of integrins, paxillin, and focal adhesion kinase, as well as their phosphorylation. In contrast, the mesenchymal marker N-cadherin and the myoepithelial marker P-cadherin were down-regulated, suggesting that 1,25D reverses the myoepithelial features associ-ated with the aggressive forms of human breast cancer. However, it is to be noted that not all breast cancer cell lines respond to 1,25D. In many cases this can be attributed to the lack of or low VDR expression or function [55, 56], but it may also be due to alterations in 1,25D-metabolizing enzymes which can reduce the levels of 1,25D below its effective concentration [57].

Among the breast cancer cell lines that do respond to 1,25D a range of phenotype alterations has been reported [58], emphasizing that the mechanistic basis for the differentiating effects of 1,25D in the breast cancer cell system will be very complex. Together with the uncertainty about whether induced differentiation of breast cancer

cells, per se, has potential clinical significance, mechanistic studies in this system have been largely directed to the antiproliferative effects of 1,25D on breast cancer cells. These studies revealed that induction of apoptosis and G1 cell cycle arrest result in inhibition of tumor cell growth in several types of breast cancer cells [20, 57, 59], but the relationship of these biological effects to differentiation is not obvious. Nonetheless, some hints did result from those studies, as exemplified below.

An interesting set of candidate 1,25D-target proteins was identified by proteomic screening of a breast cancer cell line sensitive to 1,25D (MCF-7) and from a subclone of these cells derived by resistance to 1,25D (MCF-7/DRES) [60], and some of these proteins can be related to differentiation and associated phenotypic cellular changes. Examples are Rho-GDI and Rock-DI, known to participate in the formation of focal adhesions and stress fibers which contribute to the adhesive epithelial phenotype and changes in cell shape [60]. Proteins previously linked to pathways involved in 1,25D-induced differentiation such as phospho-p38, MEK2, RAS-GAP were also noted in this screen [52]. In a tissue culture study, the JNK pathway, also known to contribute to 1,25D-induced differentiation of colon and myeloid cells [61], was shown to cooperate with the p38 pathway to transactivate VDR in breast cancer cells, but this was proposed to contribute to the anti-proliferative rather than the differentiation-inducing effects of 1,25D in these cells [38]. The antiproliferative effects of 1,25D can also be explained by the reduction in EGFR mRNA and protein, but this is seen in only some, but not all, breast cancer cell lines [62, 63].

Another suggested link to differentiation in 1,25D-treated breast cancer cells is that VDR and estrogen receptor (ER) pathways converge to regulate BRCA-1, thus controlling the balance between signaling of differentiation and proliferation [64]. Since ER is important for mammary gland differentiation, studies that pursue this concept would be very valuable, and it already appears that the overexpression of ER and VDR is not sufficient to make ER-negative breast cancer cells responsive to 1α,hydroxy-vitamin D_5, a vitamin D analog known to mediate differentiation in a manner similar to 1,25D [65, 66].

7.2.3 Prostate Cancer

Similar to breast cancer cells, prostate cancer originates in hormone-dependent epithelial cells, and, as in breast cancer cell lines, 1,25D has anti-proliferative effects in some, but not all, established prostate cancer cell lines. The anti-proliferative action of 1,25D is, to a variable degree, due to the induction of cell death by apoptosis [67] and to cell cycle arrest [68], but to what extent these are associated with differentiation is uncertain.

The evidence of prostate cancer cell differentiation includes the release of prostate specific antigen (PSA) from cells treated with a differentiating agent such as 1,25D [69–71]. This can be useful in cultured cells, but in patients the increasing PSA levels suggest progressive disease, making it difficult to acquire data on the

7 Induction of Differentiation in Cancer Cells by Vitamin D   151

role of differentiation in clinical trials [72]. A study of the role of 1,25D in the differentiation of the normal rat prostate gland was based on morphological characteristics, which included an increased abundance of cytoplasmic secretory vesicles [73]. This characteristic has been used as a differentiation marker, along with the expression of keratins 8, 17, and 18 in human prostate cancer PC-3 cells [74]. In other studies [75, 76], the increased expression of E-cadherin was used as a maker of differentiation. However, although many reports on the effects of 1,25D on prostate cancer cells include the word "differentiation," the documentation most often focuses on the anti-proliferative effects of 1,25D exposure, which may, or may not be associated with phenotypic differentiation.

In a recent microarray analysis of 1,25D regulation of gene expression in LNCaP cells, Krishman et al. [77] reported several findings that appear relevant to 1,25D-induced differentiation. In addition to the major upregulation of the expression of the insulin-like growth factor binding protein-3 (IGFBP-3), which functions to inhibit cell proliferation by upregulating p21/Cip1 [78], it was noted that among about a dozen genes upregulated by 1,25D was the "prostate differentiation factor," a member of the bone morphogenetic protein (BMP) family, which is generally involved in growth and differentiation of both embryonic and adult tissues [79]. Also interesting was the finding that in these cells 1,25D regulates those genes which are androgen-responsive, and the genes which encode the enzymes involved in androgen catabolism. Further, Feldman and colleagues showed that 1,25D upregulates the expression and activity of the androgen receptor (AR) [80, 81], raising the possibility that the differentiation effects of 1,25D on prostate cells are not direct, but are due to modifications of the level or the activity of AR. Interestingly, it has also been suggested that androgens upregulate the expression of VDR [82]; thus, a positive feedback loop that includes 1,25D activation of VDR could be a factor in inducing differentiation of cancer cells derived from the hormonally regulated tissues (Fig. 7.2), while in normal cells the sex hormone (androgen or estrogen) is sufficient to promote differentiation. Since 1,25D has an established anticancer activity in prostate cells, it can be assumed that in this scenario VDR selectively enhances the AR-mediated androgenic pro-differentiation, but not the proliferation-enhancing activity (Fig. 7.2). In addition, it is likely that nuclear receptors for retinoids, glucocorticoids, and PPAR affect the signaling pathways, directly or indirectly. Whether the demonstrated 1,25D-induced decrease in the expression of COX-2 and an increase in 15-PGDH in prostate cancer cells [77, 83] have any relationship to cell differentiation, remains to be established.

Prostate cancer cells are also known to undergo "trans-differentiation" to a neuroendocrine phenotype, and when this phenotype is found in human tumors it may indicate an aggressive form of the disease [84]. Although currently 1,25D has no known role in this form of differentiation, this may be a promising area of future research, since recent studies point to a key role of NFκB, as well as IL-6 in this process [85, 86]. This suggestion is based on the finding that in some cells 1,25D upregulates the expression of C/EBP β [87], which cooperates with NFκB in regulation of the secretion of the cytokine IL-6 in neuroendocrine human prostate cancer cells [85].

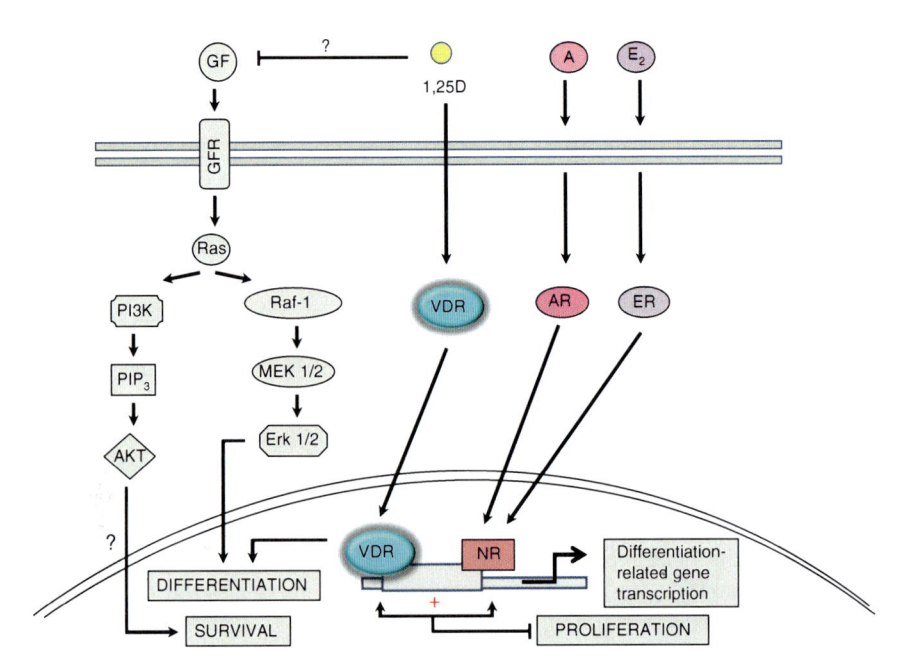

Fig. 7.2 Signaling of differentiation by 1,25D in hormone-dependent cancer cells. This schematic illustrates the hypothesis that in normal breast or prostate cells, estrogen (E_2) or androgen (*A*) is sufficient to induce differentiation, respectively. In cancer cells, the differentiation signal provided by the hormone-liganded nuclear receptor (*NR*) may need to be amplified by cooperation with 1,25D-activated VDR to induce differentiation. Since cells also receive signals from growth factors (*GF*), several of which activate Ras, the presence of a Ras-activated signaling pathways is exemplified by the AKT and extracellular-signal regulated kinase (*ERK*) cascades, though the role of these pathways in the differentiation of hormone-dependent cells is uncertain

7.2.4 *Keratinocytes and Squamous Cell Carcinoma Cells*

While there is extensive evidence of 1,25D-induced differentiation in normal keratinocytes, the studies of the induction of differentiation in squamous cell carcinomas (SCC), composed essentially of neoplastic keratinocytes, are less conclusive. Differentiation can be detected by the presence of various components of the keratinizing cells, such as cytokeratins K1 and K10, cornifin beta, involucrin, and transglutaminase, considered to be a late marker of squamous cell differentiation to normal epidermal keratinocytes [88]. The expression of target genes of 1,25D and analogs can also be taken as evidence that SCC cell lines can be driven to differentiation by these compounds [89]. Such genes include N-cadherin, which when overexpressed restores the epithelial phenotype also in prostate cancer cells [90], cystatin M, protease M, type XIII collagen, and desmoglein 3 [89]. Bikle and colleagues have presented persuasive models for induction of keratinocyte differentiation by increased calcium levels and by calcium-1,25D interactions [91, 92]. The key features of calcium-induced human keratinocyte differentiation appear to include the

7 Induction of Differentiation in Cancer Cells by Vitamin D 153

recruitment of phosphatidylinositol 3-kinase (PI3K) to a complex at the cell plasma membrane consisting of E-cadherin, β-catenin, and p120-catenin. This complex is postulated to activate PI3K leading to the accumulation of phosphatidylinositol 3,4,5-triphosphate (PIP_3), which binds to and activates phospholipase C gamma-1 (PLC-γ1) [93, 94]. The activated phospholipase generates inositol triphosphate (IP_3) which stimulates the release of calcium from the intracellular stores in the endoplasmic reticulum, and diacylglycerol, which together with increased intracellular calcium activates PKC. PKC, and perhaps calcium activation of other enzymes, then initiate signaling cascades that impinge on nuclear transcription factors such as AP-1, which lead to differentiation [95].

How much of this description applies to the 1,25D-induced differentiation is less clear, but Bikle et al. [91] presented a plausible model in which 1,25D interacts with calcium to induce keratinocyte differentiation. This model also includes a G-protein-coupled calcium-sensing surface receptor (CaR), which when activated by 1,25D leads to the activation of PKC, with consequences described above. The associated influx of calcium, which occurs in human keratinocytes after exposure to 1,25D has been recently shown to be mediated, at least in part, by the calcium-selective channel TRPV6 upregulated at the mRNA and protein levels by 1,25D [96]. A cohesive picture of 1,25D-induced keratinocyte differentiation is quite well, but perhaps not completely developed. For instance, regulation of AP-1 activity in cultured human keratinocytes by 1,25D was reported to be independent of PKC [97], in contrast to the model presented by Bikle et al. [91]. Takahashi et al. [98] reported that treatment of normal human keratinocytes with 1,25D increases the expression of cystatin A, a cysteine protease inhibitor which is a component of the cornified envelope, and that it is the suppression of the Raf-1/MEK-1/ERK signaling pathway which is responsible for this effect. However, cystatin A expression is stimulated by the Ras/MEKK-/MKK7/JNK pathway [99], consistent with the schematic model of Bikle et al. [91], and explaining why PKC activation may not be essential for AP-1 activation in this cell system. An enigmatic role of caspase-14 in keratinocyte differentiation induced by 1,25D has been reported [100], and it was suggested that the absence of caspase-14 contributes to the psoriatic phenotype. Since caspase-14 is a non-apoptotic protein, it is unclear if this is related to the report that 1,25D protects keratinocytes from apoptosis [101]. On the other hand, the identification of Kruppel-like factor 4 (KLF-4) and c-fos as 1,25D-responsive genes in gene expression profiling of 1,25D-treated keratinocytes [102] fits in well with the existing knowledge of differentiation signaling, as c-fos is a component of the AP-1 transcription factor, and KLF-4 is a transcription factor with a major role in cell fate decisions [103–105]. Recently, it was reported that yet another transcription factor, PPAR gamma, also has a major role in 1,25D-induced differentiation of keratinocytes [106]. In these studies, dominant negative (dn) PPAR gamma inhibited the expression of involucrin (a differentiation marker), suppressed AP-1 binding to DNA, and prevented the 1,25D-induced phosphorylation of p38. Thus, the keratinocyte system provided a wealth of interesting information on 1,25D as a differentiation-promoting and survival-regulating agent.

Transformed keratinocytes which give rise to SCC tend to be resistant to the differentiation-inducing action of 1,25D [107, 108], even though apoptosis and cell cycle arrest induced by 1,25D have been demonstrated in models of SCC [109, 110]. While VDR expression is required for 1,25D-induced differentiation, the resistance of SCCs to 1,25D is not due to the lack of functional VDR [111]. The possible explanations for the 1,25D resistance include the finding that the VDRE in the human PLCγ-1 gene is not functional [111]. Another explanation for the resistance is that increased serine phosphorylation of retinoid X receptor alpha (RXRα) by the Ras/MAPK pathway leads to its degradation, and thus VDR loses its heterodimeric partner for gene transactivation [112]. Yet another possibility is that VDR coactivators in SCCs are not appropriate for transactivation of differentiation-inducing genes [95]. Specifically, it was suggested that the expression of differentiation markers required a complex of VDR with the Src family of coactivators [113], but in SCC the DRIP coactivator complex is overexpressed, and there is a failure of SCCs to switch from DRIP to Src, resulting in the inability to express genes required for differentiation. It would be interesting to learn if this model has a wider applicability.

7.2.5 Osteosarcoma and Osteoblasts

Differentiation, as well as growth inhibition, has been documented in 1,25D-treated human and rat osteosarcoma cells [114, 115]. The differentiation was recognized by a morphological change to the chondrocyte phenotype, and by increased Alk Pase staining. The presence of Alk Pase or osteocalcin could also be detected at the mRNA level [115]. In human fetal osteoblastic cell line responsive to 1,25D, mineralized nodules were detected [116], demonstrating that an advanced degree of differentiation can be achieved in this cell system. Interestingly, 1,25D-induced differentiation in osteoblasts and osteocytes is accompanied by an increase in the potential for cell survival through enhanced anti-apoptotic signaling [117]. It is possible that this is mediated by EGFR-relayed signals, as in contrast to other cell types [32, 62, 118], 1,25D-treated osteoblastic cells show increased levels of EGFR mRNA [119].

Recent studies suggest that the anti-apoptotic effects of 1, 25D on osteoblasts and osteocytes are mediated by Src, PI3K, and JNK kinases [117]. The suggested mechanisms include an association of Src with VDR, though transcriptional mechanisms were required, as shown by an inhibition of the biological effect by exposure to actinomycin D or cycloheximide. The association of VDR with other proteins may be particularly important in osteoblast cells induced to differentiate by 1,25D, as another group reported that IGF-binding protein-5 (IBP-5) interacts with VDR, and blocks the RXR/VDR heterodimerization in the nuclei of MG-63 and U2-OS cells, thus attenuating the expression of bone differentiation markers [120]. Also, in ROS 17/28 cells the NFκB p65 subunit integrates into the VDR transcription complex and disrupts VDR binding to its coactivator Src-1 [121]. Although protein–protein binding between VDR and p65 has not been demonstrated, this remains a possibility, further highlighting the importance of this mode of control of VDR activity.

7.3 Leukemias

Hematological malignances are a diverse group of diseases, but can be divided into two major groups, the lymphocytic and myeloid leukemias. Although normal activated B and T lymphocytes express VDR, and 1,25D has antiproliferative effects on these cell types (e.g., [122, 123]), this does not appear to alter their differentiation state, and lymphocytic leukemia cells do not respond to 1,25D. In contrast, 1,25D has been known since 1981 to induce maturation of mouse myeloid leukemia cells [124], and this can also take place in a wide variety of human myeloid leukemia cell lines, with the exception of the lines derived from the most immature acute myeloid leukemia (AML) blast cells (e.g., [125–127]).

Differentiation induced by 1,25D usually results in a monocyte-like phenotype, but prolonged exposure to 1,25D confers cell surface changes that result in adherence to the substratum, making the differentiated cells macrophage-like [124, 128]. The monocyte characteristics are recognized by changes related to phagocytosis, such as the ability to break down esters, assayed by the "non-specific esterase" (NSE) cytochemical reaction, also known as "monocyte-specific esterase" (MSE) since in the hematopoietic cells this esterase is specific for monocytes and macrophages [129]. Also related to phagocytosis is the ability to generate reactive oxygen species (ROS) including superoxide, usually recognized by the nitroblue tetrazolium (NBT) or cytochrome reduction [130, 131]. The availability of Flow Cytometry (FC) for the recognition of surface proteins has made the study of the differentiating effects of 1,25D on myeloid leukemia cells quite simple, using CD14, a receptor for complexes of lipopolysaccharides (LPS) and LPS-binding protein [132], a near-definitive marker of the monocytic phenotype. This is usually supplemented by the FC determination of CD11b, or another subunit of the human neutrophil surface protein that mediates cellular adherence [133]. In contrast to myeloid cells induced to differentiate by the phorbol ester TPA, in 1,25D-treated cells the ability to adhere develops more slowly than the ability to phagocytose. Consequently, 1,25D treatment results in an earlier appearance of the CD14 antigen, usually accompanied in parallel by MSE positivity, than the appearance of CD11b and NBT positivity [134, 135]. Generally, at least two of the above parameters are measured to demonstrate monocytic differentiation, and FC methods require the use of paired isotypic IgG controls for each test sample to avoid obtaining false-positive data. Exposure of AML cells to 1,25D also results in G1 phase cell cycle arrest, which follows, rather than precedes, the phenotypic differentiation [134], and is often taken as the confirmatory evidence that differentiation has taken place. However, in contrast to cells from most solid tumors, monocytic differentiation of AML cells is accompanied by increased expression of anti-apoptotic proteins, and consequently 1,25D-treated myeloid cells have an increased cell survival potential [136–140].

The topic of 1,25D-induced leukemia cell differentiation has been extensively studied in many laboratories. These include several groups in Japan [141–145], and a group in Birmingham, England [146, 147], who made many valuable contributions to the field. Notably, combined basic and clinical studies of 1,25D-induced

leukemia cell differentiation were very comprehensively developed by Koeffler and his various collaborators [148–151]. Their impressive achievements are described in the preceding chapter in this volume. Accordingly, what follows in the remainder of this section is an outline of the signaling mechanisms of AML cells that have occupied the attention of the corresponding author's laboratory.

In these studies, the laboratory has focused on HL60 cells, a widely available cell line derived from a patient with promyeloblastic leukemia, with the objective of achieving with the currently available tools as clear a picture as possible of the signaling of monocytic differentiation. In this model, outlined in Figs. 7.3 and 7.4, a plausible sequence of events is presented, but it is likely that other pathways are

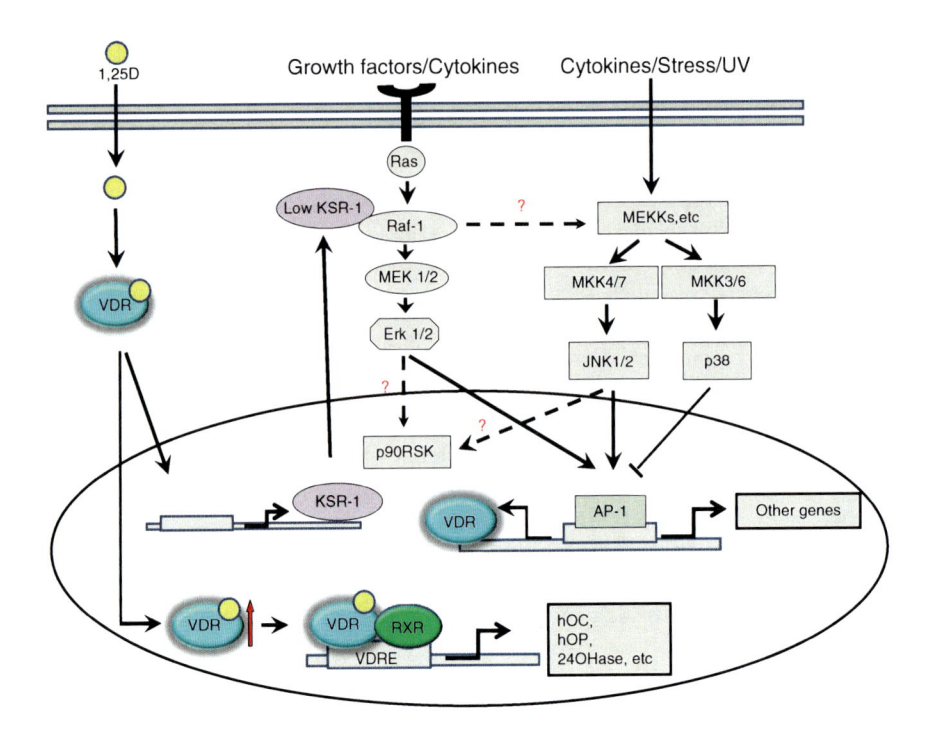

Fig. 7.3 Suggested signaling of the early stages of 1,25D-induced monocytic differentiation. Binding of 1,25D to vitamin D receptor (*VDR*) stimulates its translocation to the cell nucleus, where it heterodimerizes with retinoid X receptor (*RXR*) and in myeloid precursor cells transactivates genes containing vitamin D_3 response element (*VDREs*) in their promoter regions. These include genes which encode proteins involved in calcium homeostasis and bone integrity, such as osteocalcin (*hOC*), osteopontin (*hOP*), and the 1,25D-catabolic enzyme 24-hydroxylase (24OHase). It is postulated that the regulators of signaling pathways, e.g., KSR-1, are also upregulated in myeloid cells and alter Ras signaling from the cell membrane, so that signaling by Mitogen activated protein kinases (*MAPKs*) (MEKs, ERKs, and JNKs) increases the AP-1 activity. This can have a positive feedback effect on differentiation by increasing VDR abundance. It is also suggested that a potential negative feedback mechanism is provided by p38 MAPK, as inhibition of its signaling by SB203580 enhances 1,25D-induced monocytic differentiation

7 Induction of Differentiation in Cancer Cells by Vitamin D   157

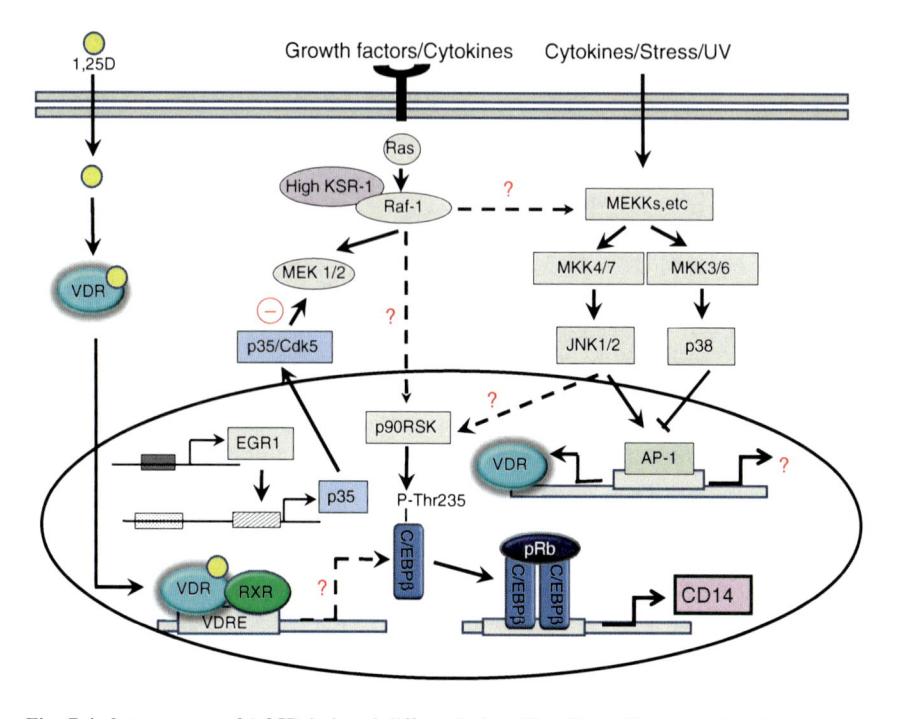

Fig. 7.4 Later stages of 1,25D-induced differentiation. This figure illustrates that the transcription factor Egr-1, known to be upregulated by 1,25D (189), can increase the expression of p35/Nck5a (p35) activator of Cdk5. Cdk5 activated by p35 then can phosphorylate MEK on Thr286, a site which inactivates it [200], as shown by the Θ symbol. This diminishes ERK1/2 activity, downstream from MEK (not shown here), but Raf-1 can activate p90RSK directly, which in turn activates the transcription factor C/EBP β, perhaps bound to pRb, and increases the expression of CD14, as part of monocytic differentiation. The activation of p90RSK may also be increased by the Jun N-terminal kinase (*JNK*) pathway, which also activates AP-1, and may lead to VDR expression. The interplay between the signaling by 1,25D, growth factor, and stress add to the overall complexity of the induction of the monocytic phenotype

also operative, but remain to be convincingly demonstrated. The details of the scheme are described below.

7.3.1 Signaling of Monocytic Differentiation by MAPK and Parallel Pathways

Early in our investigations we recognized that 1,25D-induced monocytic differentiation is not a single continuous process, but a series of events that can be divided into at least two overlapping phases. In the first phase, which lasts 24–48 h, the cells continue in the normal cell cycle while expressing markers of monocytic phenotype, such as CD14 and NSE. In the next phase, the G1 to S phase cell cycle block becomes apparent, and the expression of CD11b is also prominent, indicating a

beginning of the transition to the macrophage phenotype. The first phase is characterized by high levels of ERKs activated by phosphorylation, and these levels decrease as the cells enter the second phase, while the levels of the cell cycle inhibitor $p27^{Klp1}$ increase at that time. Serum-starved HL60 cells or cells treated with the MAPK inhibitor PD 98059 have reduced growth rate and a slower rate of differentiation, but the G1 block under these conditions also coincides with decreased levels of activated ERK1/2 [152]. Our data suggested that the MEK/ERK pathway maintains cell proliferation during the early stages of differentiation, and the consequent G1 block leads to "terminal" differentiation. Using a different experimental design similar results were obtained by Marcinkowska [153].

We also demonstrated that the JNK pathway, as shown by the increased phosphorylation of c-jun, plays a role in the induction of differentiation of HL60 cells by 1,25D. The data showed that 1,25D-induced differentiation of a stable clone of U937 cells transfected with a dominant negative construct of JNK-1 was reduced, as compared to cells transfected with a control construct [154], and potentiation of 1,25D-induced differentiation by the plant antioxidants curcumin and silibinin increased the phosphorylation of c-jun [155]. This suggested that the JNK-jun pathway is involved in 1,25D-induced differentiation, and was further established in experiments which showed that the AP-1 transcription factor complex is required for this process, since c-jun, together with ATF-2, is the principal component of this complex [140]. This appears to be of wider significance, as c-jun expression was also reported to enhance macrophage differentiation of U937 cells [156].

However, it seems clear that the ERK and JNK MAPK pathways are not the only ones involved in signaling of 1,25D-induced differentiation. For instance, compounds SB203580 and SB2902190, reported to be specific inhibitors of the signaling protein p38 MAP kinase [157], were found to markedly accelerate monocytic differentiation of HL60 cells induced by low concentrations of 1,25D [158]. Paradoxically, these compounds also induced a sustained enhancement of p38 phosphorylation and of its activity in cell extracts in the absence of added inhibitor, which raised the possibility of a lack of specificity of SB compounds in this cell system, or of an up-regulation of the upstream components of the p38 pathway. In addition, SB 203580 or SB 202190 treatment of HL60 cells resulted in a prolonged activation of the JNK and the ERK MAPK pathways [158]. Honma and colleagues also found that SB203580 treatment of HL60, HT93 and ML-1 human myeloid leukemia cell lines increased cellular ERK activity [159]. These data are consistent with the hypothesis that in HL60 cells an interruption of a negative feedback loop from a p38 target activates a common regulator of multiple MAPK pathways, but it is also possible that SB203580 has an additional, unknown, action.

Another signaling cascade known to be activated by 1,25D in human AML cells is the PI3K-AKT pathway, which is often envisaged to signal from the cell membrane to the intracellular regulators in parallel with the MAPK pathways, e.g., [160]. As first noted by Reiner and colleagues [161], monocytic leukemia cells THP-1 exposed to 1,25D in serum-free medium show a rapid and transient increase in PI3K activity, which was attributed to the formation of a VDR-PI3K protein

complex. However, it is not clear if the lack of growth factors normally provided by the serum contributes to the observed effects. The role of the PI3K pathway in 1,25-induced differentiation was further studied by Marcinkowska and colleagues [162–164], who showed that the activation of PI3K by 1,25D can also be demonstrated in HL60 cells, and that the signal is transmitted to AKT. This function of AKT may contribute to the differentiation-related increase in 1,25D-induced cell survival [139]. An additional role of PI3K, as well as of the Ras/Raf/ERK, pathway in human leukemia cells is the stimulation of steroid sulfatase activity, an enzyme that converts inactive estrogen and androgen precursors to the active sex hormones [147]. If this is also operative in breast and/or prostate tissues, it could offer an explanation for the mutual activation of VDR and the estrogen and androgen nuclear receptors, as shown in Fig. 7.2.

The mechanisms of the upregulation of MAPK pathways in the initial phase of 1,25D action on leukemia cells are still unclear. The very rapid effects of 1,25D on the MAPK pathway in intestinal cells that result in rapid calcium transport ("transcaltachia") have been attributed to a cell membrane receptor ("mVDR") [165–167], but whether direct, non-genomic action of such mVDR can initiate or enhance MAPK pathways activity in leukemia cells has not been well documented. In non-starved leukemia cells, 1,25D elicits less rapid (hours rather than minutes) activation of the MAPKs. One possibility is that this is achieved by the transcriptional upregulation of Kinase Suppressor of Ras-1 (KSR-1), a membrane-associated kinase/molecular scaffold, also known as ceramide-activated protein kinase [168, 169]. Although a kinase activity associated with KSR-1 has been reported [170–172], the best established function of KSR-1 is to provide a platform for Raf-1 kinase to phosphorylate and thus activate its downstream targets in the MAPK pathways [173, 174]. Thus, since KSR-1 has been shown to have a functional DNA element regulated by VDR (VDRE) [175], the activation of the MAPKs may be a direct, "genomic" action of 1,25D, as depicted in Fig. 7.3, rather than signaling initiating at the membrane and "non-genomic."

Our studies [169, 176] combined with those of Marcinkowska and colleagues [164, 177] suggest that leukemia cell differentiation is initiated when 1,25D promotes nuclear translocation of liganded VDR, which dimerizes with RXR and transactivates several VDRE-regulated genes, including KSR-1 and KSR-2. The latter appears to have a role in increasing the survival potential of differentiating monocytic cells [24], while KSR-1 acts as a scaffold, which by simultaneously binding to Ras and Raf-1 (and perhaps other proteins) facilitates or redirects the signaling cascade, at least initially, to MEK/ERK, and thus amplifies the signal that initiates monocytic differentiation (Fig. 7.3).

Raf-1 participation has been shown to be required for the later stages of differentiation, when an impairment in cell cycle progression becomes apparent, and at this more advanced point of the differentiation process MEK/ERK signaling does not appear to be involved [178, 179]. While this requires further study, the current model, also supported by observations in other differentiation signaling systems [180–182], suggests that Raf-1 can signal p90RSK activation independently of MEK and ERK, as outlined in Fig. 7.4.

160   E. Gocek and G.P. Studzinski

A rather speculative mechanism describing how MEK/ERK signaling is diminished in the later stages of differentiation, when cell proliferation becomes arrested, is presented below.

7.3.2 p35/Cdk5, a Protein Kinase System That May Interface Differentiation Processes with Cell Cycle Arrest

After 24–48 h of exposure of myeloid leukemia cells to moderate concentrations of 1,25D (1–10 nM), cell cycle progression becomes progressively arrested, principally due to a G1 to S phase block, though a G2 phase block can also be observed [183]. Several mechanisms could explain these cell cycle effects, and these include activation of cyclin-dependent kinase 5 (Cdk5).

Cdk5 is a proline-directed serine-threonine kinase with sequence homology to the cyclin-activated kinases which regulate cell cycle progression, but its best known function is participation in differentiation of neuronal cells [184]. When combined with a "cyclin-like" neuronal Cdk5 activator (Nck5a) 35 kDa protein (p35/Nck5a, or p35), the p35/Cdk5 complex functions in monocytic cells and has an important role in the normal, and possibly abnormal development of this hematopoietic lineage. Our initial observations were that in HL60 cells treated with 1,25D the monocytic phenotype and expression of Cdk5 appear in parallel. Both active and inactive Cdk5 was associated with cyclin D1 protein, and the inhibition of Cdk5 expression by an antisense oligonucleotide construct reduced the intensity of 1,25D-induced expression of the monocytic marker CD14 [185]. This finding demonstrated a novel (other than neuronal) cellular type for Cdk5 activity, and a concomitant enhancement of monocytic differentiation.

The above study showed that protein levels and kinase activity of Cdk5 increase in HL60 cells induced to monocytic differentiation by 1,25D, but did not establish the specificity of the association of Cdk5 with the monocytic phenotype. Therefore, we showed in a subsequent study that the upregulation of Cdk5 does not occur in granulocytic differentiation, whereas an inhibition of Cdk5 activity by the pharmacological inhibitor olomoucine, or of its expression by a plasmid construct expressing antisense Cdk5, switches the 1,25D-induced monocytic phenotype (a combination of the positive NSE reaction, the expression of the CD14 marker, and morphology) to a general myeloid phenotype (a positive NBT reaction, the CD11b marker, and morphology) [186]. These findings showed that in human myeloid cells the upregulation of Cdk5 is specifically associated with the monocytic phenotype.

The Nck5a 35 kDa protein has hitherto been considered to be exclusively expressed in neuronal cells, as its name implies [187]. However, since we had clear evidence that Cdk5 is an active kinase in human leukemia cells HL60 and U937 induced to differentiate with 1,25D, and since the "classical" cyclins (e.g., cyclin D1, cyclin E) are not known to activate Cdk5, we investigated whether p35 can be detected in cells with active Cdk5. Indeed, we demonstrated that p35 is expressed in normal human monocytes and in leukemic cells induced to differentiate toward

7 Induction of Differentiation in Cancer Cells by Vitamin D   161

the monocytic lineage, but not in lymphocytes, or cells induced to granulocytic differentiation by retinoic acid. The activator p35 is present in a complex with Cdk5 that has protein kinase activity, and when ectopically expressed together with Cdk5 in undifferentiated HL60 cells it induces the expression of CD14 and NSE markers of the monocytic phenotype [188]. These observations not only indicate a functional relationship between Cdk5 and p35, but also support a role for this complex in monocytic differentiation.

A likely link to the diminution of ERK MAPK pathway activity at the onset of phase 2 of 1,25D-induced differentiation is provided by the EGR-1 → p35/Cdk5 ---| MEK 1/2 pathway, that was elucidated in leukemia cells by this laboratory [189]. The schematic representation is shown in Fig. 7.4, and the supporting data can be summarized as follows.

7.3.2.1 Control of p35 Expression by the EGR-1 Transcription Factor

The evidence that supports a role of EGR-1 in regulating the expression of p35 includes the coordinate expression of EGR-1 along with Cdk5, and the co-inhibition of the 1,25D-induced upregulation of these proteins by PD 98059, an inhibitor of the MEK/ERK1/2 pathway [171, 190]. Further, the promoter region of human p35 has an EGR-1 binding site that overlaps with an Sp1 site, and a gel shift assay showed that a double-stranded oligonucleotide that contained this sequence bound proteins in nuclear extracts from 1,25D-treated HL60 cells. The EGR-1-site binding proteins were competed most efficiently by an anti-EGR-1 antibody, though some competition was also observed with an anti-Sp1 antibody, but no competition was observed with an irrelevant antibody, e.g., anti-VDR. The data suggested that EGR-1, and perhaps Sp1 proteins, regulate the expression of p35 and contribute to induction of the monocytic phenotype. A "decoy" EGR-1 response element oligonucleotide inhibited both 1,25D-induced p35 expression and monocytic differentiation [189].

7.3.2.2 The Cdk5/p35 Complex Phosphorylates MEK

We also found that the Cdk5/p35 can phosphorylate MEK in cell extracts [189]. If this can be demonstrated to occur in leukemia cells, it will provide a potential mechanism for the inhibition of the MAPK/ERK pathway seen in the later stages of differentiation (48 h after the addition of 1,25D to the cultures), since phosphorylation of MEK by p35/Cdk5 inhibits its kinase activity. Intriguingly, upregulation of p35 (which activates Cdk5) is observed *pari passu* as ERK 1/2 phosphorylation is waning, consistent with a cause–effect relationship. We have thus proposed a mechanism that can shut down cell proliferation, possibly by allowing p27^{Kip1} to accumulate in the cell nucleus due to a decline in ERK 1/2 activity, since it has been reported that the ERK pathway can increase nuclear export of p27 [191].

7.3.2.3 C/EBP Β Transcription Factor as an Effector of Monocytic Differentiation

One of the downstream targets of the MAPK-RSK pathway is a nuclear transcription factor, the CAAT and Enhancer Binding Protein β (C/EBP β). This transcription factor has been reported to be activated by phosphorylation both by ERK [192] and by RSK [193], and can interact directly with the promoter of CD14, one of the principal markers of monocytic differentiation [194], as illustrated in Fig. 7.4. We showed that the expression of C/EBP β is increased by 1,25D in parallel with markers of differentiation; conversely, the knockdown of its expression by antisense oligonucleotides, or of its transcriptional activity by "decoy" promoter competition, inhibited 1,25D-induced differentiation [195]. In an additional study, the data suggested that 1,25D induced phosphorylation of C/EBP β isoforms on Thr235, and that the C/EBP β-2 isoform is one of the principal differentiation-related transcription factors in this system [87].

These findings suggest that 1,25D can induce leukemic progenitor cells, which have the potential to differentiate into several hematopoietic lineages, to become nonproliferating monocyte-like cells by changing the ratio of nuclear transcription factors in a manner that permits this form of differentiation [196]. In this scenario, the event that initiates leukemic transformation, such as a mutation, alters the proper balance of transcription factor activity necessary for normal granulocytic cell differentiation. However, 1,25D-induced expression of C/EBP β then allows the cells to bypass this block to granulocytic differentiation by becoming mono-cyte-like cells instead (Fig. 7.5).

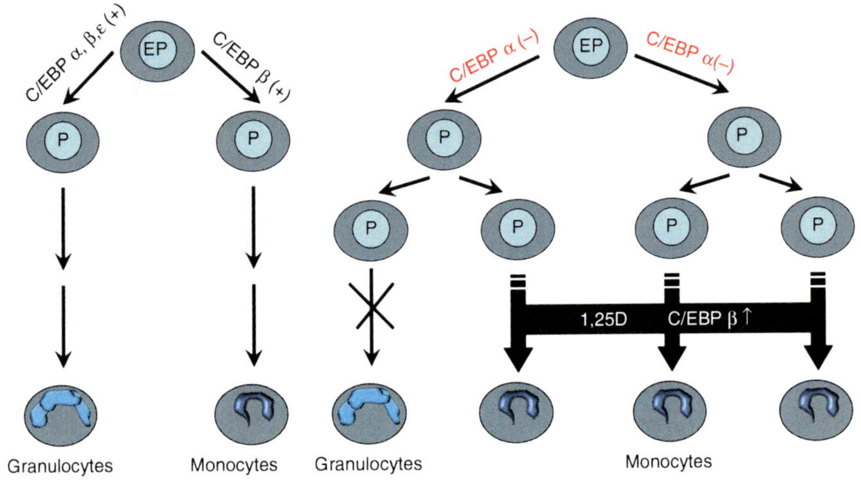

Fig. 7.5 The suggested role of CAAT/enhancer binding protein β in 1,25D-induced bypass of the differentiation block in leukemia cells. In this scenario, C/EBP α is indispensable for normal granulopoiesis, while C/EBP β regulates monocytic differentiation. When C/EBP α is mutated or inactivated and granulopoiesis is blocked, immature myeloid cells accumulate in the bone marrow and appear in the peripheral blood resulting in acute myeloid leukemia (*AML*). 1,25D-induced expression of C/EBP β may allow the cells to bypass this block to granulocytic differentiation by switching the lineage of cell differentiation from granulocytes to monocytes

7 Induction of Differentiation in Cancer Cells by Vitamin D   163

Interestingly, 1,25D has also been reported to have a negative effect on differentiation, as it inhibits IL-4/GM-CSF-induced differentiation of human monocytes into dendritic cells, and this contributes to 1,25D immunosuppressive activity [197, 198]. The data also suggested that 1,25D specifically downregulates the expression of CSF-1, and promoted spontaneous apoptosis of mature dendritic cells, further demonstrating the pleiotropic effects of 1,25D and the cell type-specificity of the outcomes.

7.4 Conclusion

The signaling pathways presented here are shown to control the activity of several transcription factors, such as the ubiquitous AP-1 complex, the nuclear receptor VDR, and the lineage-determining C/EBP family of transcription factors. While these clearly play a role in 1,25D-induced differentiation of HL60 cells, there may be redundancy of important cellular regulators, and other pathways and transcription factors are likely to be involved. The initial steps that activate the differentiation-inducing actions of 1,25D are not entirely clear, and while cell membrane-associated events have a role, these events are not necessarily rapid but are sustained. It is likely that microRNAs will be found to further control or modulate 1,25D signaling, as retinoic acid-induced differentiation of NB4 AML cells has been shown to be associated with the upregulation of a number of microRNAs, and the downregulation of microRNA 181b [199]. Thus, extensive additional investigations are warranted to provide a basis for the design of improved therapies of leukemia and solid tumors.

Acknowledgments We thank Drs. Michael Danilenko, David Goldberg, and Ewa Marcinkowska for comments on the manuscript, and Ms. Vivienne Lowe for expert secretarial assistance. The author's experimental work was supported by grants from the National Cancer Institute RO1-CA 44722–18 and RO1-CA 117942–01, and from the Polish Ministry of Science and Higher Education, grant No. 2622/P01/2006/31.

References

1. Smithers DW (1969) Maturation in human tumours. Lancet 2:949–952
2. Walton JD, Kattan DR, Thomas SK, Spengler BA, Guo HF, Biedler JL, Cheung NK, Ross RA (2004) Characteristics of stem cells from human neuroblastoma cell lines and in tumors. Neoplasia 6:838–845
3. Chen SJ, Zhu YJ, Tong JH, Dong S, Huang W, Chen Y, Xiang WM, Zhang L, Li XS, Qian GQ (1991) Rearrangements in the second intron of the RARalfa gene are present in a large majority of patients with acute promyelocytic leukemia and are used as molecular marker for retinoic acid-induced leukemic cell differentiation. Blood 78:2696–2701
4. Degos L, Wang ZY (2001) All trans retinoic acid in acute promyelocytic leukemia. Oncogene 20:7140–7145
5. Schlenk RF, Frohling S, Hartmann F, Fischer JT, Glasmacher A, del Valle F, Grimminger W, Gotze K, Waterhouse C, Schoch R, Pralle H, Mergenthaler HG, Hensel M, Koller E, Kirchen H, Preiss J, Salwender H, Biedermann HG, Kremers S, Griesinger F, Benner A, Addamo B,

Dohner K, Haas R, Dohner H (2004) Phase III study of all-trans retinoic acid in previously untreated patients 61 years or older with acute myeloid leukemia. Leukemia 18:1798–1803

6. Camerini T, Mariani L, De Palo G, Marubini E, Di Mauro MG, Decensi A, Costa A, Veronesi U (2001) Safety of the synthetic retinoid fenretinide: long-term results from a controlled clinical trial for the prevention of contralateral breast cancer. J Clin Oncol 19:1664–1670

7. Jung SJ, Lee YY, Pakkala S, de Vos S, Elstner E, Norman AW, Green J, Uskokovic M, Koeffler HP (1994) 1, 25(OH)$_2$-16ene-vitamin D3 is a potent antileukemic agent with low potential to cause hypercalcemia. Leuk Res 18:453–463

8. Jones G (2008) Pharmacokinetics of vitamin D toxicity. Am J Clin Nutr 88:582S–586S

9. Bouillon R, Okamura WH, Norman AW (1995) Structure-function relationships in the vitamin D endocrine system. Endocr Rev 16:200–257

10. Ji Y, Wang X, Donnelly RJ, Uskokovic MR, Studzinski GP (2002) Signaling of monocytic differentiation by a non-hypercalcemic analog of vitamin D3, 1, 25(OH)$_2$-5, 6 trans-16-ene-vitamin D3, involves nuclear vitamin D receptor (nVDR) and non-nVDR-mediated pathways. J Cell Physiol 191:198–207

11. Collins ED, Bishop JE, Bula CM, Acevedo A, Okamura WH, Norman AW (2005) Effect of 25-hydroxyl group orientation on biological activity and binding to the 1alpha, 25-dihydroxy vitamin D3 receptor. J Steroid Biochem Mol Biol 94:279–288

12. Aparna R, Subhashini J, Roy KR, Reddy GS, Robinson M, Uskokovic MR, Venkateswara Reddy G, Reddanna P (2008) Selective inhibition of cyclooxygenase-2 (COX-2) by 1alpha, 25-dihydroxy-16-ene-23-yne-vitamin D3, a less calcemic vitamin D analog. J Cell Biochem 104:1832–1842

13. Danilenko M, Wang X, Studzinski GP (2001) Carnosic acid and promotion of monocytic differentiation of HL60-G cells initiated by other agents. J Natl Cancer Inst 93:1224–1233

14. Danilenko M, Wang Q, Wang X, Levy J, Sharoni Y, Studzinski GP (2003) Carnosic acid potentiates the antioxidant and prodifferentiation effects of 1alpha, 25-dihydroxyvitamin D3 in leukemia cells but does not promote elevation of basal levels of intracellular calcium. Cancer Res 63:1325–1332

15. Danilenko M, Studzinski GP (2004) Enhancement by other compounds of the anti-cancer activity of vitamin D3 and its analogs. Exp Cell Res 298:339–358

16. Coffman FD, Studzinski GP (1999) Differentiation-related mechanisms which suppress DNA replication. Exp Cell Res 248:58–73

17. Harrison LE, Wang QM, Studzinski GP (1999) Butyrate-induced G2/M block in CaCo-2 colon cancer cells is associated with decreased p34cdc2 activity. Proc Soc Exp Biol Med 222:150–156

18. Harrison LE, Wang QM, Studzinski GP (1999) 1, 25-dihydroxyvitamin D3-induced retardation of the G(2)/M traverse is associated with decreased levels of p34(cdc2) in HL60 cells. J Cell Biochem 75:226–234

19. Wang QM, Studzinski GP, Chen F, Coffman FD, Harrison LE (2000) p53/56(lyn) antisense shifts the 1, 25-dihydroxyvitamin D3-induced G1/S block in HL60 cells to S phase. J Cell Physiol 183:238–246

20. Welsh J, Simboli-Campbell M, Narvaez CJ, Tenniswood M (1995) Role of apoptosis in the growth inhibitory effects of vitamin D in MCF-7 cells. Adv Exp Med Biol 375:45–52

21. Li F, Ling X, Huang H, Brattain L, Apontes P, Wu J, Binderup L, Brattain MG (2005) Differential regulation of survivin expression and apoptosis by vitamin D3 compounds in two isogenic MCF-7 breast cancer cell sublines. Oncogene 24:1385–1395

22. Myrthue A, Rademacher BL, Pittsenbarger J, Kutyba-Brooks B, Gantner M, Qian DZ, Beer TM (2008) The iroquois homeobox gene 5 is regulated by 1,25-dihydroxyvitamin D3 in human prostate cancer and regulates apoptosis and the cell cycle in LNCaP prostate cancer cells. Clin Cancer Res 14:3562–3570

23. Wang X, Studzinski GP (1997) Antiapoptotic action of 1, 25-dihydroxyvitamin D3 is associated with increased mitochondrial MCL-1 and RAF-1 proteins and reduced release of cytochrome c. Exp Cell Res 235:210–217

24. Wang X, Patel R, Studzinski GP (2008) hKSR-2, a vitamin D-regulated gene, inhibits apoptosis in arabinocytosine-treated HL60 leukemia cells. Mol Cancer Ther 7:2798–2806
25. Helander HF (1975) Enzyme patterns and protein absorption in rat colon during development. Acta Anat (Basel) 91:330–349
26. Ono K (1976) Alkaline phosphatase activity of the large intestinal principal cells in postnatal developing rats. Acta Histochem 57:312–319
27. Chen A, Davis BH, Bissonnette M, Scaglione-Sewell B, Brasitus TA (1999) 1,25-Dihydroxyvitamin D3 stimulates activator protein-1-dependent CaCo-2 cell differentiation. J Biol Chem 274:35505–35513
28. Marvin-Guy LF, Duncan P, Wagniere S, Antille N, Porta N, Affolter M, Kussmann M (2008) Rapid identification of differentiation markers from whole epithelial cells by matrix-assisted laser desorption/ionisation time-of-flight mass spectrometry and statistical analysis. Rapid Commun Mass Spectrom 22:1099–1108
29. Palmer HG, Gonzalez-Sancho JM, Espada J, Berciano MT, Puig I, Baulida J, Quintanilla M, Cano A, de Herreros AG, Lafarga M, Munoz A (2001) Vitamin D3 promotes the differentiation of colon carcinoma cells by the induction of E-cadherin and the inhibition of beta-catenin signaling. J Cell Biol 154:369–387
30. Garland CF, Comstock GW, Garland FC, Helsing KJ, Shaw EK, Gorham ED (1989) Serum 25-hydroxyvitamin D and colon cancer: eight-year prospective study. Lancet 2:1176–1178
31. Giovannucci E, Liu Y, Rimm EB, Hollis BW, Fuchs CS, Stampfer MJ, Willett WC (2006) Prospective study of predictors of vitamin D status and cancer incidence and mortality in men. J Natl Cancer Inst 98:451–459
32. Tong WM, Hofer H, Ellinger A, Peterlik M, Cross HS (1999) Mechanism of antimitogenic action of vitamin D in human colon carcinoma cells: relevance for suppression of epidermal growth factor-stimulated cell growth. Oncol Res 11:77–84
33. Belleli A, Shany S, Levy J, Guberman R, Lamprecht SA (1992) A protective role of 1, 25-dihydroxyvitamin D3 in chemically induced rat colon carcinogenesis. Carcinogenesis 13:2293–2298
34. Shabahang M, Buras RR, Davoodi F, Schumaker LM, Nauta RJ, Evans SR (1993) 1, 25-Dihydroxyvitamin D3 receptor as a marker of human colon carcinoma cell line differentiation and growth inhibition. Cancer Res 53:3712–3718
35. Vandewalle B, Adenis A, Hornez L, Revillion F, Lefebvre J (1994) 1, 25-dihydroxyvitamin D3 receptors in normal and malignant human colorectal tissues. Cancer Lett 86:67–73
36. Sheinin Y, Kaserer K, Wrba F, Wenzl E, Kriwanek S, Peterlik M, Cross HS (2000) In situ mRNA hybridization analysis and immunolocalization of the vitamin D receptor in normal and carcinomatous human colonic mucosa: relation to epidermal growth factor receptor expression. Virchows Arch 437:501–507
37. Cross HS, Bareis P, Hofer H, Bischof MG, Bajna E, Kriwanek S, Bonner E, Peterlik M (2001) 25-Hydroxyvitamin D3–1alpha-hydroxylase and vitamin D receptor gene expression in human colonic mucosa is elevated during early cancerogenesis. Steroids 66:287–292
38. Qi X, Pramanik R, Wang J, Schultz RM, Maitra RK, Han J, DeLuca HF, Chen G (2002) The p38 and JNK pathways cooperate to trans-activate vitamin D receptor via c-Jun/AP-1 and sensitize human breast cancer cells to vitamin D3-induced growth inhibition. J Biol Chem 277:25884–25892
39. Wiese RJ, Uhland-Smith A, Ross TK, Prahl JM, DeLuca HF (1992) Up-regulation of the vitamin D receptor in response to 1, 25-dihydroxyvitamin D3 results from ligand-induced stabilization. J Biol Chem 267:20082–20086
40. Alrefai WA, Scaglione-Sewell B, Tyagi S, Wartman L, Brasitus TA, Ramaswamy K, Dudeja PK (2001) Differential regulation of the expression of Na^+/H^+ exchanger isoform NHE3 by PKC-alpha in CaCo-2 cells. Am J Physiol Cell Physiol 281:C1551–C1558
41. Wali RK, Baum CL, Sitrin MD, Brasitus TA (1990) 1, 25(OH)$_2$ vitamin D3 stimulates membrane phosphoinositide turnover, activates protein kinase C, and increases cytosolic calcium in rat colonic epithelium. J Clin Invest 85:1296–1303

42. Gaschott T, Werz O, Steinmeyer A, Steinhilber D, Stein J (2001) Butyrate-induced differentiation of CaCo-2 cells is mediated by vitamin D receptor. Biochem Biophys Res Commun 288:690–696
43. Cui M, Klopot A, Jiang Y, Fleet JC (2009) The effect of differentiation on 1, 25 dihydroxyvitamin D-mediated gene expression in the enterocyte-like cell line, CaCo-2. J Cell Physiol 218(1):113–121
44. Easwaran V, Pishvaian M, Salimuddin, Byers S (1999) Cross-regulation of beta-catenin-LEF/TCF and retinoid signaling pathways. Curr Biol 9:1415–1418
45. Shah S, Islam MN, Dakshanamurthy S, Rizvi I, Rao M, Herrell R, Zinser G, Valrance M, Aranda A, Moras D, Norman A, Welsh J, Byers SW (2006) The molecular basis of vitamin D receptor and beta-catenin crossregulation. Mol Cell 21:799–809
46. Larriba MJ, Valle N, Palmer HG, Ordonez-Moran P, Alvarez-Diaz S, Becker KF, Gamallo C, de Herreros AG, Gonzalez-Sancho JM, Munoz A (2007) The inhibition of Wnt/beta-catenin signalling by 1alpha, 25-dihydroxyvitamin D3 is abrogated by Snail1 in human colon cancer cells. Endocr Relat Cancer 14:141–151
47. Pendas-Franco N, Garcia JM, Pena C, Valle N, Palmer HG, Heinaniemi M, Carlberg C, Jimenez B, Bonilla F, Munoz A, Gonzalez-Sancho JM (2008) DICKKOPF-4 is induced by TCF/beta-catenin and upregulated in human colon cancer, promotes tumour cell invasion and angiogenesis and is repressed by 1alpha, 25-dihydroxyvitamin D3. Oncogene 27:4467–4477
48. Gumbiner BM (1996) Cell adhesion: the molecular basis of tissue architecture and morphogenesis. Cell 84:345–357
49. Wilson AJ, Velcich A, Arango D, Kurland AR, Shenoy SM, Pezo RC, Levsky JM, Singer RH, Augenlicht LH (2002) Novel detection and differential utilization of a c-myc transcriptional block in colon cancer chemoprevention. Cancer Res 62:6006–6010
50. Huang YC, Chen JY, Hung WC (2004) Vitamin D3 receptor/Sp1 complex is required for the induction of p27Kip1 expression by vitamin D3. Oncogene 23:4856–4861
51. Zinser GM, Suckow M, Welsh J (2005) Vitamin D receptor (VDR) ablation alters carcinogen-induced tumorigenesis in mammary gland, epidermis and lymphoid tissues. J Steroid Biochem Mol Biol 97:153–164
52. Welsh J (2004) Vitamin D and breast cancer: insights from animal models. Am J Clin Nutr 80:1721S–1724S
53. Escaleira MT, Brentani MM (1999) Vitamin D3 receptor (VDR) expression in HC-11 mammary cells: regulation by growth-modulatory agents, differentiation, and Ha-Ras transformation. Breast Cancer Res Treat 54:123–133
54. Pendas-Franco N, Gonzalez-Sancho JM, Suarez Y, Aguilera O, Steinmeyer A, Gamallo C, Berciano MT, Lafarga M, Munoz A (2007) Vitamin D regulates the phenotype of human breast cancer cells. Differentiation 75:193–207
55. Agadir A, Lazzaro G, Zheng Y, Zhang XK, Mehta R (1999) Resistance of HBL100 human breast epithelial cells to vitamin D action. Carcinogenesis 20:577–582
56. Valrance ME, Brunet AH, Welsh J (2007) Vitamin D receptor-dependent inhibition of mammary tumor growth by EB1089 and ultraviolet radiation in vivo. Endocrinology 148:4887–4894
57. Byrne B, Welsh J (2007) Identification of novel mediators of Vitamin D signaling and 1, 25(OH)$_2$D3 resistance in mammary cells. J Steroid Biochem Mol Biol 103:703–707
58. Wang Q, Lee D, Sysounthone V, Chandraratna RAS, Christakos S, Korah R, Wieder R (2001) 1, 25-dihydroxyvitamin D3 and retinoic acid analogues induce differentiation in breast cancer cells with function- and cell-specific additive effects. Breast Cancer Res Treat 67:157–168
59. Simboli-Campbell M, Narvaez CJ, van Weelden K, Tenniswood M, Welsh J (1997) Comparative effects of 1, 25(OH)$_2$D3 and EB1089 on cell cycle kinetics and apoptosis in MCF-7 breast cancer cells. Breast Cancer Res Treat 42:31–41
60. Byrne BM, Welsh J (2005) Altered thioredoxin subcellular localization and redox status in MCF-7 cells following 1, 25-dihydroxyvitamin D3 treatment. J Steroid Biochem Mol Biol 97:57–64
61. Li QP, Qi X, Pramanik R, Pohl NM, Loesch M, Chen G (2007) Stress-induced c-Jun-dependent Vitamin D receptor (VDR) activation dissects the non-classical VDR pathway from the classical VDR activity. J Biol Chem 282:1544–1551

62. McGaffin KR, Acktinson LE, Chrysogelos SA (2004) Growth and EGFR regulation in breast cancer cells by vitamin D and retinoid compounds. Breast Cancer Res Treat 86:55–73
63. McGaffin KR, Chrysogelos SA (2005) Identification and characterization of a response element in the EGFR promoter that mediates transcriptional repression by 1, 25-dihydroxyvitamin D3 in breast cancer cells. J Mol Endocrinol 35:117–133
64. Campbell MJ, Gombart AF, Kwok SH, Park S, Koeffler HP (2000) The anti-proliferative effects of 1alpha, 25(OH)$_2$D3 on breast and prostate cancer cells are associated with induction of BRCA1 gene expression. Oncogene 19:5091–5097
65. Lazzaro G, Agadir A, Qing W, Poria M, Mehta RR, Moriarty RM, Das Gupta TK, Zhang XK, Mehta RG (2000) Induction of differentiation by 1alpha-hydroxyvitamin D$_5$ in T47D human breast cancer cells and its interaction with vitamin D receptors. Eur J Cancer 36:780–786
66. Peng X, Jhaveri P, Hussain-Hakimjee EA, Mehta RG (2007) Overexpression of ER and VDR is not sufficient to make ER-negative MDA-MB231 breast cancer cells responsive to 1alpha-hydroxyvitamin D$_5$. Carcinogenesis 28:1000–2007
67. Campbell MJ, Reddy GS, Koeffler HP (1997) Vitamin D3 analogs and their 24-oxo metabolites equally inhibit clonal proliferation of a variety of cancer cells but have differing molecular effects. J Cell Biochem 66:413–425
68. Getzenberg RH, Light BW, Lapco PE, Konety BR, Nangia AK, Acierno JS, Dhir R, Shurin Z, Day RS, Trump DL, Johnson CS (1997) Vitamin D inhibition of prostate adenocarcinoma growth and metastasis in the Dunning rat prostate model system. Urology 50:999–1006
69. Zhao XY, Feldman D (2001) The role of vitamin D in prostate cancer. Steroids 66:293–300
70. Krishnan AV, Peehl DM, Feldman D (2003) Inhibition of prostate cancer growth by vitamin D: Regulation of target gene expression. J Cell Biochem 88:363–71
71. Beer TM, Garzotto M, Park B, Mori M, Myrthue A, Janeba N, Sauer D, Eilers K (2006) Effect of calcitriol on prostate-specific antigen in vitro and in humans. Clin Cancer Res 12:2812–2816
72. Reiter W (1999) The clinical value of the Enzymun-Test for total and free PSA – a multi-centre evaluation. Anticancer Res 19:5559–5562
73. Konety BR, Schwartz GG, Acierno JS Jr, Becich MJ, Getzenberg RH (1996) The role of vitamin D in normal prostate growth and differentiation. Cell Growth Differ 7:1563–1570
74. Floryk D, Tollaksen SL, Giometti CS, Huberman E (2004) Differentiation of human prostate cancer PC-3 cells induced by inhibitors of inosine 5′-monophosphate dehydrogenase. Cancer Res 64:9049–9056
75. Campbell MJ, Elstner E, Holden S, Uskokovic M, Koeffler HP (1997) Inhibition of proliferation of prostate cancer cells by a 19-nor-hexafluoride vitamin D3 analogue involves the induction of p21waf1, p27kip1 and E-cadherin. J Mol Endocrinol 19:15–27
76. Ortel B, Sharlin D, ÓDonnell D, Sinha AK, Maytin EV, Hasan T (2002) Differentiation enhances aminolevulinic acid-dependent photodynamic treatment of LNCaP prostate cancer cells. Br J Cancer 87:1321–1327
77. Krishnan AV, Shinghal R, Raghavachari N, Brooks JD, Peehl DM, Feldman D (2004) Analysis of vitamin D-regulated gene expression in LNCaP human prostate cancer cells using cDNA microarrays. Prostate 59:243–251
78. Boyle BJ, Zhao XY, Cohen P, Feldman D (2001) Insulin-like growth factor binding protein-3 mediates 1 alpha, 25-dihydroxyvitamin D3 growth inhibition in the LNCaP prostate cancer cell line through p21/WAF1. J Urol 165:1319–1324
79. Paralkar VM, Vail AL, Grasser WA, Brown TA, Xu H, Vukicevic S, Ke HZ, Qi H, Owen TA, Thompson DD (1998) Cloning and characterization of a novel member of the transforming growth factor-beta/bone morphogenetic protein family. J Biol Chem 273:13760–13767
80. Zhao XY, Ly LH, Peehl DM, Feldman D (1999) Induction of androgen receptor by 1alpha, 25-dihydroxyvitamin D3 and 9-cis retinoic acid in LNCaP human prostate cancer cells. Endocrinology 140:1205–1212
81. Zhao XY, Peehl DM, Navone NM, Feldman D (2000) 1alpha, 25-dihydroxyvitamin D3 inhibits prostate cancer cell growth by androgen-dependent and androgen-independent mechanisms. Endocrinology 141:2548–2556

82. Tuohimaa P, Lyakhovich A, Aksenov N, Pennanen P, Syvala H, Lou YR, Ahonen M, Hasan T, Pasanen P, Blauer M, Manninen T, Miettinen S, Vilja P, Ylikomi T (2001) Vitamin D and prostate cancer. J Steroid Biochem Mol Biol 76:125–134

83. Moreno J, Krishnan AV, Swami S, Nonn L, Peehl DM, Feldman D (2005) Regulation of prostaglandin metabolism by calcitriol attenuates growth stimulation in prostate cancer cells. Cancer Res 65:7917–7925

84. Cussenot O, Villette JM, Cochand-Priollet B, Berthon P (1998) Evaluation and clinical value of neuroendocrine differentiation in human prostatic tumors. Prostate Suppl 8:43–51

85. Xiao W, Hodge DR, Wang L, Yang X, Zhang X, Farrar WL (2004) Co-operative functions between nuclear factors NFkB and CCAT/enhancer-binding protein-beta (C/EBP-beta) regulate the IL-6 promoter in autocrine human prostate cancer cells. Prostate 61:354–370

86. Mori R, Xiong S, Wang Q, Tarabolous C, Shimada H, Panteris E, Danenberg KD, Danenberg PV, Pinski JK (2009) Gene profiling and pathway analysis of neuroendocrine transdifferentiated prostate cancer cells. Prostate 69:12–23

87. Marcinkowska E, Garay E, Gocek E, Chrobak A, Wang X, Studzinski GP (2006) Regulation of C/EBPbeta isoforms by MAPK pathways in HL60 cells induced to differentiate by 1, 25-dihydroxyvitamin D3. Exp Cell Res 312:2054–2065

88. Abban G, Yildirim NB, Jetten AM (2008) Regulation of the vitamin D receptor and cornifin beta expression in vaginal epithelium of the rats through vitamin D3. Eur J Histochem 52:107–114

89. White JH (2004) Profiling 1, 25-dihydroxyvitamin D3-regulated gene expression by microarray analysis. J Steroid Biochem Mol Biol 89–90:239–244

90. Tomita K, van Bokhoven A, van Leenders GJ, Ruijter ET, Jansen CF, Bussemakers MJ, Schalken JA (2000) Cadherin switching in human prostate cancer progression. Cancer Res 60:3650–3654

91. Bikle DD, Ng D, Tu CL, Oda Y, Xie Z (2001) Calcium- and vitamin D-regulated keratinocyte differentiation. Mol Cell Endocrinol 177:161–171

92. Xie Z, Bikle DD (2007) The recruitment of phosphatidylinositol 3-kinase to the E-cadherin-catenin complex at the plasma membrane is required for calcium-induced phospholipase C-gamma1 activation and human keratinocyte differentiation. J Biol Chem 282:8695–8703

93. Falasca M, Logan SK, Lehto VP, Baccante G, Lemmon MA, Schlessinger J (1998) Activation of phospholipase C gamma by PI 3-kinase-induced PH domain-mediated membrane targeting. Embo J 17:414–422

94. Xie Z, Singleton PA, Bourguignon LY, Bikle DD (2005) Calcium-induced human keratinocyte differentiation requires src- and fyn-mediated phosphatidylinositol 3-kinase-dependent activation of phospholipase C-gamma1. Mol Biol Cell 16:3236–3246

95. Bikle DD, Oda Y, Xie Z (2005) Vitamin D and skin cancer: a problem in gene regulation. J Steroid Biochem Mol Biol 97:83–91

96. Lehen'kyi V, Beck B, Polakowska R, Charveron M, Bordat P, Skryma R, Prevarskaya N (2007) TRPV6 is a Ca^{2+} entry channel essential for Ca^{2+}-induced differentiation of human keratinocytes. J Biol Chem 282:22582–22591

97. Johansen C, Iversen L, Ryborg A, Kragballe K (2000) 1alpha, 25-dihydroxyvitamin D3 induced differentiation of cultured human keratinocytes is accompanied by a PKC-independent regulation of AP-1 DNA binding activity. J Invest Dermatol 114:1174–1179

98. Takahashi H, Ibe M, Honma M, Ishida-Yamamoto A, Hashimoto Y, Iizuka H (2003) 1, 25-dihydroxyvitamin D3 increases human cystatin A expression by inhibiting the Raf-1/MEK1/ERK signaling pathway of keratinocytes. Arch Dermatol Res 295:80–87

99. Takahashi H, Honma M, Ishida-Yamamoto A, Namikawa K, Kiyama H, Iizuka H (2001) Expression of human cystatin A by keratinocytes is positively regulated via the Ras/MEKK1/MKK7/JNK signal transduction pathway but negatively regulated via the Ras/Raf-1/MEK1/ERK pathway. J Biol Chem 276:36632–36638

100. Lippens S, Kockx M, Denecker G, Knaapen M, Verheyen A, Christiaen R, Tschachler E, Vandenabeele P, Declercq W (2004) Vitamin D3 induces caspase-14 expression in psoriatic lesions and enhances caspase-14 processing in organotypic skin cultures. Am J Pathol 165:833–841

7 Induction of Differentiation in Cancer Cells by Vitamin D   169

101. Manggau M, Kim DS, Ruwisch L, Vogler R, Korting HC, Schafer-Korting M, Kleuser B (2001) 1alpha, 25-dihydroxyvitamin D3 protects human keratinocytes from apoptosis by the formation of sphingosine-1-phosphate. J Invest Dermatol 117:1241–1249

102. Lu J, Goldstein KM, Chen P, Huang S, Gelbert LM, Nagpal S (2005) Transcriptional profiling of keratinocytes reveals a vitamin D-regulated epidermal differentiation network. J Invest Dermatol 124:778–785

103. Feinberg MW, Wara AK, Cao Z, Lebedeva MA, Rosenbauer F, Iwasaki H, Hirai H, Katz JP, Haspel RL, Gray S, Akashi K, Segre J, Kaestner KH, Tenen DG, Jain MK (2007) The Kruppel-like factor KLF4 is a critical regulator of monocyte differentiation. Embo J 26:4138–4148

104. Autieri MV (2008) Kruppel-like factor 4: transcriptional regulator of proliferation, or inflammation, or differentiation, or all three? Circ Res 102:1455–1457

105. Alder JK, Georgantas RW 3rd, Hildreth RL, Kaplan IM, Morisot S, Yu X, McDevitt M, Civin CI (2008) Kruppel-like factor 4 is essential for inflammatory monocyte differentiation in vivo. J Immunol 180:5645–5652

106. Dai X, Sayama K, Shirakata Y, Tokumaru S, Yang L, Tohyama M, Hirakawa S, Hanakawa Y, Hashimoto K (2008) PPAR gamma is an important transcription factor in 1 alpha, 25-dihydroxyvitamin D3-induced involucrin expression. J Dermatol Sci 50:53–60

107. Bikle DD, Pillai S, Gee E (1991) Squamous carcinoma cell lines produce 1, 25 dihydroxyvitamin D, but fail to respond to its prodifferentiating effect. J Invest Dermatol 97:435–441

108. Solomon C, White JH, Kremer R (1999) Mitogen-activated protein kinase inhibits 1, 25-dihydroxyvitamin D3-dependent signal transduction by phosphorylating human retinoid X receptor alpha. J Clin Invest 103:1729–1735

109. McGuire TF, Trump DL, Johnson CS (2001) Vitamin D3-induced apoptosis of murine squamous cell carcinoma cells. Selective induction of caspase-dependent MEK cleavage and up-regulation of MEKK-1. J Biol Chem 276:26365–26373

110. Ma Y, Yu WD, Kong RX, Trump DL, Johnson CS (2006) Role of nongenomic activation of phosphatidylinositol 3-kinase/Akt and mitogen-activated protein kinase/extracellular signal-regulated kinase kinase/extracellular signal-regulated kinase 1/2 pathways in 1,25D3-mediated apoptosis in squamous cell carcinoma cells. Cancer Res 66:8131–8138

111. Xie Z, Bikle DD (1998) Differential regulation of vitamin D responsive elements in normal and transformed keratinocytes. J Invest Dermatol 110:730–733

112. Goltzman D, White J, Kremer R (2001) Studies of the effects of 1, 25-dihydroxyvitamin D on skeletal and calcium homeostasis and on inhibition of tumor cell growth. J Steroid Biochem Mol Biol 76:43–47

113. Oda Y, Sihlbom C, Chalkley RJ, Huang L, Rachez C, Chang CP, Burlingame AL, Freedman LP, Bikle DD (2003) Two distinct coactivators, DRIP/mediator and SRC/p160, are differentially involved in vitamin D receptor transactivation during keratinocyte differentiation. Mol Endocrinol 17:2329–2339

114. Tokuumi Y (1995) Correlation between the concentration of 1, 25 alpha dihydroxyvitamin D3 receptors and growth inhibition, and differentiation of human osteosarcoma cells induced by vitamin D3. Nippon Seikeigeka Gakkai Zasshi 69:181–190

115. Van Auken M, Buckley D, Ray R, Holick MF, Baran DT (1996) Effects of the vitamin D3 analog 1 alpha, 25-dihydroxyvitamin D3-3 beta-bromoacetate on rat osteosarcoma cells: comparison with 1 alpha, 25-dihydroxyvitamin D3. J Cell Biochem 63:302–310

116. Harris SA, Enger RJ, Riggs BL, Spelsberg TC (1995) Development and characterization of a conditionally immortalized human fetal osteoblastic cell line. J Bone Miner Res 10:178–86

117. Vertino AM, Bula CM, Chen JR, Almeida M, Han L, Bellido T, Kousteni S, Norman AW, Manolagas SC (2005) Nongenotropic, anti-apoptotic signaling of 1alpha, 25(OH)$_2$-vitamin D3 and analogs through the ligand binding domain of the vitamin D receptor in osteoblasts and osteocytes. Mediation by Src, phosphatidylinositol 3-, and JNK kinases. J Biol Chem 280:14130–14137

118. Cordero JB, Cozzolino M, Lu Y, Vidal M, Slatopolsky E, Stahl PD, Barbieri MA, Dusso A (2002) 1, 25-Dihydroxyvitamin D down-regulates cell membrane growth- and nuclear growth-promoting signals by the epidermal growth factor receptor. J Biol Chem 277:38965–38971

119. Gonzalez EA, Disthabanchong S, Kowalewski R, Martin KJ (2002) Mechanisms of the regulation of EGF receptor gene expression by calcitriol and parathyroid hormone in UMR 106–01 cells. Kidney Int 61:1627–1634

120. Schedlich LJ, Muthukaruppan A, ÓHan MK, Baxter RC (2007) Insulin-like growth factor binding protein-5 interacts with the vitamin D receptor and modulates the vitamin D response in osteoblasts. Mol Endocrinol 21:2378–2390

121. Lu X, Farmer P, Rubin J, Nanes MS (2004) Integration of the NFkappaB p65 sub-unit into the vitamin D receptor transcriptional complex: identification of p65 domains that inhibit 1, 25-dihydroxyvitamin D3-stimulated transcription. J Cell Biochem 92:833–848

122. Lemire JM, Adams JS, Sakai R, Jordan SC (1984) 1 alpha, 25-dihydroxyvitamin D3 suppresses proliferation and immunoglobulin production by normal human peripheral blood mononuclear cells. J Clin Invest 74:657–661

123. Tsoukas CD, Provvedini DM, Manolagas SC (1984) 1, 25-dihydroxyvitamin D3: a novel immunoregulatory hormone. Science 224:1438–1440

124. Abe E, Miyaura C, Sakagami H, Takeda M, Konno K, Yamazaki T, Yoshiki S, Suda T (1981) Differentiation of mouse myeloid leukemia cells induced by 1 alpha, 25-dihydroxyvitamin D3. Proc Natl Acad Sci USA 78:4990–4994

125. Tanaka H, Abe E, Miyaura C, Kuribayashi T, Konno K, Nishii Y, Suda T (1982) 1 alpha, 25-Dihydroxycholecalciferol and a human myeloid leukaemia cell line (HL-60). Biochem J 204:713–719

126. Studzinski GP, Bhandal AK, Brelvi ZS (1985) A system for monocytic differentiation of leukemic cells HL 60 by a short exposure to 1, 25-dihydroxycholecalciferol. Proc Soc Exp Biol Med 179:288–295

127. Munker R, Norman A, Koeffler HP (1986) Vitamin D compounds. Effect on clonal proliferation and differentiation of human myeloid cells. J Clin Invest 78:424–430

128. Brackman D, Lund-Johansen F, Aarskog D (1995) Expression of cell surface antigens during the differentiation of HL-60 cells induced by 1, 25-dihydroxyvitamin D3, retinoic acid and DMSO. Leuk Res 19:57–64

129. Uphoff CC, Drexler HG (2000) Biology of monocyte-specific esterase. Leuk Lymphoma 39:257–270

130. Steiner M, Priel I, Giat J, Levy J, Sharoni Y, Danilenko M (2001) Carnosic acid inhibits proliferation and augments differentiation of human leukemic cells induced by 1, 25-dihydroxyvitamin D3 and retinoic acid. Nutr Cancer 41:135–144

131. Sharabani H, Izumchenko E, Wang Q, Kreinin R, Steiner M, Barvish Z, Kafka M, Sharoni Y, Levy J, Uskokovic M, Studzinski GP, Danilenko M (2006) Cooperative antitumor effects of vitamin D3 derivatives and rosemary preparations in a mouse model of myeloid leukemia. Int J Cancer 118:3012–3021

132. Wright SD, Ramos RA, Tobias PS, Ulevitch RJ, Mathison JC (1990) CD14, a receptor for complexes of lipopolysaccharide (LPS) and LPS binding protein. Science 249:1431–1433

133. Hickstein DD, Ozols J, Williams SA, Baenziger JU, Locksley RM, Roth GJ (1987) Isolation and characterization of the receptor on human neutrophils that mediates cellular adherence. J Biol Chem 262:5576–5580

134. Studzinski GP, Rathod B, Wang QM, Rao J, Zhang F (1997) Uncoupling of cell cycle arrest from the expression of monocytic differentiation markers in HL60 cell variants. Exp Cell Res 232:376–387

135. Ji Y, Kutner A, Verstuyf A, Verlinden L, Studzinski GP (2002) Derivatives of vitamins D_2 and D3 activate three MAPK pathways and upregulate pRb expression in differentiating HL60 cells. Cell Cycle 1:410–415

136. Sikora E, Grassilli E, Bellesia E, Troiano L, Franceschi C (1993) Studies of the relationship between cell proliferation and cell death. III. AP-1 DNA-binding activity during concanavalin A-induced proliferation or dexamethasone-induced apoptosis of rat thymocytes. Biochem Biophys Res Commun 192:386–391

137. Hewison M, Dabrowski M, Vadher S, Faulkner L, Cockerill FJ, Brickell PM, ÓRiordan JL, Katz DR (1996) Antisense inhibition of vitamin D receptor expression induces apoptosis in monoblastoid U937 cells. J Immunol 156:4391–4400
138. Marcinkowska E, Chrobak A, Wiedlocha A (2001) Evading apoptosis by calcitriol-differentiated human leukemic HL-60 cells is not mediated by changes in CD95 receptor system but by increased sensitivity of these cells to insulin. Exp Cell Res 270:119–127
139. Zhang Y, Zhang J, Studzinski GP (2006) AKT pathway is activated by 1, 25-dihydroxyvitamin D3 and participates in its anti-apoptotic effect and cell cycle control in differentiating HL60 cells. Cell Cycle 5:447–451
140. Wang X, Studzinski GP (2006) The requirement for and changing composition of the activating protein-1 transcription factor during differentiation of human leukemia HL60 cells induced by 1, 25-dihydroxyvitamin D3. Cancer Res 66:4402–4409
141. Nakagawa K, Kurobe M, Konno K, Fujishima T, Takayama H, Okano T (2000) Structure-specific control of differentiation and apoptosis of human promyelocytic leukemia (HL-60) cells by A-ring diastereomers of 2-methyl-1alpha, 25-dihydroxyvitamin D3 and its 20-epimer. Biochem Pharmacol 60:1937–1947
142. Urahama N, Ito M, Sada A, Yakushijin K, Yamamoto K, Okamura A, Minagawa K, Hato A, Chihara K, Roeder RG, Matsui T (2005) The role of transcriptional coactivator TRAP220 in myelomonocytic differentiation. Genes Cells 10:1127–1137
143. Ikezoe T, Bandobashi K, Yang Y, Takeuchi S, Sekiguchi N, Sakai S, Koeffler HP, Taguchi H (2006) HIV-1 protease inhibitor ritonavir potentiates the effect of 1, 25-dihydroxyvitamin D3 to induce growth arrest and differentiation of human myeloid leukemia cells via down-regulation of CYP24. Leuk Res 30:1005–1011
144. Takahashi E, Nakagawa K, Suhara Y, Kittaka A, Nihei K, Konno K, Takayama H, Ozono K, Okano T (2006) Biological activities of 2alpha-substituted analogues of 1alpha, 25-dihydroxyvitamin D3 in transcriptional regulation and human promyelocytic leukemia (HL-60) cell proliferation and differentiation. Biol Pharm Bull 29:2246–2250
145. Suzuki T, Tazoe H, Taguchi K, Koyama Y, Ichikawa H, Hayakawa S, Munakata H, Isemura M (2006) DNA microarray analysis of changes in gene expression induced by 1, 25-dihydroxyvitamin D3 in human promyelocytic leukemia HL-60 cells. Biomed Res 27:99–109
146. Hughes PJ, Brown G (2006) 1Alpha, 25-dihydroxyvitamin D3-mediated stimulation of steroid sulphatase activity in myeloid leukaemic cell lines requires VDRnuc-mediated activation of the RAS/RAF/ERK-MAP kinase signalling pathway. J Cell Biochem 98:590–617
147. Hughes PJ, Lee JS, Reiner NE, Brown G (2008) The vitamin D receptor-mediated activation of phosphatidylinositol 3-kinase (PI3Kalpha) plays a role in the 1alpha, 25-dihydroxyvitamin D3-stimulated increase in steroid sulphatase activity in myeloid leukaemic cell lines. J Cell Biochem 103:1551–1572
148. Munker R, Kobayashi T, Elstner E, Norman AW, Uskokovic M, Zhang W, Andreeff M, Koeffler HP (1996) A new series of vitamin D analogs is highly active for clonal inhibition, differentiation, and induction of WAF1 in myeloid leukemia. Blood 88:2201–2209
149. Muto A, Kizaki M, Yamato K, Kawai Y, Kamata-Matsushita M, Ueno H, Ohguchi M, Nishihara T, Koeffler HP, Ikeda Y (1999) 1, 25-Dihydroxyvitamin D3 induces differentiation of a retinoic acid-resistant acute promyelocytic leukemia cell line (UF-1) associated with expression of p21(WAF1/CIP1) and p27(KIP1). Blood 93:2225–2233
150. Shiohara M, Uskokovic M, Hisatake J, Hisatake Y, Koike K, Komiyama A, Koeffler HP (2001) 24-Oxo metabolites of vitamin D3 analogues: disassociation of their prominent anti-leukemic effects from their lack of calcium modulation. Cancer Res 61:3361–3368
151. ÓKelly J, Uskokovic M, Lemp N, Vadgama J, Koeffler HP (2006) Novel Gemini-vitamin D3 analog inhibits tumor cell growth and modulates the Akt/mTOR signaling pathway. J Steroid Biochem Mol Biol 100:107–16
152. Wang X, Studzinski GP (2001) Activation of extracellular signal-regulated kinases (ERKs) defines the first phase of 1, 25-dihydroxyvitamin D3-induced differentiation of HL60 cells. J Cell Biochem 80:471–482

153. Marcinkowska E (2001) Evidence that activation of MEK1, 2/Erk1, 2 signal transduction pathway is necessary for calcitriol-induced differentiation of HL-60 cells. Anticancer Res 21:499–504
154. Wang Q, Wang X, Studzinski GP (2003) Jun N-terminal kinase pathway enhances signaling of monocytic differentiation of human leukemia cells induced by 1, 25-dihydroxyvitamin D3. J Cell Biochem 89:1087–1101
155. Wang Q, Salman H, Danilenko M, Studzinski GP (2005) Cooperation between antioxidants and 1, 25-dihydroxyvitamin D3 in induction of leukemia HL60 cell differentiation through the JNK/AP-1/Egr-1 pathway. J Cell Physiol 204:964–974
156. Szabo E, Preis LH, Birrer MJ (1994) Constitutive cJun expression induces partial macrophage differentiation in U-937 cells. Cell Growth Differ 5:439–446
157. Cuenda A, Rouse J, Doza YN, Meier R, Cohen P, Gallagher TF, Young PR, Lee JC (1995) SB 203580 is a specific inhibitor of a MAP kinase homologue which is stimulated by cellular stresses and interleukin-1. FEBS Lett 364:229–233
158. Wang X, Rao J, Studzinski GP (2000) Inhibition of p38 MAP kinase activity up-regulates multiple MAP kinase pathways and potentiates 1, 25-dihydroxyvitamin D3-induced differentiation of human leukemia HL60 cells. Exp Cell Res 258:425–437
159. Ishii Y, Sakai S, Honma Y (2001) Pyridinyl imidazole inhibitor SB203580 activates p44/42 mitogen-activated protein kinase and induces the differentiation of human myeloid leukemia cells. Leuk Res 25:813–820
160. Weinberg RA (2006) The Biology of Cancer. Garland Science/Taylor & Francis/LLC, London
161. Hmama Z, Nandan D, Sly L, Knutson KL, Herrera-Velit P, Reiner NE (1999) 1alpha, 25-dihydroxyvitamin D3-induced myeloid cell differentiation is regulated by a vitamin D receptor-phosphatidylinositol 3-kinase signaling complex. J Exp Med 190:1583–1594
162. Marcinkowska E, Kutner A (2002) Side-chain modified vitamin D analogs require activation of both PI 3-K and erk1, 2 signal transduction pathways to induce differentiation of human promyelocytic leukemia cells. Acta Biochim Pol 49:393–406
163. Marcinkowska E, Wiedlocha A (2003) Phosphatidylinositol-3 kinase-dependent activation of Akt does not correlate with either high mitogenicity or cell migration induced by FGF-1. Anticancer Res 23:4071–4077
164. Gocek E, Kielbinski M, Marcinkowska E (2007) Activation of intracellular signaling pathways is necessary for an increase in VDR expression and its nuclear translocation. FEBS Lett 581:1751–1757
165. Berry DM, Antochi R, Bhatia M, Meckling-Gill KA (1996) 1, 25-Dihydroxyvitamin D3 stimulates expression and translocation of protein kinase Calpha and Cdelta via a nongenomic mechanism and rapidly induces phosphorylation of a 33-kDa protein in acute promyelocytic NB4 cells. J Biol Chem 271:16090–16096
166. Zanello SB, Collins ED, Marinissen MJ, Norman AW, Boland RL (1997) Vitamin D receptor expression in chicken muscle tissue and cultured myoblasts. Horm Metab Res 29:231–6
167. Marcinkowska E (2001) A run for a membrane vitamin D receptor. Biol Signals Recept 10:341–349
168. Okazaki T, Bielawska A, Bell RM, Hannun YA (1990) Role of ceramide as a lipid mediator of 1 alpha, 25-dihydroxyvitamin D3-induced HL-60 cell differentiation. J Biol Chem 265:15823–15831
169. Wang X, Studzinski GP (2004) Kinase suppressor of RAS (KSR) amplifies the differentiation signal provided by low concentrations 1, 25-dihydroxyvitamin D3. J Cell Physiol 198:333–342
170. Zhang Y, Yao B, Delikat S, Bayoumy S, Lin XH, Basu S, McGinley M, Chan-Hui PY, Lichenstein H, Kolesnick R (1997) Kinase suppressor of Ras is ceramide-activated protein kinase. Cell 89:63–72
171. Wang X, Studzinski GP (2001) Phosphorylation of raf-1 by kinase suppressor of ras is inhibited by "MEK-specific" inhibitors PD 098059 and U0126 in differentiating HL60 cells. Exp Cell Res 268:294–300
172. Xing HR, Kolesnick R (2001) Kinase suppressor of Ras signals through Thr269 of c-Raf-1. J Biol Chem 276:9733–9741

7 Induction of Differentiation in Cancer Cells by Vitamin D   173

173. Michaud NR, Therrien M, Cacace A, Edsall LC, Spiegel S, Rubin GM, Morrison DK (1997) KSR stimulates Raf-1 activity in a kinase-independent manner. Proc Natl Acad Sci USA 94:12792–12796
174. Morrison DK (2001) KSR: a MAPK scaffold of the Ras pathway? J Cell Sci 114:1609–1612
175. Wang X, Wang TT, White JH, Studzinski GP (2006) Induction of kinase suppressor of RAS-1(KSR-1) gene by 1, alpha25-dihydroxyvitamin D3 in human leukemia HL60 cells through a vitamin D response element in the 5″-flanking region. Oncogene 25:7078–7085
176. Garay E, Donnelly R, Wang X, Studzinski GP (2007) Resistance to 1, 25D-induced differentiation in human acute myeloid leukemia HL60–40AF cells is associated with reduced transcriptional activity and nuclear localization of the vitamin D receptor. J Cell Physiol 213:816–825
177. Gocek E, Kielbinski M, Wylob P, Kutner A, Marcinkowska E (2008) Side-chain modified vitamin D analogs induce rapid accumulation of VDR in the cell nuclei proportionately to their differentiation-inducing potential. Steroids 73:1359–1366
178. Wang X, Studzinski GP (2006) Raf-1 signaling is required for the later stages of 1, 25-dihydroxyvitamin D3-induced differentiation of HL60 cells but is not mediated by the MEK/ERK module. J Cell Physiol 209:253–260
179. Jamshidi F, Zhang J, Harrison JS, Wang X, Studzinski GP (2008) Induction of differentiation of human leukemia cells by combinations of COX inhibitors and 1, 25-dihydroxyvitamin D3 involves Raf1 but not Erk 1/2 signaling. Cell Cycle 7:917–924
180. Porras A, Muszynski K, Rapp UR, Santos E (1994) Dissociation between activation of Raf-1 kinase and the 42-kDa mitogen-activated protein kinase/90-kDa S6 kinase (MAPK/RSK) cascade in the insulin/Ras pathway of adipocytic differentiation of 3T3L1 cells. J Biol Chem 269:12741–12748
181. Kuo WL, Abe M, Rhee J, Eves EM, McCarthy SA, Yan M, Templeton DJ, McMahon M, Rosner MR (1996) Raf, but not MEK or ERK, is sufficient for differentiation of hippocampal neuronal cells. Mol Cell Biol 16:1458–1470
182. Dhillon AS, Pollock C, Steen H, Shaw PE, Mischak H, Kolch W (2002) Cyclic AMP-dependent kinase regulates Raf-1 kinase mainly by phosphorylation of serine 259. Mol Cell Biol 22:3237–3246
183. Godyn JJ, Xu H, Zhang F, Kolla S, Studzinski GP (1994) A dual block to cell cycle progression in HL60 cells exposed to analogues of vitamin D3. Cell Prolif 27:37–46
184. Cicero S, Herrup K (2005) Cyclin-dependent kinase 5 is essential for neuronal cell cycle arrest and differentiation. J Neurosci 25:9658–9668
185. Chen F, Studzinski GP (1999) Cyclin-dependent kinase 5 activity enhances monocytic phenotype and cell cycle traverse in 1, 25-dihydroxyvitamin D3-treated HL60 cells. Exp Cell Res 249:422–428
186. Chen F, Rao J, Studzinski GP (2000) Specific association of increased cyclin-dependent kinase 5 expression with monocytic lineage of differentiation of human leukemia HL60 cells. J Leukoc Biol 67:559–566
187. Tsai LH, Delalle I, Caviness VS Jr, Chae T, Harlow E (1994) p35 is a neural-specific regulatory subunit of cyclin-dependent kinase 5. Nature 371:419–423
188. Chen F, Studzinski GP (2001) Expression of the neuronal cyclin-dependent kinase 5 activator p35Nck5a in human monocytic cells is associated with differentiation. Blood 97:3763–3767
189. Chen F, Wang Q, Wang X, Studzinski GP (2004) Up-regulation of Egr1 by 1, 25-dihydroxyvitamin D3 contributes to increased expression of p35 activator of cyclin-dependent kinase 5 and consequent onset of the terminal phase of HL60 cell differentiation. Cancer Res 64:5425–5433
190. Dudley DT, Pang L, Decker SJ, Bridges AJ, Saltiel AR (1995) A synthetic inhibitor of the mitogen-activated protein kinase cascade. Proc Natl Acad Sci US A 92:7686–7689
191. Foster JS, Fernando RI, Ishida N, Nakayama KI, Wimalasena J (2003) Estrogens down-regulate p27Kip1 in breast cancer cells through Skp2 and through nuclear export mediated by the ERK pathway. J Biol Chem 278:41355–41366

192. Park BH, Qiang L, Farmer SR (2004) Phosphorylation of C/EBPbeta at a consensus extracellular signal-regulated kinase/glycogen synthase kinase 3 site is required for the induction of adiponectin gene expression during the differentiation of mouse fibroblasts into adipocytes. Mol Cell Biol 24:8671–8680
193. Buck M, Poli V, Hunter T, Chojkier M (2001) C/EBPbeta phosphorylation by RSK creates a functional XEXD caspase inhibitory box critical for cell survival. Mol Cell 8:807–816
194. Pan Z, Hetherington CJ, Zhang DE (1999) CCAAT/enhancer-binding protein activates the CD14 promoter and mediates transforming growth factor beta signaling in monocyte development. J Biol Chem 274:23242–23248
195. Ji Y, Studzinski GP (2004) Retinoblastoma protein and CCAAT/enhancer-binding protein beta are required for 1, 25-dihydroxyvitamin D3-induced monocytic differentiation of HL60 cells. Cancer Res 64:370–377
196. Studzinski GP, Wang X, Ji Y, Wang Q, Zhang Y, Kutner A, Harrison JS (2005) The rationale for deltanoids in therapy for myeloid leukemia: role of KSR-MAPK-C/EBP pathway. J Steroid Biochem Mol Biol 97:47–55
197. Penna G, Adorini L (2000) 1 Alpha, 25-dihydroxyvitamin D3 inhibits differentiation, maturation, activation, and survival of dendritic cells leading to impaired alloreactive T cell activation. J Immunol 164:2405–2411
198. Zhu K, Glaser R, Mrowietz U (2002) Vitamin D3 and analogues modulate the expression of CSF-1 and its receptor in human dendritic cells. Biochem Biophys Res Commun 297:1211–1217
199. Garzon R, Pichiorri F, Palumbo T, Visentini M, Aqeilan R, Cimmino A, Wang H, Sun H, Volinia S, Alder H, Calin GA, Liu CG, Andreeff M, Croce CM (2007) MicroRNA gene expression during retinoic acid-induced differentiation of human acute promyelocytic leukemia. Oncogene 26:4148–4157
200. Sharma P, Veeranna SM, Amin ND, Sihag RK, Grant P, Ahn N, Kulkarni AB, Pant HC (2002) Phosphorylation of MEK1 by cdk5/p35 down-regulates the mitogen-activated protein kinase pathway. J Biol Chem 277:528–534

Chapter 8
Vitamin D and Cancer Chemoprevention

Sarah A. Mazzilli, Mary E. Reid, and Barbara A. Foster

Abstract Epidemiological evidence suggests that there is an inverse relationship between vitamin D and cancer. To investigate this relationship, a number of preclinical studies have been conducted focusing on the chemopreventive nature of dietary intake of vitamin D_3 and the administration of the active metabolite of vitamin D_3 (1,25(OH)2 D_3) and analogs of various forms of D_3. In addition, clinical studies have also have begun to assess the role of vitamin D in cancer prevention focusing on the administration of vitamin D_3. For colorectal and breast cancers, preclinical studies in a number of animal models suggest that diets containing sufficient levels of vitamin D_3 and calcium may slow tumor progression. Additionally, studies in examining the use of 1,25(OH)2 D_3 and/or analogs of vitamin D in animal models of colorectal, prostate, lung, and breast cancers further support the chemopreventive potential for vitamin D in these cancers, when administered during early stage disease. Overall the preclinical studies support the chemopreventive role of vitamin D in cancer, however further studies are required to understand how to effectively utilize vitamin D in the clinic. Clinical studies have not strongly supported the use of vitamin D as a chemopreventive agent potentially due to study design. However, new trails are currently on-going to further assess the clinical benefits of vitamin D in reducing cancer incidence and mortality.

Keywords Chemoprevention • Vitamin D_3 • 1,25(OH)2 D_3 • Colorectal cancer • Breast cancer • Prostate cancer and lung cancer

---

B.A. Foster (✉)
Pharmacology & Therapeutics,
Roswell Park Cancer Institute,
Elm & Carlton Streets, Buffalo, NY 14263, USA
e-mail: barbara.foster@roswellpark.org

D.L. Trump and C.S. Johnson (eds.), *Vitamin D and Cancer*,
DOI 10.1007/978-1-4419-7188-3_8, © Springer Science+Business Media, LLC 2011

8.1 Introduction

Current epidemiological data suggest that Vitamin D may act as a chemopreventive agent to reduce cancer incidence and mortality. The hypothesis that there is an inverse relationship between sunlight, vitamin D and cancer was first noted in 1937 by Peller, who proposed that those exposed to more sunlight had fewer internal cancers [1]. Following Nixon's declaration of war on cancer in 1970, maps were created to examine the geographical distribution of cancer mortality. It was these maps that lead the Garlands' to publish a study in 1980, proposing that vitamin D and calcium protected against colon cancer [2]. This study caught the attention of many, leading to further research into the potential preventive nature of vitamin D against cancer.

It has been proposed that the serum $25(OH)D_3$ levels needed to obtain a preventive effect is in the range of 30–60 ng/mL. However, a large percentage of individuals have serum $25(OH)D_3$ levels far below that level and are thought to be vitamin D deficient. Vitamin D deficiencies are associated with lifestyle and environmental factors that result in inadequate sun exposure and dietary intake of vitamin D. The amount of vitamin D that is able to be synthesized in the skin by UV-B exposure is determined by a number of variables including: geographic latitude, weather, time of day, pollution and use of sun protection lotions or sprays [3, 7]. In addition, campaigns to control sun exposure due to its association with skin cancer may also play a role in the growing number of individuals with low vitamin D levels [4]. In the US, dietary vitamin D is responsible for only a small percent of serum $25(OH)D_3$ levels, as the American diet does not include many foods that are naturally high in vitamin D. Although many foods in the American diet are supplemented with vitamin D, such as milk, yogurt, select juices and bread products, the contributions are less than that of multi-vitamins [5]. Currently the majority of multi-vitamins only contain 400 international units (IU) of Vitamin D_3. This is based on the 1997 recommendations of the Food and Nutrition Board (FNB) at the Institute of Medicine of The National Academies for adequate vitamin D_3 intake [6]. However, due to changes in lifestyle that have resulted in reduced sun exposure it is now being suggested that daily intake recommendations be increased to $\geq 1,000$ IU [8]. Increasing the recommended daily vitamin D intake particularly during the winter months may reduce the number of people deficient in vitamin D [4].

Current epidemiological studies have examined the relationship between serum $25(OH)D_3$ levels and both incidence and mortality rates in cancers of the colon, breast, prostate, ovarian, renal and lung. A recent review by Garland et al. stated that raising serum $25(OH)D_3$ levels to 40–60 ng/mL may prevent 58,000 new cases of breast cancer and 49,000 new cases of colorectal cancer, in addition to potentially reducing the mortality rates of individuals with colon, breast and prostate cancer by as much as 50% [8]. An inverse relationship between sunlight exposure and lung cancer incidence has been proposed by Mohr et al. after examining data from patients in 111 countries [9].

8   Vitamin D and Cancer Chemoprevention

In addition to the large body of epidemiological evidence that supports the use of vitamin D as a chemopreventive agent for cancer, *in vitro* tissue culture studies have elucidated many of the mechanisms by which vitamin D and its active metabolites act to inhibit the growth of malignant cells [10–13]. These studies have set the foundation for *in vivo* preclinical and clinical studies. As the epidemiological and molecular mechanisms of vitamin D have previously been discussed, this chapter will focus on the preclinical and clinical evidence that support the use of vitamin D as a chemopreventive agent across different cancer subtypes.

8.2   Pre-clinical Studies

Pre-clinical studies evaluating the chemopreventive effects of vitamin D are essential for establishing the rational for designing clinical trials. Here we summarize the pre-clinical studies that have been conducted in animal models of colon, prostate, breast, and lung and briefly discuss other tumor subtypes that are currently under investigation. These studies have examined not only differences in tumor growth associated with changes in dietary vitamin D levels but also through administration of the active metabolite of vitamin D $(1,25(OH)_2D_3)$ or vitamin D analogs. *In vitro* assays of $1,25(OH)_2D_3$ have demonstrated that $1,25(OH)_2D_3$ is responsible for the most potent anticancer effects as measured by proliferation, apoptosis, differentiation and cell cycle arrest [10–13]. However, administration of $1,25(OH)_2D_3$ can cause hypercalcemia and associated toxicities; therefore analogs of $1,25(OH)_2D_3$ are also being examined in efforts to maintain anticancer responses while lowering toxicity [14].

8.2.1   Colorectal Cancer

Studies by Lipkin et al. and Newmark et al. examined the effects of a diet high in fat and low in vitamin D and calcium (Western-style diet) on the induction of neoplasms in the colons of C57Bl/6 mice with and without the adenomatous polyposis coli (APC) gene mutations [15–17]. Comparisons are made between mice fed the Western-style diet containing 20% fat (corn oil)/g, 0.5 mg calcium (Ca)/g and 0.11 IU vitamin D_3/g/diet and the AIN-76A diet containing 5% fat/g, 5 mg Ca/g and 1 IU vitamin D_3/g/diet for various amounts of time ranging from several weeks to 2 years. These studies demonstrated that the C57Bl/6 mice on a Western diet developed hyperproliferative colon crypt hyperplasia while APC mice on a Western diet had an increased incidence of carcinoma with more invasive disease. However, mice that were fed diets high in vitamin D_3 and calcium did not develop lesions.

Tangpricha et al. performed additional studies that further examined the effect of low vitamin D_3 and calcium in the diet [18, 19]. In these studies, two cohorts of

MC-26 tumor bearing mice were used to examine the chemopreventive effects of dietary vitamin D_3 through the administration 0 IU (vitamin D deficient cohort) or 50,000 IU of vitamin D_3 (vitamin D_3 sufficient cohort) in the diet. When mice on the vitamin D_3 deficient diet had a mean serum 25(OH)D_3 level of ≤ 5 ng/mL, all mice in both cohorts received 10,000 MC-26 cells subcutaneously. The vitamin D sufficient cohort maintained a mean 25(OH)D_3 serum level of 26 ng/mL. This study also demonstrated that a diet deficient in vitamin D results in larger tumor volumes as compared to a vitamin D sufficient diet.

In addition to examining the effects of dietary intake of vitamin D_3, studies have been performed to examine the chemopreventive effects of the active metabolite of vitamin D, 1,25(OH)$_2D_3$ on the formation and the progression of colorectal cancers. Fichera et al. examined the chemopreventive effect of a 1,25(OH)$_2D_3$ analog (1α,25-dihydroxy-16,23(Z)-diene-26,27-hexafluoro-19-nor-cholecalciferol) (Ro26–2198) on colon carcinogenesis in A/J mice treated with the carcinogens azoxymethane (AOM) and Dextran sulfate sodium (DSS) [20]. The AOM/DSS carcinogen-induced mouse model recapitulates many aspects of human colon cancer via the induction of colitis that progresses into carcinoma. The AOM/DSS mice received Ro26–2198 (0.01 μg/kg body weight/day $\times$ 28 days) or vehicle by mini-osmotic pump 1 week prior to treatment with carcinogen. Subsequently, AOM/SDS mice are treated with a single dose of 5 mg/kg AOM and receive 3% DSS in their water for 7 days at the beginning of week 3. Mice receiving Ro26–2198 treatment had a delayed onset of colitis and those not treated with Ro26–2198 had several dysplastic foci. These results support a chemopreventive effect of vitamin D in colorectal cancer.

To further support that vitamin D has chemopreventive properties, Kallay et al. compared hyperproliferation and oxidative damage in mice with wild-type vitamin D receptor (VDR) (VDR$^{+/+}$), heterozygote VDR (VDR$^{+/-}$) and knock out of the VDR (VDR$^{-/-}$) mice [21]. An inverse relationship was found between VDR expression and proliferation in the colon, with the VDR $^{-/-}$ mice having a higher rate of proliferation. These studies demonstrate a significant role for vitamin D in modulating proliferation. Additionally it was demonstrated that there was an increase in the expression of 8-hydroxy-20-deoxyguanosine (8-OHdG), a marker of oxidative stress in the VDR$^{-/-}$ mice resulting in the VDR$^{-/-}$ mice having a higher amount of oxidative damage. Over all this study demonstrated that the genomic action of 1,25(OH)$_2D_3$ that is modulated by VDR expression is required to protect against the nutritional linked hyperproliferation and oxidative damage.

By and large, the preclinical studies examining the effect of vitamin D_3 in the diet and/or administration of the active metabolite or its analogs support the notion that there is chemopreventive potential for vitamin D in colorectal cancers. The rationale that vitamin D has chemopreventive potential is further reinforced by the demonstration that there is a relationship between VDR status and proliferation. More studies may be required to further elucidate the impact of dose and timing for clinical studies; however, it is plausible that vitamin D can alter the course of progression of colorectal cancers.

8   Vitamin D and Cancer Chemoprevention

8.2.2   Prostate Cancer

In a study by Banach-Petrosky et al. the chemopreventive activity of $1,25(OH)_2D_3$ was investigated in the Nkx3.1;Pten mutant model of prostate cancer [18]. This model has a loss of function of Nkx3.1 and the tumor suppressor Pten. With time, Nkx3.1;Pten mutant mice develop progressive prostate cancer with histopathology ranging from intraepithelial neoplasia (PIN) to adenocarcinoma. In this study, wild-type litter mates (Nkx3.1+/+;Pten+/+) were compared with mutant mice (Nkx3.1−/−;Pten+/−). An osmotic pump was used to give a continuous dose of $1,25(OH)_2D_3$ to the animals at a rate of 0.25 µL/h for a dose of 46 ng/kg/day. Treatment was initiated prior to the formation of cancerous lesions or after cancer had been established. Disease status was evaluated by histological evaluation. Treatment with $1,25(OH)_2D_3$ had no effect on the wild-type litter mates. However, mutant mice displayed a reduction in high-grade PIN lesions when treatment was administered prior to the onset of cancer. In the precancerous cohort treated with vehicle alone 0/8 animals had low-grade PIN and 8/8 had high-grade PIN with invasion compared with the $1,25(OH)_2D_3$ treatment cohort that had 10/12 with low-grade PIN and 2/12 with high-grade PIN. In contrast when $1,25(OH)_2D_3$ was administered to animals with established disease no preventive effect was observed. This study demonstrated a clear preventive effect of $1,25(OH)_2D_3$ when treatment was administered in the precancerous stage.

Perez-Stable et al. used the Gγ/T-15 model of prostate cancer to examine the chemopreventive activity of the $1,25(OH)2D_3$ analog EB1089 [22]. The Gγ/T-15 model is a transgenic mouse model that uses the human fetal the globin promoter to express SV40 T antigen. These mice rapidly develop prostate cancer with expression of the transgene detectable by 11 weeks of age and tumors present by 16 weeks of age. The transgene is expressed in the cells in the basal layer of the prostate. The tumors that develop are refractory to androgens and have a more neuroendocrine phenotype. Mice were administered EB1089 by IP injection three times a week at 0.5, 2, 3, 5, or 10 µg/kg starting at 14 weeks of age, 0.5, 2, 3, or 4 µg/kg at 12 weeks of age and 2 µg/kg at 9 weeks of age. Animals were palpated for tumors 3× week and tissues collected 21 days post detection of a palpable tumor or at 24 weeks of age. Prostatic tissues were collected and evaluated for the presence of tumors. In this model, no difference in the tumor incidence was observed at any treatment dose or timing of initiation of treatment. However, tumor size was decreased in animals treated with higher doses of EB1089 (>4 µg/kg) and the number of metastatic lesions was decreased in animals receiving the 10 µg/kg dose. The authors demonstrate that EB1089 inhibits growth in BPH-1 cells expressing SV40 T antigen. Thus, the expression of the transgene does not render the cells unresponsive to EB1089. The authors contend that the target cells in the model may be insensitive to vitamin D. This is supported by the low level of VDR expression in target cells that undergo carcinogenesis in this model. It should be noted that the most effective doses at inhibiting tumor size were not administered at the early time point (9 weeks of age). The doses given at the earliest time point were not effective

at inhibiting tumor growth and may have been too low to be effective. So while this study does not demonstrate a chemopreventive effect of the vitamin D analog, there are several factors that may contribute to the lack of response.

The studies in the Nkx3.1;Pten model indicate that vitamin D may elicit different responses when administered in early versus late stage disease, with the preventive benefits being greatest when $1,25(OH)_2D_3$ is administered prior to established disease. Elevated $1,25(OH)_2D_3$ levels prevented/reduced disease progression when administered early, while $1,25(OH)_2D_3$ had an antiproliferative effect on established disease. This study supports the use of vitamin D for the prevention of early stage prostate cancer. The studies using the Gγ/T-15 model did not demonstrate a preventive effect of the vitamin D analog, but did demonstrate an antiproliferative response for the primary tumor and the metastatic lesions. The lack of a preventive effect in the Gγ/T-15 model compared to Nkx3.1;Pten model could be due to several compounding factors. The target cells may not be able to respond to VDR as suggested by the lack of VDR; the dose of vitamin D used at the early stage disease was much lower than that used in the Nkx3.1;Pten model and was a dose that was not sufficient to reduce proliferation in the model; and the phenotype of the disease was different. The Nkx3.1;Pten model develops adenocarcinoma which retains androgen responsiveness while the Gγ/T-15 model develops prostate cancer from the basal cells that is hormone refractory and has a more neuroendocrine phenotype. These studies suggest that vitamin D may be more effective as a chemopreventive agent against adenocarcinoma and less effective against hormone refractory disease. However, both studies support a role for vitamin D to prevent/limit the growth of prostate cancer at both early and late stage disease.

8.2.3 Breast Cancer

To examine the chemopreventive effects of vitamin D in breast cancer similar methods seen in the examination of colorectal cancer were employed. Jacobson et al. used a carcinogen-induced rat model of breast cancer to examine the effects of a high fat combined with low vitamin D and low calcium diet on formation of tumors compared to a low fat and calcium and vitamin D sufficient diet [23]. The rat model utilized in this study was a female Sprague-Dawley rat treated with dimethylbenz(a) anthracene (DMBA). At 43 days of age the rats received a starter diet consisting of 7% sunflower seed oil (SF)/kcal, 1.5 mg calcium (Ca)/kcal, and 0.5 IU vitamin D_3 (D)/kcal. Subsequently, the rats were treated with 2.5 mg DMBA via gastric gavage and maintained on the starter diet for a second week. The rats were then split into six cohorts receiving: (I) 38.5% SF/kcal, 1.5 mg Ca/kcal and 0.5 IU D/kcal per diet; (II) 38.5% SF/kcal, 0.25 mg Ca/kcal and 0.05 IU D/kcal per diet; (III) 38.5% SF/kcal, 0.1 mg Ca/kcal & 0.05 IU D/kcal per diet; (IV) 7% SF/kcal, 1.5 mg Ca/kcal and 0.5 IU D/kcal per diet; (V) 7% SF/kcal, 0.25 mg Ca/kcal and 0.05 IU D/kcal per diet; and (VI) 7% SF/kcal, 0.1 mg Ca/kcal and 0.05 IU D/kcal per diet. The rats

8 Vitamin D and Cancer Chemoprevention

were maintained on these diets for 24 weeks. At the end of 24 weeks the cohorts that were fed a high fat and low calcium and low vitamin D diet (Cohorts II and III) had a greater number of mammary lesions and tumors as compared to the low fat groups (Cohorts IV, V, VI). There were more tumors in the low fat, low vitamin D, low calcium cohort (Cohort VI) compared to the other low fat diet cohorts (Cohorts IV, V). Thus, suggesting that the combination of low vitamin D and low calcium results in enhanced mammary tumorigenesis, especially when combined with a high fat diet.

Similar to Jacobson's study, Xue et al. examined the effects of low vitamin D and low calcium in combination with high fat diets on the number of terminal ducts in mouse mammary glands (NTDMG) in C57BL/6 J mice [24]. Terminal ducts are the cancer prone region in the mammary tissue of both mice and humans. An increase in the NTDMG increases the risk of developing mammary tumors; therefore this study used NTDMG to evaluate the effects of low vitamin D and low calcium in a high fat diet. Mice were split in to two cohorts, one received standard AIN-76A diet containing 12% Fat/kcal, 1.4 mg calcium (Ca)/kcal and 0.3 IU vitamin D_3 (D)/kcal/diet the other cohort received a high fat diet containing 40% Fat/kcal, 0.11 mg Ca/kcal and 0.05 IU D/kcal per diet. The NTDMG were determined at 8, 14 and 20 weeks of diet administration. The authors further demonstrated that a diet high in fat and low in both vitamin D and calcium resulted in an increased risk for tumorigenesis as demonstrated by the increased NTFMG in mice on the high fat diet for 14 and 20 week. Furthermore, the increased NTFMG was also associated with increased proliferation in animals on high fat and low vitamin D and low calcium diets.

In addition to examining the effects of vitamin D_3, Anzano et al. examined the chemopreventive nature of the $1,25(OH)_2D_3$ analog, $1\alpha,25$-dihydroxy-16-ene-23-yne-26,27-hexafluorocholecalciferol (Ro24–5531) in a carcinogen-induced rat model [14]. The carcinogen-induced N-nitroso-N-methyl urea (NMU) rat model used in this study forms invasive mammary adenocarcinoma in rats treated with a single intervenous injection of 15 mg/kg NMU [25]. The rats in this study were treated with NMU and a week following were put on a diet with either 0, 2.5, or 1.25 nmol Ro24–5531/kg. The rats were followed for 6 months and palpable tumors were measured. The rats on the both Ro24–5531 supplemented diets had similar effect, in that there was ~24% reduction in tumor incidence compared to the diet with no Ro24–5531. Thus, these studies demonstrate a chemopreventive effect of Ro24–5531 against breast cancer in this model.

Murillo et al. also examined the chemopreventive effects associated with a different vitamin D analog, 1α (OH)D_5 [26]. The authors sought to not only examine overall changes in incidence and multiplicity, but also examined stage specific effects of treating animals with 1α (OH)D_5. To examine tumor incidence and multiplicity Sprauge–Dawley rats were treated with an intervenous injection of 50 mg/kg of the carcinogen, N-methyl-N-nitrosourea (MNU) to induce mammary tumors. The stage specific studies were conducted in Sprauge–Dawley rats that were treated with 15 mg of dimethylbenz(a)anthracene (DMBA) in 1 mL of corn oil intragastrically. In the tumor incidence and multiplicity studies, the rats were given diets

containing either 0, 25, or 50 µg/kg 1α (OH)D$_5$/diet beginning 2 weeks prior to MNU injections and followed for an additional 120 days. These studies demonstrated that 1α (OH)D$_5$ reduced both tumor incidence by 26.7% and 33.4% and tumor multiplicity by 25% and 50% in the 25 and 50 µg/kg 1α (OH)D$_5$/diet groups respectively as compared to the untreated group. To examine the stage specific effects, three treatment cohorts were created all of which received 40 µg/kg 1α (OH)D$_5$ in the diet beginning at the following times: I. prior to initiation/promotion at 2 weeks prior to DMBA treatment; II. during initiation at the time of DMBA; or III. during promotion at 1 week post DMBA treatment. A fourth group was treated with rat chow containing no 1α (OH)D$_5$. The results of this study demonstrated no significant effects of 1α (OH)D$_5$ in the diet during the initiation phase; however, tumor incidence was reduced by 37.5% in rats receiving 1α (OH)D$_5$ during the promotional stage.

To further investigate vitamin D's chemopreventive effect, Zinser et al. conducted a study to examine the role of $1,25(OH)_2D_3$ on mammary gland development during puberty [27]. After demonstrating that VDR was present in a number of mouse mammary cell lines the authors compared the mammary development in VDR$^{-/-}$ mice on a high calcium diet to VDR$^{wt/wt}$ mice. The study showed that the mammary glands in the VDR$^{-/-}$ mice were heavier, had enhance ductal growth and increased secondary branch points and had an increased number of terminal end buds compared to the VDR$^{wt/wt}$ mice.

Overall the examination of vitamin D$_3$ in the diet and the administration of $1,25(OH)_2D_3$ or its analogs demonstrated a reduction in tumor incidence in a number of animal models. Additionally, the illustration that $1,25(OH)_2D_3$ is involved in the control of mammary gland growth and development furthers the rationale that vitamin D$_3$ may be useful in altering the course of mammary tumorigenesis. Together these studies provide rationale for continued exploration into the clinical application for vitamin D$_3$ as a chemopreventive agent to potentially reduce the incidence and mortality of breast cancer.

8.2.4 Lung Cancer

There are currently no published studies examining the preventive effects of dietary vitamin D$_3$ in lung cancer animal models. However, Mernitz et al. examined the active metabolite, $1,25(OH)_2D_3$ for its potential to inhibit lung carcinogenesis in the 4-(methynitrosamino)-1-(3-pyridyl)-1-butanone (NNK) carcinogen-induced animal model [28]. The mice in this study were fed a diet with 2.5, or 5 µg $1,25(OH)_2D_3$/kg diet (0.5 and 1.0 µg $1,25(OH)_2D_3$/kg body weight/day) for 20 weeks. A single administration of 100 mg/kg body weight of NNK was injected 3 weeks from the start of the $1,25(OH)_2D_3$ diet. Following 20 weeks on the diet, the lungs of treated animals were analyzed and lung lesions were quantified to determine the incidence and multiplicity of pulmonary surface tumors. Lung tumor incidence was reduced by 36% in the mice treated with 2.5 µg/kg diet of $1,25(OH)_2D_3$ and by 82% in those

treated with 5 μg/kg diet of $1,25(OH)_2D_3$. The tumor multiplicity was reduced by 85% in the 2.5 μg/kg diet of $1,25(OH)_2D_3$ cohort and by 98% in the 5 μg/kg diet of $1,25(OH)_2D_3$ cohort. Although there was a reduction in both the tumor incidence and multiplicity, both groups had toxicities associated with treatment including weight loss and kidney calcium deposits. However, the authors demonstrated that the toxicities were ameliorated when 9-*cis* retinoic acid (15 mg/kg diet) was added to the diet.

In addition to examining how vitamin D effects tumor progression, Nakagawa et al. published a study examining $1,25(OH)_2D_3$'s ability to prevent metastasis [29]. The ability of Lewis lung carcinoma (LCC) cells to metastasize to the lungs following intravenous injection were evaluated in syngenic vitamin D receptor (VDR) null mutant (VDR$^{-/-}$) mice and VDR wild-type (VDR$^{+/+}$) mice. VDR$^{-/-}$ mice on a normal diet (1.2% calcium, 0.6% phosphorus and 108 IU vitamin D_3/100 g diet) exhibit hypocalcemia and had extremely high serum levels of $1,25(OH)_2D_3$. The authors hypothesized that the high serum levels would inhibit metastatic growth of the LCC cells. To test this hypothesis the hypocalcemia, and/or hypervitaminosis D were corrected in the VDR$^{-/-}$ mice using dietary manipulations. The results demonstrated that the metastatic growth of LCC cells was greatly reduced in the VDR$^{-/-}$ in response to the high serum levels of $1,25(OH)_2D_3$, suggesting high serum levels of $1,25(OH)_2D_3$ may act to prevent lung metastasis. Although these studies do demonstrate that vitamin D has the potential to act as a chemopreventive agent in lung cancer, further studies are required to elucidate optimal formulation, dosing and administration methods to translate its usefulness in the clinic. In addition, more information about how vitamin D deficient versus sufficient diets effect the progression of lung cancer will also aid in elucidating the chemopreventive nature of vitamin D.

8.2.5   All Other Cancers

The chemopreventive nature of vitamin D is starting to be investigated in a number of other cancer subtypes that are less commonly studied, however few published studies exist to date. This section will summarize the one or two published studies that are available for melanoma, and retinoblastoma.

There is strong evidence that UV-B radiation that results in the synthesis of vitamin D in the skin also contributes to the development of melanoma [30]. Although UV-B exposure is a major contributor to vitamin D status, supplementation with dietary vitamin D is being suggested as a safer approach for populations at risk of melanoma. However, more recently studies are being conducted to examine if vitamin D may play a role in reducing some of the damaging effects associated with UV-B exposure, for example a study by Dixon et al. examined the use of a topical treatment of 0.33 μM $1,25(OH)_2D_3$ in Skh:HR1 mice [31]. The Skh:HR1 mice are hairless mice that form skin cancer following UV-B radiation. Mice were either untreated, treated with $1,25(OH)_2D_3$ pre and post UV-B exposure or treated

with $1,25(OH)_2D_3$ post UV-B exposure only. The treatment of $1,25(OH)_2D_3$ pre and post UV-B exposure appeared to reduce the amount of DNA damage as measured by the number of cyclobutane pyrimidine dimers (CPDs) formed. Further examination into the efficacy of vitamin D as a preventive agent is required, however the current study begins to shed a positive light for a preventive mechanism for melanoma.

Retinoblastoma is common in children that has relatively high cure rates [32]. However, although treatments are successful they are often destructive and may cause visual impairment, thus finding methods to prevent progression may reduce the impairments associated with treatment. A study by Albert et al. examines the potential for the use of $1,25(OH)_2D_3$ in the prevention of retinoblastoma in a transgenic mouse model of retinoblastoma [33]. The retinoblastoma transgenic mice express SV40 T antigen in the retina, which inactivates the p 105^{Rb} protein resulting in the formation of ocular tumors beginning at 14 weeks of age [34]. 8–10 week old mice were treated with either 0.05 mg or 0.025 mg of $1,25(OH)_2D_3$ five times a week for 5 weeks then sacrificed at 5 months age. In mice treated with high dose $1,25(OH)_2D_3$, 20% had no evidence of disease while the remaining had organ confined disease. In the mice treated with low dose $1,25(OH)_2D_3$, 13% had no evidence of disease. In contrast, all untreated mice formed bilateral disease that involved large invading tumors. This model clearly demonstrates that $1,25(OH)_2D_3$ inhibits the growth and local extension of retinoblastoma, suggesting a potential preventive role for vitamin D for retinoblastoma.

8.2.6 Summary

Overall the preclinical studies support a chemopreventive role for vitamin D in cancer. More studies are needed to understand the impact of vitamin D deficiency on cancer initiation and progression. Likewise, more information is needed to define sufficient levels of vitamin D necessary to achieve an anticancer benefit as well as defining the optimal levels for achieving the greatest anticancer benefit. A greater understanding of the molecular mechanism by which vitamin D exerts its chemopreventive effects and defining the molecular phenotype of the target cells that respond to vitamin chemoprevention therapy will enhance our ability to effectively utilize vitamin D and its analogs to reduce the incidence and impact of cancers in the clinic.

8.3 Clinical Prevention Trials

While the epidemiology of vitamin D status has been associated with lower cancer rates, supported by preclinical research, there have been only a few clinical prevention trials in humans that appear in the literature. These trials are included in the

8   Vitamin D and Cancer Chemoprevention

following discussion. While the doses have varied, all use the vitamin D in the form of vitamin D_3 (cholecalciferol).

8.3.1   Results from the Women's Health Initiative

The Women's Health Initiative (WHI) CaD trial was a double-blind, placebo control factorial trial of 36,282 postmenopausal women treated with 1,000 mg/day of calcium and 400 IU/day of vitamin D, in the form of vitamin D_3 [35–37]. The primary endpoint for this trial was hip fracture with colon cancer as an established secondary endpoint. Women were excluded from the trial if they had a predicted survival of less than 3 years, current use of corticosteroids, a history of renal stones, and regular intake of vitamin D supplements of 600 IU/day. Adherence between the treatment groups was comparable as was the frequency of sigmoidoscopy.

Colorectal Cancer (CRC) Endpoint: After an average of 7 years of follow-up, 168 were diagnosed with colon cancer in the treatment group and 154 were diagnosed in the placebo group. These results showed a non-significant difference in the rates of colorectal cancer, with a hazard ratio (HR) of 1.08 (95% confidence interval (CI) 0.86–1.34). The association between the treatments and colorectal cancer did not change when women with prior CRC were excluded.

Breast Cancer (BC) Endpoint: The hazard ratio (HR) for invasive breast cancer was 0.96 (95% CI=0.85–1.09) between the treatment group (n=528) and the placebo group (n=546), after an average of 7 years of follow-up. No significant interactions were noted with physical activity or BMI, both independent risk factors for breast cancer. Breast cancer histology and stage were not significant factors in breast cancer rates between the two treatment groups, however the tumors found in the treated patients were significantly smaller, with a p=.05. Mortality endpoint: another secondary endpoint evaluated in the WHI Calcium-Vitamin D trial was total mortality. A total of 744 deaths were reported in the treatment group versus 807 in the placebo group. The HR for total mortality was 0.91 (95% CI=0.83–1.01). Additional HRs calculated for stroke and cancer were consistently non-significant. Age and seasonality did not show significant interactions with the mortality outcome.

Toxicity and Safety of the Interventions: As reported in 2006, there was no significant association with the treatment groups and death (HR=0.91, 95% CI=0.83–1.01), total cancer risk (HR=0.98, 95% CI=0.91–1.05), cancer death or colorectal polyps (HR=0.94, 95% CI=0.83–1.01). The major toxicity of vitamin D supplementation is related to increased serum calcium and renal stones. There was a significant increase in the reports of kidney stones in women in the treatment versus placebo groups (HR=1.17, 95% CI=1.02–1.34, p=.02). While there were no obvious benefits of supplementation with calcium and vitamin D_3, there was an increase in reported toxicities, even at a dose that is now considered low by current supplementation levels.

There are limitations to this trial. The dose of vitamin D_3 may not have been large enough to substantively change vitamin D status, to the range seen in epidemiologic studies. Since the start of the WHI, other supplementation studies have used doses more in the range of 800–2,000 IU vitamin D/day. This is compounded by compliance issues. Particularly in the colorectal analysis, the authors suggest that since participants were not discouraged from taking additional calcium and vitamin D supplements and reported an increased in supplementation that was greater than the national average, the drop-in to the treatment would make differences in CRC between treatment groups more difficult to detect. Finally, the length of follow-up was in the range of 7 years. It may be that the length of treatment to change to course of CRC progression may be more in the range of 10–20 years. Designing a chemoprevention trial with treatment phases of 1–2 decades is not feasible. To continue to investigate vitamin D and calcium for CRC prevention, alternative designs could include forms of these agents that have greater expected effect sizes or the use of intermediate biomarkers as endpoints, such as colorectal adenomas or genetic changes, may be employed.

8.3.2   Colon Cancer Prevention

A study by Lappe et al. was a 4-year double-blind, placebo-controlled randomized trial of 1,179 postmenopausal women from rural Nebraska [38]. These women were randomized to either 1,400–1,500 mg/day of calcium alone, calcium plus 1,100 IU of vitamin D_3, or matched placebo. Subjects were 55 years or older, no history of cancer and capable of 4 years of participation. Mean age was 66.7 years, mean body mass index was 29 (±5.7) nmol/L and baseline 25(OH)D_3 was 71.8 (±20.3) nmol/L. The primary outcome was fracture but colon cancer was a formal secondary endpoint. The vitamin D intervention was sufficient to raise the serum 25(OH)D_3 to >80 nmol/L. Overall, both the calcium alone and calcium with vitamin D groups showed a significant difference (Chi-square=7.3; p value<0.03) and the calcium plus vitamin D group showed a relative risk (RR) of 0.40 (95% CI=0.20–0.82; $p=0.013$). When participants with cancers that developed within the first year were excluded, the calcium plus vitamin D group showed a relative risk (RR) of 0.23 (95% CI=0.09–0.60; $p=0.013$). This restriction of cases had no effect on the risk estimates for the calcium alone group, suggesting that the benefit of the vitamin D supplements on new cancers was attenuated by cancers that were most likely preclinical at the time of randomization.

Serious adverse events and toxicities: No serious adverse events were reported. There was no difference in the reports of renal calculi between treatment groups.

A study by Fedirko et al. reported a pilot, randomized, double-blind trial with a factorial design to evaluate the effects of vitamin D_3 (800 IU/day) and calcium (2 g/day) on biomarkers in the normal colorectal mucosa [39, 40]. Ninety-two women and men were recruited and treated for a period of 6 months. Several markers, including p21, MIB-1, hTERT, Bcl-2 and Bax were evaluated in the colonic crypts. Results showed

8 Vitamin D and Cancer Chemoprevention

that p21 significantly increased by 242% in the vitamin D alone and calcium alone groups. The combined treatment group showed a non-significant increase of 25%. There were no significant changes in MIB-1 or hTERT markers. Bax increased significantly by 56% along the full length of the crypts ($p=0.02$) in the vitamin D alone group and not significantly in the other two intervention groups. The changes in Bax expression were seen predominantly in the differentiation zone of the crypts while Bcl-2 did not change throughout treatment.

8.3.3 On-going Clinical Trials

Three additional studies, supported by NIH funding are currently underway. The VITAL trial (PI: JE Manson) will enroll 20,000 men and women and is designed to test in a randomized, factorial study the independent and combined effects of vitamin D (1,600 IU/D) and omega-3 fatty acids (1 g/D) on cancer and cardiovascular endpoints. Another study, which is using oral calcitriol (1,25 dihydroxycholecalciferol) to prevent the recurrence and progression of premalignant bronchoepithelial lesions in high risk lung cancer patients (PI: ME Reid). Finally, topical vitamin D is being tested for the prevention of basal cell carcinoma (BCC) in a pilot study of high risk skin cancer patients.

References

1. Mohr SB (2009) A brief history of vitamin D and cancer prevention. Ann Epidemiol 19:79–83
2. Garland CF, Garland FC (1980) Do sunlight and vitamin D reduce the likelihood of colon cancer? Int J Epidemiol 9:227–231
3. Holick MF (1994) McCollum Award Lecture, 1994: vitamin D – new horizons for the 21st century. Am J Clin Nutr 60:619–630
4. Ginde AA, Liu MC, Camargo CA Jr (2009) Demographic differences and trends of vitamin D insufficiency in the US population, 1988–2004. Arch Intern Med 169:626–632
5. Ovesen L, Brot C, Jakobsen J (2003) Food contents and biological activity of 25-hydroxyvitamin D: a vitamin D metabolite to be reckoned with? Ann Nutr Metab 47:107–113
6. Institute of Medicine FaNB (1997) Dietary reference intakes: calcium, phosphorus, magnesium, vitamin D, and fluoride. National Academy Press, Washington, DC
7. Pilz S, Tomaschitz A, Obermayer-Pietsch B, Dobnig H, Pieber TR (2009) Epidemiology of vitamin D insufficiency and cancer mortality. Anticancer Res 29:3699–3704
8. Garland CF, Gorham ED, Mohr SB, Garland FC (2009) Vitamin D for cancer prevention: global perspective. Ann Epidemiol 19:468–483
9. Mohr SB, Garland CF, Gorham ED, Grant WB, Garland FC (2008) Could ultraviolet B irradiance and vitamin D be associated with lower incidence rates of lung cancer? J Epidemiol Commun Health 62:69–74
10. Colston KW, Chander SK, Mackay AG, Coombes RC (1992) Effects of synthetic vitamin D analogues on breast cancer cell proliferation in vivo and in vitro. Biochem Pharmacol 44:693–702

11. Deeb KK, Trump DL, Johnson CS (2007) Vitamin D signaling pathways in cancer: potential for anticancer therapeutics. Nat Rev Cancer 7:684–700

12. Peehl DM, Skowronski RJ, Leung GK, Wong ST, Stamey TA, Feldman D (1994) Antiproliferative effects of 1, 25-dihydroxyvitamin D_3 on primary cultures of human prostatic cells. Cancer Res 54:805–810

13. Shabahang M, Buras RR, Davoodi F, Schumaker LM, Nauta RJ, Uskokovic MR, Brenner RV, Evans SR (1994) Growth inhibition of HT-29 human colon cancer cells by analogues of 1, 25-dihydroxyvitamin D_3. Cancer Res 54:4057–4064

14. Anzano MA, Smith JM, Uskokovic MR, Peer CW, Mullen LT, Letterio JJ, Welsh MC, Shrader MW, Logsdon DL, Driver CL et al (1994) 1 alpha, 25-Dihydroxy-16-ene-23-yne-26, 27-hexafluorocholecalciferol (Ro24–5531), a new deltanoid (vitamin D analogue) for prevention of breast cancer in the rat. Cancer Res 54:1653–1656

15. Lipkin M, Yang K, Edelmann W, Xue L, Fan K, Risio M, Newmark H, Kucherlapati R (1999) Preclinical mouse models for cancer chemoprevention studies. Ann N Y Acad Sci 889:14–19

16. Newmark HL, Yang K, Kurihara N, Fan K, Augenlicht LH, Lipkin M (2009) Western-style diet-induced colonic tumors and their modulation by calcium and vitamin D in C57Bl/6 mice: a preclinical model for human sporadic colon cancer. Carcinogenesis 30:88–92

17. Yang K, Edelmann W, Fan K, Lau K, Leung D, Newmark H, Kucherlapati R, Lipkin M (1998) Dietary modulation of carcinoma development in a mouse model for human familial adenomatous polyposis. Cancer Res 58:5713–5717

18. Banach-Petrosky W (2006) Vitamin D inhibits the formation of prostatic intraepithelial neoplasia in Nkx3.1;Pten mutant mice. Clinical Cancer Research 12;5895

19. Tangpricha V, Spina C, Yao M, Chen TC, Wolfe MM, Holick MF (2005) Vitamin D deficiency enhances the growth of MC-26 colon cancer xenografts in Balb/c mice. J Nutr 135:2350–2354

20. Fichera A, Little N, Dougherty U, Mustafi R, Cerda S, Li YC, Delgado J, Arora A, Campbell LK, Joseph L, Hart J, Noffsinger A, Bissonnette M (2007) A vitamin D analogue inhibits colonic carcinogenesis in the AOM/DSS model. J Surg Res 142:239–245

21. Kallay E, Bareis P, Bajna E, Kriwanek S, Bonner E, Toyokuni S, Cross HS (2002) Vitamin D receptor activity and prevention of colonic hyperproliferation and oxidative stress. Food Chem Toxicol 40:1191–1196

22. Perez-Stable CM, Schwartz GG, Farinas A, Finegold M, Binderup L, Howard GA, Roos BA (2002) The G gamma/T-15 transgenic mouse model of androgen-independent prostate cancer: target cells of carcinogenesis and the effect of the vitamin D analogue EB 1089. Cancer Epidemiol Biomarkers Prev 11:555–563

23. Jacobson EA, James KA, Newmark HL, Carroll KK (1989) Effects of dietary fat, calcium, and vitamin D on growth and mammary tumorigenesis induced by 7, 12-dimethylbenz(a) anthracene in female Sprague-Dawley rats. Cancer Res 49:6300–6303

24. Xue L, Newmark H, Yang K, Lipkin M (1996) Model of mouse mammary gland hyperproliferation and hyperplasia induced by a western-style diet. Nutr Cancer 26:281–287

25. Moon RC, Thompson HJ, Becci PJ, Grubbs CJ, Gander RJ, Newton DL, Smith JM, Phillips SL, Henderson WR, Mullen LT, Brown CC, Sporn MB (1979) N-(4-hydroxyphenyl) retinamide, a new retinoid for prevention of breast cancer in the rat. Cancer Res 39:1339–1346

26. Murillo G, Mehta RG (2005) Chemoprevention of chemically-induced mammary and colon carcinogenesis by 1alpha-hydroxyvitamin D5. J Steroid Biochem Mol Biol 97:129–136

27. Zinser G, Packman K, Welsh J (2002) Vitamin D(3) receptor ablation alters mammary gland morphogenesis. Development 129:3067–3076

28. Mernitz H, Smith DE, Wood RJ, Russell RM, Wang XD (2007) Inhibition of lung carcinogenesis by 1alpha, 25-dihydroxyvitamin D_3 and 9-cis retinoic acid in the A/J mouse model: evidence of retinoid mitigation of vitamin D toxicity. Int J Cancer 120:1402–1409

29. Nakagawa K, Kawaura A, Kato S, Takeda E, Okano T (2005) 1 alpha, 25-Dihydroxyvitamin D(3) is a preventive factor in the metastasis of lung cancer. Carcinogenesis 26:429–440

8 Vitamin D and Cancer Chemoprevention

30. Egan KM (2009) Vitamin D and melanoma. Ann Epidemiol 19:455–461
31. Dixon KM, Deo SS, Wong G, Slater M, Norman AW, Bishop JE, Posner GH, Ishizuka S, Halliday GM, Reeve VE, Mason RS (2005) Skin cancer prevention: a possible role of 1, 25dihydroxyvitamin D_3 and its analogs. J Steroid Biochem Mol Biol 97:137–143
32. Mastrangelo D, De Francesco S, Di Leonardo A, Lentini L, Hadjistilianou T (2007) Does the evidence matter in medicine? The retinoblastoma paradigm. Int J Cancer 121:2501–2505
33. Albert DM, Marcus DM, Gallo JP, ÓBrien JM (1992) The antineoplastic effect of vitamin D in transgenic mice with retinoblastoma. Invest Ophthalmol Vis Sci 33:2354–2364
34. Windle JJ, Albert DM, ÓBrien JM, Marcus DM, Disteche CM, Bernards R, Mellon PL (1990) Retinoblastoma in transgenic mice. Nature 343:665–669
35. LaCroix AZ, Kotchen J, Anderson G, Brzyski R, Cauley JA, Cummings SR, Gass M, Johnson KC, Ko M, Larson J, Manson JE, Stefanick ML, Wactawski-Wende J (2009) Calcium plus vitamin D supplementation and mortality in postmenopausal women: the Women's Health Initiative calcium-vitamin D randomized controlled trial. J Gerontol A Biol Sci Med Sci 64:559–567
36. Wactawski-Wende J, Kotchen JM, Anderson GL, Assaf AR, Brunner RL, ÓSullivan MJ, Margolis KL, Ockene JK, Phillips L, Pottern L, Prentice RL, Robbins J, Rohan TE, Sarto GE, Sharma S, Stefanick ML, Van Horn L, Wallace RB, Whitlock E, Bassford T, Beresford SA, Black HR, Bonds DE, Brzyski RG, Caan B, Chlebowski RT, Cochrane B, Garland C, Gass M, Hays J, Heiss G, Hendrix SL, Howard BV, Hsia J, Hubbell FA, Jackson RD, Johnson KC, Judd H, Kooperberg CL, Kuller LH, LaCroix AZ, Lane DS, Langer RD, Lasser NL, Lewis CE, Limacher MC, Manson JE (2006) Calcium plus vitamin D supplementation and the risk of colorectal cancer. N Engl J Med 354:684–696
37. Chlebowski RT, Johnson KC, Kooperberg C, Pettinger M, Wactawski-Wende J, Rohan T, Rossouw J, Lane D, ÓSullivan MJ, Yasmeen S, Hiatt RA, Shikany JM, Vitolins M, Khandekar J, Hubbell FA (2008) Calcium plus vitamin D supplementation and the risk of breast cancer. J Natl Cancer Inst 100:1581–1591
38. Lappe JM, Travers-Gustafson D, Davies KM, Recker RR, Heaney RP (2007) Vitamin D and calcium supplementation reduces cancer risk: results of a randomized trial. Am J Clin Nutr 85:1586–1591
39. Fedirko V, Bostick RM, Flanders WD, Long Q, Shaukat A, Rutherford RE, Daniel CR, Cohen V, Dash C (2009) Effects of vitamin D and calcium supplementation on markers of apoptosis in normal colon mucosa: a randomized, double-blind, placebo-controlled clinical trial. Cancer Prev Res (Phila Pa) 2:213–223
40. Fedirko V, Bostick RM, Flanders WD, Long Q, Sidelnikov E, Shaukat A, Daniel CR, Rutherford RE, Woodard JJ (2009) Effects of vitamin d and calcium on proliferation and differentiation in normal colon mucosa: a randomized clinical trial. Cancer Epidemiol Biomarkers Prev 18:2933–2941

Chapter 9
Molecular Biology of Vitamin D Metabolism and Skin Cancer

Florence S.G. Cheung and Juergen K.V. Reichardt

Abstract It is well known that UV exposure is essential for subcutaneous vitamin D synthesis, which is important in maintaining mineral and bone homeostasis. In this chapter, we discuss findings in recent epidemiologic, in vitro and in vivo studies that suggest vitamin D has an additional role, skin cancer prevention. With accumulating evidence on the neoplastic effects of vitamin D, studies on vitamin D analogs have shown promising results. Thus we are currently faced with the dilemma in seeking a fine balance between the amount of sun exposure needed to produce sufficient vitamin D to maintain its function in bone health and possible anticancer effects, while avoiding excessive exposure that can increase the risk of skin cancer development. This is further complicated by the fact that the amount of vitamin D synthesized from UV exposure is influenced by age, culture, and existing medical conditions of the individual. The designing of vitamin D analogs and appropriate recommendations on sun exposure requires further understanding of the vitamin D pathway and its actions, as well as any genetic factors that may influence the therapeutic outcome.

Keywords Skin cancer • Solar UV radiation • Vitamin D • Epidemiology • Prevention • Vitamin D receptor • 1,25-dihydroxyvitamin D • Keratinocytes • Differentiation • Photoprotection • Vitamin D analogs

Abbreviations

Aa	Amino acids
AC	Adenylyl cyclase

J.K.V. Reichardt (✉)
Plunket Chair of Molecular Biology (Medicine),
Bosch Institute, The University of Sydney,
Medical Foundation Building (K25), 92–94 Parramatta Road,
Camperdown, NSW 2006, Australia
and
School of Pharmacy and Molecular Sciences, James Cook University,
Townsville, Qld 4811, Australia
e-mail: jreichardt@med.usyd.edu.au

BCC	Basal cell carcinoma
Ca^{2+}	Calcium
CaR	Calcium receptor
cAMP	Cyclic AMP
CBP	CREB binding protein
CPDs	Cyclobutane pyrimidine dimmers
DAG	Diacylglycerol
7-DHC	7-Dehydrocholesterol
DRIP	Vitamin D receptor-interacting protein
ERK	Extracellular signal-regulated kinase
HATs	Histone acetyltransferases
HDACs	Histone deacetylases
IP_3	Inositol 1,4,5-triphosphate
JNK	c-Jun NH2-terminal kinase
MAPK	Mitogen-activated protein kinase
MARRS	Membrane associated rapid response steroid-binding
MEK	MAPK/ERK kinase
MM	Malignant melanoma
NCoA62-SKIP	Nuclear co-activator 62 kDa-SKI-interacting protein
NCoR	Nuclear receptor co-repressor
nVDRE	Negative vitamin D response element
NO	Nitric oxide
1a-OHase	1a-Hydroxylase
24-OHase	24-Hydroxylase
25-OHase	25-Hydroxylases
$25OHD_3$	25-HydroxyvitaminD$_3$
$1,25(OH)_2D_3$	1,25Dihydroxyvitamin D$_3$
OPG	Osteoprotegrin
PI3K	Phosphatidylinositide 3-kinase
PIP_2	Phosphatidylinositol 4,5-bisphosphate
PKA	Protein kinase A
PKC	Protein kinase C
PLC	Phospholipase C
PTH	Parathyroid hormone
Pol	Polymerase
RANKL	Receptor activator of NF-kB ligand
RXR	Retinoid X receptor
SCC	Squamous cell carcinomas
SMRT	Silencing mediator for retinoid and thyroid hormone receptors
SRC	Steroid receptor co-activators
TD	Thymine dimmers
TF2B	Transcription factor 2B
TRPV	Transient receptor potential vanilliod
UVA	Ultraviolet A
UVB	Ultraviolet B

9 Molecular Biology of Vitamin D Metabolism and Skin Cancer   193

UVC   Ultraviolet C
UVR   UV radiation
VDIR   VDR-interacting repressor
VDR   Vitamin D receptor
VDRE   Vitamin D response element

9.1 Introduction

Incidence and mortality rates of skin cancer in most developed countries have experienced a steady increase over the past 25 years [57]. In the past few decades, the 5 year survival has improved to over 90% in some developed countries including the United States, Sweden and Australia [57], but survival rates in many nations remain low [36]. Therefore, it is important to understand the cellular and molecular events involved in skin cancer pathogenesis to provide new approaches to reduce the incidence and mortality of skin cancer.

It is long known that ultraviolet B (UVB) (280–315 nm) irradiation is a major cause of skin cancer. Cyclobutane pyrimidine dimers (CPDs) constitute the major DNA photoproducts upon exposure to UVB light [140]. If not repaired, these can become initiating mutations in skin cancer [140] or if the DNA damage is irreparable, the cell may undergo apoptosis [144]. Skin chronically exposed to UV radiation (UVR) may also suffer irreversible suppression of cell-mediated immunity promoting skin cancer outgrowth [45].

UVR is also essential in the synthesis of pre-vitamin D from 7-dehydrocholesterol (7-DHC) in the skin. Pre-vitamin D_3 then undergoes further hydroxylation reactions in the liver and kidneys to form 25-hydroxyvitamin D_3 (25OHD$_3$) and 1,25-dihydroxyvitamin D_3 (1,25(OH)$_2$D$_3$) respectively [69]. The 1,25(OH)$_2$D$_3$ formed from the kidney is essential in maintaining mineral and bone homeostasis (Fig. 9.1a). Vitamin D deficiency can arise in older individuals as a result of age related factors including reduced capacity to produce vitamin D, reduced sunlight exposure, lower vitamin D intake and decline in renal function [116].

Interestingly, epidemiologic studies have shown seasonal melanoma fatality patterns, with fatality rates lower during summer than in winter [17]. In addition, fatality from melanoma is lower in people with a history of higher sun exposure than in people with low sun exposure [9]. Together with the knowledge that UV exposure is important for vitamin D synthesis, this raised the idea of a possible relationship between melanoma and vitamin D. The effect of sun exposure on vitamin D status appears to be important in protecting against a number of non-cutaneous cancers, including cancers of the breast, colon and prostate and non-Hodgkin lymphoma [17, 55, 56, 87, 101].

Much of the knowledge of the connection between vitamin D and the epidemiological data on cancer have been contributed by investigations into the role of vitamin D in extra-renal tissues, initiated by the discovery of the vitamin D receptor (VDR) in breast cancer cells [44]. Other experiments have also demonstrated the

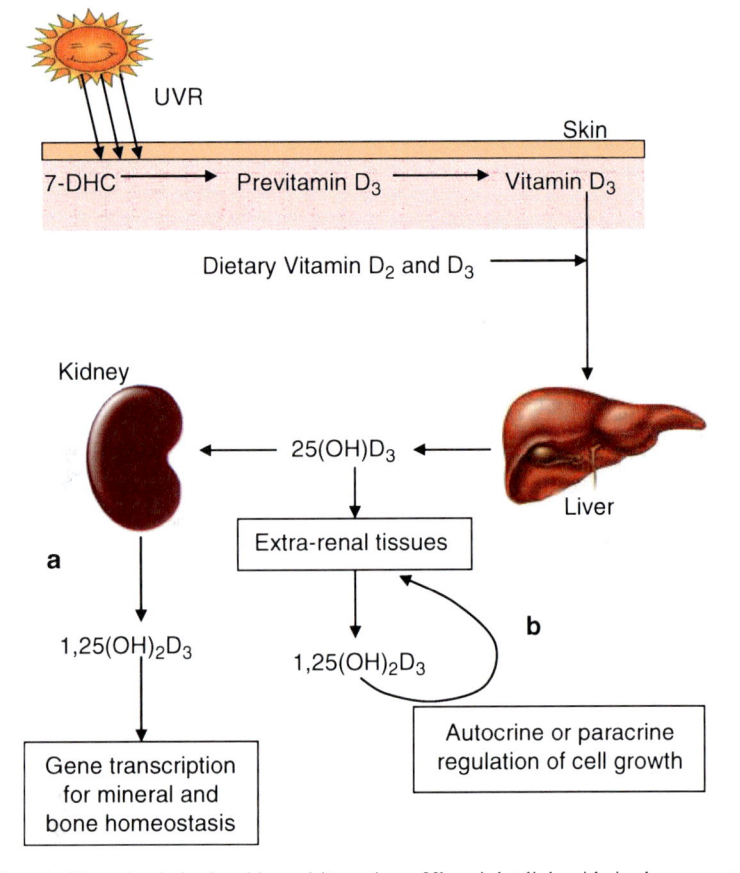

Fig. 9.1 Vitamin D synthesis in the skin and its actions. Ultraviolet light aids in the conversion of 7-DHC to previtamin D_3 which thermally isomerizes to vitamin D_3. Both synthesized and ingested vitamin D are hydroxylated in the liver to form $25(OH)D_3$ and the kidneys (**a**) or extra renal tissues (**b**) to form $1,25(OH)_2D_3$ which acts on target cells to elicit a biological response

presence of VDR in various cancer cell lines [51]. More importantly, growth inhibitory effects of $1,25(OH)_2D_3$ have been demonstrated in breast cancer [28], prostate [106, 121], colon [31, 141] and melanoma cell lines in culture [29]. There is also accumulating evidence on $1,25(OH)_2D_3$ having anti-proliferative, pro-differentiation and photoprotective properties in keratinocytes which makes it potentially very attractive as an anti-cancer agent [38].

Therefore, UVR has a dual effect on skin cancer development and vitamin D synthesis that is important in maintaining the health of the body as well as preventing cancer development. Considering that solar dependant vitamin D synthesis contributes to 90% of the body's vitamin D requirement [133], when determining vitamin D recommendation levels, we face the dilemma in seeking a fine balance between the amount of sun exposure to produce sufficient vitamin D while avoiding excessive exposure that can increase the risk of skin cancer development.

9.2 The Induction of Skin Cancer by UV Radiation

Solar UV spectrum is composed of ultraviolet A (UVA) (315–400 nm), UVB (280–315 nm) and ultraviolet C (UVC) (<280 nm). The harmful short wavelength UVC and most of the UVB (up to 310 nm) is absorbed by the ozone layer and is therefore not physiologically significant. On the other hand, UVA reaches the earth's surface and up to 50% of UVA energy penetrates to the dermis. The effects of UVA include DNA oxidative damage, solar elastosis and skin ageing [130]. The remaining UVB is the most energetic component of the solar UV spectrum and is almost completely absorbed by the outer layer of the skin, the epidermis [130].

DNA is the predominant chromophore in the epidermis and absorbs most strongly at 260 nm with decreasing absorption from the UVB to UVA spectra. The major type of damage to DNA upon UVB absorption is the cycloaddition of the C5–C6 double bonds of adjacent pyrimidines to cause the formation of cyclobutane pyrimidine dimmers (CPD), e.g., thymine dimers (TD) [26, 32, 132]. If not repaired, these can become initiating mutations in skin cancer [140] or if the DNA damage is irreparable, the cell may undergo apoptosis [144] which is the situation with sun burn cells. If the DNA damage escapes the gene repair system and is in a gene involved in DNA repair, apoptosis, proliferation or cell cycle control, tumor growth can arise [130]. In fact, in squamous cell carcinoma (SCC) and basal cell carcinoma (BCC), the p53 gene, an essential transcription factor regulating cell cycle control and apoptosis, bears point mutations with the features of UVB-induced point mutations. These UVB signature mutations are C to T or CC to TT transitions that are associated with di-pyrimidinic sites [19]. In addition, skin cells chronically exposed to UVR may also suffer irreversible cell-mediated immunity suppression [45], which may generate immune tolerance against immunogenic skin tumors and exacerbate cancer outgrowth.

9.3 The Vitamin D_3 Metabolic Pathway and Its Actions

9.3.1 UV Radiation Induced Vitamin D_3 Synthesis in the Skin

Apart from the genotoxic effect of UVR, UVR also plays an important role in the synthesis of vitamin D. The term vitamin D generally refers to two molecules, vitamin D_2 and D_3. Vitamin D is obtained through two sources. A small proportion of vitamin D_2 and D_3 can be obtained from the diet (Fig. 9.1). Vitamin D_3 can be obtained from fatty fish or fish liver oil [70] while vitamin D_2 (ergocalciferol), is the form of vitamin D produced by plants through the irradiation of the plant steroid, ergosterol [116]. Majority of the vitamin D_3 required is synthesized subcutaneously. The synthesis of vitamin D_3 in human and animals begins via a photolysis reaction in which ultraviolet light converts 7-dehydrocholesterol (7-DHC) to previtamin D_3, which then isomerizes to vitamin D_3 (cholecalciferol). Both vitamin D_2 and vitamin

D_3, either ingested or synthesized enter the liver where they are metabolized by liver mitochondrial and microsomal 25-hydroxylase (25-OHase), the gene product of *CYP27A1*, to $25OHD_3$. This is the main circulating form of vitamin D_3 [116]. Further hydroxylations occur in the proximal tubules of the kidneys where $1,25(OH)_2D_3$ (calcitriol) is produced via kidney 1α-hydroxylase (1α-OHase), the gene product of *CYP27B1* (Fig. 9.1a). It has been also shown that the entire pathway to forming $1,25(OH)_2D_3$ from 7-dehydrocholesterol can occur in the human skin [93, 104], demonstrating the importance of the human skin in the synthesis of vitamin D.

The $1,25(OH)_2D_3$ produced in the kidney is then transported in the blood and is mostly bound to the vitamin D binding protein with only a very small amount of its free form being able to elicit a biological response [116].

Serum level of $1,25(OH)_2D_3$ is regulated by 25-hydrodxyvitamin D 24-hydroxylase (24-OHase) which is encoded by the *CYP24A1* gene. The *CYP24A1* gene is strongly induced by $1,25(OH)_2D_3$ [118]. With adequate levels of $1,25(OH)_2D_3$ the 24-OHase acts on $25OHD_3$ and $1,25(OH)_2D_3$ to form the inactive metabolites $24,25(OH)_2D_3$ and $1\alpha,24,25(OH)_2D_3$. The expression of *CYP27B1* is also down regulated by its own gene product $1,25(OH)_2D_3$ [109]. Thus by inducing *CYP24A1* and down regulating *CYP27B1*, $1,25(OH)_2D_3$ possesses its own feedback regulation via these two genes.

9.3.2   *Genomic Actions of 1,25-Dihydroxyvitamin D_3*

The genomic actions of $1,25(OH)_2D_3$ is depicted in Fig. 9.2. This is initiated by the uptake of free $1,25(OH)_2D_3$ into the target cells. In the cell, $1,25(OH)_2D_3$ can bind to the vitamin D receptor (VDR). The VDR belongs to the nuclear hormone receptor superfamily and is a ligand activated transcription factor that recognize and binds to distinctive sequences, known as vitamin D response elements (VDRE), located in the promoter of vitamin D responsive genes [38]. VDREs typically contain two hexanucleotide repeats separated by varying number of nucleotides of any base, for example GGTTCA-NNN-GGTTCA [154]. The binding of $1,25(OH)_2D_3$ with VDR induces a significant conformation change that is essential for a number of downstream events including phosphorylation, dimerisation with the retinoid X receptor (RXR) and most importantly, the recruitment of co-activators and transcription machinery to the promoter, reviewed in [38].

In the absence of a ligand, the VDR is only loosely bound to the RXR. Binding of the $1,25(OH)_2D_3$ to VDR induces conformation changes to expose the surfaces for co- activating factor binding and high affinity dimerization with the RXR [63]. The heterodimerisation with the RXR allows the VDR to bind with higher affinity to the promoter of target genes. This high affinity interaction is achieved by binding of the VDR and the RXR to the 3' and 5' strand of the VDRE sequence respectively [89].

DNA in the non-active state is coiled tightly around the histones to form nucleosomes. The initiation of replication and transcription requires the acetylation of lysines in the N-terminal tails of histone by histone acetyltransferases (HATs) to "loosen" the nucleosome core to allow access of DNA binding sites to proteins mediating transcription. This acetylation can be reversed by the removal of acetyl groups by histone

9   Molecular Biology of Vitamin D Metabolism and Skin Cancer   197

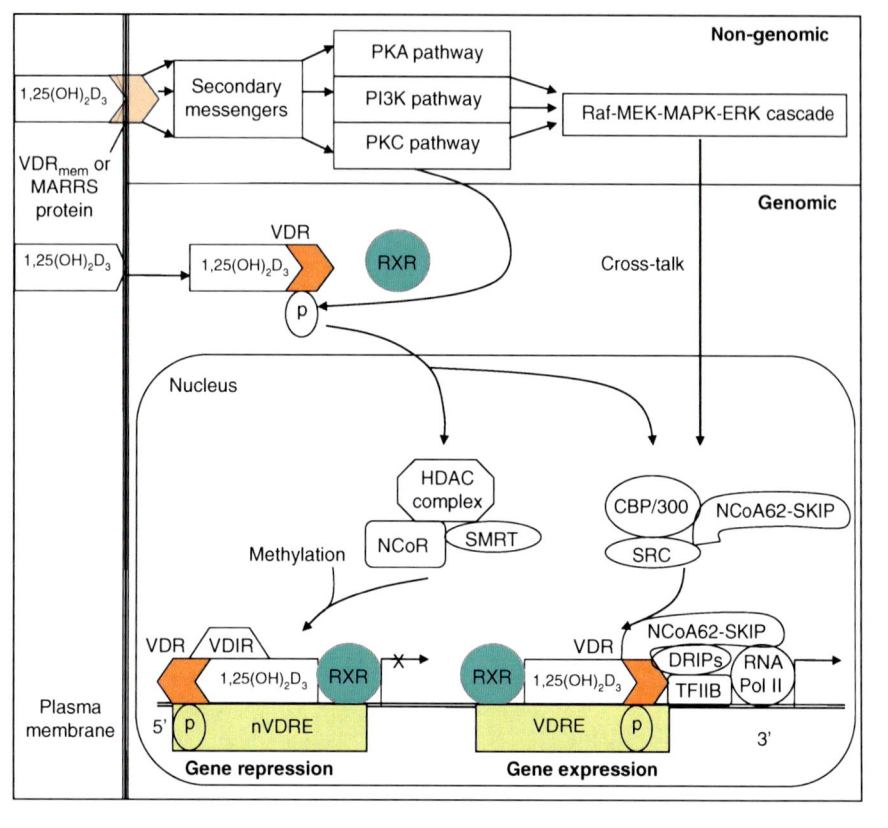

Fig. 9.2 Genomic and non-genomic actions of $1,25(OH)_2D_3$. Gene expression by $1,25(OH)_2D_3$ via the genomic pathway is mediated by the uptake of $1,25(OH)_2D_3$ into the target cell and binding to vitamin D receptor (*VDR*). The $1,25(OH)_2D_3$-VDR complex dimerizes with the retinoid X receptor (*RXR*) to bind onto the VDRE with RXR and VDR on the 5′and 3′ half site of the vitamin D response element (*VDRE*) respectively. Upon $1,25(OH)_2D_3$ binding, conformation changes of the VDR allows the VDR to bind co-activators such as SRC, NCoA62-SKIP and CBP/300 which relax and de-repress the chromatin. The vitamin D receptor-interacting protein (*DRIPs*) complex is then recruited to aid the entry of the transcription machinery TFIIB and RNA Pol II. Gene repression by $1,25(OH)_2D_3$ involves the binding of VDR and the RXR to the 5′and 3′ site of the nVDRE respectively. The association of VDIR with the VDR recruits the SMRT-HDAC complex and NCoR, together with methylation activity, keeps the chromatin in a repressed state. The non-genomic pathway is characterized by $1,25(OH)_2D_3$ binding to a membrane receptor possibly the VDR_{mem} or MARRS protein which activates secondary messengers that in turn can activate the PKA, PI3K and the protein kinase C (*PKC*) pathway to ultimately lead to the activation of extracellular signal-regulated kinase (*ERK*) in the Raf-MEK-MAPK-ERK cascade. Both PKC and ERK modulate the transcriptional activity VDR through phosphorylation, providing cross-talk between the genomic and non-genomic pathways

deacetylases (HDACs) [95]. To initiate gene expression, the induced conformational change of VDR by $1,25(OH)_2D_3$ binding aids in the disassociation of co-repressors such as nuclear receptor co-repressor (NCoR) and silencing mediator for retinoid and thyroid hormone receptors (SMRT) [152]. SMRT brings deacetylation activities to the site by binding to a repressive complex containing histone binding proteins and HDACs [95]. This de-repression of the DNA allows the recruitment of co-activators.

The AF-2 domain of the VDR becomes exposed upon $1,25(OH)_2D_3$ binding and serves as a binding platform for transcriptional activators [110]. Kim et al. investigated the recruitment of co-factors in $1,25(OH)_2D_3$ induced gene expression. These co-factors possessing HAT activity and include members of the p160 co-activators (steroid receptor co-activators (SRC)-1, SRC-2 and SRC-3), CREB binding protein (CBP)/p300 co-activators [83] and nuclear co-activator 62 kDa-SKI-interacting protein (NCoA62-SKIP) [7]. After the chromatin is relaxed by acetylation, the vitamin D receptor interacting proteins (DRIPs) complex at the AF-2 region facilitates the entry of transcription machinery proteins, such as RNA polymerase (Pol) II [52] and transcription factor 2B (TF2B) [94]. Different nuclear hormone receptors may direct tissue specific gene regulation by recruiting various members of the HAT proteins/co-activators [148].

On the other hand, $1,25(OH)_2D_3$ can also repress gene expression. The repression is mediated by the binding of $1,25(OH)_2D_3$ to VDR to induce the interaction of the VDR to VDR-interacting repressor (VDIR) which can bind to a negative VDRE (nVDRE). Binding of VDIR to this motif leads to the replacement of HAT with HDAC [108]. It has recently been found that this VDIR-VDR co-repressor complex together with HDAC recruits the DNA methyltransferase which methylates CpG sites [82]. At this stage MeCP2 can bind to the methylated CpG sequences and repress transcription by interacting with the HDAC complex [95]. Therefore, the HDAC and methylation activities work in parallel to mediate $1,25(OH)_2D_3$ induced trans-repression of VDR target genes.

9.3.3 Non-genomic Actions of 1,25-Dihydroxyvitamin D_3

In $1,25(OH)_2D_3$, the single bond between the A ring and the fused C-D rings allows rotation of the A ring around the C-D fused rings. This flexibility creates the formation of *trans* and *cis* conformations of the molecule that dictates the type of response elicited by the molecule [113, 114]. Apart from the genomic effect described earlier, in the mid-1980s, a rapid, nongenomic response was recognized and is mediated by the *cis*-$1,25(OH)_2D_3$ [112]. This molecule has the ability to activate multiple cell-signalling cascades and bring about a broad range of effects in cell survival and proliferation [115].

Non-genomic actions of $1,25(OH)_2D_3$ involves the binding of $1,25(OH)_2D_3$ to a cell surface membrane receptor. It has been well documented that non genomic pathway involves the activation of the Raf-mitogen-activated protein kinase extracellular signal-regulated kinase kinase (MEK)-mitogen-activated protein kinase (MAPK)-extracellular signal-regulated kinase (ERK) cascade. However, the receptor and the exact pathways that lead to the activation of Raf are still to be confirmed. Candidates for this putative surface membrane receptor include the classical cytosolic VDR (called VDR_{mem}) [75, 84] and the $1,25(OH)_2D_3$-membrane associated rapid response steroid-binding ($1,25(OH)_2D_3$–MARRS) protein [111].

It has been proposed that the binding of $1,25(OH)_2D_3$ to G protein coupled receptors or protein-tyrosine kinase receptors [100] is an essential part of the non-genomic

9 Molecular Biology of Vitamin D Metabolism and Skin Cancer

action of this molecule. The stimulation of phospholipase C (PLC) β and PLCγ by G proteins and protein-tyrosine kinase receptors respectively, leads to the hydrolysis of phosphatidylinositol 4,5-bisphosphate (PIP_2) in the inner layer of the plasma membrane to form the second messengers diacylglycerol (DAG) and inositol 1,4,5-triphosphate (IP_3). DAG remains at the plasma membrane and activates kinases in the protein kinase C (PKC) family. On the other hand, IP_3 is release to the cytoplasm to stimulate the release of Ca^{2+} from intracellular stores to increase cytosolic calcium (Ca^{2+}) levels. The Ca^{2+} released can either act on protein kinases (some members of the PKC need both DAG and Ca^{2+} to be activated) or cause the opening of calcium channels in the plasma membrane to allow the influx of extracellular Ca^{2+} for a more sustained response. PIP_2 can also initiate another second messenger signaling pathway when it is phosphorylated by phosphatidylinositide 3-kinase (PI3K) to produce PIP_3. PIP_3 acts to recruit the protein kinases Akt and PDK1 to the plasma membrane. Akt is subsequently phosphorylated and activated to phosphorylate downstream targets such as regulators proteins for cell survival, transcription factors and other protein kinases. Additionally, activation of the G protein can also stimulate adenylyl cyclase (AC) activity. AC synthesizes cyclic AMP (cAMP) from ATP. cAMP then binds to the regulatory subunits of protein kinase A (PKA) to release the catalytic subunits which are now able to phosphorylate their target proteins [30].

Activation of the PKC and PKA in the non-genomic pathway can phosphorylate the VDR involved in the genomic pathway to modulate its activity (Fig. 9.2) [38]. This suggests that kinase activation on the non-genomic pathway may have a role in determining the functional outcome of the VDR in the genomic pathway.

In addition to the VDR, target proteins of PKC, PI3K and PKA pathways also include proteins involved in the Raf-MEK-MAPK-ERK pathway (Fig. 9.2). This is initiated by the activation of Ras which in turn activates the Raf protein serine/threonine kinase and subsequently the MEK-MAPK-ERK cascade. This ultimately allows ERK to phosphorylate a range of targets such as other protein kinases and transcription factors. Thus, the PKA, PKC and ERK signaling pathway intersects with the classical genomic pathway to provide "cross-talk" between the non-classical membrane receptor pathway and the classical genomic pathway (Fig. 9.2). This allows a complex fine tune regulatory mechanism to action of $1,25(OH)_2D_3$ in regulating mineral and bone homeostasis, cellular proliferation and differentiation, that are important in healthy and diseased states.

9.3.4 Classical Roles of 1,25-Dihydroxyvitamin D_3

The most well known and classical role of $1,25(OH)_2D_3$ is its function in calcium and phosphate homeostasis and bone mineral metabolism [67]. The vitamin D endocrine system maintains mineral homeostasis and bone metabolism by the appropriate transcriptional activation of genes or repression of target genes in cells that are involved in these processes [98]. The importance of this role is shown in studies using 1α-hydroxylase, vitamin D receptor and a combination of

1α-hydroxylase and VDR knock out mice [120]. These experiments showed that $1,25(OH)_2D_3$ and VDR are both crucial for calcium absorption, longitudinal bone growth and normal bone remodeling. Cloning of the *CYP27B1* gene [54], showed that patients with vitamin D-dependant rickets type I had defects in the *CYP27B1* gene and are unable to convert $25OHD_3$ to $1,25(OH)_2D_3$. On the other hand, patients with vitamin D-dependant rickets type II (hereditary vitamin D resistant rickets) do not have a functioning VDR [49].

The formation of $1,25(OH)_2D_3$ from the hydroxylation of $25OHD_3$ by 1α-hydroxylase in the kidneys is regulated by parathyroid hormone (PTH) which in turn is regulated by Ca^{2+} levels. The Ca^{2+} sensing receptor in the parathyroid cell regulates the secretion of PTH. Secreted PTH then binds to the PTH membrane receptor of the renal proximal tubular cell to induce cAMP and PIP_2 signaling pathways (described in Sect. 9.3.3), which leads to the transcriptional activation and upregulation of *CYP27B1* [3]. In addition, the enhanced expression of *CYP27B1* can also be mediated by Ca^{2+} independent of the PTH pathway but the mechanism involved in this process is still not well understood [63]. Upregulation of *CYP27B1* causes the increased synthesis of the 1α-hydroxylase enzyme which acts on the intestinal cell through the genomic pathway (described in Sect. 9.3.2) to upregulate the expression of transient receptor potential vanilliod (TRPV) 5, TRPV 6, calbindins, Ca^{2+} pump and the Na^+/Ca^{2+} exchanger (Table 9.1). These proteins all take part in the transcellular pathway in the uptake of Ca^{2+} from diet [122]. TRPV5 and TRPV6 (more abundant in the intestinal cell) are Ca^{2+} channel proteins on the apical surface of the intestine that mediate the entry of Ca^{2+} [156]. Upon entry of the Ca^{2+},

Table 9.1 The effects of $1,25(OH)_2D_3$ in various tissues

Tissue	Protein	Gene regulation	Effect
Small intestine	TRPV 5	Upregulation	Entry of Ca^{2+} into the intestinal cell
	TRPV 6	Upregulation	
	Calbindins	Upregulation	Ca^{2+} transport from entry size to basolateral membrane
	Ca^{2+} pump	Upregulation	Ca^{2+} exit from intestinal cell
	Na^+/Ca^{2+} exchanger	Upregulation	
Bone	RANKL	Upregulation	Osteoclastogenesis
	OPA	Downregualtion	Bone cell differentiation
Skin	Involucrin	Upregulation	Keratinocyte differentiation
	Transglutaminase K	Upregulation	
	Loricrin	Upregulation	
	filaggrin	Upregulation	
	CaR	Upregulation	
	PLCγ	Upregulation	
Others (immune system, prostate, breast, colon)	–	–	Regulation of proliferation and differentiation

9 Molecular Biology of Vitamin D Metabolism and Skin Cancer   201

calbindin carries Ca^{2+} from the entry side to the basolateral membrane of the intestinal cell where it exits to the lamina propria via the plasma membrane Ca^{2+} pump and the Na^+/Ca^{2+} exchanger. Apart from this mechanism, $1,25(OH)_2D_3$ can also cause rapid absorption of calcium (called transcaltachia) via binding of a membrane receptor to activate the rapid non genomic pathway [113], also described earlier in Sect. 9.3.3. A negative feedback loop exists through high levels of calcium and $1,25(OH)_2D_3$ levels to regulate and decrease the level of PTH [98].

Calcium homeostasis is also important in maintaining bone health. The normal bone remodeling cycle begins with the resorption of existing bone by osteoclasts followed by the synthesis of unmineralized bone by osteoblasts (osteoid). With adequate levels of $1,25(OH)_2D_3$ and mineral, the osteoblast mineralizes the osteoid [116]. The differentiation, development, activation and survival of the osteoclast depend on the binding of the receptor activator of NF-κB ligand (RANKL) on the surface of preosteoblastic cells to RANK on the osteoclastic precursor cells. On the other hand, this process can be blocked by the binding of osteoprotegrin (OPG) to RANK to inhibit its binding to RANKL [18]. $1,25(OH)_2D_3$ plays a role in osteoclastogenesis by upregulating and repressing of RANKL and OPG expression respectively [147] (Table 9.1). PTH also increases RANKL and decreases OPG production [92], thus, $1,25(OH)_2D_3$ may also indirectly enhance osteoclastogenesis by its influence on PTH levels. Therefore, PTH can enhance osteoclastogenesis to release bone minerals into the circulation to maintain calcium homeostasis. During times of adequate/high calcium in the circulation, PTH decreases and bone mineralization occurs by utilizing the mineral in the circulation. Thus PTH and $1,25(OH)_2D_3$ co-operate to coordinately regulate bone remodeling and calcium homeostasis. Vitamin D is well known for its role in mineral and bone homeostasis; however, epidemiological studies seem to suggest another role for this hormone.

9.4 Epidemiological Evidence on the Relationship of Sun exposure and Cancer

9.4.1 Epidemiologic Evidence on the Role of 1,25-Dihydroxyvitamin D_3 in Skin Cancer

The three common types of skin cancers include melanoma and two nonmelanocytic skin cancers, squamous cell carcinoma (SCC) and basal cell carcinoma (BCC). It is clear that UVR produces harmful photoproducts in DNA (Sect. 9.2) and increase in sun exposure leading to increase in skin cancer risk has been supported by many studies [4, 119]. Migrant studies have examined the effect of migration from an area of low ambient solar UV radiation to one of high ambient solar radiation. The risk of each type of skin cancer was greater for native-born Australians than for migrants [47, 86]. The rates were similar in people who migrated in Australia (a high ambient solar radiation area) before 10 years of age

compared to those who were born in Australia, whereas migration after the age of 10 had a quarter of the rate of native-born Australians [71]. Risk for all three types of skin cancer also showed a positive correlation with ambient solar radiation and increasing average annual hours of bright sunlight though the extent of this correlation seems to vary depending on the type of skin cancer [4]. The frequencies of all three cancers were generally the greatest on high sun-exposed body sites such as the face, ears and neck and low on the rarely exposed sites [5, 59]. Interestingy, the densities for melanoma and BCC are higher on the more intermittently exposed shoulders and back while SCC has a lower density on these sites and is higher on the back of the hands. This association is consistent with results of the study on the relationship of personal sun exposure with skin cancer risk. SCC is strongly related to total sun exposure and occupational sun exposure (continuous pattern of exposure), while melanoma and to a lesser extent, BCC, show significant associations with non-occupational/recreational (intermittent) exposure and sunburn (intense intermittent exposure) [46]. Thus, with the evidence that SCC, BCC and melanoma is caused by sun exposure, it is of no surprise that a latitude gradient of skin cancer exists, with increasing incidence and mortality rates corresponding with increasing proximity to the equator [25, 91]. The magnitude of the latitude gradient was approximately 65% and 50% greater in incidence and mortality of melanoma respectively, for body areas most intermittently exposed compared with those with a least intermittent pattern of exposure [24].

Although there is a vast amount of persuasive evidence that support the classical belief that sun exposure causes skin cancer, a recent study by [17] provided a new school of thought on the relationship of sun exposure and skin cancer development. A number of previous studies have shown that that the incidence of cutaneous melanoma varies by season with a peak in summer [16, 20, 126, 136, 137]. It has been hypothesized that if the higher incidence in summer is due to increased awareness and detection of lesions on exposed skin, thinner lesions will be seen; whereas a late stage promotion effect from the summer sun will yield thick lesions with worse prognosis independent of Breslow thickness. Although increased thinner and less aggressive lesions were indeed found in younger women during summer which seems to correlate increased incidence with awareness, there was still a significant increase of 18% in incidence for the constantly exposed head and neck. Thus, the data do not exclude the possibility of greater awareness in summer or a late-stage promotional effect of sun exposure (consistent with the classical belief). Interestingly, the same study also found a significant 20% of reduced fatality for melanomas diagnosed in summer to those diagnosed in winter. These rates were independent of seasonal thickness variation, age, sex, anatomical site and histologic type of the melanoma [17]. Therefore, these results are suggestive of a more complex pathway in the development or progression of melanoma that is not restricted to the classical effects of direct sun exposure [17].

Consistent with the results obtained by Boniol et al. were the results found by [9] who conducted a study to investigate the effect of sun exposure on melanoma fatality. This study showed that solar elastosis, sunburns and intermitted sun exposure were inversely associated with melanoma fatality. This finding was also

9 Molecular Biology of Vitamin D Metabolism and Skin Cancer

independent from confounding factors including sex, age, Breslow thickness, anatomic site, social class, skin awareness, skin self examination and physician examination [9]

Thus, such epidemiological studies yield interesting results that imply a complex process in the development of melanoma. Knowing that vitamin D synthesis is dependant on UV exposure, the effect of sun exposure with increased melanoma survival raised the possibility of a link between vitamin D and skin cancer.

9.4.2 Polymorphisms of the Vitamin D Receptor

The involvement of $1,25(OH)_2D_3$ in skin cancer is also supported by genetic evidence. As the $1,25(OH)_2D_3$ must act via the VDR to elicit the genomic effect and a possible VDR_{mem} to elicit the non-genomic pathway, it is expected that any changes in the genetic sequence and expression of VDR will have an effect in $1,25(OH)_2D_3$ action, and in turn on skin cancer outcome.

The most well known polymorphisms in the VDR include the polymorphism at the 5'*FokI* restriction site in exon 2; an alteration in intron 8 to generate the *Bsm*I and *Apa*I restriction sites; a synonymous polymorphism in exon 9, generating a *Taq*I (*t*) restriction site and a poly-A microsatellite in the 3'untranslated region [8]. The 5' *FokI* restriction site does not seem to show any linkage to the other polymorphisms, whereas the latter four polymorphisms are in strong linkage disequilibrium [48]. Thus, in the studies of [76, 97], the analysis of the *Taq*I was assumed to represent the 3' cluster of polymorphisms. The 5'*FokI* polymorphism involves a T to C transition at the ATG start site, producing two variants of the protein, a shorter protein (*F*) of 424 amino acids (aa) and a longer protein of 427aa (*f*) [8]. In one study [76], it was found that a significant reduction in risk of malignant melanoma (MM) was associated with the *FF* phenotype. It has previously been reported that the *F* allele with the shorter protein of 424 aa had higher transcriptional activation activity and *FokI* polymorphism has a functional significance [76]. This was consistent with the finding that the *f* being a risk allele [97]. The same study found that the *t* allele was protective against melanoma with a *tt* genotype reducing the risk by 29%. Interestingly, [76] did not find a significant association with melanoma risk but showed the genotype combination *ttff* was significantly associated with tumors of increased Breslow thickness and which raised the idea that genetic variants of VDR can be a determinant of melanoma outcome. The role of VDR polymorphisms have also been studied in other cancers such as the breast and colon, however, the results were not always consistent [77, 79, 153]. These controversies may be due to differences in vitamin D serum levels and sample variations [97]. It is also thought that polymorphisms in the 3'UTR may have cell type specific effect that can play a role in altered VDR transcription [155]. Furthermore, the possible interactions of VDR polymorphism haplotypes with other known risk factors can have an impact on melanoma risk [96].

Understanding the functional effects of the VDR variants can aid us in understanding the action of vitamin D in the presence of a particular VDR variant. Genotyping patients for the VDR gene can help us predict the action of vitamin D for each individual. This in turn may be useful in advising high risk individuals to take precautions for preventing skin cancer development. Furthermore, patients carrying different VDR variants may also cause them to respond differently to therapies and knowledge on the functional effects of VDR variants should allow the development of drugs that will act most efficiently on the patient with minimal side effects.

Taken together, the epidemiological data from these studies show that there is a link between sun exposure and skin cancer. More importantly, these evidences suggest a possible link between the role of vitamin D and skin cancer.

9.5 Vitamin D and Skin Cancer

9.5.1 The Role and Expression of 1,25-Dihydroxyvitamin D_3 in Extra Renal Sites

Apart from the classical role of $1,25(OH)_2D_3$ in maintaining mineral homeostasis via the intestine, parathyroid, bone and kidney. $1,25(OH)_2D_3$ also has non classical functions in extra-renal tissues. The idea of extra-renal synthesis of $1,25(OH)_2D_3$ started when it was observed that the administration of vitamin D in anephric patients led to an significant increase of serum $1,25(OH)_2D_3$ levels compared to controls and this increase of $1,25(OH)_2D$ had significant correlation with the precursor 25OHD levels [90]. This observation was confirmed in another study by the oral administration of 25OHD to uremic mongrel dogs and anephric patients which also found a similar significant correlation between serum levels of 25OHD and $1,25(OH)_2D_3$ [42]. The enzyme expressed in extra-renal tissues acts locally in an autocrine/paracrine manner (Fig. 9.1b) which serves to complement the endocrine circulating $1,25(OH)_2D_3$ produced by the kidneys [80]. This locally elevated concentration of $1,25(OH)_2D_3$ can alter gene expression in a tissue specific manner that eventually limit proliferation and induces differentiation. These effects of proliferation and differentiation regulation by $1,25(OH)_2D_3$ has been described in various tissues including the cells of the immune system [53, 65], prostate, breast [160], colon, bone as well as the skin [124] (Table 9.1). In fact, the *CYP27B1* gene has recently been expressed in the transgenic mouse and it has been shown that the 5′ flanking region itself provides sufficient information for directing cell and tissue specific expression [2]. This is in agreement with the idea mentioned earlier in Sect. 9.3.2, that the presence of different transcription factors in different tissues and cell differentiation state allows nuclear receptors to regulate gene transcription in a tissue and time specific manner. These exciting findings of tissue specific proliferation and differentiation regulation by $1,25(OH)_2D_3$ in extra-renal sites provide an important and direct link on the actions of $1,25(OH)_2D_3$ in various cancers including skin cancer.

9.5.2 The Role of 1,25-Dihydroxyvitamin D_3 in Normal Skin

Before investigating the role of $1,25(OH)_2D_3$ in the skin in healthy and diseased state, it is important to know the process of keratinocyte differentiation in the epidermis. The epidermis is composed of four layers. Directly on top of the basal lamina, the basal layer of the epidermis is the stratum basale, followed by the stratum spinosum, then the stratum granulosum and finally the most superficial layer, the stratum corneum [134]. As the cells differentiate, they gradually migrate up from the base layer, stratum basale, to the stratum spinosum then granulosum to finally become completely differentiated keratinocytes in the stratum corneum. Proliferating keratinocytes found in the stratum basale express keratin 5 and 14. Upon entering the stratum spinosum, the cell expresses keratin 1 and 10 instead of 5 and 14 and the synthesis of involucrin and transglutaminase-K, an enzyme cross linking the involucrin with other substrates for the formation of the cornified envelope is now evident. By the time the cells reach the stratum granulosum, granules containing loricrin and the keratin filaments bundling protein precursor, profilaggrin are present. Lamella bodies in this layer, which secretes fatty acid, ceramide, and cholesterol, fill the intercorneocyte space to bind the corneocytes together in the stratum corneum providing the skin its elasticity and barrier function [10].

The fact that keratinocytes are the only cells that supports the complete vitamin D metabolic pathway from 7-DHC to $1,25(OH)_2D_3$ [93, 104] and the observation of $1,25(OH)_2D_3$ induces keratinocyte differentiation [73] together with the fact that the expression and levels of VDR and $1,25(OH)_2D_3$ vary with differentiation [72] strongly suggest that $1,25(OH)_2D_3$ is an autocrine/paracrine factor for keratinocyte differentiation [12]. In experiments with 1αOHase knockout mice [11], it was observed that there were no gross epidermal phenotype differences between the knockout and their wild type littermates, however, there is a reduction of the differentiation markers involucrin, filaggrin and loricrin. It was also found that $1,25(OH)_2D_3$ and calcium act together in a synergistic manner to elicit prodifferentiation effects including the activation of involucrin and transglutaminase gene expression (Table 9.1) [150]. A plausible explanation of the observed synergistic effect of $1,25(OH)_2D_3$ and calcium arises from the close proximity of the calcium and VDR elements in the promoter of the involucrin gene although the mechanism of this synergistic effect is still unknown for the transglutaminase gene [13].

The calcium signaling pathway for keratinocyte differentiation is very similar to the rapid non genomic/surface membrane pathway of $1,25(OH)_2D_3$ signaling (described in detail in Sect. 9.3.3). The binding of extracellular calcium to the calcium receptor (CaR) activates the receptor to stimulate PLC activity which leads to the formation of DAG and IP_3 that eventually causes the release of intracellular calcium stores from the endoplasmic reticulum and the golgi. This initial and sustained increase of IP_3 through PLCβ and γ respectively allows the sustained increase of intacellular calcium to induce genes necessary for differentiation [78, 165]. During the differentiation process, apart from inducing the expression of involucrin and transglutaminase, $1,25(OH)_2D_3$ also induces CaR [131] and PLCγ expression [164] (Table 9.1). Thus, the requirement for $1,25(OH)_2D_3$ to induce the proteins needed for differentiation is consistent with

in vitro findings that the stratum basale with the least differentiated keratinocytes have the highest levels of *CYP27B1* and VDR [149, 166]. Therefore, disturbance to the process of $1,25(OH)_2D_3$ mediated expression of these essential proteins for differentiation can lead to diseases of the skin including cancer.

9.5.3 The Role of Vitamin D in Regulating Proliferation and Differentiation in Skin Cancer

Transformed keratinocytes in squamous cell carcinomas (SCC) are not responsive to the differentiation and proliferation effects of $1,25(OH)_2D_3$ [138]. The vitamin D receptor interacting protein (DRIP) (DRIP205 is the major subunit for anchoring the complex to the VDR) and steroid receptor co-activators (SRC) including SRC 2 and 3 are the two main co-activator complexes that interact with the VDR in keratinocytes to initiate the transcription of the differentiation markers [43]. A model was initially proposed that DRIP205 complex dominates in binding with the VDR during the proliferation/early differentiation stages and SRC complex is the one dominating in late differentiation stages [117]. It was also found that SCC overexpresses DRIP205 and hence it was thought that this elevation of DRIP205 levels inhibited the switch to SRC maintaining these transformed cells in a proliferation state [15]. However, a follow up study [64] proved that this proposed model of switching from DRIP205 to SRC is inadequate. The results from the follow up study suggested that knock down of VDR, DRIP205 and SRC significantly decreased the early marker keratin 1 and late markers loricrin and filaggrin. However, only the knock down DRIP205 significantly reduced the early marker keratin 10 and the intermediate marker involucrin. Thus, this latest study show that VDR, DRIP and SRC are all required for induction of both early and late differentiation markers. Also, the recruitment of the appropriate co-activator by the $1,25(OH)_2D_3$-VDR complex is gene specific and not differentiation stage specific. Further investigations are required to fully elucidate the keratinocyte differentiation process in order to suggest targets for drug treatments.

It is known that activated Ras oncogenes can contribute to the development of SCC and basal cell carcinoma (BCC) [41, 146, 157]. An immortalized squamous cell line with activated Ras oncogene, HPK1A Ras, was compared to the original immortalized squamous cell line (HPK1A) to investigate how keratinocytes can exhibit $1,25(OH)_2D_3$ resistance in growth with respect to the Ras oncogene [58]. It was found that the ability of $1,25(OH)_2D_3$ to induce trans-activation for growth inhibition was significantly decrease in HPK1A Ras compared to HPK1A cells. The growth inhibition by $1,25(OH)_2D_3$ on HPK1A Ras cells was restored by the addition of a MAPK kinase inhibitor. These results were reproducible when tested with a reporter gene containing an upstream VDRE. An antibody to the binding domain for the RXR yield a super shift only in HPK1A cells and follow up experiments using anti-phosphothreonine and anti-phosphoserine antibody demonstrated serine phosphorylation of RXR only in HPK1A Ras cells. In addition, serine

9 Molecular Biology of Vitamin D Metabolism and Skin Cancer

phosphorylation with control HPK1A cells was detected with over expression of active MAPK kinase and these cells failed to drive reporter activity. The reverse was then tested by using the $1,25(OH)_2D_3$ resistant HPK1A Ras cells expressing a mutant RXR of serine to alanine at the relevant position. Indeed the restoration of reporter activity and the detection of serine phosphorylation confirmed that an activated Ras/MAPK signaling pathway in tumor cells can cause the phosphorylation of the RXR, which in turn may interfere with $1,25(OH)_2D_3$ transactivation mediated growth inhibition. Further understanding of the exact mechanism of how RXR phosphorylation can lead to the disturbance of its interaction with proteins required for $1,25(OH)_2D_3$ transactivation, which could yield important ideas for chemoprevention therapies.

9.5.4 The Role of Vitamin D in Photoprotection

The most well known consequence of UVB radiation is the appearance of apoptotic or sunburn cells [88]. Cellular stresses including UV irradiation activates c-Jun NH2-terminal kinase (JNK) [74] and there is evidence that upregulation of stress activated protein kinases (SAPKs) promotes apoptosis [158, 163]. The tumor suppressor gene, p53, can either induce cell cycle arrest by upregulating cyclin dependant kinase inhibitor P21 [144] or inducing apoptosis if the damage is extensive and cannot be repaired [37]. The interaction between JNK and p53, and the precise pathway of JNK mediated apoptosis and carcinogenesis is not yet fully elucidated. The interaction of p53 with JNK could conceivably prevent the interaction of p53 to the p21 promoter to inhibit cell cycle arrest and thus favors apoptosis [142]. It has been demonstrated that JNK2 knockout mice have a lower number of papillomas and malignant tumors induced by 12-O-tetradecanoylphorbol-13-acetate compared to wild type mice, suggesting that JNK2 is critical in tumor promotion [27].

De Haes et al. found that pretreating keratinocytes for 24 h prior to UVB radiation with pharmacological dose of $1,25(OH)_2D_3$ (1 μM) reduced apoptosis by 55–70%. Moreover, a reduction of UVB stimulated JNK activation of more than 30% was also found together with a 90% inhibition of mitochondrial cytochrome c release [33]. This can possibly be explained by the recent finding of the ability of p53 to protect cells against UV induced apoptosis via the binding and inactivation of JNK pathway, which is responsible for the induction of mitochondrial death signaling [99].

It has also been noted [33] that the culture conditions in terms of dose and preincubation time of $1,25(OH)_2D_3$ were very similar to those used to conduct growth inhibition experiments on proliferating keratinocytes [14, 139]. It is hypothesized that the observed accumulation of keratinocytes in the G_1 phase of these experiments may have protected the DNA from the genotoxic effects of UVB, as the unfolded structure of DNA in the S phase will render it more susceptible to UVB induced DNA damage [123]. This hypothesis is in agreement with the findings of

p53 having a dual role of JNK inactivation while in the same time activate cell cycle arrest related genes to protect cells from apoptosis upon UVB irradiation [99].

If $1,25(OH)_2D_3$ could prevent the apoptosis of UVB irradiated cells, the next concern is the danger of allowing cells with increased DNA damage to survive [61]. Gupta et al. tested whether $1,25(OH)_2D_3$ enhanced cell survival would lead to an accumulation of UV induced DNA damage. Cells treated at physiological dose of $1,25(OH)_2D_3$ (10^{-9} M) 24 h prior to irradiation not only showed significant dose dependant increase of cell survival, but a dose dependant decrease in TD was also observed. Such effects can be reproduced by treating cells with $1,25(OH)_2D_3$ immediately after irradiation. More importantly, there was a corresponding increase in p53 with decreasing TD. As it is known that UV induced increases in nitric oxide (NO) products [22] can enhance DNA damage by UVR [151] and inhibit CPD repair [6], the levels of nitrite were also measured and a significant reduction of nitrite in $1,25(OH)_2D_3$ treated cells was found. Therefore these experiments [61] suggest that the reduction of TD or DNA damage by $1,25(OH)_2D_3$, is due to the increase of p53 along with a decrease of NO products that results in increased DNA repair. Taken together, the effect of $1,25(OH)_2D_3$ on UV irradiated cells is to reduce the number of apoptotic cells and enhance cell survival by improving UVB induced DNA damage repair. The protection of $1,25(OH)_2D_3$ against the formation of CPD was also supported by another study [34], however, these effects were only seen using pharmacological doses and a suppression in p53 was obtained. It is argued that the suppression of CPD formation by $1,25(OH)_2D_3$ may have prevented the need for p53 accumulation for DNA repair. However, such discrepancies may also be due to the difference in cell culture and irradiation conditions.

The fact that the photoprotective effects of adding $1,25(OH)_2D_3$ immediately after irradiation was comparable to those with 24 h $1,25(OH)_2D_3$ pretreatment, prompted studies to investigate the mechanism of $1,25(OH)_2D_3$ in producing such effects. A series of elegant studies [39, 40, 162] found that the photoprotective effects of $1,25(OH)_2D_3$ described above can be reproduced by three low- calcemic analogs of vitamin D both in vitro and in vivo. It was described in Chapter 2 (Sect. 2.2) that the existence of *trans* and *cis* isomers allows $1,25(OH)_2D_3$ to mediate genomic as well as rapid, non genomic responses. Rapid response signaling is mediated by the *cis* conformers. These experiments showed that *cis*-locked, low calcemic rapid response agonists, $1,25(OH)_2$lumisterol$_3$ (JN) and $1,25(OH)_2$–7-dehydrocholesterol (JM) added immediately after irradiation, displayed similar protective effects to that of $1,25(OH)_2D_3$ at physiological doses. A rapid response antagonist (HL) completely blocked the photoprotective effects [162] while a genomic response antagonist (TEI-9647) had no effect [40]. In fact, the protective effects of the low calcemic rapid response agonist, JN, has been confirmed recently in vivo [39]. Furthermore, the low calcemic homo hybrid analog (QW) with some transcriptional capacity, was also able to reduce pyrimidine dimmers as well as immunosuppression in the same level of effectiveness as $1,25(OH)_2D_3$ when topically applied to the epidermis of irradiated hairless Skh:HR1 mice [40]. These results show QW to be a potential candidate in skin cancer prevention (see Sect. 9.5.5). Therefore, the data from these studies provide strong evidence for vitamin D photoprotection via the rapid response pathway.

9 Molecular Biology of Vitamin D Metabolism and Skin Cancer

9.5.5 *Vitamin D Analogs as Potential Agent for Skin Cancer Prevention*

Accumulating evidence of the pro differentiating, anti proliferating and photo-protective effects of $1,25(OH)_2D_3$ from in vitro, in vivo and epidemiologic studies have raised a growing interest in the possibility of making vitamin D a therapeutic agent [21, 107]. Most clinical trials are impeded by the severe hypercalcemia effect of $1,25(OH)_2D_3$ and the problem of 24-OHase degradation. This raises the idea that the development of vitamin D analogs with more specific actions to minimize current side effects will have a much greater clinical potential [62, 103].

The hypercalcemic vitamin D analog QW was described earlier in Sect. 9.5.4 in respect to its photoprotective effects in reducing CPDs [40]. In fact, QW has undergone some intense pre-clinical trials and its therapeutic effects was compared to $1,25(OH)_2D_3$ as well as Paricalcitol, another hypercalcemic vitamin D analog. It was found that QW was 80–100 times less calciuric than the classical $1,25(OH)_2D_3$ [127]. Both QW and Paricalcitol were tested in SCC models and the molecular mechanism were shown to involve a number of pathways, such as those induced in growth cycle arrest, DNA synthesis inhibition, as well as apoptosis promotion and pro survival actions. To elicit their anti-tumor effects, both QW and Paricalcitol decreased the positive cell cycle regulator cyclin dependant kinase 2 and inhibited the pro-survival/pro-growth pathway mediators such as phospho-Akt, phospho-MEK and phospho-ERK. More importantly, apart from its low calcemic properties, the ability of QW to induce the cell cycle inhibitor p27 and inhibit phospho-ERK was not seen in $1,25(OH)_2D_3$. In summary, QW was proved to be a more potent compound in SCC inhibition [1] and test results for QW and Paricalcitol to date are very promising. Testing of other low calcemic vitamin D analogs such as TX527 and TX522 of its photoprotective effects against UV irradiation are also underway. With a potency of 100 times more than $1,25(OH)_2D_3$ the results demonstrated that $1,25(OH)_2D_3$ analogs have great promise in chemoprevention therapies for UVB induced skin cancer [35].

9.6 Future Perspectives: Current Controversies on Sun Exposure and Vitamin D Recommendations

Given the detrimental role of UVR in the development of skin cancer, during the last decades, health campaigns and prevention programs have recommended the use of sunscreens, protective clothing and the avoidance of sunlight [133]. However, there is also accumulating evidence from epidemiological, in vitro and in vivo studies on the benefits of vitamin D. Thus, scientists face the dilemma of how much UVR is needed to produce an adequate amount of vitamin D to maintain everyday functions, and more interestingly, anticancer effects (summarized in Table 9.1), while

preventing the development of skin cancer due to the over exposure of UVR [68]. This urges careful evaluation of these current recommendations to the public.

Some parts of the population with medical conditions such as patients with xeroderma pigmentosum who are defective in DNA repair [85] and patients receiving organ transplants that are on immunosuppressive drugs are extremely sensitive to UVR induced skin cancer [23], and are already taking these precautions. With the avoidance of sun exposure, it was found that patients in these two groups had significantly lower 25(OH)D$_3$ serum levels compared to controls [128, 129]. Also, it was found that over 80% of veiled women had significantly low blood 25(OH)D$_3$ levels [60]. This problem of insufficient or even deficient in vitamin D levels were more apparent when studies in the southern states in Australia which has relatively low levels of sunlight, revealed that 42% of women were vitamn D insufficient and 8% of 20–59 year old women were vitamin D deficient in vitctoria. In addition, in Hobart, up tp 10% of healthy 8 year old children were insufficient in vitamin D [81]. Thus, inadequate vitamin D levels are a problem in all age groups.

Vitamin D status is characterized by bone health and PTH levels. This is because increased PTH induces the expression of *CYP27B1* to maintain vitamin D and in turn Ca^{2+} concentrations in the blood to ensure sufficient levels are available for bone mineralization. Vitamin D sufficiency is defined by the absence of bone disease with a PTH level of less than 65 pg/mL and a serum 25(OH)D$_3$ concentration of equal to or above 50 nmol/L. Vitamin insufficiency is accompanied with normal but high bone turnover and is characterized by PTH levels less than 65 pg/mL but can be reduced by vitamin D supplementation. Vitamin D insufficiency occurs when serum 25(OH)D$_3$ concentration is between 25 and 50 nmol/L. People with vitamin deficiency have a PTH level of more than 65 pg/mL and a serum 25(OH)D$_3$ concentration of less than 25 nmol/L [116, 161]. These patients have a high bone turnover, and in more severe cases with serum 25(OH)D$_3$ concentration less than 12.5 nmol/L, osteomalacia results, where newly formed bone cannot be mineralized. Osteomalacia may be asymptomatic, but the patient may also experience a diffuse bone and muscle pain, and skeletal weakness [125]. These vitamin D status and characteristics are summarized in Table 9.2.

Table 9.2 Indicators of vitamin D status

Vitamin D status	Serum PTH concentrations (pg/mL)	Serum 25(OH)D$_3$ concentration (nmol/L)	Characteristics
Sufficiency	<65	>50	No bone disease
Insufficiency	<65 but can be reduced by vitamin D supplementation	25–50	High but normal bone turnover
Deficiency	>65	12.5–25 (moderate) <12.5 (severe)	High bone turnover, rickets or osteomalacia

9 Molecular Biology of Vitamin D Metabolism and Skin Cancer   211

Knowing the consequences of vitamin D inadequacy, key parties in Australia involved in skin cancer control have decided to provide more updated guidelines to the puiblic regarding the importance of UVR in vitamin D synthesis [143]. Apart from reminding people of the harmful effects of UVR on skin cancer, the new message to the public stepped away from the idea of needing protection against the sun at all times and stressed the importance of maintaining adequate vitamin D levels by encouraging outdoor activities (Cancer Council Australia, 2007). However, appropriate precautions needs to be taken during outdoor activities For incidental sun exposure of less than 10 min, the application of sunscreen may not be necessary, but sunscreen application is recommended if periods of sun exposure sufficient to produce erythema (redness) are intended [116]. Although it has been found that the use of sunscreen can have a negative effect on vitamin D synthesis [105], other clinical studies on long term use of sunscreens showed that normal vitamin D levels can still be maintained [102, 145]. The use of sunscreen is also encouraged by the fact that once previtamin and vitamin D_3 has been formed, further exposure to sunlight will cause their degradation into inert over irradiation products [66] and this further UVR exposure will only lead to increases in DNA damage. Based on this, it has been pragmatically decided that exposure of hands, face and arms to a third to a half of a minimum erythemal dose for 5–15 min four to six times a week with the dark skinned and elderly population needing the greatest exposure of these recommended values [68, 116]. However, if sun exposure is limited by medical or cultural reasons, a tailored vitamin D supplementation plan may be necessary [143].

Currently, there are still no recommended dietary intake levels in place in Australia but the daily vitamin D intake recommended by the Food and Nutrition Board of the US Institute are 200 IU, 400 IU and 600–800 IU for ages 0–50 years, 51–70 years and 71+ years respectively [50]. Yet, these recommended values have been challenged by the findings that 200 IU/day has no effect on bone status and the recommendation of 1,000 IU has been suggested to adequately prevent bone disease, fractures and possibly protect against some cancers [159]. Moreover, it has even been reported that 800 IU/day supplemented vitamin D did not reduce osteoporotic fractures in some vitamin D replete individuals [135].

In conclusion, much research is needed to further understand the health benefits that accompanying sun exposure. More importantly, it is essential to further elucidate the molecular mechanisms underlying the actions of vitamin D in preventing classical diseases relating to bone health as well as non classical diseases such as cancer. Such investigations should take into consideration not only different age and racial groups, but also their health status including genetical variations in key vitamin D metabolizing genes (Fig. 9.3). The findings in these future studies will yield invaluable knowledge to aid appropriate recommendations for sun exposure and vitamin D intake. These sun exposure levels will also have to take into account of keeping the fine balance between UV exposure derived health benefits and preventing skin cancer. Ultimately, this knowledge can be translated into the development of improved vitamin D analogs to efficiently treat vitamin D related diseases with minimal side effects.

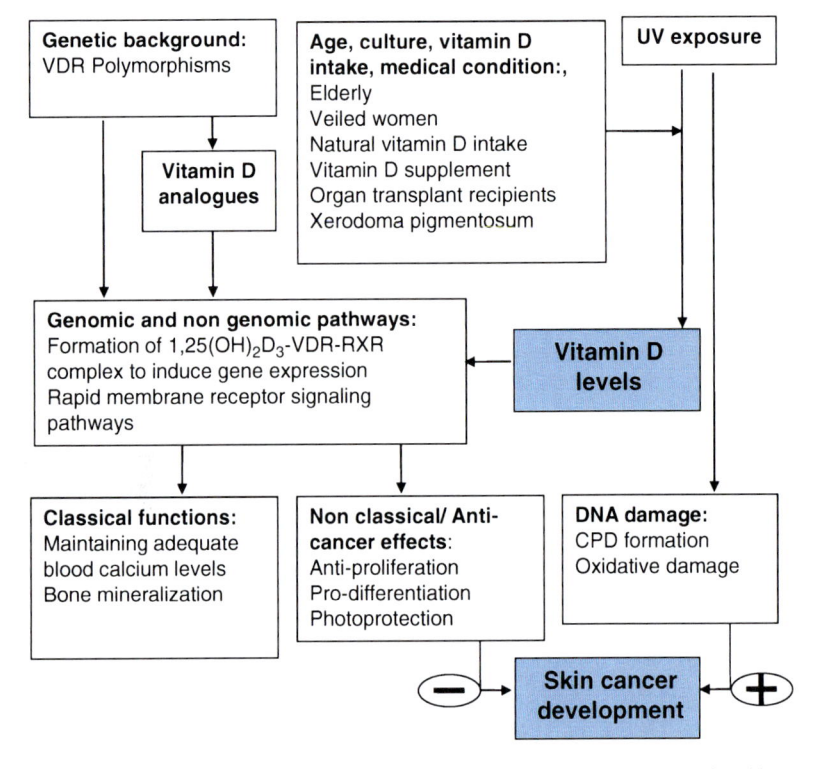

Fig. 9.3 Vitamin D and skin cancer. UV radiation can cause DNA damage that can lead to skin cancer. However, the vitamin D level in our body is also dependant on UV radiation induced vitamin D synthesis as well as our vitamin intake from food sources. The amount of vitamin D synthesized from UV exposure is influenced by the age, culture, and existing medical conditions of the individual. Genetic variants of the vitamin D receptor (*VDR*) can influence the ability of vitamin D to elicit its biological actions. Vitamin D in the body can carry out classical functions such as mineral and bone homeostasis as well as non classical functions that can result in anti-cancer effects. Thus the balance of UV exposure in causing DNA damage and vitamin D synthesis determines the risk of skin cancer

Achknowledgments We thank Lucia Musumeci and Cheng Li (The University of Sydney) for their helpful contributions and Bruce Armstrong (The University of Sydney) for his helpful discussions and comments on this chapter. The work in the lab of JKVR was supported by NCI grant P01 CA108964 (Project 1). JKVR is also a Medical Foundation Fellow at The University of Sydney.

References

1. Alagbala AA, Johnson CS, Trump DL et al (2006) Antitumor effects of two less-calcemic vitamin D analogs (Paricalcitol and QW-1624F2–2) in squamous cell carcinoma cells. Oncology 70(6):483–492
2. Anderson PH, Hendrix I, Sawyer RK et al (2008) Co-expression of CYP27B1 enzyme with the 1.5 kb CYP27B1 promoter-luciferase transgene in the mouse. Mol Cell Endocrinol 285(1–2):1–9

3. Armbrecht HJ, Boltz MA, Hodam TL (2003) PTH increases renal 25(OH)D3–1alpha -hydroxylase (CYP1alpha) mRNA but not renal 1, 25(OH)2D3 production in adult rats. Am J Physiol Renal Physiol 284(5):F1032–F1036

4. Armstrong BK, Kricker A (2001) The epidemiology of UV induced skin cancer. J Photochem Photobiol B 63(1–3):8–18

5. Armstrong BK, Kricker A, English DR (1997) Sun exposure and skin cancer. Australas J Dermatol 38(Suppl 1):S1–S6

6. Bau DT, Gurr JR, Jan KY (2001) Nitric oxide is involved in arsenite inhibition of pyrimidine dimer excision. Carcinogenesis 22(5):709–716

7. Baudino TA, Kraichely DM, Jefcoat SC Jr et al (1998) Isolation and characterization of a novel coactivator protein, NCoA-62, involved in vitamin D-mediated transcription. J Biol Chem 273(26):16434–16441

8. Berndt SI, Dodson JL, Huang WY et al (2006) A systematic review of vitamin D receptor gene polymorphisms and prostate cancer risk. J Urol 175(5):1613–1623

9. Berwick M, Armstrong BK, Ben-Porat L et al (2005) Sun exposure and mortality from melanoma. J Natl Cancer Inst 97(3):195–199

10. Bikle DD (2004) Vitamin D and skin cancer. J Nutr 134(12 Suppl):3472S–3478S

11. Bikle DD, Chang S, Crumrine D et al (2004) 25 Hydroxyvitamin D 1 alpha-hydroxylase is required for optimal epidermal differentiation and permeability barrier homeostasis. J Invest Dermatol 122(4):984–992

12. Bikle DD, Nemanic MK, Gee E et al (1986) 1, 25-Dihydroxyvitamin D3 production by human keratinocytes. Kinetics and regulation. J Clin Invest 78(2):557–566

13. Bikle DD, Ng D, Oda Y et al (2002) The vitamin D response element of the involucrin gene mediates its regulation by 1, 25-dihydroxyvitamin D3. J Invest Dermatol 119(5):1109–1113

14. Bikle DD, Pillai S (1993) Vitamin D, calcium, and epidermal differentiation. Endocr Rev 14(1):3–19

15. Bikle DD, Xie Z, Ng D et al (2003) Squamous cell carcinomas fail to respond to the prodif-ferentiating actions of 1, 25(OH)2D: why? Recent Results Cancer Res 164:111–122

16. Blum A, Ellwanger U, Garbe C (1997) Seasonal patterns in the diagnosis of cutaneous malignant melanoma: analysis of the data of the German Central Malignant Melanoma Registry. Br J Dermatol 136(6):968–969

17. Boniol M, Armstrong BK, Dore JF (2006) Variation in incidence and fatality of melanoma by season of diagnosis in new South Wales, Australia. Cancer Epidemiol Biomarkers Prev 15(3):524–526

18. Boyce BF, Xing L (2008) Functions of RANKL/RANK/OPG in bone modeling and remod-eling. Arch Biochem Biophys 473(2):139–146

19. Brash DE, Rudolph JA, Simon JA et al (1991) A role for sunlight in skin cancer: UV-induced p53 mutations in squamous cell carcinoma. Proc Natl Acad Sci USA 88(22):10124–10128

20. Braun MM, Tucker MA, Devesa SS et al (1994) Seasonal variation in frequency of diagnosis of cutaneous malignant melanoma. Melanoma Res 4(4):235–241

21. Brown AJ, Slatopolsky E (2008) Vitamin D analogs: Therapeutic applications and mecha-nisms for selectivity. Mol Aspects Med 29(6):433–452

22. Bruch-Gerharz D, Ruzicka T, Kolb-Bachofen V (1998) Nitric oxide in human skin: current status and future prospects. J Invest Dermatol 110(1):1–7

23. Buell JF, Hanaway MJ, Thomas M et al (2005) Skin cancer following transplantation: the Israel Penn International Transplant Tumor Registry experience. Transplant Proc 37(2):962–963

24. Bulliard JL (2000) Site-specific risk of cutaneous malignant melanoma and pattern of sun exposure in New Zealand. Int J Cancer 85(5):627–632

25. Bulliard JL, Cox B, Elwood JM (1994) Latitude gradients in melanoma incidence and mortality in the non-Maori population of New Zealand. Cancer Causes Control 5(3):234–240

26. Cadet J, Voituriez L, Grand A et al (1985) Recent aspects of the photochemistry of nucleic acids and related model compounds. Biochimie 67(3–4):277–292

27. Chen N, Nomura M, She QB et al (2001) Suppression of skin tumorigenesis in c-Jun NH(2)-terminal kinase-2-deficient mice. Cancer Res 61(10):3908–3912

28. Chouvet C, Vicard E, Devonec M et al (1986) 1, 25-Dihydroxyvitamin D3 inhibitory effect on the growth of two human breast cancer cell lines (MCF-7, BT-20). J Steroid Biochem 24(1):373–376
29. Colston K, Colston MJ, Feldman D (1981) 1, 25-dihydroxyvitamin D3 and malignant melanoma: the presence of receptors and inhibition of cell growth in culture. Endocrinology 108(3):1083–1086
30. Cooper GM, Hausman RE (2004) The cell: a molecular approach, 3rd edn. Sinaeur, Sunderland
31. Cross HS, Pavelka M, Slavik J et al (1992) Growth control of human colon cancer cells by vitamin D and calcium in vitro. J Natl Cancer Inst 84(17):1355–1357
32. de Gruijl FR, van Kranen HJ, Mullenders LH (2001) UV-induced DNA damage, repair, mutations and oncogenic pathways in skin cancer. J Photochem Photobiol B 63(1–3):19–27
33. De Haes P, Garmyn M, Degreef H et al (2003) 1, 25-Dihydroxyvitamin D3 inhibits ultraviolet B-induced apoptosis, Jun kinase activation, and interleukin-6 production in primary human keratinocytes. J Cell Biochem 89(4):663–673
34. De Haes P, Garmyn M, Verstuyf A et al (2005) 1, 25-Dihydroxyvitamin D3 and analogues protect primary human keratinocytes against UVB-induced DNA damage. J Photochem Photobiol B 78(2):141–148
35. De Haes P, Garmyn M, Verstuyf A et al (2004) Two 14-epi analogues of 1, 25-dihydroxyvitamin D3 protect human keratinocytes against the effects of UVB. Arch Dermatol Res 295(12):527–534
36. de Vries E, Coebergh JW (2004) Cutaneous malignant melanoma in Europe. Eur J Cancer 40(16):2355–2366
37. Decraene D, Agostinis P, Pupe A et al (2001) Acute response of human skin to solar radiation: regulation and function of the p53 protein. J Photochem Photobiol B 63(1–3):78–83
38. Deeb KK, Trump DL, Johnson CS (2007) Vitamin D signalling pathways in cancer: potential for anticancer therapeutics. Nat Rev Cancer 7(9):684–700
39. Dixon KM, Deo SS, Norman AW et al (2007) In vivo relevance for photoprotection by the vitamin D rapid response pathway. J Steroid Biochem Mol Biol 103(3–5):451–456
40. Dixon KM, Deo SS, Wong G et al (2005) Skin cancer prevention: a possible role of 1, 25dihydroxyvitamin D3 and its analogs. J Steroid Biochem Mol Biol 97(1–2):137–143
41. Dlugosz A, Merlino G, Yuspa SH (2002) Progress in cutaneous cancer research. J Investig Dermatol Symp Proc 7(1):17–26
42. Dusso A, Lopez-Hilker S, Rapp N et al (1988) Extra-renal production of calcitriol in chronic renal failure. Kidney Int 34(3):368–375
43. Eelen G, Verlinden L, De Clercq P et al (2006) Vitamin D analogs and coactivators. Anticancer Res 26(4A):2717–2721
44. Eisman JA, Martin TJ, MacIntyre I et al (1979) 1, 25-dihydroxyvitamin-D-receptor in breast cancer cells. Lancet 2(8156–8157):1335–1336
45. Elsner P, Holzle E, Diepgen T et al (2007) Recommendation: daily sun protection in the prevention of chronic UV-induced skin damage. J Dtsch Dermatol Ges 5(2):166–173
46. Elwood JM, Jopson J (1997) Melanoma and sun exposure: an overview of published studies. Int J Cancer 73(2):198–203
47. English DR, Armstrong BK, Kricker A et al (1998) Demographic characteristics, pigmentary and cutaneous risk factors for squamous cell carcinoma of the skin: a case-control study. Int J Cancer 76(5):628–634
48. Fang Y, van Meurs JB, D'Alesio A et al (2005) Promoter and 3′-untranslated-region haplotypes in the vitamin d receptor gene predispose to osteoporotic fracture: the rotterdam study. Am J Hum Genet 77(5):807–823
49. Feldman GF, Pike WJ (2005) Vitamin D, 2nd edn. Academic, San Diego
50. Food and Nutrition Board (1997) Dietary reference intakes for Ca, P, Mg, vitamin D and F. National Academic Press, Wiahington, DC
51. Frampton RJ, Suva LJ, Eisman JA et al (1982) Presence of 1, 25-dihydroxyvitamin D3 receptors in established human cancer cell lines in culture. Cancer Res 42(3):1116–1119

9 Molecular Biology of Vitamin D Metabolism and Skin Cancer

52. Freedman LP (1999) Multimeric coactivator complexes for steroid/nuclear receptors. Trends Endocrinol Metab 10(10):403–407
53. Fritsche J, Mondal K, Ehrnsperger A et al (2003) Regulation of 25-hydroxyvitamin D3–1 alpha-hydroxylase and production of 1 alpha, 25-dihydroxyvitamin D3 by human dendritic cells. Blood 102(9):3314–3316
54. Fu GK, Lin D, Zhang MY et al (1997) Cloning of human 25-hydroxyvitamin D-1 alpha-hydroxylase and mutations causing vitamin D-dependent rickets type 1. Mol Endocrinol 11(13):1961–1970
55. Garland CF, Garland FC (1980) Do sunlight and vitamin D reduce the likelihood of colon cancer? Int J Epidemiol 9(3):227–231
56. Garland FC, Garland CF, Gorham ED et al (1990) Geographic variation in breast cancer mortality in the United States: a hypothesis involving exposure to solar radiation. Prev Med 19(6):614–622
57. Geller AC, Swetter SM, Brooks K et al (2007) Screening, early detection, and trends for melanoma: current status (2000–2006) and future directions. J Am Acad Dermatol 57(4):555–572, quiz 73–6
58. Goltzman D, White J, Kremer R (2001) Studies of the effects of 1, 25-dihydroxyvitamin D on skeletal and calcium homeostasis and on inhibition of tumor cell growth. J Steroid Biochem Mol Biol 76(1–5):43–47
59. Green A, MacLennan R, Youl P et al (1993) Site distribution of cutaneous melanoma in Queensland. Int J Cancer 53(2):232–236
60. Grover SR, Morley R (2001) Vitamin D deficiency in veiled or dark-skinned pregnant women. Med J Aust 175(5):251–252
61. Gupta R, Dixon KM, Deo SS et al (2007) Photoprotection by 1, 25 dihydroxyvitamin D3 is associated with an increase in p53 and a decrease in nitric oxide products. J Invest Dermatol 127(3):707–715
62. Hansen CM, Hamberg KJ, Binderup E et al (2000) Seocalcitol (EB 1089): a vitamin D analogue of anti-cancer potential. Background, design, synthesis, pre-clinical and clinical evaluation. Curr Pharm Des 6(7):803–828
63. Haussler MR, Whitfield GK, Haussler CA et al (1998) The nuclear vitamin D receptor: biological and molecular regulatory properties revealed. J Bone Miner Res 13(3):325–349
64. Hawker NP, Pennypacker SD, Chang SM et al (2007) Regulation of human epidermal keratinocyte differentiation by the vitamin D receptor and its coactivators DRIP205, SRC2, and SRC3. J Invest Dermatol 127(4):874–880
65. Hewison M, Freeman L, Hughes SV et al (2003) Differential regulation of vitamin D receptor and its ligand in human monocyte-derived dendritic cells. J Immunol 170(11):5382–5390
66. Holick MF (1994) McCollum Award Lecture, 1994: vitamin D–new horizons for the 21st century. Am J Clin Nutr 60(4):619–630
67. Holick MF (1996) Vitamin D and bone health. J Nutr 126(4 Suppl):1159S–1164S
68. Holick MF (2001) Sunlight "D"ilemma: risk of skin cancer or bone disease and muscle weakness. Lancet 357(9249):4–6
69. Holick MF (2003) Evolution and function of vitamin D. Recent Results Cancer Res 164:3–28
70. Hollis BW (2005) Circulating 25-hydroxyvitamin D levels indicative of vitamin D sufficiency: implications for establishing a new effective dietary intake recommendation for vitamin D. J Nutr 135(2):317–322
71. Holman CD, Armstrong BK (1984) Cutaneous malignant melanoma and indicators of total accumulated exposure to the sun: an analysis separating histogenetic types. J Natl Cancer Inst 73(1):75–82
72. Horiuchi N, Clemens TL, Schiller AL et al (1985) Detection and developmental changes of the 1, 25-(OH)2–D3 receptor concentration in mouse skin and intestine. J Invest Dermatol 84(6):461–464
73. Hosomi J, Hosoi J, Abe E et al (1983) Regulation of terminal differentiation of cultured mouse epidermal cells by 1 alpha, 25-dihydroxyvitamin D3. Endocrinology 113(6):1950–1957

74. Huang C, Ma W, Ding M et al (1997) Direct evidence for an important role of sphingomyelinase in ultraviolet-induced activation of c-Jun N-terminal kinase. J Biol Chem 272(44):27753–27757
75. Huhtakangas JA, Olivera CJ, Bishop JE et al (2004) The vitamin D receptor is present in caveolae-enriched plasma membranes and binds 1 alpha, 25(OH)2-vitamin D3 in vivo and in vitro. Mol Endocrinol 18(11):2660–2671
76. Hutchinson PE, Osborne JE, Lear JT et al (2000) Vitamin D receptor polymorphisms are associated with altered prognosis in patients with malignant melanoma. Clin Cancer Res 6(2):498–504
77. Ingles SA, Ross RK, Yu MC et al (1997) Association of prostate cancer risk with genetic polymorphisms in vitamin D receptor and androgen receptor. J Natl Cancer Inst 89(2):166–170
78. Jaken S, Yuspa SH (1988) Early signals for keratinocyte differentiation: role of Ca2 + -mediated inositol lipid metabolism in normal and neoplastic epidermal cells. Carcinogenesis 9(6):1033–1038
79. John EM, Schwartz GG, Koo J et al (2005) Sun exposure, vitamin D receptor gene polymorphisms, and risk of advanced prostate cancer. Cancer Res 65(12):5470–5479
80. Jones G (2007) Expanding role for vitamin D in chronic kidney disease: importance of blood 25-OH-D levels and extra-renal 1alpha-hydroxylase in the classical and nonclassical actions of 1alpha, 25-dihydroxyvitamin D(3). Semin Dial 20(4):316–324
81. Jones G, Blizzard C, Riley MD et al (1999) Vitamin D levels in prepubertal children in Southern Tasmania: prevalence and determinants. Eur J Clin Nutr 53(10):824–829
82. Kim MS, Fujiki R, Kitagawa H et al (2007) 1alpha, 25(OH)2D3-induced DNA methylation suppresses the human CYP27B1 gene. Mol Cell Endocrinol 265–266:168–173
83. Kim S, Shevde NK, Pike JW (2005) 1, 25-Dihydroxyvitamin D3 stimulates cyclic vitamin D receptor/retinoid X receptor DNA-binding, co-activator recruitment, and histone acetylation in intact osteoblasts. J Bone Miner Res 20(2):305–317
84. Kim YS, MacDonald PN, Dedhar S et al (1996) Association of 1 alpha, 25-dihydroxyvitamin D3-occupied vitamin D receptors with cellular membrane acceptance sites. Endocrinology 137(9):3649–3658
85. Kraemer KH, Lee MM, Scotto J (1987) Xeroderma pigmentosum. Cutaneous, ocular, and neurologic abnormalities in 830 published cases. Arch Dermatol 123(2):241–250
86. Kricker A, Armstrong BK, English DR et al (1991) Pigmentary and cutaneous risk factors for non-melanocytic skin cancer–a case-control study. Int J Cancer 48(5):650–662
87. Kricker A, Armstrong BK, Hughes AM et al (2008) Personal sun exposure and risk of non Hodgkin lymphoma: a pooled analysis from the Interlymph Consortium. Int J Cancer 122(1):144–154
88. Kulms D, Schwarz T (2000) Molecular mechanisms of UV-induced apoptosis. Photodermatol Photoimmunol Photomed 16(5):195–201
89. Kurokawa R, Yu VC, Naar A et al (1993) Differential orientations of the DNA-binding domain and carboxy-terminal dimerization interface regulate binding site selection by nuclear receptor heterodimers. Genes Dev 7(7B):1423–1435
90. Lambert PW, Stern PH, Avioli RC et al (1982) Evidence for extrarenal production of 1 alpha, 25-dihydroxyvitamin D in man. J Clin Invest 69(3):722–725
91. Lee JA, Scotto J (1993) Melanoma: linked temporal and latitude changes in the United States. Cancer Causes Control 4(5):413–418
92. Lee SK, Lorenzo JA (1999) Parathyroid hormone stimulates TRANCE and inhibits osteoprotegerin messenger ribonucleic acid expression in murine bone marrow cultures: correlation with osteoclast-like cell formation. Endocrinology 140(8):3552–3561
93. Lehmann B, Genehr T, Knuschke P et al (2001) UVB-induced conversion of 7-dehydrocholesterol to 1alpha, 25-dihydroxyvitamin D3 in an in vitro human skin equivalent model. J Invest Dermatol 117(5):1179–1185
94. Leong GM, Wang KS, Marton MJ et al (1998) Interaction between the retinoid X receptor and transcription factor IIB is ligand-dependent in vivo. J Biol Chem 273(4):2296–2305
95. Lewin B (2004) Genes VIII. Prentice Hall, Upper Saddle River
96. Li C, Liu Z, Wang LE et al (2008) Haplotype and genotypes of the VDR gene and cutaneous melanoma risk in non-Hispanic whites in Texas: a case-control study. Int J Cancer 122(9):2077–2084

97. Li C, Liu Z, Zhang Z et al (2007) Genetic variants of the vitamin D receptor gene alter risk of cutaneous melanoma. J Invest Dermatol 127(2):276–280

98. Lips P (2006) Vitamin D physiology. Prog Biophys Mol Biol 92(1):4–8

99. Lo PK, Huang SZ, Chen HC et al (2004) The prosurvival activity of p53 protects cells from UV-induced apoptosis by inhibiting c-Jun NH2-terminal kinase activity and mitochondrial death signaling. Cancer Res 64(23):8736–8745

100. Losel R, Wehling M (2003) Nongenomic actions of steroid hormones. Nat Rev Mol Cell Biol 4(1):46–56

101. Luscombe CJ, French ME, Liu S et al (2001) Outcome in prostate cancer associations with skin type and polymorphism in pigmentation-related genes. Carcinogenesis 22(9):1343–1347

102. Marks R, Foley PA, Jolley D et al (1995) The effect of regular sunscreen use on vitamin D levels in an Australian population. Results of a randomized controlled trial. Arch Dermatol 131(4):415–421

103. Masuda S, Jones G (2003) Vitamin D analogs–drug design based on proteins involved in vitamin D signal transduction. Curr Drug Targets Immune Endocr Metabol Disord 3(1):43–66

104. Matsumoto K, Azuma Y, Kiyoki M et al (1991) Involvement of endogenously produced 1, 25-dihydroxyvitamin D-3 in the growth and differentiation of human keratinocytes. Biochim Biophys Acta 1092(3):311–318

105. Matsuoka LY, Ide L, Wortsman J et al (1987) Sunscreens suppress cutaneous vitamin D3 synthesis. J Clin Endocrinol Metab 64(6):1165–1168

106. Miller GJ, Stapleton GE, Hedlund TE et al (1995) Vitamin D receptor expression, 24-hydroxylase activity, and inhibition of growth by 1alpha, 25-dihydroxyvitamin D3 in seven human prostatic carcinoma cell lines. Clin Cancer Res 1(9):997–1003

107. Muindi JR, Modzelewski RA, Peng Y et al (2004) Pharmacokinetics of 1alpha, 25-dihydroxyvitamin D3 in normal mice after systemic exposure to effective and safe antitumor doses. Oncology 66(1):62–66

108. Murayama A, Kim MS, Yanagisawa J et al (2004) Transrepression by a liganded nuclear receptor via a bHLH activator through co-regulator switching. Embo J 23(7):1598–1608

109. Murayama A, Takeyama K, Kitanaka S et al (1998) The promoter of the human 25-hydroxyvitamin D3 1 alpha-hydroxylase gene confers positive and negative responsiveness to PTH, calcitonin, and 1 alpha, 25(OH)2D3. Biochem Biophys Res Commun 249(1):11–16

110. Nakajima S, Hsieh JC, MacDonald PN et al (1994) The C-terminal region of the vitamin D receptor is essential to form a complex with a receptor auxiliary factor required for high affinity binding to the vitamin D-responsive element. Mol Endocrinol 8(2):159–172

111. Nemere I (2005) The 1, 25D3-MARRS protein: contribution to steroid stimulated calcium uptake in chicks and rats. Steroids 70(5–7):455–457

112. Nemere I, Yoshimoto Y, Norman AW (1984) Calcium transport in perfused duodena from normal chicks: enhancement within fourteen minutes of exposure to 1, 25-dihydroxyvitamin D3. Endocrinology 115(4):1476–1483

113. Norman AW (2006) Minireview: vitamin D receptor: new assignments for an already busy receptor. Endocrinology 147(12):5542–5548

114. Norman AW, Bouillon R, Farach-Carson MC et al (1993) Demonstration that 1 beta, 25-dihydroxyvitamin D3 is an antagonist of the nongenomic but not genomic biological responses and biological profile of the three A-ring diastereomers of 1 alpha, 25-dihydroxyvitamin D3. J Biol Chem 268(27):20022–20030

115. Norman AW, Henry HL, Bishop JE et al (2001) Different shapes of the steroid hormone 1alpha, 25(OH)(2)-vitamin D(3) act as agonists for two different receptors in the vitamin D endocrine system to mediate genomic and rapid responses. Steroids 66(3–5): 147–158

116. Nowson CA, Diamond TH, Pasco JA et al (2004) Vitamin D in Australia. Issues and recommendations. Aust Fam Physician 33(3):133–138

117. Oda Y, Sihlbom C, Chalkley RJ et al (2003) Two distinct coactivators, DRIP/mediator and SRC/p160, are differentially involved in vitamin D receptor transactivation during keratinocyte differentiation. Mol Endocrinol 17(11):2329–2339

118. Ohyama Y, Ozono K, Uchida M et al (1996) Functional assessment of two vitamin D-responsive elements in the rat 25-hydroxyvitamin D3 24-hydroxylase gene. J Biol Chem 271(48):30381–30385

119. Oliveria SA, Saraiya M, Geller AC et al (2006) Sun exposure and risk of melanoma. Arch Dis Child 91(2):131–138

120. Panda DK, Miao D, Bolivar I et al (2004) Inactivation of the 25-hydroxyvitamin D 1alpha-hydroxylase and vitamin D receptor demonstrates independent and interdependent effects of calcium and vitamin D on skeletal and mineral homeostasis. J Biol Chem 279(16):16754–16766

121. Peehl DM, Skowronski RJ, Leung GK et al (1994) Antiproliferative effects of 1, 25-dihydroxyvitamin D3 on primary cultures of human prostatic cells. Cancer Res 54(3):805–810

122. Perez AV, Picotto G, Carpentieri AR et al (2008) Minireview on regulation of intestinal calcium absorption. Emphasis on molecular mechanisms of transcellular pathway. Digestion 77(1):22–34

123. Peterson CL, Logie C (2000) Recruitment of chromatin remodeling machines. J Cell Biochem 78(2):179–185

124. Pillai S, Bikle DD, Elias PM (1988) 1, 25-Dihydroxyvitamin D production and receptor binding in human keratinocytes varies with differentiation. J Biol Chem 263(11):5390–5395

125. Plotnikoff GA, Quigley JM (2003) Prevalence of severe hypovitaminosis D in patients with persistent, nonspecific musculoskeletal pain. Mayo Clin Proc 78(12):1463–1470

126. Polednak AP (1984) Seasonal patterns in the diagnosis of malignant melanoma of skin and eye in upstate New York. Cancer 54(11):2587–2594

127. Posner GH, Jeon HB, Sarjeant A et al (2004) Low-calcemic, efficacious, 1alpha, 25-dihydroxyvitamin D3 analog QW-1624F2–2: calcemic dose-response determination, preclinical genotoxicity testing, and revision of A-ring stereochemistry. Steroids 69(11–12):757–762

128. Querings K, Girndt M, Geisel J et al (2006) 25-hydroxyvitamin D deficiency in renal transplant recipients. J Clin Endocrinol Metab 91(2):526–529

129. Querings K, Reichrath J (2004) A plea for the analysis of Vitamin-D levels in patients under photoprotection, including patients with xeroderma pigmentosum (XP) and basal cell nevus syndrome (BCNS). Cancer Causes Control 15(2):219

130. Rass K, Reichrath J (2008) UV damage and DNA repair in malignant melanoma and non-melanoma skin cancer. Adv Exp Med Biol 624:162–178

131. Ratnam AV, Bikle DD, Cho JK (1999) 1, 25 dihydroxyvitamin D3 enhances the calcium response of keratinocytes. J Cell Physiol 178(2):188–196

132. Ravanat JL, Douki T, Cadet J (2001) Direct and indirect effects of UV radiation on DNA and its components. J Photochem Photobiol B 63(1–3):88–102

133. Reichrath J (2006) The challenge resulting from positive and negative effects of sunlight: how much solar UV exposure is appropriate to balance between risks of vitamin D deficiency and skin cancer? Prog Biophys Mol Biol 92(1):9–16

134. Ross MH, Kaye GI, Pawlina W (2003) Histology: a text and atlas, 4th edn. Lippincott Williams & Wilkins, Philadelphia

135. Sambrook P (2005) Vitamin D and fractures: quo vadis? Lancet 365(9471):1599–1600

136. Schwartz SM, Armstrong BK, Weiss NS (1987) Seasonal variation in the incidence of cutaneous malignant melanoma: an analysis by body site and histologic type. Am J Epidemiol 126(1):104–111

137. Scotto J, Nam JM (1980) Skin melanoma and seasonal patterns. Am J Epidemiol 111(3):309–314

138. Sebag M, Henderson J, Rhim J et al (1992) Relative resistance to 1, 25-dihydroxyvitamin D3 in a keratinocyte model of tumor progression. J Biol Chem 267(17):12162–12167

139. Segaert S, Garmyn M, Degreef H et al (1997) Retinoic acid modulates the anti-proliferative effect of 1, 25-dihydroxyvitamin D3 in cultured human epidermal keratinocytes. J Invest Dermatol 109(1):46–54

140. Setlow RB (1966) Cyclobutane-type pyrimidine dimers in polynucleotides. Science 153(734):379–386

141. Shabahang M, Buras RR, Davoodi F et al (1993) 1, 25-Dihydroxyvitamin D3 receptor as a marker of human colon carcinoma cell line differentiation and growth inhibition. Cancer Res 53(16):3712–3718

9 Molecular Biology of Vitamin D Metabolism and Skin Cancer

142. Shaulian E, Schreiber M, Piu F et al (2000) The mammalian UV response: c-Jun induction is required for exit from p53-imposed growth arrest. Cell 103(6):897–907
143. Sinclair C (2006) Risks and benefits of sun exposure: implications for public health practice based on the Australian experience. Prog Biophys Mol Biol 92(1):173–178
144. Smith ML, Chen IT, Zhan Q et al (1995) Involvement of the p53 tumor suppressor in repair of u.v.-type DNA damage. Oncogene 10(6):1053–1059
145. Sollitto RB, Kraemer KH, DiGiovanna JJ (1997) Normal vitamin D levels can be maintained despite rigorous photoprotection: six years' experience with xeroderma pigmentosum. J Am Acad Dermatol 37(6):942–947
146. Spencer JM, Kahn SM, Jiang W et al (1995) Activated ras genes occur in human actinic keratoses, premalignant precursors to squamous cell carcinomas. Arch Dermatol 131(7):796–800
147. St-Arnaud R (2008) The direct role of vitamin D on bone homeostasis. Arch Biochem Biophys 473(2):225–230
148. Struhl K, Moqtaderi Z (1998) The TAFs in the HAT. Cell 94(1):1–4
149. Stumpf WE, Sar M, Reid FA et al (1979) Target cells for 1, 25-dihydroxyvitamin D3 in intestinal tract, stomach, kidney, skin, pituitary, and parathyroid. Science 206(4423):1188–1190
150. Su MJ, Bikle DD, Mancianti ML et al (1994) 1, 25-Dihydroxyvitamin D3 potentiates the keratinocyte response to calcium. J Biol Chem 269(20):14723–14729
151. Suzuki T, Inukai M (2006) Effects of nitrite and nitrate on DNA damage induced by ultraviolet light. Chem Res Toxicol 19(3):457–462
152. Tagami T, Lutz WH, Kumar R et al (1998) The interaction of the vitamin D receptor with nuclear receptor corepressors and coactivators. Biochem Biophys Res Commun 253(2):358–363
153. Taylor JA, Hirvonen A, Watson M et al (1996) Association of prostate cancer with vitamin D receptor gene polymorphism. Cancer Res 56(18):4108–4110
154. Turunen MM, Dunlop TW, Carlberg C et al (2007) Selective use of multiple vitamin D response elements underlies the 1 alpha, 25-dihydroxyvitamin D3-mediated negative regulation of the human CYP27B1 gene. Nucleic Acids Res 35(8):2734–2747
155. Uitterlinden AG, Fang Y, Van Meurs JB et al (2004) Genetics and biology of vitamin D receptor polymorphisms. Gene 338(2):143–156
156. van Abel M, Hoenderop JG, Bindels RJ (2005) The epithelial calcium channels TRPV5 and TRPV6: regulation and implications for disease. Naunyn Schmiedebergs Arch Pharmacol 371(4):295–306
157. van der Schroeff JG, Evers LM, Boot AJ et al (1990) Ras oncogene mutations in basal cell carcinomas and squamous cell carcinomas of human skin. J Invest Dermatol 94(4):423–425
158. Verheij M, Bose R, Lin XH et al (1996) Requirement for ceramide-initiated SAPK/JNK signalling in stress-induced apoptosis. Nature 380(6569):75–79
159. Vieth R (1999) Vitamin D supplementation, 25-hydroxyvitamin D concentrations, and safety. Am J Clin Nutr 69(5):842–856
160. Welsh J (2004) Vitamin D and breast cancer: insights from animal models. Am J Clin Nutr 80(6 Suppl):1721S–1724S
161. Wolpowitz D, Gilchrest BA (2006) The vitamin D questions: how much do you need and how should you get it? J Am Acad Dermatol 54(2):301–317
162. Wong G, Gupta R, Dixon KM et al (2004) 1, 25-Dihydroxyvitamin D and three low-calcemic analogs decrease UV-induced DNA damage via the rapid response pathway. J Steroid Biochem Mol Biol 89–90(1–5):567–570
163. Xia Z, Dickens M, Raingeaud J et al (1995) Opposing effects of ERK and JNK-p38 MAP kinases on apoptosis. Science 270(5240):1326–1331
164. Xie Z, Bikle DD (1997) Cloning of the human phospholipase C-gamma1 promoter and identification of a DR6-type vitamin D-responsive element. J Biol Chem 272(10):6573–6577
165. Xie Z, Bikle DD (1999) Phospholipase C-gamma1 is required for calcium-induced keratinocyte differentiation. J Biol Chem 274(29):20421–20424
166. Zehnder D, Bland R, Williams MC et al (2001) Extrarenal expression of 25-hydroxyvitamin d(3)-1 alpha-hydroxylase. J Clin Endocrinol Metab 86(2):888–894

Chapter 10
Vitamin D and Prostate Cancer

Christine M. Barnett and Tomasz M. Beer

Abstract Following epidemiological observations that suggested links between low vitamin D exposure and increased risk of prostate cancer, interest in clarifying a potential role of this steroid hormone in prostate cancer has grown. While the results have been mixed, epidemiologic studies have suggested that severe vitamin D deficiency may increase the risk of clinically important prostate cancer. Laboratory investigation provides clear evidence of the potential of vitamin D receptor (VDR) ligands to induce growth arrest and promote apoptosis in a variety of cancer models. Because there are hundreds of vitamin D responsive genes, multiple mechanisms for these observations have been proposed.

Prompted by clear evidence of dose-dependent antitumor effects, efforts to harness this knowledge to improve patient outcomes has focused primarily on the development of high dose calcitriol, often in combination with other anti-neoplastic agents. After encouraging phase II results, the phase III effort failed when excess deaths were reported in the experimental arm of a trial that compared calcitriol with docetaxel to prednisone with docetaxel. In addition to targeting the vitamin D receptor, the two arms of this study differed with respect to the dose, schedule, and dose intensity of the chemotherapy agent and steroids, making definitive conclusions about the potential of vitamin D receptor targeted therapy difficult. No prospective randomized studies aimed at prostate cancer prevention have been reported.

Continued efforts to target vitamin D signaling for prostate cancer prevention and treatment are needed in light of the strong preclinical evidence supporting the importance of this signaling pathway. Better understanding of the human prostate cancer's biologic heterogeneity in vitamin D sensitivity may allow for more robust identification of ways in which vitamin D can be harnessed to help men who suffer from this disease.

T.M. Beer (✉)
OHSU, Knight Cancer Institute, 3303 SW Bond Ave, CH14R,
Portland, OR 97239, USA
e-mail: beert@ohsu.edu

D.L. Trump and C.S. Johnson (eds.), *Vitamin D and Cancer*,
DOI 10.1007/978-1-4419-7188-3_10, © Springer Science+Business Media, LLC 2011

Keywords Skin cancer • Solar UV radiation • Vitamin D • Epidemiology • Prevention • Vitamin D receptor • 1,25-dihydroxyvitamin D • Keratinocytes • Differentiation • Photoprotection • Vitamin D analogs • Prostate cancer

Disclosure OHSU and Dr. Beer have a significant financial interest in Novacea a company that may have a commercial interest in the results of this research and technology. This potential conflict of interest has been reviewed and managed by OHSU and the Integrity Program Oversight Council.

Abbreviations

AIPC	Androgen independent prostate cancer
ASCENT	AIPC Study of Calcitriol Enhancing Taxotere
AUC	Area under the concentration curve
C_{max}	Peak blood calcitriol concentrations
EGFR	Epidermal growth factor receptors
NMU	N-nitroso-N-methylurea
NSAIDS	Non-steroidal anti-inflammatory agents
RXR	Retinoid-X receptor
VDR	Vitamin D receptors
VDRE	Vitamin D response element

10.1 Introduction

Stimulated by epidemiological observations that suggest links between low vitamin D exposure and increased risk of prostate cancer [1, 2], a number of investigators have sought to examine the hypothesis that vitamin D receptor (VDR) signaling may impact prostate cancer risk, progression, outcomes, and treatment. This work continues to this day and has yielded encouraging but also conflicting results.

10.2 Vitamin D Physiology

Vitamin D is an important regulatory hormone in the human body that belongs to the steroid receptor superfamily. Its calcium regulatory activity is well known, but additional roles for vitamin D are being increasingly recognized. The principal hormonally active form of vitamin D, 1,25-OH$_2$ vitamin D, is synthesized through a number of steps starting with conversion of 7-deoxycholesterol to pre-vitamin D catalyzed by UV-B sunlight. Pre-vitamin D is then converted to 25-OH vitamin D in the liver by the enzyme 25-hydroxylase. The enzyme 1-alpha-hydroxylase is needed for the final conversion of 25-OH vitamin D to 1,25-OH$_2$ vitamin D.

10 Vitamin D and Prostate Cancer

This endocrine enzyme is located predominately in the kidney, but has also been found in other tissues such as the colon and the prostate [3–7]. The circulating levels of $1,25\text{-OH}_2$ vitamin D are tightly regulated by calcium levels and parathyroid hormone. Renal 1-alpha-hydroxylase activity is enhanced by hypocalcemia through transcriptional regulation. The expression of the CYP27B1 gene, which encodes 1-alpha-hydroxylase, is upregulated by parathyroid hormone [8]. $1,25\text{-OH}_2$ vitamin D in turn inhibits transcription of 1-alpha-hydroxylase creating a regulatory feedback loop [9, 10]. In contrast, non-renal 1-alpha-hydroxylase, that is responsible for autocrine and paracrine, but not endocrine vitamin D activation, is thought to be constitutively active [7, 11]. Unlike its renal counterpart, extra-renal 1-alpha-hydroxylase is not down-regulated by its downstream product, $1,25\text{-OH}_2$ vitamin D [12]. Thus, tissues that express 1-alpha-hydroxylase, including potentially certain tumors, may experience tissue $1,25\text{-OH}_2$ vitamin D levels that reflect circulating levels of the substrate (25-OH vitamin D). $1,25\text{-OH}_2$ vitamin D also induces the CYP27A1 gene that encodes 24-hydroxylase. This enzyme catalyses 24-hydroxylation of 25-OH vitamin D, creating, $24,25\text{-OH}_2$ vitamin D, a hormonally inactive alternative to $1,25\text{-OH}_2$ vitamin D [9, 11, 13]. Local activity of the competing 24-hydroxylase in some cancer tissues, may also impact on tissue $1,25\text{-OH}_2$ vitamin D concentrations by diverting the substrate [11, 14].

Prostate carcinoma cell lines express vitamin D receptors (VDR) [15–17]. VDR expression in human prostate cancer specimens has also been reported [18]. Interestingly, prostate cell lines also express 1-alpha-hydroxylase [3, 5]. However, it has been shown in cell culture that prostate cancer cells have reduced 1-alpha-hydroxylase activity when compared to normal prostate epithelial cells, [3, 19]. As a consequence, prostate cancer cells may lose the ability to convert 25-OH vitamin D to $1,25\text{-OH}_2$ vitamin D. Loss of the ability to locally produce activated vitamin D may result in the loss of an important break on cancer cell proliferation. This hypothesis has led to the suggestion that 1-alpha-hydroxylase may act as a tumor suppressor gene [20]. Because VDR exerts predominantly growth inhibitory effects on prostate cancer cell lines, it is plausible that loss of the autocrine vitamin D loop with reduced 1-alpha-hydroxylase activity contributes to the progression of prostate cancer [21]. Also, VDR activity has been shown to be altered in prostate cancer cells, with decreased ligand-inducible DNA binding activity, altered recruitment of coregulators SRC-1 and CBP, and increased recruitment of SMRT corepressor [22]. These alterations may further exacerbate the effects of a relative deficiency of $1,25\text{-OH}_2$ vitamin D concentrations in prostate cancer.

10.3 The Biologic Activity of Vitamin D in Prostate Cancer

Vitamin D activity involves both rapid induction of cell signaling pathways, and genomic receptor-mediated pathways. The vitamin D receptor is an intracellular steroid receptor that acts as a ligand activated transcription factor [23].

When VDR is activated it binds to the promoter regions of specific genes and regulates the transcription of mRNA of these genes. The VDR (once activated by vitamin D) forms a heterodimer with the retinoid-X receptor (RXR) and then binds to the regulatory region of the gene in the presence of a coactivator and corepressor complex. Many genes involving calcium and bone metabolism including osteoclastin [24] and osteopontin [25] are regulated this way. In addition, other genes regulating the cell cycle, apoptosis, and cell proliferation have been found to have a vitamin D response element (VDRE) and are induced or down-regulated by vitamin D. Some genes with vitamin D response elements that are activated by vitamin D include p21 [26] and GADD45 [27], which play an important role in cell cycle regulation, and CYP2A1, [11, 28] which encodes 24-hydroxylase. Notable genes down-regulated by vitamin D include PTH [29] and CYP2B1 [8], which regulate 1-alpha-hydroxylase production. Also, vitamin D has been shown to down-regulate insulin-like growth factor [30] and Bcl-2 [31]. Through the regulation of these genes as well many others, vitamin D can shift the balance of cell survival signals in favor of apoptosis and growth arrest. There are many other vitamin D-regulated genes and a partial list of these is provided in Table 10.1. Notably, many of these genes are important regulators of cell growth and apoptosis.

In addition to VDR-mediated activities of vitamin D, there are rapid non-genomic signals induced by vitamin D. Examples include rapid intestinal absorption of calcium induced by vitamin D [32] as well as the induction of signaling cascades such as Raf-MEK-MAPK-ERK signaling pathway [11, 33–35] and protein kinase C [36] among others. These rapid signals may be mediated by translocation of the VDR to the plasma membrane [11, 37] (Table 10.2).

Because the VDR regulates so many genes including those effecting cell growth and cancer development, many recent studies have been devoted to looking at different genetic variants of the VDR and their relation to prostate cancer risk. Most of these studies have been focused on five VDR gene polymorphisms, the *poly-A* microsatellite, and four restriction sites: *FokI, BsmI, ApaI,* and *TaqI.* Much like epidemiologic studies with serum levels of vitamin D, some studies involving these polymorphisms have shown strong associations with increased prostate cancer risk, but overall results between different studies are inconclusive [13, 38–42].

10.4 Mechanisms of Anti-neoplastic Activity

Because there are so many different genes affected by vitamin D, different anti-neoplastic activity mechanisms predominate under different experimental conditions, and in different tumor models. Nevertheless, vitamin D activity against prostate cancer is seen across a range of tumor models.

Not surprisingly, given that multiple cell cycle regulatory genes are regulated by vitamin D, a number of investigators have demonstrated vitamin D-induced growth arrest in G1 [11, 26, 43–46]. This has been attributed, at least in part, to transcriptional

10 Vitamin D and Prostate Cancer

Table 10.1 Selected genes found to have a functional VDRE

Calcium/bone metabolism:
Osteoclastin [24]
Osteopopontin [25]
Bone sialoprotein [155]
PTH (repression) [29]
PTHrp [156]
Calcium binding proteins (calbindin, D28-k, dak) [157]
RANKL [158]
Cell cycle regulators:
p21 [26]
GADd45 [27]
IGFBP3 [159, 160]
Cell adhesion:
Fibronectin [161]
Beta-3 integrin [162]
Involucrin [163]
Cell signaling:
cfos [164]
Phospholipase C [165]
EGFR [166]
TNF-alpha [65]
Vitamin D metabolism and others:
Runx2/Cbfa1 [167]
Insulin receptor [30]
Carbonic anhydrase II [168]
Human growth hormone [169]
Fructose 1,6 bisphosphatase [170]
CYP2A1 [11, 28, 171]
CYP2B1 (repression) [8]
25(OH)D3 24-hydroxylase [172]

Table 10.2 Non-genomic signals regulated by vitamin D

Protein kinase C [33, 36]
Raf-MEK-MAPK-ERK pathway [33–35]
Protein lipase A [173]
Protein kinase A [174]
Phosphatidyl inositol 3-kinase/Akt [11, 32]
Rapid intestinal calcium absorption [11, 32]
Bcl-2 downregulation [31]
Interruption of IL-8 [63]

activation of cyclin-dependent kinase inhibitors p21 (directly) and p27 (indirectly) [26, 45]. While vitamin D regulates the transcription of these cell cycle regulators, it also inhibits some mitosis signaling pathways. These include, but are not limited to, epidermal growth factor receptors (EGFR), [47] c-myc, [48, 49] and ERK/MAPK [35, 50, 51] (see Table 10.1).

Specific to prostate cancer, multiple studies have shown the antiproliferative effects of vitamin D on prostate cancer cells in cell lines, [17] human primary culture, [52] and in rodent models [53].

While normal prostate cells express 1-alpha-hydroxylase, this activity can be lost when prostate cancer develops, [3, 19, 54] perhaps reducing the cell's ability to produce 1,25-OH$_2$ vitamin D from its circulating precursor, 25-OH vitamin D. Loss of local 1-alpha-hydroxylase activity may render cancer cells dependent on circulating 1,25-OH$_2$ vitamin D for growth suppression activity. Indeed, restoring LNCaP cells 1-alpha-hydroxylase activity with gene transfer [3, 54] has been shown to restore effect of 25-OH vitamin D on cell proliferation. Interestingly, colon cancer cells rarely lose 1-alpha-hydroxylase activity and sometimes even have increased activity, [4] perhaps making colon cancer more responsive to the effects of circulating 25-OH vitamin D than prostate cancer [21]. These biologic differences may have significant clinical implications. Because circulating 1,25-OH$_2$ vitamin D levels are tightly regulated and remain relatively stable during mild deficiency states, tissues that rely on renally activated vitamin D for VDR signaling would remain relatively unaffected by vitamin D deficiency until it is severe. In contrast, tissues with significant local production of 1,25-OH$_2$ vitamin D would see differences in VDR signaling with changes in circulating 25-OH vitamin D levels, which more closely mirror the overall vitamin D status.

In animal models of cancer, the antineoplastic activity of vitamin D has been shown to translate into a reduction in metastatic potential. In rodent models, there has been demonstration of reduction in metastases with vitamin D therapy [55–57] and slowed growth of the prostate cancer [58, 59]. Reduced prostate cancer cell invasiveness with vitamin D therapy has been demonstrated in vitro by several investigators [37, 60–62]. 1,25 Vitamin D also decreases IL-8 signaling in prostate cancer, thus inhibiting endothelial migration and therefore inhibiting growth and invasion of the cancer [63].

In addition to growth inhibition, vitamin D induced apoptosis has been shown in several prostate cancer cell culture models. To explain this, vitamin D has been shown to down-regulate Bcl–2, [31] an important protein in anti-apoptotic pathways in prostate cancer cells, and other cancer cell lines. Vitamin D also upregulates expression of pro-apoptotic proteins BAK and BAX [64]. Down-regulation of insulin-like growth factor receptor in response to vitamin D has also been shown, [30] along with up-regulation of TNF-alpha, [65] all important in apoptotic pathways.

10.5 Epidemiology

10.5.1 UV Exposure and Prostate Cancer Risk

The hypothesis that vitamin D plays a role in prostate cancer biology was formulated after geographic studies showed that prostate cancer-related mortality was geographically dependent, with the greatest mortality in northern regions [66]. This geographic distribution is consistent with an inverse relationship between prostate cancer risk and UV exposure, and presumably, vitamin D levels [67]. After the initial

10 Vitamin D and Prostate Cancer

study by Hanchette, et al. other studies have also shown the correlation between living in areas characterized by low UV exposure and increased risk of prostate cancer diagnosis and death due to prostate cancer [68, 69]. One study measured exposure to UV radiation, a sunbathing score, and correlated low exposure to an increased risk of prostate cancer with an OR 3.03 for men with the lowest quartile of UV exposure [69]. Two recent studies have supported the hypothesis for a protective effect of sunlight [70, 71]. Two other studies done recently in Norway interestingly did not find a geographic or latitude dependent increased risk of prostate cancer mortality [72, 73] after correcting for season of diagnosis. Notably, these studies examined a limited range of latitudes as they considered only the Norwegian population (Table 10.3).

One possible explanation for the incomplete concordance among these studies may be rooted in the populations that were examined. If prostate cancer indeed relies on circulating 1,25-OH$_2$ vitamin D levels for VDR signaling, these would only be altered in states of relatively severe vitamin D deficiency. Normal homeostatic regulatory mechanisms maintain 1,25-OH$_2$ vitamin D levels across a fairly broad range of 25-OH vitamin D concentrations. Luscombe's study was done in the UK where there is a high prevalence of vitamin D deficiency and therefore changes in 25-OH vitamin D levels would have the most effect on tissue 1,25-OH$_2$ vitamin D levels. Another possible confounder in analyses of UV exposure is the seasonal nature of UV availability. Interestingly, several recent studies have linked the season of diagnosis and cancer mortality [72–74]. Patients diagnosed in the summer and fall had greater survival than patients diagnosed in the winter. Zhou et al. found that patients diagnosed and undergoing surgery for early stage lung cancer in the summer had a longer relapse-free survival than those that were diagnosed and underwent surgery in the winter (HR 0.33). Robsahm et al. found similar results for prostate cancer with a summer diagnosis of prostate cancer conferring a 20–30% reduction in risk of death when compared to other seasons of diagnosis. Recently, Lagunova et al. showed that patients diagnosed with prostate cancer in the summer and autumn had a better prognosis than those patients diagnosed in winter or spring with a relative risk of death of 0.8. This study was done in Norway where there is a relatively high prevalence of vitamin D deficiency and the seasonal variation in UV exposure is extreme. While the prostate cancer studies did not include measurement of vitamin D levels in the patients, a follow-up of the Harvard School of Public Health lung cancer study did. They reported that patients with early stage lung cancer whose vitamin D blood levels and vitamin D intake was above the median had a significantly lower risk of recurrence and death when compared to patients below the median for both of these measures (HR 0.67 and 0.64, respectively) [75].

10.5.2 Dietary Vitamin D and Calcium Intake and Prostate Cancer Risk

Relevant studies of diet and prostate cancer risk have focused not only on dietary intake of vitamin D, but also on calcium intake. High dietary calcium

Table 10.3 Studies of UV exposure and prostate cancer

Study	Location	Methods	Results
Hanchette et al. (1992) [66]	United States	Measured "epidemiology index" with cloud cover and latitude taken into account, "UV count" with latitude and altitude taken into account	Inverse correlation with areas of lower UV exposure and increased prostate cancer mortality
Luscombe, et al. (2001) [69]	UK	Cases and controls measured "sunbathing score" and UV exposure from questionnaire data	Increased risk of prostate cancer mortality with decreased UV exposure: OR 3.03 with lowest quartile of exposure
Grant et al. (2002) [68]	United States	UV exposure data from UVB radiation exposure map and USDA UV measuring stations, combined with cancer mortality rates 1970–1994	Inverse correlation with lower UV exposure and increased prostate cancer mortality
Bodialwa et al. (2003) [71]	UK	Cases and controls, measured months of cumulative UV exposure per year and "sunbathing score"	Cumulative UV exposure and sunbathing score lower in cancer group
Robsahm et al. (2004) [73]	Norway	Norway divided into eight regions based on latitude and climate. Mean annual "erythmogenic UV radiation" averaged for each region. Combined with cancer mortality data from 1964 to 1992	No geographic variation in prostate cancer mortality Diagnosis in autumn associated with decreased prostate cancer mortality (OR 0.83)
John EM et al. (2005) [70]	San Fransisco Bay Area	Compared measured skin pigmentation in non-sun exposed areas and sun exposed areas	Reduced risk of advanced prostate cancer with highest sun exposure (biggest difference in pigmentation) (OR 0.51)
Schwartz et al. (2006) [67]	United States	Prostate cancer data combined with UV index data from NOAA	Inverse correlation with areas of lower UV exposure and increased prostate cancer mortality, especially high risk north of 40° north latitude
Lagunova et al. (2007) [72]	Norway	Norway divided into groups based on annual ambient UV exposure and SCC of skin incidence	No latitude variation in prostate cancer mortality. Diagnosis in summer and autumn associated with decreased prostate cancer mortality (RR 0.8)

would be expected to reduce renal 1-alpha hydroxylation of 25-OH vitamin D [76, 77]. One limitation of these dietary studies is the lack of consistent concurrent measurement of blood calcium or vitamin D levels. A recent study, [78] however, did measure serum calcium levels and found that with serum calcium greater than 10.2 mg/dL there was an increased risk of mortality from prostate cancer. This was only statistically significant in sub-groups with high BMI and when separated out for race. Along these same lines, a 2004 meta-analysis reported an increased risk of prostate cancer with high milk consumption with an odds ratio of 1.68 [79]. Other studies have supported the association between high calcium intake and increased prostate cancer risk [77, 80, 81]. Consistent with these findings, some studies have shown that high milk consumption is associated with a reduction in circulating 1,25 vitamin D levels [76, 77]. However, these findings have not been either universal or completely consistent. There have been multiple studies that do not show an increased risk of prostate cancer with increased calcium intake [82–86]. Interestingly, a recent study correlated dietary, but not supplemental calcium intake to an increase in prostate cancer risk [87].

Overall, dietary studies that evaluate vitamin D intake have not shown a consistent protective effect for prostate cancer, [76, 80, 84, 88] as has been demonstrated for colon cancer. This observation is consistent with the hypothesis that loss of 1-alpha-hydroxylase activity in prostate cancer renders the tumor less susceptible to modest fluctuations of serum 25-OH vitamin D that occurs with variations in dietary intake. To the extent that circulating $1,25\text{-OH}_2$ vitamin D may be important in prostate cancer, only severe vitamin D deficiency states where renal 1-alpha-hydroxylation is reduced would be expected to result in adverse cancer outcomes.

10.5.3 Vitamin D Blood Levels and the Risk of Prostate Cancer

There are only a handful of epidemiologic studies that have measured vitamin D levels and examined the association with risk of prostate cancer. These results have been mixed but, in general, studies done in areas with a high prevalence of vitamin D deficiency have shown an association between low levels of vitamin D and subsequent development of prostate cancer. There have been 11 case–control studies that measured vitamin D and examined prostate cancer risk (see Table 10.4).

Overall, four of the studies showed an association between decreased vitamin D levels and increased prostate cancer risk [89–92]. Three of these studies included subjects with a high (>50%) prevalence of vitamin D deficiency (defined as 25-OH vitamin D<20 ng/mL). In contrast, all of the studies that showed no association between vitamin D blood levels and prostate cancer risk examined populations with a much lower prevalence of vitamin D deficiency, mostly less than 20% [93–98] and even one at zero [99].

Table 10.4 Case-control studies of vitamin D level and prostate cancer risk

Study	Population	Number of subjects	% Vitamin D deficient	Conclusions
Corder (1993) [90]	African-American and Caucasian men in CA	181 cases, 181 controls	~50%	Decreased risk of prostate cancer in men older than 57yo with higher levels of 1,25-OH$_2$, especially in those men with low 25-OH levels
Braun (1995) [94]	Caucasians in MD	61 cases, 122 controls	~10%	Null
Gann (1996) [96]	US physicians	232 cases, 414 controls	~20%	High 1,25-OH$_2$ associated with non-significant reduction in prostate cancer risk
Nomura (1998) [99]	Japanese Americans in HI	136 cases, 136 controls	None	Null
Ahohen (2000) [89]	Finnish men	149 cases, 566 controls	>60%	Low levels of 25-OH are associated with increased risk of earlier and more aggressive prostate cancer in men less than 52 years
Tuohimaa (2004) [92]	Scandanavian men	622 cases, 1451 controls	~50%	Both high and low levels of 25-OH are associated with an increased risk of prostate cancer
Platz (2004) [98]	US health professionals	460 cases, 460 controls	~20%	Null
Jacobs (2004) [97]	Eastern US Caucasians	83 cases, 166 controls	20%	Null
Li et al. (2007) [91]	US Physicians	492 cases, 644 controls	19%	Higher levels of 1,25-OH$_2$ were associated with decreased risk of aggressive prostate cancer in older (>65 years) men. Also low 1,25-OH$_2$ in combination with low 25-OH was associated with highest risk of aggressive prostate cancer
Faupel-Badger et al. (2007) [95]	Finnish men	296 cases, 297 controls	~50%	Null
Ahn et al. (2008) [93]	Caucasian Americans	749 cases, 781 control	<15%	No association with low levels of 25-OH vitamin D and risk of prostate cancer, possible increased risk of aggressive prostate cancer with high 25-OH vitamin D levels

10 Vitamin D and Prostate Cancer

There were important differences between the four positive studies. Two of the positive studies showed that higher 1,25-OH$_2$ vitamin D levels were associated with a protective effect against prostate cancer [90, 91]. Corder et al. had a large number of vitamin D deficient subjects (approximately 50%). In Li, et al. the protective effect of the higher 1,25-OH$_2$ vitamin D levels was a reduction in the risk of aggressive prostate cancers. Two other studies showed a link between low 25-OH vitamin D levels and increased risk of prostate cancer [89, 92]. Tuohimaa et al. showed an increased risk of prostate cancer with extreme 25-OH vitamin D deficiency (<7.6 ng/mL) but also showed an increased risk of prostate cancer with highest 25-OH vitamin D levels suggesting a U-shaped relationship between vitamin D status and prostate cancer risk [92].

This suggestion of an increased risk of prostate cancer at higher 25-OH vitamin D levels was reproduced in one recent study [93]. Ahn et al. found a statistically significant increase in risk of aggressive prostate cancers (Gleason > 7) with higher 25-OH vitamin D levels. This possible increased risk at higher vitamin D levels has not been fully explained and requires further investigation.

In the aforementioned 2007 study by Li et al., there was an increased risk of aggressive prostate cancer when both 1,25-OH$_2$ vitamin D and 25-OH vitamin D levels were low, but not solely with low 25-OH levels. This additive effect of low levels of both forms of vitamin D was also shown by Corder et al. Three of the studies that had positive results, demonstrating increased risk of aggressive prostate cancer, but not necessarily an increased risk of lower grade cancers [89, 91, 93]. Of the studies that had null results, two did not analyze risk based on aggressiveness [94, 99] and three had relatively small numbers of aggressive cases [95, 96, 98]. This may support a hypothesis that vitamin D deficient states will increase the risk of aggressive prostate cancers, rather than all grades of prostate cancers.

Thus, epidemiologic evidence is mixed, but generally consistent with the hypothesis that circulating 1,25-OH$_2$ vitamin D levels, and factors that influence them (i.e., oral calcium intake, severe vitamin D deficiency) play a role in prostate cancer development and its course [100]. There are multiple preclinical observations involving vitamin D and prostate cancer risk and mortality that still need further investigation with humans. In addition to the ongoing trials with vitamin D analogs in treating prostate cancer, the observation that 1-alpha-hydroxylase is reduced or lost in prostate cancer tissue needs further confirmation and study in humans.

10.6 Therapeutic Applications of Vitamin D

10.6.1 Vitamin D in Combination with Other Antineoplastic Agents in Preclinical Models

Experiments in preclinical models suggest that VDR ligands enhance the activity of a broad range of antineoplastic agents.

10.6.1.1 Steroids

In preclinical models, the steroid, dexamethasone, enhances the antineoplastic activity of vitamin D [101, 102]. It has been shown to increase vitamin D induced cell cycle arrest and apoptosis and increase vitamin D-mediated suppression of phospho-Erk 1/2, phospho-Akt levels and tumor derived endothelial cell growth [101–104]. Dexamethasone has also been shown to directly increase VDR protein levels and ligand binding in the squamous cell carcinoma model SCC [102].

10.6.1.2 Cytotoxic Chemotherapy

Combining of VDR ligands with several classes of chemotherapy drugs has shown to result in additive and supra-additive activity in several preclinical models of cancer. Specifically, docetaxel [105], paclitaxel, [106] platinum compounds [107], and mitoxantrone [108] have been rendered more active by combinations with vitamin D in preclinical in vitro models of prostate cancer. Confirmation in in vivo models has been reported for paclitaxel and mitoxantrone [106, 108]. Studies in models of other neoplasms yield similar observations [109–112], but further study is required to fully clarify the mechanisms of these interactions.

10.6.1.3 Retinoid Receptor Ligands

As previously mentioned, after ligand binding, VDR forms heterodimers with the retinoid X receptor (RXR), thus interactions between these two receptor systems would be expected [113, 114]. Both apoptosis [114] and angiogenesis inhibition is synergistically enhanced when VDR and RXR ligands are co-administered in preclinical models [114]. Several overlapping mechanisms of anticancer activity, including modulation of IGFBP-3 expression, [115] inhibition of telomerase reverse transcriptase in prostate cancer cells [116] as well as induction of cell cycle checkpoint proteins like p21 may explain these observations.

10.6.1.4 Tamoxifen

A study in Sprague-Dawley rats reports that there was a significant increase in the inhibition of N-nitroso-N-methylurea (NMU) induced mammary carcinogenesis when VDR ligands are co-administered with tamoxifen [117]. Enhanced apoptosis was seen in MCF-7 cells in vitro and in vitro when this combination was evaluated [118, 119]. It maybe that MCF-7 cells are inversely sensitive to vitamin D and antiestrogens [120]. While these findings originate from breast cancer models, they may have relevance to prostate cancer biology as well.

10 Vitamin D and Prostate Cancer

10.6.1.5 Non-steroidal Anti-inflammatory Agents (NSAIDS)

In LNCaP cells, VDR ligands and ibuprofen acted synergistically to inhibit growth [121]. Both decreased G1-S transition and enhanced apoptosis were noted when the two agents were used together [122]. Expression of prostaglandin synthesizing COX-2 gene was decreased by calcitriol in LNCaP cells. At the same time, the prostaglandin inactivating 15-prostaglandin dehydrogenase gene was upregulated [121].

10.6.1.6 Radiation

Radiation sensitivity is enhanced by p21 expression, which in turn is a known VDR target [123]. In several tumor models, radiation induced apoptosis was also enhanced with VDR ligands [124, 125]. One explanation for this interaction maybe increased ceramide generation [126].

Thus, in addition to single agent activity, VDR ligands appear to enhance the activity of a broad collection of antineoplastic agents. These pre-clinical data have served as the basis for the examination of clinical activity of VDR ligands. Calcitriol, the natural VDR ligand, has been most extensively studied.

10.6.2 Clinical Trials of Calcitriol in Prostate Cancer

Calcitriol (1,25-dihydroxycholecalciferol, 1,25-OH$_2$ vitamin D) is approved for the treatment of kidney failure patients where it serves as a replacement for the inability to activate vitamin D. Nearly all pre-clinical studies suggest that the antineoplastic activity of VDR ligands, and calcitriol specifically, is dose dependent and most pronounced at supraphysiologic concentrations (typically at or above 1 nM). Consequently, studies in cancer have generally sought to examine higher doses than those required for replacement in patients with end-stage renal disease.

10.6.2.1 Phase I Studies of Single Agent Calcitriol

Daily Administration

Initial studies of calcitriol in prostate cancer patients sought to increase the dose administered on the standard daily replacement schedule. Osborn, et al. used daily administration and examined doses that ranged from 0.5 to 1.5 μg daily in 11 hormone-refractory prostate cancer patients. No PSA responses were seen in this study [127]. A similar approach was taken in a pilot study carried out in 7 hormone-naïve patients who had a rising serum PSA without metastases [128]. While there were no PSA responses, the PSA doubling time appeared to be lengthened compared to the pre-treatment PSA doubling time. Subsequent studies with other agents have clearly demonstrated variability in PSA kinetics in this

clinical setting, and therefore illustrate the need for a control arm to interpret these results, nevertheless the observation is suggestive of a treatment effect. Dose escalation was not carried out in the Gross et al. study beyond doses of 2.5 µg/day due to concern about hypercalciuria.

Every Other Day Subcutaneous Administration

The hypotheses that an alternative route of administration and dosing schedule may allow greater dose escalation by reducing the calcemic toxicity of calcitriol was examined in a clinical trial of subcutaneous administration every other day. Significant escalation was indeed possible with doses of 10 µg reached and peak calcitriol concentrations of approximately 0.7 nM at the 8 µg dose. Hypercalcemia precluded further dose escalation [129].

Weekly Oral Dosing

In the initial phase I study, weekly oral dosing demonstrated both significant potential with regard to dose escalation and revealed a formulation-specific absorption ceiling. Doses as high as 2.8 µg/kg were examined. In this study, peak blood calcitriol concentrations (C_{max}) of 3.7–6.0 nM were observed without dose limiting toxicity, but above 0.48 µg/kg, C_{max} and the area under the concentration curve (AUC) did not increase linearly [130]. Mundi et al. later confirmed that the commercially available formulation of calcitriol had non-linear pharmacokinetics [131] and later showed a similar pattern with a liquid calcitriol formulation [132].

A new formulation of calcitriol has been developed to overcome the limitation of the pharmacokinetics and the large quantity of pills required for treatment (calcitriol is only commercially available as 0.25 and 0.5 µg capsules). DN-101 (Novacea, Inc. South San Francisco), given as a single dose capsule, demonstrated dose-proportional increases in both C_{max} and AUC when studied over a range of doses (15–165 µg). Peak calcitriol concentrations (14.9 nM at the 165 µg dose) were higher than any previously reported [133]. While single dose administration was free from dose-limiting toxicity, grade 2 hypercalcemia was seen with repeat weekly dosing in the 60 µg group [134]. It is likely that a higher weekly dose would have been achievable if a more conventional grade 3 toxicity criterion were utilized or if DN-101 had been co-administered with agent(s) that have potential to reduce hypercalcemia (i.e., bisphosphonates or steroids).

10.6.2.2 Early Stage Studies of Calcitriol in Combination with Other Agents

Daily Administration

One study examined daily calcitriol, dosed at 0.5 µg daily with daily dexamethasone and weekly carboplatin in 34 patients with androgen independent prostate cancer

(AIPC) [135]. PSA response was noted in 38% of patients. The interpretation of this result is challenging because both dexamethasone and carboplatin have some activity in prostate cancer. Nonetheless, the response rate is respectable.

Dosing 3 of Every 7 Days

Dosing calcitriol for 3 consecutive days, every 7 days was evaluated in two studies in combination with other drugs. The first trial was a phase I combination with paclitaxel, with daily doses up to 38 μg on three consecutive days. C_{max} ranges of 1.4–3.5 nM at the highest doses did not produce dose limiting toxicity [131]. The second study was in combination with zoledronate with dexamethasone added upon progression [136]. Calcitriol was administered on the same schedule as it was on the previous study at doses of 30 μg. While there were not dose limiting toxicities, three patients did have dose reductions due to laboratory abnormalities. The only patient responses to this regimen were observed when dexamethasone was added upon patient progression.

Intravenous Calcitriol

Having observed an absorption-related pharmacokinetic ceiling, the Roswell Park group examined weekly intravenous calcitriol in a phase I study that included patients with a range of solid tumors [137]. In this study, gefitinib was given as the partner drug. Dose limiting hypercalcemia was reached in two patients who were receiving 96 μg of calcitriol/week (Table 10.5).

In a series of studies, intermittent dosing has been shown to result in significant dose escalation. A novel formulation, DN-101 circumvented the previously described non-linear pharmacokinetics, and in doing so provided evidence that the phenomenon is likely to be related to the formulation rather than the parent compound. DN-101 also allowed for much more convenient dosing that required one or several capsules instead of dozens if not more than 100. As a result, the development of DN-101 allowed large scale trials of high dose calcitriol.

10.6.2.3 Phase II Studies

Weekly Dosing

Patients who had a biochemical progression after prostatectomy or radiation therapy were enrolled in a non-randomized study of weekly calcitriol of 0.5 μg/kg [138]. Patients were treated for a median of 10 months demonstrating the long-term safety of this approach. Lengthening of the PSA doubling time when compared to pre-treatment and a handful of minor PSA reductions with treatment were seen. Absent a control arm, it would be difficult to be certain whether these observations indicate true anti-tumor activity.

Table 10.5 Dose escalation studies of calcitriol in cancer

Investigator	Dose of calcitriol	Schedule and route	Companion drugs	Dose limiting toxicity	Peak concentrations
Osborn et al. (1995) [127]	0.5–1.5 μg	Daily/orally	None	Hypercalcemia	NR
Gross et al. (1998) [128]	0.5–2.5 μg	Daily/orally	None	Hypercalciuria	NR
Smith et al. (1999) [129]	2.0–10 μg	Every other day/subcutaneous	None	Hypercalcemia	NR
Beer (2001) [130]	0.06–2.8 μg/kg	Weekly/orally	None	Not determined	3.7–6.0 nM
Muindi (2002) [131]	4–38 μg	Daily for 3 days every 7 days/orally	Paclitaxel 80 mg/m^2 weekly	Not determined	1.4–3.5 nM
Morris (2004) [136]	4–30 μg	Daily for 3 days every 7 days/orally	Zoledronate 4 mg IV monthly; dexamethosone 0.75 mg BID added at progression	Not determined	0.9–2.3 nM
Fakih (2007) [137]	15–96 μg	Weekly/I.V.	Gefitinib 250 mg daily	Hypercalcemia	6.68±1.42 ng/mL

NR not reported

10 Vitamin D and Prostate Cancer

Building on the pre-clinical evidence of synergy with taxanes, the next effort involved combining weekly calcitriol with docetaxel. Chemotherapy-naïve metastatic androgen-independent prostate cancer patients received oral calcitriol 0.5 µg/kg on day 1, followed by docetaxel 36 mg/m² intravenously on day 2 weekly for 6 consecutive weeks, repeated every 8 weeks in a phase II single institution clinical trial [139]. Of the 37 patients, 81% had a confirmed PSA response, while toxicity was similar to what would be expected with docetaxel alone. RECIST criteria for response was met in 53% of the 15 patients with measurable disease. The median overall survival was 19.5 months. These results were quite encouraging when contrasted with contemporary results seen with docetaxel alone and stimulated the development of a larger effort.

ASCENT (AIPC Study of Calcitriol Enhancing Taxotere) was launched to more robustly examine the possibility that weekly calcitriol enhances the activity of weekly docetaxel. This placebo-controlled international multi-institutional randomized study that compared weekly DN-101 + docetaxel to placebo + docetaxel in 250 patients with chemotherapy-naïve AIPC enrolled at 48 sites in the US and Canada. For 3 consecutive weeks out of 4, 45 µg of DN-101 was given 24 hours before docetaxel 36 mg/m². Although the study did not meet its primary endpoint of PSA response rate improvement, the observed trend favored the experimental arm with an overall PSA response rate of 63% compared to 52%, $p=0.07$. Overall survival, a secondary endpoint, was better in the experimental arm than in the docetaxel arm (HR 0.67, $p=0.035$). Interestingly, calcitriol did not appear to add toxicity to docetaxel and exploratory analyses suggested a lower incidence of thrombotic and gastrointestinal toxicity in the experimental arm. The overall results of ASCENT were thought to be sufficiently encouraging to warrant a phase III program [140].

The 3 days out of 7 schedule was also examined further in a phase II study with dexamethasone [141]. In this study, calcitriol was given at 8–12 µg/day for 3 consecutive days repeated every week. Four milligrams of dexamethasone was given for 4 of every 7 days. Nineteen percent of the 37 patients enrolled had a PSA response and treatment was well tolerated. While encouraging, this response rate is difficult to interpret with confidence because the activity of this dose and schedule of dexamethasone is not known (Table 10.6).

Less Frequent Dosing

A dose de-escalation study of 60 µg of calcitriol was administered to AIPC patients every 3 weeks 24 hours before chemotherapy with docetaxel and estramustine [142]. Although this study was not designed to test efficacy, responses were seen in 55% of chemotherapy naïve patients and 9% of patients previously treated with docetaxel-containing chemotherapy, while at the same time showing that 60 µg of calcitriol can be safely administered.

Calcitriol 0.5 µg/kg dosed every 4 weeks was evaluated in combination with carboplatin dosed at AUC of 7 (6 in patients with prior radiation) in a small phase II study of patients with AIPC [143]. Seventeen patients had a response rate of less

Table 10.6 Phase II studies of high dose calcitriol combinations

Investigator	Dose	Schedule of oral calcitriol	Companion drugs	Number of patients	Efficacy results
Beer (2004) [143]	0.5 µg/kg	24 h prior to carboplatin every 4 weeks	Carboplatin AUC of 7 (6 in patients with prior radiation) every 28 days	17	1 of 17 patients had a PSA response
Beer (2003) [139]	0.5 µg/kg	24 h prior to each dose of docetaxel weekly	Docetaxel 36 mg/m^2 weekly for 6 consecutive weeks repeated every 8 weeks	37	81% had PSA response. 8 of 15 responded in measurable disease. Median overall survival 19.5 months
Beer (2007) [140]	45 µg (DN-101)	24 h prior to each dose of docetaxel weekly	Docetaxel 36 mg/m^2 weekly for 3 consecutive weeks repeated every 4 weeks	250 (125 DN-101, 125 placebo)	Improved survival with DN-101 (HR 0.67, $p=0.035$), trend favoring DN-101 with respect to PSA response rates (overall 63% vs 52%, $p=0.07$, within 6 months 58% vs 49%, $p=0.16$)
Tiffany (2005) [142]	45 µg (DN-101)	24 h prior to each dose of docetaxel 3-weekly	Estramustine 280 mg on days 1–5, and docetaxel 70 mg/ m^2 on day 2	24	PSA decline in 55% of chemotherapy naïve patients and 9% of patients previously treated with docetaxel-containing chemotherapy
Trump (2006) [141]	8–12 µg	Daily for 3 days every 7 days/orally	Dexamethasone 4 mg for 4 of every 7 days	37	PSA decline seen in 19% of patients
Chen (in press)	180 µg (DN-101)	24 h prior to each dose of mitoxantrone 3-weekly	mitoxantrone 12 mg/m^2 every 3 weeks	19	Five of 19 patients (26%; 95 CI 9–51%) achieved a PSA decline and 47% (95% CI 21–73%) achieved an analgesic response

than 10% with unremarkable toxicity. It is unclear if the infrequent dosing or the platinum resistance of prostate cancer had an impact on these results.

Nineteen patients with metastatic AIPC received DN-101 180 μg p.o. on day 1 and mitoxantrone 12 mg/m² i.v. on day 2 every 21 days with continuous daily prednisone 10 mg p.o. for a maximum of 12 cycles. This trial examined the highest dose of calcitriol evaluated in a phase II study, but used an infrequent dosing schedule. Five of 19 patients (26%; 95 CI 9%–51%) achieved a PSA decline and 47% (95% CI 21%–73%) achieved an analgesic response (BJU International, in press).

Overall, the phase II studies of infrequently given high dose calcitriol, even using very high doses, did not produce remarkable results, suggesting that weekly dosing maybe a more promising strategy.

10.6.2.4 Phase III Studies

With encouraging results from the ASCENT study in hand, Novacea, Inc. pursued phase III development of DN-101. The ASCENT-2 study sought to determine if the addition of DN-101 to docetaxel improved overall survival. The design of this study faced several important challenges. While much of the high dose calcitriol program, and the encouraging results from the ASCENT study were derived from a program of weekly administration of high dose calcitriol along with weekly chemotherapy, a 3-weekly regimen of docetaxel and prednisone had become the standard of care. Tannock et al. reported that docetaxel 75 mg/m² with low dose daily prednisone improve the overall survival of AIPC patients over the prior standard of mitoxantrone and prednisone. At the same time, a weekly regimen of 30 mg/m² administered for 5 of every 6 weeks, designed to be equal in dose intensity to the 3-weekly arm, but distinct from all previously studies weekly regimens of docetaxel in prostate cancer, did not produce a survival improvement.

The phase III program, with the primary endpoint of survival, compared the winning arm of ASCENT that consisted of 45 μg of DN-101 + docetaxel at 36 mg/m² weekly for 3 of every 4 weeks to the FDA approved standard of docetaxel 75 mg/m² with daily prednisone. This large study was halted early due to excess deaths in the experimental arm. Recently, the Food and Drug Administration lifted the resulting hold on studies of DN-101. This disappointing result is difficult to interpret due to the multiple differences between the two arms of the study. In addition to the presence or absence of high dose calcitriol, the two arms differ with respect to: (1) the dose and schedule of docetaxel, (2) the dose intensity of docetaxel, (3) the use of prednisone, and (4) the dose and schedule of dexamethasone. Thus, this unblinded study did not directly examine the contribution of high dose calcitriol to the safety and efficacy of chemotherapy. Rather, it was designed to meet the regulatory requirements for drug approval. The failure of this study leaves us uncertain about the potential of high dose calcitriol as a useful cancer treatment.

10.6.3 Clinical Trials of Calcitriol Analogs in Prostate Cancer

An alternative to calcitriol, calcitriol analogs have been developed in the hope of overcoming calcemic toxicity, while maintaining antineoplastic activity. Many compounds have been chemically synthesized, primarily with side chain modifications. It is hoped that reduced calcemic toxicity may be a result of differences in protein binding, VDR affinity, and drug metabolism [144–146].

After phase I studies in pancreatic and hepatocellular carcinoma, [147] Seocalcitol (EB 1089, Leo Pharmaceuticals, Ballerup, Denmark) 10 µg entered phase II studies. Results in unresectable hepatocellular carcinoma show that 2 of 33 evaluable patients had a complete remission enduring beyond 29 months (last point of analysis), [148] while no responses were seen in pancreatic cancer [149]. Another analog, topical calcipotriol, had observed responses in 3 of 14 patients with locally advanced or cutaneous metastatic adenocarcinoma of the breast [150].

In a phase I study of 1-alpha-hydroxyvitamin D_2 [151] 12.5 µg was identified as the safe dose due to dose limiting hypercalcemia and renal insufficiency. Two of 25 androgen independent prostate cancer patients had objective responses, which lead to the development of a phase II study. In this follow-up study, 26 patients were enrolled to evaluate progression free survival. One patient had stable disease for more than 2 years, while the median time to progression was 12 weeks (mean 19 weeks). In a randomized phase II study, 70 chemotherapy-naïve men with AIPC were treated with weekly docetaxel with or without 1-alpha-hydroxyvitamin D_2 given at a dose of 10 µg/day. The response rates, time to disease progression, and toxicity were similar in both arms of the study [152].

Another analog, ILX23–7553, was evaluated in a phase I clinical trial. It was found that doses up to 45 µg/m²/day for 5 consecutive days repeated every 14 days was safe, but the number of capsules required prompted early closure. The authors conclude that a reformulation at a higher dose may be a more feasible study in the future [153].

19-Nor-1alpha-25-dihydroxyvitamin D2 (paricalcitol) was examined in a phase I study in 18 patients with androgen-independent prostate cancer. Paricalcitol was given i.v. three times per week with doses between 5 and 25 µg tested [154]. While some PSA declines were seen, no patient had a sustained PSA response. One episode of hypercalcemia was noted at the highest dose tested. Interestingly, serum parathyroid hormone levels, elevated at study entry in 41% of patients, were reduced with therapy.

Vitamin D remains an exciting area of investigation in prostate cancer epidemiology, prevention, and therapy. Despite compelling biology and supportive epidemiology, to date, definitive results have not been reported. There are sufficient data to expect that with further work, a role for vitamin D in reducing the risk of prostate cancer diagnosis and death, as well as improved outcomes in prostate cancer treatment will be identified. It is tempting to consider that human biologic heterogeneity in vitamin D sensitivity has not been fully considered in the studies conducted to

date. Increased attention to the underlying molecular defects in individual prostate cancer may allow for more robust identification of ways in which vitamin D can be harnessed to help men who suffer from this disease.

References

1. Garland CF, Garland FC (1980) Do sunlight and vitamin D reduce the likelihood of colon cancer? Int J Epidemiol 9:227–231
2. Schwartz GG, Hulka BS (1990) Is vitamin D deficiency a risk factor for prostate cancer? (Hypothesis). Anticancer Res 10:1307–1311
3. Chen TC, Wang L, Whitlatch LW et al (2003) Prostatic 25-hydroxyvitamin D-1alpha-hydroxylase and its implication in prostate cancer. J Cell Biochem 88:315–322
4. Cross HS, Bareis P, Hofer H et al (2001) 25-Hydroxyvitamin D(3)-1alpha-hydroxylase and vitamin D receptor gene expression in human colonic mucosa is elevated during early cancerogenesis. Steroids 66:287–292
5. Schwartz GG, Whitlatch LW, Chen TC et al (1998) Human prostate cells synthesize 1, 25-dihydroxyvitamin D3 from 25-hydroxyvitamin D3. Cancer Epidemiol Biomarkers Prev 7:391–395
6. Tangpricha V, Flanagan JN, Whitlatch LW et al (2001) 25-hydroxyvitamin D-1alpha-hydroxylase in normal and malignant colon tissue. Lancet 357:1673–1674
7. Zehnder D, Bland R, Williams MC et al (2001) Extrarenal expression of 25-hydroxyvitamin d(3)-1 alpha-hydroxylase. J Clin Endocrinol Metab 86:888–894
8. Brenza HL, DeLuca HF (2000) Regulation of 25-hydroxyvitamin D3 1alpha-hydroxylase gene expression by parathyroid hormone and 1, 25-dihydroxyvitamin D3. Arch Biochem Biophys 381:143–152
9. Haussler MR, Whitfield GK, Haussler CA et al (1998) The nuclear vitamin D receptor: biological and molecular regulatory properties revealed. J Bone Miner Res 13:325–349
10. Takeyama K, Kitanaka S, Sato T et al (1997) 25-Hydroxyvitamin D3 1alpha-hydroxylase and vitamin D synthesis. Science 277:1827–1830
11. Deeb KK, Trump DL, Johnson CS (2007) Vitamin D signalling pathways in cancer: potential for anticancer therapeutics. Nat Rev Cancer 7:684–700
12. Hewison M, Zehnder D, Bland R et al (2000) 1alpha-Hydroxylase and the action of vitamin D. J Mol Endocrinol 25:141–148
13. Holick CN, Stanford JL, Kwon EM et al (2007) Comprehensive association analysis of the vitamin D pathway genes, VDR, CYP27B1, and CYP24A1, in prostate cancer. Cancer Epidemiol Biomarkers Prev 16:1990–1999
14. Cross HS, Bises G, Lechner D et al (2005) The vitamin D endocrine system of the gut – its possible role in colorectal cancer prevention. J Steroid Biochem Mol Biol 97:121–128
15. Hedlund TE, Moffatt KA, Miller GJ (1996) Vitamin D receptor expression is required for growth modulation by 1 alpha, 25-dihydroxyvitamin D3 in the human prostatic carcinoma cell line ALVA-31. J Steroid Biochem Mol Biol 58:277–288
16. Miller GJ, Stapleton GE, Ferrara JA et al (1992) The human prostatic carcinoma cell line LNCaP expresses biologically active, specific receptors for 1 alpha, 25-dihydroxyvitamin D3. Cancer Res 52:515–520
17. Skowronski RJ, Peehl DM, Feldman D (1993) Vitamin D and prostate cancer: 1, 25 dihydroxyvitamin D3 receptors and actions in human prostate cancer cell lines. Endocrinology 132:1952–1960
18. Beer TM, Myrthue A, Garzotto M et al (2004) Randomized study of high-dose pulse calcitriol or placebo prior to radical prostatectomy. Cancer Epidemiol Biomarkers Prev 13: 2225–2232

19. Hsu JY, Feldman D, McNeal JE et al (2001) Reduced 1alpha-hydroxylase activity in human prostate cancer cells correlates with decreased susceptibility to 25-hydroxyvitamin D3-induced growth inhibition. Cancer Res 61:2852–2856

20. Chen TC (2008) 25-Hydroxyvitamin D-1 alpha-hydroxylase (CYP27B1) is a new class of tumor suppressor in the prostate. Anticancer Res 28:2015–2017

21. Giovannucci E (2005) The epidemiology of vitamin D and cancer incidence and mortality: a review (United States). Cancer Causes Control 16:83–95

22. Hidalgo AA, Paredes R, Garcia VM et al (2007) Altered VDR-mediated transcriptional activity in prostate cancer stroma. J Steroid Biochem Mol Biol 103:731–736

23. Mangelsdorf DJ, Thummel C, Beato M et al (1995) The nuclear receptor superfamily: the second decade. Cell 83:835–839

24. Nanes MS, Kuno H, Demay MB et al (1994) A single up-stream element confers responsiveness to 1, 25-dihydroxyvitamin D3 and tumor necrosis factor-alpha in the rat osteocalcin gene. Endocrinology 134:1113–1120

25. Koszewski NJ, Reinhardt TA, Horst RL (1996) Vitamin D receptor interactions with the murine osteopontin response element. J Steroid Biochem Mol Biol 59:377–388

26. Liu M, Lee MH, Cohen M et al (1996) Transcriptional activation of the Cdk inhibitor p21 by vitamin D3 leads to the induced differentiation of the myelomonocytic cell line U937. Genes Dev 10:142–153

27. Jiang F, Li P, Fornace AJ Jr et al (2003) G2/M arrest by 1, 25-dihydroxyvitamin D3 in ovarian cancer cells mediated through the induction of GADD45 via an exonic enhancer. J Biol Chem 278:48030–48040

28. Murayama A, Kim MS, Yanagisawa J et al (2004) Transrepression by a liganded nuclear receptor via a bHLH activator through co-regulator switching. EMBO J 23:1598–1608

29. Hawa NS, O'Riordan JL, Farrow SM (1996) Functional analysis of vitamin D response elements in the parathyroid hormone gene and a comparison with the osteocalcin gene. Biochem Biophys Res Commun 228:352–357

30. Maestro B, Davila N, Carranza MC et al (2003) Identification of a Vitamin D response element in the human insulin receptor gene promoter. J Steroid Biochem Mol Biol 84:223–230

31. Guzey M, Kitada S, Reed JC (2002) Apoptosis induction by 1alpha, 25-dihydroxyvitamin D3 in prostate cancer. Mol Cancer Ther 1:667–677

32. Nemere I, Yoshimoto Y, Norman AW (1984) Calcium transport in perfused duodena from normal chicks: enhancement within fourteen minutes of exposure to 1, 25-dihydroxyvitamin D3. Endocrinology 115:1476–1483

33. Beno DW, Brady LM, Bissonnette M et al (1995) Protein kinase C and mitogen-activated protein kinase are required for 1, 25-dihydroxyvitamin D3-stimulated Egr induction. J Biol Chem 270:3642–3647

34. Morelli S, Buitrago C, Boland R et al (2001) The stimulation of MAP kinase by 1, 25(OH)(2)-vitamin D(3) in skeletal muscle cells is mediated by protein kinase C and calcium. Mol Cell Endocrinol 173:41–52

35. Rossi AM, Capiati DA, Picotto G et al (2004) MAPK inhibition by 1alpha, 25(OH)2-Vitamin D3 in breast cancer cells. Evidence on the participation of the VDR and Src. J Steroid Biochem Mol Biol 89–90:287–290

36. Wali RK, Baum CL, Sitrin MD et al (1990) 1, 25(OH)2 vitamin D3 stimulates membrane phosphoinositide turnover, activates protein kinase C, and increases cytosolic calcium in rat colonic epithelium. J Clin Invest 85:1296–1303

37. Larsson D, Hagberg M, Malek N et al (2008) Membrane initiated signaling by 1, 25alpha-dihydroxyvitamin D3 in LNCaP prostate cancer cells. Adv Exp Med Biol 617:573–579

38. Berndt SI, Dodson JL, Huang WY et al (2006) A systematic review of vitamin D receptor gene polymorphisms and prostate cancer risk. J Urol 175:1613–1623

39. Cicek MS, Liu X, Schumacher FR et al (2006) Vitamin D receptor genotypes/haplotypes and prostate cancer risk. Cancer Epidemiol Biomarkers Prev 15:2549–2552

40. Ma J, Stampfer MJ, Gann PH et al (1998) Vitamin D receptor polymorphisms, circulating vitamin D metabolites, and risk of prostate cancer in United States physicians. Cancer Epidemiol Biomarkers Prev 7:385–390

41. Mikhak B, Hunter DJ, Spiegelman D et al (2007) Vitamin D receptor (VDR) gene polymorphisms and haplotypes, interactions with plasma 25-hydroxyvitamin D and 1, 25-dihydroxyvitamin D, and prostate cancer risk. Prostate 67:911–923

42. Whitfield GK, Remus LS, Jurutka PW et al (2001) Functionally relevant polymorphisms in the human nuclear vitamin D receptor gene. Mol Cell Endocrinol 177:145–159

43. Campbell MJ, Elstner E, Holden S et al (1997) Inhibition of proliferation of prostate cancer cells by a 19-nor-hexafluoride vitamin D3 analogue involves the induction of p21waf1, p27kip1 and E-cadherin. J Mol Endocrinol 19:15–27

44. Sheikh MS, Rochefort H, Garcia M (1995) Overexpression of p21WAF1/CIP1 induces growth arrest, giant cell formation and apoptosis in human breast carcinoma cell lines. Oncogene 11:1899–1905

45. Wang QM, Jones JB, Studzinski GP (1996) Cyclin-dependent kinase inhibitor p27 as a mediator of the G1-S phase block induced by 1, 25-dihydroxyvitamin D3 in HL60 cells. Cancer Res 56:264–267

46. Zhuang SH, Burnstein KL (1998) Antiproliferative effect of 1alpha, 25-dihydroxyvitamin D3 in human prostate cancer cell line LNCaP involves reduction of cyclin-dependent kinase 2 activity and persistent G1 accumulation. Endocrinology 139:1197–1207

47. Tong WM, Kallay E, Hofer H et al (1998) Growth regulation of human colon cancer cells by epidermal growth factor and 1, 25-dihydroxyvitamin D3 is mediated by mutual modulation of receptor expression. Eur J Cancer 34:2119–2125

48. Matsumoto K, Hashimoto K, Nishida Y et al (1990) Growth-inhibitory effects of 1, 25-dihydroxyvitamin D3 on normal human keratinocytes cultured in serum-free medium. Biochem Biophys Res Commun 166:916–923

49. Reitsma PH, Rothberg PG, Astrin SM et al (1983) Regulation of myc gene expression in HL-60 leukaemia cells by a vitamin D metabolite. Nature 306:492–494

50. Capiati DA, Rossi AM, Picotto G et al (2004) Inhibition of serum-stimulated mitogen activated protein kinase by 1alpha, 25(OH)2-vitamin D3 in MCF-7 breast cancer cells. J Cell Biochem 93:384–397

51. Park WH, Seol JG, Kim ES et al (2000) Induction of apoptosis by vitamin D3 analogue EB1089 in NCI-H929 myeloma cells via activation of caspase 3 and p38 MAP kinase. Br J Haematol 109:576–583

52. Peehl DM, Skowronski RJ, Leung GK et al (1994) Antiproliferative effects of 1,25-dihydroxyvitamin D3 on primary cultures of human prostatic cells. Cancer Res 54:805–810

53. Oades GM, Dredge K, Kirby RS et al (2002) Vitamin D receptor-dependent antitumour effects of 1, 25-dihydroxyvitamin D3 and two synthetic analogues in three in vivo models of prostate cancer. BJU Int 90:607–616

54. Whitlatch LW, Young MV, Schwartz GG et al (2002) 25-Hydroxyvitamin D-1alpha-hydroxylase activity is diminished in human prostate cancer cells and is enhanced by gene transfer. J Steroid Biochem Mol Biol 81:135–140

55. Getzenberg RH, Light BW, Lapco PE et al (1997) Vitamin D inhibition of prostate adenocarcinoma growth and metastasis in the Dunning rat prostate model system. Urology 50:999–1006

56. Lokeshwar BL, Schwartz GG, Selzer MG et al (1999) Inhibition of prostate cancer metastasis in vivo: a comparison of 1, 23-dihydroxyvitamin D (calcitriol) and EB1089. Cancer Epidemiol Biomarkers Prev 8:241–248

57. Yudoh K, Matsuno H, Kimura T (1999) 1alpha, 25-dihydroxyvitamin D3 inhibits in vitro invasiveness through the extracellular matrix and in vivo pulmonary metastasis of B16 mouse melanoma. J Lab Clin Med 133:120–128

58. Schwartz GG, Hill CC, Oeler TA et al (1995) 1, 25-Dihydroxy-16-ene-23-yne-vitamin D3 and prostate cancer cell proliferation in vivo. Urology 46:365–369

59. Schwartz GG, Oeler TA, Uskokovic MR et al (1994) Human prostate cancer cells: inhibition of proliferation by vitamin D analogs. Anticancer Res 14:1077–1081

60. Bao BY, Yeh SD, Lee YF (2006) 1alpha, 25-dihydroxyvitamin D3 inhibits prostate cancer cell invasion via modulation of selective proteases. Carcinogenesis 27:32–42

61. Schwartz GG, Wang MH, Zang M et al (1997) 1 alpha, 25-Dihydroxyvitamin D (calcitriol) inhibits the invasiveness of human prostate cancer cells. Cancer Epidemiol Biomarkers Prev 6:727–732

62. Sung V, Feldman D (2000) 1, 25-Dihydroxyvitamin D3 decreases human prostate cancer cell adhesion and migration. Mol Cell Endocrinol 164:133–143

63. Bao BY, Yao J, Lee YF (2006) 1alpha, 25-dihydroxyvitamin D3 suppresses interleukin-8-mediated prostate cancer cell angiogenesis. Carcinogenesis 27:1883–1893

64. Ylikomi T, Laaksi I, Lou YR et al (2002) Antiproliferative action of vitamin D. Vitam Horm 64:357–406

65. Hakim I, Bar-Shavit Z (2003) Modulation of TNF-alpha expression in bone marrow macrophages: involvement of vitamin D response element. J Cell Biochem 88:986–998

66. Hanchette CL, Schwartz GG (1992) Geographic patterns of prostate cancer mortality. Evidence for a protective effect of ultraviolet radiation. Cancer 70:2861–2869

67. Schwartz GG, Hanchette CL (2006) UV, latitude, and spatial trends in prostate cancer mortality: all sunlight is not the same (United States). Cancer Causes Control 17:1091–1101

68. Grant WB (2002) An estimate of premature cancer mortality in the U.S. due to inadequate doses of solar ultraviolet-B radiation. Cancer 94:1867–1875

69. Luscombe CJ, Fryer AA, French ME et al (2001) Exposure to ultraviolet radiation: association with susceptibility and age at presentation with prostate cancer. Lancet 358:641–642

70. John EM, Schwartz GG, Koo J et al (2005) Sun exposure, vitamin D receptor gene polymorphisms, and risk of advanced prostate cancer. Cancer Res 65:5470–5479

71. Bodiwala D, Luscombe CJ, French ME et al (2003) Susceptibility to prostate cancer: studies on interactions between UVR exposure and skin type. Carcinogenesis 24:711–717

72. Lagunova Z, Porojnicu AC, Dahlback A et al (2007) Prostate cancer survival is dependent on season of diagnosis. Prostate 67:1362–1370

73. Robsahm TE, Tretli S, Dahlback A et al (2004) Vitamin D3 from sunlight may improve the prognosis of breast-, colon- and prostate cancer (Norway). Cancer Causes Control 15:149–158

74. Zhou W, Suk R, Liu G et al (2005) Vitamin D is associated with improved survival in early-stage non-small cell lung cancer patients. Cancer Epidemiol Biomarkers Prev 14:2303–2309

75. Zhou W, Heist RS, Liu G et al (2007) Circulating 25-hydroxyvitamin D levels predict survival in early-stage non-small-cell lung cancer patients. J Clin Oncol 25:479–485

76. Giovannucci E, Rimm EB, Wolk A et al (1998) Calcium and fructose intake in relation to risk of prostate cancer. Cancer Res 58:442–447

77. Rodriguez C, McCullough ML, Mondul AM et al (2003) Calcium, dairy products, and risk of prostate cancer in a prospective cohort of United States men. Cancer Epidemiol Biomarkers Prev 12:597–603

78. Skinner HG, Schwartz GG (2008) Serum calcium and incident and fatal prostate cancer in the National Health and Nutrition Examination Survey. Cancer Epidemiol Biomarkers Prev 17:2302–2305

79. Qin LQ, Xu JY, Wang PY et al (2004) Milk consumption is a risk factor for prostate cancer: meta-analysis of case-control studies. Nutr Cancer 48:22–27

80. Chan JM, Giovannucci E, Andersson SO et al (1998) Dairy products, calcium, phosphorous, vitamin D, and risk of prostate cancer (Sweden). Cancer Causes Control 9:559–566

81. Grant WB (1999) An ecologic study of dietary links to prostate cancer. Altern Med Rev 4:162–169

82. Baron JA, Beach M, Wallace K et al (2005) Risk of prostate cancer in a randomized clinical trial of calcium supplementation. Cancer Epidemiol Biomarkers Prev 14:586–589

83. Berndt SI, Carter HB, Landis PK et al (2002) Calcium intake and prostate cancer risk in a long-term aging study: the Baltimore Longitudinal Study of Aging. Urology 60:1118–1123

84. Chan JM, Pietinen P, Virtanen M et al (2000) Diet and prostate cancer risk in a cohort of smokers, with a specific focus on calcium and phosphorus (Finland). Cancer Causes Control 11:859–867

85. Park Y, Mitrou PN, Kipnis V et al (2007) Calcium, dairy foods, and risk of incident and fatal prostate cancer: the NIH-AARP Diet and Health Study. Am J Epidemiol 166:1270–1279

86. Tavani A, Gallus S, Franceschi S et al (2001) Calcium, dairy products, and the risk of prostate cancer. Prostate 48:118–121

10 Vitamin D and Prostate Cancer

87. Ahn J, Albanes D, Peters U et al (2007) Dairy products, calcium intake, and risk of prostate cancer in the prostate, lung, colorectal, and ovarian cancer screening trial. Cancer Epidemiol Biomarkers Prev 16:2623–2630
88. Kristal AR, Cohen JH, Qu P et al (2002) Associations of energy, fat, calcium, and vitamin D with prostate cancer risk. Cancer Epidemiol Biomarkers Prev 11:719–725
89. Ahonen MH, Tenkanen L, Teppo L et al (2000) Prostate cancer risk and prediagnostic serum 25-hydroxyvitamin D levels (Finland). Cancer Causes Control 11:847–852
90. Corder EH, Guess HA, Hulka BS et al (1993) Vitamin D and prostate cancer: a prediagnostic study with stored sera. Cancer Epidemiol Biomarkers Prev 2:467–472
91. Li H, Stampfer MJ, Hollis JB et al (2007) A prospective study of plasma vitamin D metabolites, vitamin D receptor polymorphisms, and prostate cancer. PLoS Med 4:e103
92. Tuohimaa P, Tenkanen L, Ahonen M et al (2004) Both high and low levels of blood vitamin D are associated with a higher prostate cancer risk: a longitudinal, nested case-control study in the Nordic countries. Int J Cancer 108:104–108
93. Ahn J, Peters U, Albanes D et al (2008) Serum vitamin D concentration and prostate cancer risk: a nested case-control study. J Natl Cancer Inst 100:796–804
94. Braun MM, Helzlsouer KJ, Hollis BW et al (1995) Prostate cancer and prediagnostic levels of serum vitamin D metabolites (Maryland, United States). Cancer Causes Control 6: 235–239
95. Faupel-Badger JM, Diaw L, Albanes D et al (2007) Lack of association between serum levels of 25-hydroxyvitamin D and the subsequent risk of prostate cancer in Finnish men. Cancer Epidemiol Biomarkers Prev 16:2784–2786
96. Gann PH, Ma J, Hennekens CH et al (1996) Circulating vitamin D metabolites in relation to subsequent development of prostate cancer. Cancer Epidemiol Biomarkers Prev 5: 121–126
97. Jacobs ET, Giuliano AR, Martinez ME et al (2004) Plasma levels of 25-hydroxyvitamin D, 1, 25-dihydroxyvitamin D and the risk of prostate cancer. J Steroid Biochem Mol Biol 89–90:533–537
98. Platz EA, Leitzmann MF, Hollis BW et al (2004) Plasma 1, 25-dihydroxy- and 25-hydroxy-vitamin D and subsequent risk of prostate cancer. Cancer Causes Control 15:255–265
99. Nomura AM, Stemmermann GN, Lee J et al (1998) Serum vitamin D metabolite levels and the subsequent development of prostate cancer (Hawaii, United States). Cancer Causes Control 9:425–432
100. Giovannucci E, Liu Y, Rimm EB et al (2006) Prospective study of predictors of vitamin D status and cancer incidence and mortality in men. J Natl Cancer Inst 98:451–459
101. Bernardi RJ, Trump DL, Yu WD et al (2001) Combination of 1alpha, 25-dihydroxyvitamin D(3) with dexamethasone enhances cell cycle arrest and apoptosis: role of nuclear receptor cross-talk and Erk/Akt signaling. Clin Cancer Res 7:4164–4173
102. Yu WD, McElwain MC, Modzelewski RA et al (1998) Enhancement of 1, 25-dihydroxyvitamin D3-mediated antitumor activity with dexamethasone. J Natl Cancer Inst 90:134–141
103. Hershberger PA, Modzelewski RA, Shurin ZR et al (1999) 1, 25-Dihydroxycholecalciferol (1, 25–D3) inhibits the growth of squamous cell carcinoma and down-modulates p21(Waf1/Cip1) in vitro and in vivo. Cancer Res 59:2644–2649
104. Bernardi RJ, Johnson CS, Modzelewski RA et al (2002) Antiproliferative effects of 1alpha, 25-dihydroxyvitamin D(3) and vitamin D analogs on tumor-derived endothelial cells. Endocrinology 143:2508–2514
105. Beer TM, Hough KM, Garzotto M et al (2001) Weekly high-dose calcitriol and docetaxel in advanced prostate cancer. Semin Oncol 28:49–55
106. Hershberger PA, Yu WD, Modzelewski RA et al (2001) Calcitriol (1, 25-dihydroxychole-calciferol) enhances paclitaxel antitumor activity in vitro and in vivo and accelerates paclitaxel-induced apoptosis. Clin Cancer Res 7:1043–1051
107. Moffatt KA, Johannes WU, Miller GJ (1999) 1Alpha, 25dihydroxyvitamin D3 and platinum drugs act synergistically to inhibit the growth of prostate cancer cell lines. Clin Cancer Res 5:695–703

108. Ahmed S, Johnson CS, Rueger RM et al (2002) Calcitriol (1, 25-dihydroxycholecalciferol) potentiates activity of mitoxantrone/dexamethasone in an androgen independent prostate cancer model. J Urol 168:756–761

109. Light BW, Yu WD, McElwain MC et al (1997) Potentiation of cisplatin antitumor activity using a vitamin D analogue in a murine squamous cell carcinoma model system. Cancer Res 57:3759–3764

110. Wieder R, Wang Q, Uytingco M et al (1998) 1, 25-dihydroxyvitamin D3 and all-trans retinoic acid promote apoptosis and sensitize breast cancer cells to the effects of chemotherapeutic agents. Proc Am Soc Clin Oncol 17:107a

111. Sundaram S, Chaudhry M, Reardon D et al (2000) The vitamin D3 analog EB 1089 enhances the antiproliferative and apoptotic effects of adriamycin in MCF-7 breast tumor cells. Breast Cancer Res Treat 63:1–10

112. Torres R, Calle C, Aller P et al (2000) Etoposide stimulates 1, 25-dihydroxyvitamin D3 differentiation activity, hormone binding and hormone receptor expression in HL-60 human promyelocytic cells. Mol Cell Biochem 208:157–162

113. Koga M, Sutherland RL (1991) Retinoic acid acts synergistically with 1, 25-dihydroxyvitamin D3 or antioestrogen to inhibit T-47D human breast cancer cell proliferation. J Steroid Biochem Mol Biol 39:455–460

114. Guzey M, Sattler C, DeLuca HF (1998) Combinational effects of vitamin D3 and retinoic acid (all trans and 9 cis) on proliferation, differentiation, and programmed cell death in two small cell lung carcinoma cell lines. Biochem Biophys Res Commun 249:735–744

115. Peehl DM, Feldman D (2004) Interaction of nuclear receptor ligands with the Vitamin D signaling pathway in prostate cancer. J Steroid Biochem Mol Biol 92:307–315

116. Ikeda N, Uemura H, Ishiguro H et al (2003) Combination treatment with 1alpha, 25-dihydroxyvitamin D3 and 9-cis-retinoic acid directly inhibits human telomerase reverse transcriptase transcription in prostate cancer cells. Mol Cancer Ther 2:739–746

117. Anzano MA, Smith JM, Uskokovic MR et al (1994) 1 alpha, 25-Dihydroxy-16-ene-23-yne-26, 27-hexafluorocholecalciferol (Ro24–5531), a new deltanoid (vitamin D analogue) for prevention of breast cancer in the rat. Cancer Res 54:1653–1656

118. Welsh J (1994) Induction of apoptosis in breast cancer cells in response to vitamin D and antiestrogens. Biochem Cell Biol 72:537–545

119. Abe-Hashimoto J, Kikuchi T, Matsumoto T et al (1993) Antitumor effect of 22-oxa-calcitriol, a noncalcemic analogue of calcitriol, in athymic mice implanted with human breast carcinoma and its synergism with tamoxifen. Cancer Res 53:2534–2537

120. Christensen GL, Jepsen JS, Fog CK et al (2004) Sequential versus combined treatment of human breast cancer cells with antiestrogens and the vitamin D analogue EB1089 and evaluation of predictive markers for vitamin D treatment. Breast Cancer Res Treat 85:53–63

121. Moreno J, Krishnan AV, Swami S et al (2005) Regulation of prostaglandin metabolism by calcitriol attenuates growth stimulation in prostate cancer cells. Cancer Res 65:7917–7925

122. Gavrilov V, Steiner M, Shany S (2005) The combined treatment of 1, 25-dihydroxyvitamin D3 and a non-steroid anti-inflammatory drug is highly effective in suppressing prostate cancer cell line (LNCaP) growth. Anticancer Res 25:3425–3429

123. Hsiao M, Tse V, Carmel J et al (1997) Functional expression of human p21(WAF1/CIP1) gene in rat glioma cells suppresses tumor growth in vivo and induces radiosensitivity. Biochem Biophys Res Commun 233:329–335

124. Dunlap N, Schwartz GG, Eads D et al (2003) 1alpha, 25-dihydroxyvitamin D(3) (calcitriol) and its analogue, 19-nor-1alpha, 25(OH)(2)D(2), potentiate the effects of ionising radiation on human prostate cancer cells. Br J Cancer 89:746–753

125. Polar MK, Gennings C, Park M et al (2003) Effect of the vitamin D3 analog ILX 23–7553 on apoptosis and sensitivity to fractionated radiation in breast tumor cells and normal human fibroblasts. Cancer Chemother Pharmacol 51:415–421

126. DeMasters GA, Gupta MS, Jones KR et al (2004) Potentiation of cell killing by fractionated radiation and suppression of proliferative recovery in MCF-7 breast tumor cells by the vitamin D3 analog EB 1089. J Steroid Biochem Mol Biol 92:365–374

10 Vitamin D and Prostate Cancer

127. Osborn JL, Schwartz GG, Smith DC et al (1995) Phase II trial of oral 1, 25-dihydroxyvitamin D (calcitriol) in hormone refractory prostate cancer. Urol Onc 1:195–198

128. Gross C, Stamey T, Hancock S et al (2039) Treatment of early recurrent prostate cancer with 1,25-dihydroxyvitamin D3 (calcitriol). J Urol 159:2035, discussion 2039–2040

129. Smith DC, Johnson CS, Freeman CC et al (1999) A Phase I trial of calcitriol (1, 25-dihydroxy-cholecalciferol) in patients with advanced malignancy. Clin Cancer Res 5:1339–1345

130. Beer TM, Munar M, Henner WD (2001) A Phase I trial of pulse calcitriol in patients with refractory malignancies: pulse dosing permits substantial dose escalation. Cancer 91:2431–2439

131. Muindi JR, Peng Y, Potter DM et al (2002) Pharmacokinetics of high-dose oral calcitriol: results from a phase 1 trial of calcitriol and paclitaxel. Clin Pharmacol Ther 72:648–659

132. Muindi JR, Potter DM, Peng Y et al (2005) Pharmacokinetics of liquid calcitriol formulation in advanced solid tumor patients: comparison with caplet formulation. Cancer Chemother Pharmacol 56:492–496

133. Beer TM, Javle M, Lam GN et al (2005) Pharmacokinetics and tolerability of a single dose of DN-101, a new formulation of calcitriol, in patients with cancer. Clin Cancer Res 11:7794–7799

134. Beer TM, Javle M, Henner WD et al (2004) Pharmacokinetics (PK) and tolerability of DN-101, a new formulation of calcitriol, in patients with cancer. Proc Am Assn Cancer Res 45:91

135. Flaig TW, Barqawi A, Miller G et al (2006) A phase II trial of dexamethasone, vitamin D, and carboplatin in patients with hormone-refractory prostate cancer. Cancer 107:266–274

136. Morris MJ, Smaletz O, Solit D et al (2004) High-dose calcitriol, zoledronate, and dexamethasone for the treatment of progressive prostate carcinoma. Cancer 100:1868–1875

137. Fakih MG, Trump DL, Muindi JR et al (2007) A phase I pharmacokinetic and pharmacodynamic study of intravenous calcitriol in combination with oral gefitinib in patients with advanced solid tumors. Clin Cancer Res 13:1216–1223

138. Beer TM, Lemmon D, Lowe BA et al (2003) High-dose weekly oral calcitriol in patients with a rising PSA after prostatectomy or radiation for prostate carcinoma. Cancer 97:1217–1224

139. Beer TM, Eilers KM, Garzotto M et al (2003) Weekly high-dose calcitriol and docetaxel in metastatic androgen-independent prostate cancer. J Clin Oncol 21:123–128

140. Beer TM, Ryan CW, Venner PM et al (2007) Double-blinded randomized study of high-dose calcitriol plus docetaxel compared with placebo plus docetaxel in androgen-independent prostate cancer: a report from the ASCENT Investigators. J Clin Oncol 25:669–674

141. Trump DL, Potter DM, Muindi J et al (2006) Phase II trial of high-dose, intermittent calcitriol (1, 25 dihydroxyvitamin D3) and dexamethasone in androgen-independent prostate cancer. Cancer 106:2136–2142

142. Tiffany NM, Ryan CW, Garzotto M et al (2005) High dose pulse calcitriol, docetaxel and estramustine for androgen independent prostate cancer: a phase I/II study. J Urol 174:888–892

143. Beer TM, Garzotto M, Katovic NM (2004) High-dose calcitriol and carboplatin in metastatic androgen-independent prostate cancer. Am J Clin Oncol 27:535–541

144. Kissmeyer AM, Binderup E, Binderup L et al (1997) Metabolism of the vitamin D analog EB 1089: identification of in vivo and in vitro liver metabolites and their biological activities. Biochem Pharmacol 53:1087–1097

145. Bouillon R, Verstuyf A, Verlinden L et al (1995) Non-hypercalcemic pharmacological aspects of vitamin D analogs. Biochem Pharmacol 50:577–583

146. Bouiloon R, Okamura WH, Norman AW (1995) Structure-function relationships in the vitamin D endocrine system. Endocr Rev 16:200–257

147. Gulliford T, English J, Colston KW et al (1998) A phase I study of the vitamin D analogue EB 1089 in patients with advanced breast and colorectal cancer. Br J Cancer 78:6–13

148. Dalhoff K, Dancey J, Astrup L et al (2003) A phase II study of the vitamin D analogue Seocalcitol in patients with inoperable hepatocellular carcinoma. Br J Cancer 89:252–257

149. Evans TR, Colston KW, Lofts FJ et al (2002) A phase II trial of the vitamin D analogue Seocalcitol (EB1089) in patients with inoperable pancreatic cancer. Br J Cancer 86:680–685

150. Bower M, Colston KW, Stein RC et al (1991) Topical calcipotriol treatment in advanced breast cancer. Lancet 337:701–702
151. Liu G, Oettel K, Ripple G et al (2002) Phase I trial of 1alpha-hydroxyvitamin d(2) in patients with hormone refractory prostate cancer. Clin Cancer Res 8:2820–2827
152. Attia S, Eickhoff J, Wilding G et al (2008) Randomized, double-blinded phase II evaluation of docetaxel with or without doxercalciferol in patients with metastatic, androgen-sindependent prostate cancer. Clin Cancer Res 14:2437–2443
153. Wieder R, Novick SC, Hollis BW et al (2003) Pharmacokinetics and safety of ILX23–7553, a non-calcemic-vitamin D3 analogue, in a phase I study of patients with advanced malignancies. Invest New Drugs 21:445–452
154. Schwartz GG, Hall MC, Stindt D et al (2005) Phase I/II study of 19-nor-1alpha-25-dihydroxyvitamin D2 (paricalcitol) in advanced, androgen-insensitive prostate cancer. Clin Cancer Res 11:8680–8685
155. Kim RH, Li JJ, Ogata Y et al (1996) Identification of a vitamin D3-response element that overlaps a unique inverted TATA box in the rat bone sialoprotein gene. Biochem J 318(Pt 1):219–226
156. Falzon M (1996) DNA sequences in the rat parathyroid hormone-related peptide gene responsible for 1, 25-dihydroxyvitamin D3-mediated transcriptional repression. Mol Endocrinol 10:672–681
157. Gill RK, Christakos S (1993) Identification of sequence elements in mouse calbindin-D28k gene that confer 1, 25-dihydroxyvitamin D3- and butyrate-inducible responses. Proc Natl Acad Sci USA 90:2984–2988
158. Kitazawa R, Kitazawa S (2002) Vitamin D(3) augments osteoclastogenesis via vitamin D-responsive element of mouse RANKL gene promoter. Biochem Biophys Res Commun 290:650–655
159. Matilainen M, Malinen M, Saavalainen K et al (2005) Regulation of multiple insulin-like growth factor binding protein genes by 1alpha, 25-dihydroxyvitamin D3. Nucleic Acids Res 33:5521–5532
160. Peng L, Malloy PJ, Feldman D (2004) Identification of a functional vitamin D response element in the human insulin-like growth factor binding protein-3 promoter. Mol Endocrinol 18:1109–1119
161. Polly P, Carlberg C, Eisman JA et al (1996) Identification of a vitamin D3 response element in the fibronectin gene that is bound by a vitamin D3 receptor homodimer. J Cell Biochem 60:322–333
162. Cao X, Ross FP, Zhang L et al (1993) Cloning of the promoter for the avian integrin beta 3 subunit gene and its regulation by 1, 25-dihydroxyvitamin D3. J Biol Chem 268:27371–27380
163. Bikle DD, Ng D, Oda Y et al (2002) The vitamin D response element of the involucrin gene mediates its regulation by 1, 25-dihydroxyvitamin D3. J Invest Dermatol 119:1109–1113
164. Candeliere GA, Jurutka PW, Haussler MR et al (1996) A composite element binding the vitamin D receptor, retinoid X receptor alpha, and a member of the CTF/NF-1 family of transcription factors mediates the vitamin D responsiveness of the c-fos promoter. Mol Cell Biol 16:584–592
165. Xie Z, Bikle DD (1998) Differential regulation of vitamin D responsive elements in normal and transformed keratinocytes. J Invest Dermatol 110:730–733
166. McGaffin KR, Chrysogelos SA (2005) Identification and characterization of a response element in the EGFR promoter that mediates transcriptional repression by 1, 25-dihydroxyvitamin D3 in breast cancer cells. J Mol Endocrinol 35:117–133
167. Drissi H, Pouliot A, Koolloos C et al (2002) 1, 25-(OH)2-vitamin D3 suppresses the bone-related Runx2/Cbfa1 gene promoter. Exp Cell Res 274:323–333
168. Quelo I, Machuca I, Jurdic P (1998) Identification of a vitamin D response element in the proximal promoter of the chicken carbonic anhydrase II gene. J Biol Chem 273:10638–10646
169. Seoane S, Alonso M, Segura C et al (2002) Localization of a negative vitamin D response sequence in the human growth hormone gene. Biochem Biophys Res Commun 292:250–255

10   Vitamin D and Prostate Cancer

170. Fujisawa K, Umesono K, Kikawa Y et al (2000) Identification of a response element for vitamin D3 and retinoic acid in the promoter region of the human fructose-1, 6-bisphosphatase gene. J Biochem 127:373–382
171. Murayama A, Takeyama K, Kitanaka S et al (1998) The promoter of the human 25-hydroxyvitamin D3 1 alpha-hydroxylase gene confers positive and negative responsiveness to PTH, calcitonin, and 1 alpha, 25(OH)2D3. Biochem Biophys Res Commun 249:11–16
172. Zierold C, Darwish HM, DeLuca HF (1994) Identification of a vitamin D-response element in the rat calcidiol (25-hydroxyvitamin D3) 24-hydroxylase gene. Proc Natl Acad Sci U S A 91:900–902
173. Vazquez G, Boland R, de Boland AR (1995) Modulation by 1, 25(OH)2-vitamin D3 of the adenylyl cyclase/cyclic AMP pathway in rat and chick myoblasts. Biochim Biophys Acta 1269:91–97
174. Santillan GE, Boland RL (1998) Studies suggesting the participation of protein kinase A in 1, 25(OH)2-vitamin D3-dependent protein phosphorylation in cardiac muscle. J Mol Cell Cardiol 30:225–233

Chapter 11
Vitamin D and Hematologic Malignancies

Ryoko Okamoto, Tadayuki Akagi, and H. Phillip Koeffler

Abstract The biologically active form of vitamin D, 1,25-dihydroxyvitamin D_3 [1,25(OH)$_2$D$_3$], has multiple anticancer activities including growth arrest, induction of apoptosis, and differentiation. Here, the actions of vitamin D compounds are addressed from normal to malignant hematopoietic cells. The effects are driven by binding of vitamin D to vitamin D receptor in either genomic and/or nongenomic fashions. However, its application as a therapeutic agent is limited by its side effect, hypercalcemia. 1,25(OH)$_2$D$_3$ analogs have been synthesized to obtain anti-tumor activity with less calcemic toxicity. Limited clinical studies using vitamin D compounds have had only minor clinical success for patients with leukemia or myelodysplasia syndrome. Nevertheless, preclinical studies suggest that the combination of vitamin D compounds with other agents can have additive or synergistic anticancer activities, renewing hope for future clinical studies.

Keywords Hematopoiesis • Vitamin D • Vitamin D receptor • Leukemia • Molecular mechanisms • Vitamin D analogs • Combination therapy

11.1 Overview of Hematopoiesis

Hematopoiesis is the process that leads to the formation of the highly specialized circulating blood cells from pluripotent hematopoietic stem cells (HSCs) in the bone marrow. The HSCs are the most primitive blood cells, and they have the ability for both self-renewal and pluripotency. They differentiate to more mature "committed" cells including the common lymphoid progenitor (CLP) and the common myeloid progenitor (CMP); and the latter differentiates to megakaryocyte-erythroid progenitors

---

R. Okamoto (✉)
Division of Hematology and Oncology,
Cedars-Sinai Medical Center, UCLA School of Medicine,
8700 Beverly Blvd, Los Angeles, CA 90048, USA
e-mail: ryoko.okamoto@cshs.org

D.L. Trump and C.S. Johnson (eds.), *Vitamin D and Cancer*,
DOI 10.1007/978-1-4419-7188-3_11, © Springer Science+Business Media, LLC 2011

(MEP) and granulocyte-macrophage progenitors (GMP). The MEP eventually differentiates into functional red blood cells and platelets. The GMP gives rise to mature mast cells, eosinophils, neutrophils, and monocytes/macrophages. The CLP population produces either mature T or B lymphocytes (Fig. 11.1).

The differentiation and proliferation of hematopoietic stem cells, as well as, their more mature precursor cells are highly controlled by stimulation of cytokines from the extracellular environment. Each of these stem cells has cell surface receptors for specific cytokines. Binding of cytokines to these receptors stimulates secondary intracellular signals that deliver a message to the nucleus to enhance proliferation, differentiation, and/or activation. The growth factors acting primarily on the granulocyte-macrophage pathway are granulocyte-macrophage colony-stimulating factor (GM-CSF), granulocyte colony-stimulating factor (G-CSF), and macrophage colony-stimulating factor (M-CSF). The GM-CSF also stimulates eosinophils, enhances megakaryocytic colony formation, and increases erythroid colony formation in the presence of erythropoietin (Epo). *In vivo,* the cytokine causes an increase in granulocytes, monocytes, and eosinophils. The GM-CSF can activate the monocytes and granulocytes to kill efficiently invading microbes. The G-CSF stimulates the formation of granulocyte colonies *in vitro.* It is able to act synergistically

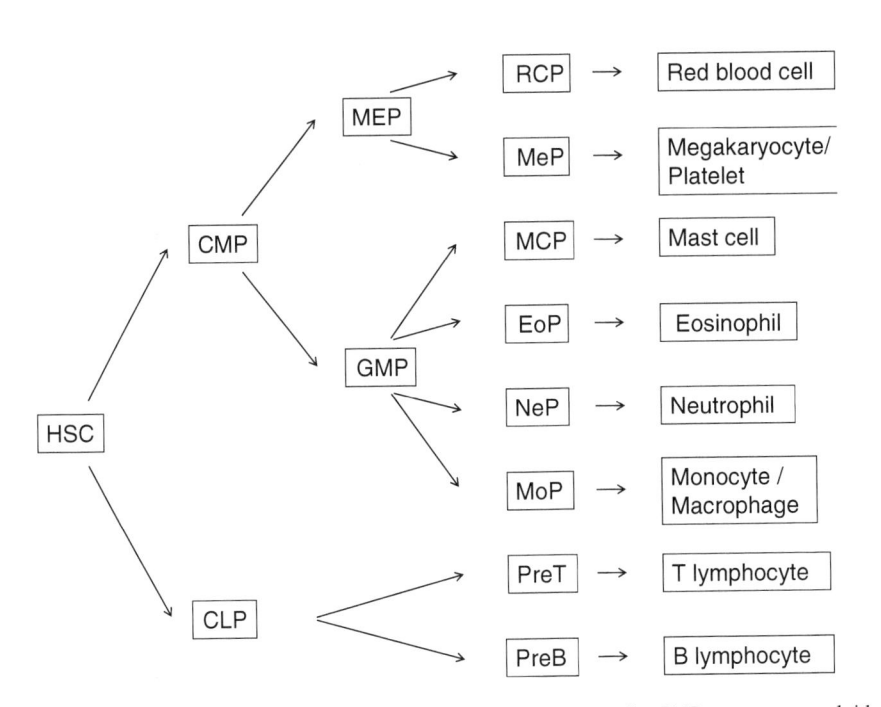

Fig. 11.1 Scheme of hematopoiesis. *HSC* hematopoietic stem cell, *CMP* common myeloid progenitor, *CLP* common lymphoid progenitor, *MEP* megakaryocyte-erythroid progenitor, *GMP* granulocyte-macrophage progenitor, *RCP* red blood cell precursor, *MeP* megakaryocyte precursor, *MCP* mast cell precursor, *EoP* eosinophil precursor, *NeP* neutrophil precursor, *MoP* monocyte-macrophage precursor, *PreT* precursor of T lymphocyte, *PreB* precursor of B lymphocyte

with interleukin (IL)-3, GM-CSF, and M-CSF. This cytokine is active *in vivo,* stimulating an increase of peripheral blood granulocytes. The M-CSF stimulates the formation of macrophage colonies *in vitro.* It maintains the survival of differentiated macrophages and increases their antitumor activities and secretion of oxygen reduction products as well as plasminogen activators. This cytokine binds to a receptor that is the product of the protooncogene c-fms. IL-3 has multilineage stimulating activity and acts directly on the granulocyte-macrophage pathway, but also enhances the development of erythroid, megakaryocytic, and mast cells, and possibly T lymphocytes. In synergy with Epo, IL-3 stimulates the formation of early erythroid stem cells, promoting the formation of colonies of red cells in soft gel culture known as BFU-E. In addition, it supports the formation of early multilineage cells *in vitro.* IL-3 also induces early progenitor cells to enter the cell cycle, and in combination with other growth factors, stimulates the production of all the myeloid cells *in vivo.* Stem cell factor (SCF) promotes survival, proliferation and differentiation of hematopoietic progenitor cells. It synergizes with other growth factors such as IL-3, GM-CSF, G-CSF and Epo to support the clonogenic growth *in vitro.* SCF is a ligand for the c-kit receptor, a tyrosine kinase receptor that is expressed in hematopoietic progenitor cells. The growth factor Epo stimulates the formation of erythroid colonies (CFU-E) *in vitro* and is the primary hormone of erythropoiesis in animals and humans. It binds to a specific receptor (Epo-R). Production of erythroblasts is regulated by Epo which is modulated by the amount of tissue oxygenation of Epo-producing cells in the kidney. Oxygen-carrying hemoglobin in the red blood cells is the physiologic rheostat determining the amounts of circulating Epo. Anemia causes tissue hypoxia, resulting in an increase of serum Epo levels.

11.2 Vitamin D Receptors in Blood Cells

The genomic actions of $1,25(OH)_2D_3$ are mediated by the intracellular vitamin D receptor (VDR), which belongs to a large family of nuclear receptors [1]. VDR forms a heterodimer with the retinoid X receptor (RXR); this complex regulates expression of target genes by binding to vitamin D responsive elements (VDREs) in the promoter regions of their target genes [2]. Patients with hereditary vitamin D-resistant rickets type II (HVDRR) have various mutations of the VDR resulting in prominent skeletal abnormalities and hematopoietic abnormalities [3, 4]. Expression of VDR has been detected in bone marrow-derived stromal cells, as well as various normal and leukemic hematopoietic cells [5, 6].

11.2.1 Vitamin D Receptors in Myeloid Cells

VDR is expressed constitutively in monocytes, neutrophils and antigen-presenting cells such as macrophages and dendritic cells [5, 7–9]. Circulating monocytes have

higher levels of VDR than tissue macrophages [10]. VDR protein levels of peripheral blood monocytes have been reported to be two-fold higher in patients with idiopathic hypercalciuria with normal serum $1,25(OH)_2D_3$ levels compared to monocytes from normal individuals [11]. On the other hand, fewer receptors have been detected in the peripheral blood mononuclear cells of patients with X-linked hypophosphatemic rickets [12]. These individuals have a significant positive correlation between VDR concentration in their mononuclear cells and their serum phosphate levels ($p < 0.05$).

Examination of a large number of myeloid leukemia cell lines blocked at various stages of maturation showed that they all expressed VDR, albeit at different levels [5]. Treatment of HL-60 myeloblastic leukemia cells with $1,25(OH)_2D_3$ (10^{-7} M) decreases their VDR protein levels by 50% at 24 h and levels return to normal after 72 h. No change of VDR mRNA expression occurred in the cells [5, 13], suggesting that one of the major sites of regulation of expression of VDR occurs at the post transcriptional level. Exposure to $1,25(OH)_2D_3$ induces the VDR to move from the cytoplasm to the nucleus, and this translocation is prevented by treatment with inhibitors of the PI3-K (LY294002) and the MAPK (PD98059) pathways [14]. Their monocyte-like differentiation of HL-60 cells treated with $1,25(OH)_2D_3$ may require functional activator protein-1 (AP-1) complexes which bind to the TRE of the promoter region in human VDR [15] (Sect. 11.4.2.1).

11.2.2 Vitamin D Receptors in Lymphoid Cells

Subsets of thymocytes, resting T lymphocytes especially those expressing either CD8+ or CD4+ and activated T lymphocytes express VDR [5, 16, 17]. VDR mRNA expression increases when these cells are stimulated to proliferate, for example after their exposure to phytohemagglutinin-A (PHA) for 24 h in vitro. Another major site of regulation of VDR expression in these cells is at the transcriptional level [5, 16]. No VDR mRNA or protein was detected in resting B lymphocytes, but VDR was up-regulated via cellular activation in vitro and in vivo, for example in normal human B cells from tonsils [16, 18]. $1,25(OH)_2D_3$ inhibits the synthesis of immunoglobulins (Ig) synthesized by B lymphocytes in vitro [19]. Their inhibition may be mediated through activation of VDR/RXR in these cells, and/or through the inhibition of T-helper activity [20]. Production of lymphokines, including IL-2, is markedly decreased by $1,25(OH)_2D_3$ in activated T lymphocytes, and this could cause the suppression of Ig synthesis [21–24]. The effects of vitamin D on the immune system are discussed in Chapter 6.

Levels of VDR mRNA in leukocytes from healthy individuals after an oral administration of $1,25(OH)_2D_3$ increased an average of 1.2 to 11.1-fold [25]. The maximum increase of VDR mRNA levels occur over 1 and 5 h, with a mean of 3.6 h. Expression of VDR is induced in the lymphocytes of patients with rheumatoid arthritis and in pulmonary lymphocytes of patients with tuberculosis and sarcoidosis [26–28]. Moreover, low levels of VDR expression were detected in

11 Vitamin D and Hematologic Malignancies   255

low-grade, non-Hodgkin's lymphoma (NHL) and in the follicular lymphoma B-cell lines SU-DHL4 and SU-DHL5 [29].

11.2.3 Hematopoiesis in VDR Knockout Mice

Studies by us using VDR knockout (KO) mice indicated that expression of VDR is dispensable for normal hematopoiesis [30]. No difference in the numbers and percentages of red and white blood cells were found between VDR KO and wild-type (WT) mice. Committed myeloid stem cells from the bone marrow cultured in methylcellulose formed similar numbers of colonies when grown in the presence of various cytokines including GM-CSF, G-CSF, M-CSF either alone or in combination with IL-3. Furthermore, bone marrow progenitor cells from VDR KO and WT mice formed a similar number and percentage of granulocyte, macrophage and granulocyte/macrophage mixed colonies when cultured in methylcellulose with GM-CSF and IL-3. Under these conditions, treatment with $1,25(OH)_2D_3$ dramatically increased the percentage of macrophage colonies derived from WT but not VDR KO bone marrow cultures. This observation demonstrates the requirement of VDR expression for $1,25(OH)_2D_3$ -induction of bone marrow progenitors into monocytes/macrophages. The proportions of T- and B-cells were normal in the VDR KO mice. However, the antigen-stimulated spleen cells from VDR KO mice produced less IFNγ and more IL-4 than those from WT mice, indicating impaired Th1 differentiation. Additionally, IL-12 stimulation induced a weaker proliferative response in VDR KO splenocytes as compared to those in WT mice, and expression of STAT4 was reduced. These results suggest that VDR plays an important role in the Th1-type immune response but not T cell development. Interestingly, another report using VDR KO mice showed that VDR is required for normal development and function of Vα14 invariant natural killer T (iNKT) cells which are involved in immune regulation, host defense against pathogens and tumor surveillance [31].

11.3 Effects of Vitamin D Compounds on Normal Hematopoiesis

$1,25(OH)_2D_3$ modulates the differentiation of normal hematopoietic progenitors. Normal human bone marrow committed stem cells cultured in either soft agar or liquid culture with $1,25(OH)_2D_3$ differentiate into macrophages. Likewise, monocytes cultured in serum-free medium with $1,25(OH)_2D_3$ become macrophages within 7 days [32–37]. These macrophages are functionally competent [35]. Concentrations of $1,25(OH)_2D_3$ causing this differentiation ranges between 10^{-10} M (slightly higher than physiological serum level) to 10^{-7} M. On the other hand, $1,25(OH)_2D_3$ (10^{-9} to 10^{-7} M) can inhibit the differentiation into CD1α + dendritic cells [38].

As mentioned earlier, $1,25(OH)_2D_3$ is able to inhibit both the synthesis of IL-2 and the proliferation of peripheral blood lymphocytes [20–23]. Indeed, $1,25(OH)_2D_3$ can regulate the expression of many lymphokines, such as GM-CSF, IFN-γ and IL-12 [20, 39, 40]. For example, Tobler et al. showed that expression of the lymphokine GM-CSF is regulated by $1,25(OH)_2D_3$ through VDR by a process independent of IL-2 production. In particular, $1,25(OH)_2D_3$ was able to inhibit both mRNA and protein expression of GM-CSF in PHA-activated normal human peripheral blood lymphocytes (PBL). The former occurred at least in part by destabilizing and shortening the half-life of the GM-CSF mRNA [39]. The down-regulation of GM-CSF was obtained at concentrations similar to those reached *in vivo*, with a 50% reduction of GM-CSF activity occurring at 10^{-10} M of $1,25(OH)_2D_3$. In addition, IL-2 did not affect the modulation of GM-CSF production by PBL which were co-cultured with $1,25(OH)_2D_3$ (10^{-10} to 10^{-7} M).

11.4   Effects of Vitamin D Compounds on Leukemic Cells

The $1,25(OH)_2D_3$ was first noted to induce leukemia cell differentiation in the M1 murine myeloid cell line [41]. Moreover, $1,25(OH)_2D_3$ extended the survival of mice inoculated with the M1 leukemia cells [42]. In spite of the promising data obtained from *in vitro* and animal studies, results of clinical trials of $1,25(OH)_2D_3$ in leukemia are limited in scope and thus far have exhibited only mediocre results. A disease that can evolve in leukemia is myelodysplastic syndrome (MDS). It is a clonal hematopoietic stem cell disorder; these individuals often have anemia, thrombocytopenia, and/or leukopenia as well as an increased number of myeloid progenitor cells in their bone marrow. $1,25(OH)_2D_3$ has had less than spectacular results as a therapy for MDS (Table 11.1) [43]. Furthermore, vitamin D_3 analogs [19-nor-$1,25(OH)_2D_3$ (Paricalcitol) or $1(OH)D_2$ (Doxercalciferol)] have had minor responses at best [44, 45].

11.4.1   Cellular Effects of Vitamin D Compounds on Leukemic Cells

A number of human AML cell lines can be inhibited in their proliferation and/or induced to undergo differentiation by $1,25(OH)_2D_3$, including HL-60, U937, THP-1, HEL and to a lesser extent NB4 cells [46, 47]. In contrast, many immature myeloid leukemia cell lines such as HL-60 blasts, KG 1, KGla and K562 are not responsive to vitamin D compounds.

Vitamin D analogs inhibit leukemic cell growth by inducing cell cycle arrest. The cells accumulate in the G0/G1 and G2/M phase of the cell cycle, with a concomitant decrease in the proportion of cells in S-phase [48–50].

Table 11.1 Trials of vitamin D compounds in myelodysplastic syndrome (MDS)

Compound	Another name	Hyper-calcemia	Dose/day (µg)	Treatment duration (months)	No. of Patients[a]	Efficiency	Reference
1,25(OH)$_2$D$_3$	Calcitriol	+ (50%)[b]	2	3	18	Occasional minor response	[53]
1(OH)D$_3$	Alfacalcidol	+ (13%)[b]	4–6	17[c]	15	Markedly decreased progression to AML	[134]
1,25(OH)$_2$D$_3$	Calcitriol	—	0.25 – 0.75	9–27	14	10/14 (71%) responded[d]	[156]
19-nor-1,25(OH)$_2$D$_2$	Paricalcitol[e]	—	8 – 56	≤ 9	12	Occasional minor response	[44]
Vitamin D$_3$	Cholecalciferol	—	50 – 100	5[c]	26	No therapeutic effect	[157]
1(OH)D$_2$	Doxercalciferol[f]	—	12.5	3	15	No therapeutic effect	[45]

[a] Only trials with ≥12 patients are listed
[b] Percentage individuals who developed hypercalcemia while on trial
[c] Median
[d] Criteria for response was not stringent
[e] Abbott Laboratories code name is Paricalcitol
[f] Bone Care Int. code name is Hectorol

HL-60 myeloblastic cell line cultured with $1,25(OH)_2D_3$ (10^{-10}–10^{-7} M, for 7 days) morphologically and functionally differentiate toward macrophages, becoming adherent to charged surfaces, developing pseudopodia, staining positively for nonspecific esterase (NSE), reducing nitroblue tetrazolium (NBT), and acquiring the ability to phagocytose yeast [36, 51, 52]. In addition, these cells have the ability to degrade bone marrow matrix *in vitro,* raising the possibility that the cells may have acquired some osteoclast-like characteristics. Leukemic cells from AML patients respond to vitamin D compounds when cultured *in vitro;* however, they are often less sensitive to this seco-steroid than are the HL-60 cell lines. They frequently undergo partial monocytic differentiation as assessed by NBT reduction, morphology, and phagocytic ability. Furthermore, their clonal growth is often inhibited [36, 53].

11.4.2 Molecular Mechanisms of Action of Vitamin D Compounds Against Leukemic Cells

Vitamin D compounds can exert their biological effects by genomic (Sect. 11.4.2.1) and/or nongenomic (Sect. 11.4.2.2) pathways. Both pathways require ligand binding to the VDR. The former pathway relies on a $1,25(OH)_2D_3$ activated VDR/RXR complex binding to VDREs in order to modulate the transcription of various target genes. The latter increases rapid intracellular Ca^{2+} influxes resulting in activation of kinases within seconds to minutes [54]. It is still unknown whether the nongenomic actions are mediated through the classical nuclear VDR, a membrane-associated VDR or other proteins. Exposure of hematopoietic cells to $1,25(OH)_2D_3$ controls myriad of genes, including those responsible for the regulation of cellular proliferation, differentiation, apoptosis and angiogenesis. Modulation of these genes by $1,25(OH)_2D_3$ may not always be a direct effect on transcription of target genes, but can reflect the entire process of differentiation associated with a series of interacting transcription factors. Nonetheless, $1,25(OH)_2D_3$-activated intracellular signaling pathways require the presence of VDR to stimulate monocyte/macrophage differentiation, as demonstrated by studies on bone marrow cells from VDR KO mice [30] and cells from patients with vitamin D-dependent rickets type II [55, 56]. The rapid, nongenomic activities of vitamin D are described in detail in Chapter xx. The molecular targets of vitamin D compounds in leukemic cells are summarized in Table 11.2.

11.4.2.1 Molecular Mechanisms of Genomic Action of $1,25(OH)_2D_3$ in Leukemic Cells

Myeloid leukemic cell lines cultured with $1,25(OH)_2D_3$ undergo an initial proliferative burst, which is followed by growth inhibition, terminal differentiation and subsequent apoptosis [57, 58]. Levels of cyclin A, D1 and E increase in the U937

11 Vitamin D and Hematologic Malignancies

Table 11.2 Molecular effects of vitamin D compounds in leukemic cells[a]

Cell cycle/apoptosis	Oncogenes	Transcription factors
Cyclin A ↑	c-myc ↓	C/EBP β ↑
Cyclin D1 ↑	Dek ↓	PU.1 ↑
Cyclin E ↑	Fli ↓	IRF8 β ↑
CDKN1A (p21) ↑	**Protooncogenes**	HoxA10 ↑
CDKN1B (p27) ↑	c-fms ↓	HoxB4 ↑
CDKN2A (p16-INK4A) ↑	**Tumor Suppressors**	AP-1[b] ↑
CDKN2B (p15-INK4B) ↑	PTEN ↑	junD binding activity ↑
CDKN2C (p18-INK4C) ↑	BTG ↑	TRAP ↑
Bcl-2 ↓	**Kinases**	TEL2 ↓
Differentiation Markers	PKC levels ↑	**Feedback Control**
CD11b ↑	PI3-K activity ↑	Cyp24 ↑
CD14 ↑	AKT activity ↑	**Immunity**
Protein synthesis and transport	MAPK activity ↑	CAMP ↑
eIF-2 ↓	ERK 1/2 activity ↑	
Importins ↓	KSR-1,-2 activity ↑	
Exportins -1,-5,-7, -t ↓		

[a]Regulation of expression or activity may occur either directly or as a consequence of differentiation. See text for details
[b]Putative components of AP-1 complex are c-jun, ATF-2, jun-B and fos-B

myelomonoblastic leukemia cells within 24 h of $1,25(OH)_2D_3$ -treatment and then expression decreases after 48 h [57]. The cyclin-dependent kinase (CDK) inhibitors CDKN1A (p21) and CDKN1B (p27) are important regulators of the cell cycle which are elevated during periods of both proliferation and growth inhibition. Expression of these proteins, as well as CDKN2A (p16-INK4A), CDKN2B (p15-INK4B) and CDKN2C (p18-INK4C) CDK inhibitors are increased in a time-dependent manner after exposure to $1,25(OH)_2D_3$ [59].

A strong correlation exists between early induction of p21 and the beginning of the differentiation program. The up-regulation of p21 mRNA occurred within 4 h of the exposure to $1,25(OH)_2D_3$ independent of *de novo* protein synthesis, suggesting a direct transcriptional activation by VDR [59]. Indeed, the p21 promoter contains a vitamin D response element, and induction requires the presence of VDR. Nevertheless, some data suggested that the marked increase of p21 protein expression in response to $1,25(OH)_2D_3$ may also be due to enhanced posttranscriptional stabilization of p21 mRNA [60]. The transcription factor p53 is a strong inducer of p21; but $1,25(OH)_2D_3$ can elevate p21 levels independently of p53 activity.

A strong up-regulation of p27 protein expression was evident after 72 h of exposure to the compound, and levels of the protein were dependent on the concentration of $1,25(OH)_2D_3$ [61]. This up-regulation was also associated with increased levels of Cyclins D1 and E, coinciding with a G1 arrest. These results suggested a prominent role of p27 in mediating the antiproliferative activity of $1,25(OH)_2D_3$ in this cell line. The $1,25(OH)_2D_3$ has a protective effect against apoptosis in HL-60 cells [62, 63]. In other cell types, inhibition of apoptosis correlates with elevated

levels of Bcl-2, but this may not be the case with myeloid cells. A down-regulation of Bcl-2 was observed both at the mRNA and protein levels after HL-60 cells were exposed to $1,25(OH)_2D_3$ [62].

Activation of the proto-oncogene c-myc is a typical feature of human leukemias. The HL-60 leukemia cell line is characterized by high levels of expression of c-myc due to gene amplification [64, 65]. Treatment of this cell line with $1,25(OH)_2D_3$ results in a down-regulation of expression of this oncogene associated with cell differentiation [66]. Suppression of c-myc by $1,25(OH)_2D_3$ and its non-calcemic analogs occurs at the transcriptional level in HL-60 cells [67, 68]. Exposure of HL-60 cells to $1,25(OH)_2D_3$ induces the expression of the proto-oncogene c-fms, which encodes the receptor for M-CSF. It occurs in parallel with the induction of CD14 expression and a block of the cell cycle in the G_0/G_1 phase [69].

$1,25(OH)_2D_3$ up-regulates the protein coding for the homeobox gene, *HOXB4*, that binds to the first exon/intron border of *MYC* to prevent transcriptional elongation, a process dependent on activation of PKC-β [70, 71]. Another homeobox gene, *HOXA10*, was found by differential display to be a gene transcriptionally induced by $1,25(OH)_2D_3$ through binding to the VDRE in the promoter during differentiation of U937 cells [72, 73].

Besides *MYC* and *HOX* genes, $1,25(OH)_2D_3$ can induce other transcription factors and coactivators to regulate gene expression. For example, exposure of U937 cells to $1,25(OH)_2D_3$ induced the expression of PU.1, IRF8 and C/EBPβ [74]. In contrast, exposure of U937 cells to $1,25(OH)_2D_3$ (10^{-8} M) down-regulated the expression of TEL2, which is a member of the ETS family [75]. Interestingly, forced overexpression of TEL2 inhibited $1,25(OH)_2D_3$ -induced differentiation.

The ligand-activated VDR can bind to the AP-1 complex. Exposure of the chronic myelogenous leukemia (CML) cell line RWLeu-4 to $1,25(OH)_2D_3$ inhibited their proliferation and enhanced the binding activity of the proto-oncogene junD to VDRE. This binding activity decreased in a $1,25(OH)_2D_3$-resistant variant $JMRD_3$ cells. Although these cells exhibit no detectable differences in the VDR, alterations in the interaction with the VDRE were important [76]. Exposure of HL-60 cells to $1,25(OH)_2D_3$, up-regulated expression of genes that code for the AP-1 complex including c-jun, ATF-2, jun-B and fos-B [15, 77]. Moreover, $1,25(OH)_2D_3$ (10^{-7}M) was also able to induce expression of the subunits of the transcriptional coactivator, Thyroid hormone Receptor-Associated Polypeptide (TRAP, also called DRIP) as early as 6 h in the HL-60 cells [78]. The TRAP complex plays a role in direct communication between the nuclear receptors and the general transcriptional machinery through direct interaction with RNA polymerase II [79]. The murine Trap220(-/-) yolk sac hematopoietic progenitor cells, as well as, TRAP knockdown HL-60 cells are resistant to induction of differentiation by $1,25(OH)_2D_3$.

Fusion proteins involving the retinoic acid receptor alpha (RARα) with either the PML or PLZF nuclear proteins are the genetic markers of acute promyelocytic leukemias (APLs). APL cells expressing PML-RARα are sensitive to retinoid induced differentiation to granulocytes in the presence of retinoic acid. In contrast, forced expression of either PML-RARα or PLZF-RARα in either U937 or HL-60 cells blocks their terminal differentiation after exposure to $1,25(OH)_2D_3$ [80]. Both PML-RARα

11 Vitamin D and Hematologic Malignancies   261

and PLZF-RARα can bind to VDR in U937 cells and sequester VDR away from activation of its normal DNA targets localization [81]. Overexpression of VDR overcomes the block in $1,25(OH)_2D_3$-stimulated differentiation caused by the fusion proteins. Of note, PLZF itself can interact directly with VDR, and overexpression of PLZF can inhibit the $1,25(OH)_2D_3$ -induced differentiation of U937 cells [82].

The HL-60 and U937 cell lines have been used to attempt to identify early response genes directly regulated by VDR. Expression of fructose 1,6-biphosphatase is up-regulated by $1,25(OH)_2D_3$ in HL-60 cells and peripheral blood monocytes [83]. cDNA microarray analysis showed that at early times, the putative oncogenes Dek and Fli-1 were down-regulated and the antiproliferative gene, BTG1 was up-regulated [84]. After exposure of HL-60 to $1,25(OH)_2D_3$, similar experiments were also noted with the importin and exportin family members which were down-regulated; these proteins mediate transportation between the nucleus and the cytoplasm [85]. Also, $1,25(OH)_2D_3$ suppressed the expression of eIF-2 in HL-60; this protein is involved in the regulation of protein synthesis [86].

About 160 years ago, sunlight or cod liver oil (both abundant source of vitamin D) was used as treatment of tuberculosis [87, 88]. *In vitro* studies suggested that $1,25(OH)_2D_3$ can have a role in activating human macrophages in host defenses against mycobacterium tuberculosis (MTB) [89]. Moreover, screening of the human genome for VDREs showed that the human cathelicidin antimicrobial peptide *(CAMP)* gene has a VDRE in its promoter; and exposure of myeloid cells to $1,25(OH)_2D_3$ and its analogs induced expression of CAMP [90–92]. Induction of CAMP by $1,25(OH)_2D_3$ has been described in hematopoietic cell lines including myeloid leukemias (U937, HL60, NB4, K562, KG-1 and THP-1) and primary hematopoietic cells including leukocytes (monocytes, neutrophils and macrophages) and bone marrow cells of both normal and leukemic individuals [93]. Interestingly, the VDRE for *CAMP* only appears in a transposable short-interspersed nuclear element (SINE), and these sequences occur only in primates [91]. Induction of this antimicrobial agent by $1,25(OH)_2D_3$ may provide significant protection against various microbes.

11.4.2.2 Molecular Mechanisms of Kinase Activities of $1,25(OH)_2D_3$ in Leukemic Cells

Data suggest that both the antiproliferative and differentiation-inducing effects of vitamin D compounds require the modulation of the intracellular kinase pathways, including PKC, PI3-K, AKT, p38 MAPK and ERK. This modulation probably occurs too quickly to be attributed to the genomic actions of vitamin D. Activation of PKC by the phorbol diesters such as TPA, promotes monocyte differentiation of leukemic cell lines [94, 95]. Differentiation of HL-60 cells in response to $1,25(OH)_2D_3$ is accompanied by increased levels of PKC-β; and this differentiation can be inhibited by a specific PKC inhibitor, chelerythrine chloride [96]. Other vitamin D analogs have been shown to stimulate expression and translocation of PKC-α and -δ during NB4 monocytic differentiation [97].

1,25(OH)$_2$D$_3$ probably activates the PI3-K/AKT pathway in both a nongenomic and genomic fashions. Activation of PI3-K may be required for the 1,25(OH)$_2$D$_3$ -stimulated myeloid differentiation, as monitored by induction of CD14 expression [98]. For example, PI3-K was activated by 1,25(OH)$_2$D$_3$ in THP-1 cells within 20 min. Pre-treatment with the PI3-K inhibitors, LY 294004 or wortmanin, inhibited monocytic differentiation in response to 1,25(OH)$_2$D$_3$ in HL-60 and THP-1 cells, as well as peripheral blood monocytes [98, 99]. Antisense oligonucleotides against PI3-K blocked induction of CD14 expression in THP-1 and HL-60 cells. Expression of the VDR was required for activation of PI3-K; and interestingly, VDR was found to be associated with the active form of the kinase. PI3-K activates (phosphorylates) AKT, as well as of its downstream targets, were activated within 6–48 h of exposure to 1,25(OH)$_2$D$_3$ in HL-60 cells [100]. PI3-K inhibitors synergized with 1,25(OH)$_2$D$_3$ to induce cell cycle arrest of HL-60 cells, associated with a synergistic up-regulation of p27. On the other hand, treatment with 1,25(OH)$_2$D$_3$ for 4 days induced the expression of PTEN, which could block the PI3-K/AKT pathway, resulting in differentiation, cell death or inhibition of growth of HL-60 cells [50].

The MAPK pathway can also be activated by 1,25(OH)$_2$D$_3$, and this also probably involves genomic and nongenomic mechanisms. Exposure of either HL-60 or NB4 cells to differentiation-inducing concentrations of vitamin D compounds cause activation and nuclear translocation of MAPK [101–103]. Rapid changes of MAPK phosphorylation occurred within 30 s of exposure to 1,25(OH)$_2$D$_3$ in NB4 cells [102]. In addition, the vitamin D$_3$ analog EB1089 was demonstrated to induced apoptosis of B-cell chronic lymphocytic leukemia cells from patients, an event preceded by stimulation of p38 MAPK and suppression of ERK activity [104]. 1,25(OH)$_2$D$_3$ stimulated the transient (24–48 h) phosphorylation of ERK1/2 in HL-60 cells. After 24 h, the level of phosphorylated ERK decreased to basal levels, while differentiation continued over an additional 48 h [105]. Furthermore, PD98059, an ERK1/2 inhibitor, blocked the 1,25(OH)$_2$D$_3$ -stimulated differentiation of HL-60 cells [106]. Kinase Suppressor of Ras-1 and -2 (KSR-1, -2) which phosphorylate Raf-1 and act as scaffolds increases the efficiency of signaling by Raf-1 [107, 108]. These two genes have an upstream promoter containing a functional VDRE motif. Knocked-down of KSR-2 blocked 1,25(OH)$_2$D$_3$ induced myeloid differentiation. Signaling by Raf-1 is required for the later stage of 1,25(OH)$_2$D$_3$ -induced differentiation and requires p90 RSK which is either directly or indirectly phosphorylated by Raf-1 [109].

11.4.3 *Vitamin D Compounds in Combination with Other Agents*

Because of the potential toxicity of 1,25(OH)$_2$D$_3$ and its analogs at the concentrations required *in vivo* to inhibit proliferation and/or induce differentiation of leukemia cells. Various attempts have been made to use them with other compounds that might act synergistically and that have an acceptable toxicity. The mechanism of action and toxicity (hypercalcemia) of vitamin D compounds differ

11 Vitamin D and Hematologic Malignancies

from chemotherapeutic agents. A variety of agents, including ATRA, arsenic trioxide(As_2O_3), Non-steroidal anti-inflammatory drugs (NSAIDs), carnosic acid, MAPK inhibitor, cisplatin, taxol, paclitacel, doxorubicin, a HIV-protease inhibitor as well as a demethylating agent have been combined with vitamin D compounds in a variety of cancers including leukemia.

Our group and others have shown that the combination of a vitamin D compound and either all-trans-retinoic acid (ATRA) or 9-cis-retinoic acid (9-cis-RA) can potentiate the terminal monocytic differentiation of HL-60, NB4 and U937 cells [110–113]. These combinations included ATRA (10^{-9} M) and either $1,25(OH)_2$–16-ene-23-yne D_3, $1,25(OH)_2$–23-yne D_3 (10^{-9} to 10^{-10} M), or 9-cis-RA and KH1060 (a 20-epi-vitamin D_3 analog) [47, 114–117]. These cells often differentiate atypically, having a neutrophilic morphology, but acquiring other properties typical of monocytes (e.g., CD14 expression); ability to bind to bacterial LPS, and express lineage specific enzymes like nonspecific acid esterase [112, 113]. The combination enhanced the decrease expression of c-myc. Interestingly, U937 cells exposed to a moderate thermal stress responded with increased differentiation after the addition of $1,25(OH)_2D_3$ and ATRA suggesting that induction of heat-shock protein may be sequestering a protein that may favors proliferation or differentiation [118].

19-nor-1,25-dihydroxyvitamin D_2 (paricalcitol) and As_2O_3 are both approved Food and Drug Administration drugs. Their combination resulted in a strong anti-proliferative effect on HL-60, NB4 and PML-RARα over-expressing U937 cells [119]. As_2O_3 decreased the levels of both the repressive PML-RARα fusion protein and the vitamin D metabolizing protein, which had been increased by paricalcitol. This combination may be effective for ATRA-resistant APL patients, as well as those with other types of AML.

NSAIDs enhance the differentiation of HL-60 cells in response to $1,25(OH)_2D_3$ and its analogs [120, 121]. This effect may occur because of a block of NF-κB activation. Bhatia et al. showed that the combination of $1,25(OH)_2D_3$ and TPA with M-CSF resulted in a synergistic response in NB4 cells, causing a complete differentiation to fully functional adherent macrophages with a rapid arrest of cell growth in the first 24 h [122].

Vitamin D compounds have also been combined successfully with naturally occurring plant products. One of these is carnosic acid, a plant-derived polyphenol antioxidant which can potentiate the pro-differentiative effects of $1,25(OH)_2D_3$ [123–125]. Differentiation was correlated with antioxidant activity, and was associated with activation of the Raf-ERK pathway and increased binding of the AP-1 transcription factor to the promoter of VDR. A p38 MAPK inhibitor (SB202190) enhanced the ability of $1,25(OH)_2D_3$ to induce differentiation of HL-60 cells [126]. In addition, the combination of the three agents (SB202190, carnosic acid and $1,25(OH)_2D_3$) further potentiated the antileukemic activity against HL-60 cells and primary AML blasts [127]. This augmented potency was associated with increased activation of the JNK-MAP kinase pathway.

Combining vitamin D compounds with traditional chemotherapy agents such as cisplatin, etoposide and doxorubicin reduces the concentration of chemotherapy

required for their antileukemic effects. Also, studies have suggested that the sequential order that the compounds are given, may be important [128, 129].

For example, pretreatment with etoposide enhanced the subsequent action of $1,25(OH)_2D_3$, but pretreatment with $1,25(OH)_2D_3$ had little effect on the activity of etoposide. The explanation for this observation is unclear now.

One of the human immunodeficiency virus type I protease inhibitors, ritonavir can enhance the antileukemic potency of $1,25(OH)_2D_3$ [130]. Ritonavir inhibits Cyp24 expression. This enzyme normally metabolizes $1,25(OH)_2D_3$ resulting in decreased levels of the active seco-steroid. By blocking this enzyme, ritonavir increases the amount of active, intracellular $1,25(OH)_2D_3$.

The combination of a demethylating agent with a vitamin D compound can have enhanced activity [74]. For example, when the demethylating agent, decitabine was combined with $1,25(OH)_2D_3$, they synergistically induced monocytic differentiation of U937 cells and primary patient AML blast cells *in vitro*.

Valproic acid (VPA) is an inhibitor of histone deacetylase which can also change the epigenetic landscape by acetylating histones and other proteins. This compound can induce myeloid differentiation [131]. In one clinical study of 19 MDS patients treated with the combination of VPA, 9-*cis*-RA and $1,25(OH)_2D_3$, 3 patients (16%) responded to treatment. A cautionary note, eight patients (42%) had suffered toxicity from the combination [132]. The investigators did not find any correlation between histone acetylation and clinical response. Clearly, further studies are required using less toxic histone deacetylating agents.

In summary, treatment of leukemia or MDS with vitamin D compound is unlikely, by itself, to be successful; but when given either in the maintenance phase of therapy after the leukemic patient is placed into remission or combined with other agents, these agents may be useful therapeutically. Furthermore, $1,25(OH)_2D_3$ can induce the expression of the antimicrobial peptide, CAMP (in Sect. 11.4.2.1), which may afford the cancer patient some protection from life-threatening infections while receiving aggressive chemotherapy.

11.5 Vitamin D Analogs Effective against Leukemic Cells

A major drawback in using $1,25(OH)_2D_3$ is its calcemic effect, which prevents pharmacological doses of the compound from being given. Vitamin D analogs have been synthesized that have enhanced potency to inhibit proliferation and promote differentiation of cancer cells, with less calcemic activity as compared with $1,25(OH)_2D_3$ (see Chapter 1). Many of these analogs *in vitro* are between 10- and 1,000-fold more active than the parental $1,25(OH)_2D_3$ in their growth suppressive activity. These novel analogs can provide a larger therapeutic window for the treatment of hematologic malignancies. A comparison of the relative antileukemic potencies of some of these vitamin D compounds is provided in Table 11.3.

The first attempts using analogs focused on 1*a*-hydroxyvitamin D_3 (1αOHD$_3$), a vitamin D_3 analog that is efficiently converted to $1,25(OH)_2D_3$ *in vivo* by

11 Vitamin D and Hematologic Malignancies

Table 11.3 Effect of vitamin D compounds on clonal proliferation of HL-60 cells in soft agar and calcium levels in mice

Compound	ED_{50}[a] (x 10^{-9} mol/l)	MTD[b] (μg)	Reference
$1,25(OH)_2D_3$	4–900	0.0625	[141–149]
$1,25(OH)_2$–16-ene-D_3	0.015	0.125	[141]
$1,25(OH)_2$–16-ene-23-yne-D_3	3	2	[141, 143]
$1,25(OH)_2$–16-ene-5,6-trans-D_3	0.03	4	[142]
$1,25(OH)_2$–16-ene-24-oxo-D_3	0.2	ND[c]	[147]
$1,25(OH)_2$–16-ene-19-nor-D_3	0.8	0.5	[147]
$1,25(OH)_2$–16-ene-24-oxo-19-nor-D_3	0.1	6	[146]
$1,25(OH)_2$–20-epi-D_3	0.006	0.00125	[143, 148, 149]
$1,25(OH)_2$–20-epi-22-oxa-24,26,27-trishomo-D_3[d]	0.001	0.0125	[148]
$1,25(OH)_2$-diene-24,26,27-trihomo-D_3[e]	0.23	0.25	[148]
19-nor-$1,25(OH)_2D_2$[f]	2.4	0.1	[49]
$1,25(OH)_2$–21-(3-methyl-3-hydroxy-butyl)-19-nor D_3[g]	0.17	ND[c]	[50]
$1,25(OH)2$–20 S-21(3-methyl-3-hydroxy-butyl)-23-yne-26,27-hexafluoro-D_3[h]	4	0.0625[i]	[155]

[a]ED_{50} represents the effective dose achieving 50% growth inhibition of HL-60 cells
[b]MTD Maximally tolerated dose; highest dose reported that did not produce hypercalcemia or other noticeable toxicities in mice when injected intraperitoneally, three times per week
[c]ND not done
[d]Leo Pharmaceutical code name is KH 1060
[e]Leo Pharmaceutical code name is EB 1089
[f]Abbott Laboratories code name is Paricalcitol
[g]Gemini-19-nor D_3
[h]Gemini-23-yne-26,27-hexafluoro-D_3
[i]At least mice that received the 0.0625 μg/mouse of Gemini-23-yne-26,27-hexafluoro-D_3 had normal serum calcium levels

D_3–25-hydroxylase. This compound was administered to mice previously inoculated with the M1 leukemia cell line, and it showed greater activity than $1,25(OH)_2D_3$ [42]. Its conversion to the active form resulted in a more prolonged elevation of plasma levels of $1,25(OH)_2D_3$, and the dose (25 pmol, every other day) produced only a slight elevation of the serum calcium. In addition, survival of the leukemic mice was increased by 50–60%; nevertheless, the more effective doses did cause hypercalcemia. Also, the administration of $1\alpha(OH)D_3$ produced tumor regression in follicular NHLs in rats, but hypercalcemia was the dose-limiting factor [29]. In one study, six patients with MDS were treated with $1\alpha(OH)D_3$ at 1 μg/day for a minimum of 3 months, but neither a good clinical response nor toxicity was observed in these individuals [133]. In another clinical study, 30 MDS patients were divided into two different groups: one group received $1\alpha(OH)D_3$ at 4–6 μg/day and the other group received placebo; the patients were treated for a median of 17 weeks [134]. An improvement of hematologic parameters was detected in only one patient, and the investigators believed that the treated group had a greater proportion of patients who did not progress to leukemia as compared to the control

group. $1\alpha(OH)D_2$ doxercalciferol is in clinical use for the treatment of secondary hyperparathyroidism for reduction of elevated parathyroid hormone levels with acceptable mild hypercalcemia and hyperphosphatemia [135]. Recently, a phase II trial of doxercalciferol was conducted in 15 patients with MDS [45]. Each received 12.5 μg/day of $1\alpha(OH)D_2$ for 12 weeks; the individuals did not develop hypercalcemia, but they also did not obtain a clinical response.

A case has been reported of an individual with chronic myelomonocytic leukemia (subtype of MDS) who achieved complete remission with 25-hydroxyvitamin D_3 [$25(OH)D_3$] for 15 months; and remission continued for 15 months after the end of the treatment [136]. These results are surprising because $25(OH)D_3$ has low activity by itself and *in vitro* has little antileukemic activity.

Calcipotriol (MC903) has a cyclopropyl group at the end of the side chain formed by the fusion of C-26 and C-27, a hydroxyl group at C-24, and a double bond at C-22. This compound is equipotent to $1,25(OH)_2D_3$ in inhibiting the proliferation and inducing the differentiation of the monoblastic cell line U937 [110, 137]. In bone marrow cultures, the analog promotes the formation of multi-nucleated osteoclast-like cells, a vitamin D function. The effects of this compound on the immune system are very similar to those produced by $1,25(OH)_2D_3$. By interfering with T-helper cell activity, calcipotriol reduces immunoglobulin production and blocks the proliferation of thymocytes induced by IL-1 [138, 139]. Exposure of the follicular NHL B-cell lines SU-DUL4 and SU-DUL5, carrying the t(14;18) translocation characteristic of the disease, to calcipotriol inhibited their proliferation, but only at high concentrations of the compound (10^{-7} M) [29]. At the same time, calcipotriol was 100-fold less active than $1,25(OH)_2D_3$ in inducing hypercalcemia and mobilizing bone calcium in rats [140]. However, the analog is rapidly inactivated in the intact animal, and therefore has been developed as a topical agent for skin diseases like psoriasis.

Introduction of a double bond at carbon 16 (C-16 ene) has proved to be an effective modification, particularly when combined with other motifs to generate a series of analogs with potent antiproliferative and pro-differentiation promoting activities, with decreased calcemic effects. Prior studies by us have shown that vitamin D_3 analogs having the C-16-ene motif were almost 100-fold more potent than $1,25(OH)_2D_3$ at inhibiting growth of HL-60 leukemia cells, but the calcemic activity was the same or markedly less than $1,25(OH)_2D_3$ [141, 142]. Combination of the C-16-double bond and the C-23-triple bond (C-23-yne) [$1,25(OH)_2$-16-ene-23-yne-D_3] produces a compound that is a more potent inducer of growth inhibition and differentiation of HL-60 cells than $1,25(OH)_2D_3$, and is 15-fold less hypercalcemic in mice. This analog has potent antiproliferative and pro-differentiating effects on leukemic cells *in vitro* [143]. In blocking HL-60 clonal growth, $1,25(OH)_2$-16-ene-23-yne D_3 has a potency of about four times higher than $1,25(OH)_2D_3$. This compound administered to vitamin D-deficient chicks is about 30 times less effective than $1,25(OH)_2D_3$ in stimulating intestinal calcium absorption and about 50 times less effective in inducing bone calcium mobilization [144]. Further experiments have demonstrated the therapeutic potential of $1,25(OH)_2$-16-ene-23-yne D_3 by its ability to prolong markedly the survival of mice that had been

11 Vitamin D and Hematologic Malignancies

inoculated with the myeloid leukemic cell line WEHI 3BD+ and treated with a high dose (1.6 μg every other day) of the compound [145]. The $1,25(OH)_2$–16-ene-19-nor-24-oxo-D_3 was synthesized as a result of previous studies that isolated 24-oxo metabolites of potent vitamin D_3 analogs, which were formed in a rat kidney perfusion system [146]. We found that these 24-oxo-metabolites had markedly reduced calcemic activity, but possessed at least an equal ability as the unmetabolized analogs to inhibit the clonal growth of breast and prostate cancer cells and myeloid leukemia cells *in vitro*. Taken together, these findings prompted the chemical synthesis of a series of vitamin D_3 analogs with $1,25(OH)_2$–16-ene-19-nor-24-oxo-D_3 being one of the more exciting compounds, having the ability to inhibit acute myeloid leukemia cells in the range of 10^{-10} M [147]. Remarkably, this compound had very little calcemic activity even when 6 μg was administered intraperitoneally to the mice, three times a week.

The compound $1,25(OH)_2$–20-epi D_3 is characterized by an inverted stoichiometry at C-20 of the side chain. The monoblastic cell line U937 cultured with this compound showed a strong induction of differentiation [148]. It was also a potent modulator of cytokine-mediated T lymphocyte activation and exerted calcemic effects comparable to $1,25(OH)_2D_3$ in rats. A study by ourselves suggested that $1,25(OH)_2$–20-epi D_3 is a potent vitamin D_3 compound at inhibiting the clonal growth of HL-60 cells and at inducing cell differentiation. In fact, it is about 2,600-fold more potent than $1,25(OH)_2D_3$ in inhibiting the clonal growth of HL-60 cells and about 5,000-fold more effective in preventing clonal proliferation of fresh human leukemic myeloid cells [149]. $1,25(OH)_2$–20-epi D_3 exerts its effects by binding directly to VDR as shown by a T lymphocytic cell line established from a patient with HVDRR. These cells with a dysfunctional VDR no longer were able to have a biologic effect. KH1060 is a potent vitamin D_3 20-epi analog with an oxygen in place of C-22 and three additional carbons in the side chain. It is about 14,000-fold more potent than $1,25(OH)_2D_3$ in inhibiting the clonal growth of the monoblastic cell line U937 [148]. It also has a powerful effect on other leukemic cells [113, 149]. However, it has the same hypercalcemic activity and the same receptor binding affinity as $1,25(OH)_2D_3$.

Paricalcitol (19-nor-1,25-dihydroxyvitamin D_2) has been approved by the Food and Drug Administration for the clinical treatment of secondary hyperparathyroidism. Clinical trials have demonstrated that it possesses very low calcemic activity [150, 151]. Studies by us and another group have demonstrated that paricalcitol has antiproliferative, pro-differentiation activities against myeloid leukemia and myeloma cell lines at a clinically achievable concentration [49, 152, 153]. Paricalcitol activity was dependent on the presence of VDR, as it was unable to induce differentiation of mononuclear bone marrow cells from VDR knockout mice, whereas cells from WT mice were differentiated toward monocytes/macrophages [49]. Furthermore, paricalcitol was able to inhibit tumor growth without causing hypercalcemia in immunodeficient mice. These observations prompted us to begin a clinical trial to treat patients with MDS. A clinical trial of oral paricalcitol was conducted on 12 MDS patients. Although paricalcitol was well-tolerated in all patients, it had only minimal activity against MDS [44].

We have found that 1α,25-dihydroxy-2 1-(3-hydroxy-3-methylbutyl)vitamin D_3 (Gemini) compounds having two side chains attached to carbon-20, increases the antitumor activities against HL-60 and NB4 compared to $1,25(OH)_2D_3$ [50, 154]. Gemini-19-nor stimulated expression of the potential tumor suppressor, PTEN [50]. Gemini-23-yne-26, 27-hexafluoro-D_3 was approximately 225-fold more potent than $1,25(OH)_2D_3$ in inhibiting the clonal growth of HL-60 [155]. This compound produces hypercalcemia at the same concentrations as $1,25(OH)_2D_3$ in mice with a maximal tolerated dose (MTD) of 0.0625 μg per mouse (intraperitoneally injections, three times per week) which is the same MTD as $1,25(OH)_2D_3$. Therefore, the Gemini compounds possess greater antiproliferative activity than $1,25(OH)_2D_3$, but produce a similar amount of hypercalcemic resulting in a larger therapeutic window.

Potential mechanisms by which vitamin D analogs have increased biological activity compared to $1,25(OH)_2D_3$ include: reduced affinity to the serum vitamin D binding protein; decreased catabolism by 24-hydroxylase; retention of biological activities by metabolic products of vitamin D analogs; increased stability of the ligand-VDR complex; increased dimerization with RXR associated with increased affinity for its VDRE in the region of target genes; and enhanced recruitment of the DRIP coactivator complex. These topics are covered in detail in Chapter 9.

In conclusion, new vitamin D analogs have enhanced antileukemic activity and decreased hypercalcemic effects compared to $1,25(OH)_2D_3$, and should be considered for the selected trials in hematologic malignancies either alone or in combination with other therapies.

11.6 Summary and Conclusions

The hormone $1,25(OH)_2D_3$ plays a role in normal hematopoiesis, enhancing the activity of monocytes-macrophages and inhibiting cytokine production by T lymphocytes. It can also inhibit proliferation and induce differentiation of various myeloid leukemia cell lines. Its activity occurs through both genomic and nongenomic pathway(s); the former action is mediated by activation of vitamin D receptors that modulates the transcription of various target genes; and the latter activity is probably mediated by rapid intracellular Ca^{2+} influxes resulting in activation of kinases within seconds to minutes. The antiproliferative effects of $1,25(OH)_2D_3$ in vivo require supraphysiological levels of the seco-steroid which causes hypercalcemia. Limited clinical trials have been performed for the treatment of preleukemia myelodysplastic syndrome with $1,25(OH)_2D_3$, but the in vitro effective dose caused hypercalcemia in vivo. Since the mid-1980s, many vitamin D analogs have been synthesized that possess reduced hypercalcemic activity and increased ability to induce cell differentiation and to inhibit proliferation of leukemic cells. Further studies have been performed in vitro and in vivo using these analogs with other differentiating and/or antiproliferative agents in the hopes that their combination,

11 Vitamin D and Hematologic Malignancies

working through different pathways, could lead to synergistic activity. Proof of principle that $1,25(OH)_2D_3$ and its analogs are beneficial in cancer has been validated in experiments conducted *in vitro* and in laboratory animals. However, to date their therapeutic value in patients is unproven.

References

1. Mangelsdorf DJ, Thummel C, Beato M, Herrlich P, Schütz G, Umesono K, Blumberg B, Kastner P, Mark M, Chambon P, Evans RM (Dec 15,1995) The nuclear receptor superfamily: the second decade. Cell 83(6):835–9
2. Christakos S, Raval-Pandya M, Wernyj RP, Yang W (June 1, 1996) Genomic mechanisms involved in the pleiotropic actions of 1,25-dihydroxyvitamin D3. Biochem J 316(Pt 2):361–71
3. Koren R, Ravid A, Liberman UA, Hochberg Z, Weisman Y, Novogrodsky A (Nov,1985) Defective binding and function of 1,25-dihydroxyvitamin D3 receptors in peripheral mononuclear cells of patients with end-organ resistance to 1,25-dihydroxyvitamin D. J Clin Invest 76(5):2012–5
4. Koeffler HP, Bishop JE, Reichel H, Singer F, Nagler A, Tobler A, Walka M, Norman AW (Mar 26, 1990) Lymphocyte cell lines from vitamin D-dependent rickets type II show functional defects in the 1 alpha,25-dihydroxyvitamin D3 receptor. Mol Cell Endocrinol 70(1):1–11
5. Kizaki M, Norman AW, Bishop JE, Lin CW, Karmakar A, Koeffler HP (Mar 15,1991) 1,25-Dihydroxyvitamin D3 receptor RNA: expression in hematopoietic cells. Blood 77(6):1238–47
6. Langub MC, Reinhardt TA, Horst RL, Malluche HH, Koszewski NJ (Sep,2000) Characterization of vitamin D receptor immunoreactivity in human bone cells. Bone 27(3):383–7
7. Zerwekh JE, Yu XP, Breslau NA, Manolagas S, Pak CY (Oct,1993) Vitamin D receptor quantitation in human blood mononuclear cells in health and disease. Mol Cell Endocrinol 96(1–2):1–6
8. Takahashi K, Nakayama Y, Horiuchi H, Ohta T, Komoriya K, Ohmori H, Kamimura T (Aug,2002) Human neutrophils express messenger RNA of vitamin D receptor and respond to 1alpha,25-dihydroxyvitamin D3. Immunopharmacol Immunotoxicol 24(3):335–47
9. Hewison M, Freeman L, Hughes SV, Evans KN, Bland R, Eliopoulos AG, Kilby MD, Moss PA, Chakraverty R (June 1, 2003) Differential regulation of vitamin D receptor and its ligand in human monocyte-derived dendritic cells. J Immunol 170(11):5382–90
10. Kreutz M, Andreesen R, Krause SW, Szabo A, Ritz E, Reichel H (Aug 15,1993) 1,25-dihydroxyvitamin D3 production and vitamin D3 receptor expression are developmentally regulated during differentiation of human monocytes into macrophages. Blood 82(4):1300–7
11. Favus MJ, Kamauskas AJ, Parks JH, Coe FL (Oct,2004) Peripheral blood monocyte vitamin D receptor levels are elevated in patients with idiopathic hypercalciuria. J Clin Endocrinol Metab 89(10):4937–43
12. Nakajima S, Yamaoka K, Yamamoto T, Okada S, Tanaka H, Seino Y (Sep,1990) Decreased concentration of 1,25-dihydroxyvitamin D3 receptors in peripheral mononuclear cells of patients with X-linked hypophosphatemic rickets: effect of phosphate supplementation. Bone Miner 1 0(3):20 1–9
13. Lee Y, Inaba M, DeLuca HF, Mellon WS (Aug 15,1989) Immunological identification of 1,25-dihydroxyvitamin D3 receptors in human promyelocytic leukemic cells (HL-60) during homologous regulation. J Biol Chem 264(23):13701–5
14. Gocek E, Kielbiński M, Marcinkowska E (May 1,2007) Activation of intracellular signaling pathways is necessary for an increase in VDR expression and its nuclear translocation. FEBS Lett 581(9):1751–7

15. Wang X, Studzinski GP (Apr 15,2006) The requirement for and changing composition of the activating protein-1 transcription factor during differentiation of human leukemia HL60 cells induced by 1,25-dihydroxyvitamin D3. Cancer Res 66(8):4402–9
16. Provvedini DM, Tsoukas CD, Deftos LJ, Manolagas SC (Sep 16,1983) 1,25-dihydroxyvitamin D3 receptors in human leukocytes. Science 221(4616):1181–3
17. Veldman CM, Cantorna MT, DeLuca HF (Feb 15,2000) Expression of 1,25-dihydroxyvitamin D(3) receptor in the immune system. Arch Biochem Biophys 374(2):334–8
18. Morgan JW, Sliney DJ, Morgan DM, Maizel AL (Jan,1999) Differential regulation of gene transcription in subpopulations of human B lymphocytes by vitamin D3. Endocrinology 140(1):38 1–91
19. Iho S, Takahashi T, Kura F, Sugiyama H, Hoshino T (June 15, 1986) The effect of 1,25-dihydroxyvitamin D3 on *in vitro* immunoglobulin production in human B cells. J Immunol 1 36(12):4427–31
20. Lemire JM, Adams JS, Kermani-Arab V, Bakke AC, Sakai R, Jordan SC (May, 1985) 1,25-Dihydroxyvitamin D3 suppresses human T helper/inducer lymphocyte activity *in vitro*. J Immunol 134(5):3032–5
21. Rigby WF, Stacy T, Fanger MW (Oct, 1984) Inhibition of T lymphocyte mitogenesis by 1,25-dihydroxyvitamin D3 (calcitriol). J Clin Invest 74(4):1451–5
22. Tsoukas CD, Provvedini DM, Manolagas SC (June 29, 1984) 1,25-dihydroxyvitamin D3: a novel immunoregulatory hormone. Science 224(4656):1438–40
23. Reichel H, Koeffler HP, Tobler A, Norman AW (May, 1987) 1 alpha,25-Dihydroxyvitamin D3 inhibits gamma-interferon synthesis by normal human peripheral blood lymphocytes. Proc Natl Acad Sci USA 84(10):3385–9
24. Tobler A, Miller CW, Norman AW, Koeffler HP (June 1988) 1,25-Dihydroxyvitamin D3 modulates the expression of a lymphokine (granulocyte-macrophage colony-stimulating factor) posttranscriptionally. J Clin Invest 81(6):1819–23
25. Soldati L, Adamo D, Bianchin C, Arcidiacono T, Terranegra A, Bianchi ML, Mora S, Cusi D, Vezzoli G (Aug, 2004) Vitamin D receptor mRNA measured in leukocytes with the TaqMan fluorogenic detection system: effect of calcitriol administration. Clin Chem 50(8):1315–21
26. Manolagas SC, Werntz DA, Tsoukas CD, Provvedini DM, Vaughan JH (Dec, 1986) 1,25-Dihydroxyvitamin D3 receptors in lymphocytes from patients with rheumatoid arthritis. J Lab Clin Med 108(6):596–600
27. Provvedini DM, Rulot CM, Sobol RE, Tsoukas CD, Manolagas SC (June 1987) 1 alpha,25-Dihydroxyvitamin D3 receptors in human thymic and tonsillar lymphocytes. J Bone Miner Res 2(3):239–47
28. Biyoudi-Vouenze R, Cadranel J, Valeyre D, Milleron B, Hance AJ, Soler P (June 1991) Expression of 1,25(0H)2D3 receptors on alveolar lymphocytes from patients with pulmonary granulomatous diseases. Am Rev Respir Dis 143(6):1376–80
29. Hickish T, Cunningham D, Colston K, Millar BC, Sandle J, Mackay AG, Soukop M, Sloane J (Oct, 1993) The effect of 1,25-dihydroxyvitamin D3 on lymphoma cell lines and expression of vitamin D receptor in lymphoma. Br J Cancer 68(4):668–72
30. O'Kelly J, Hisatake J, Hisatake Y, Bishop J, Norman A, Koeffler HP (Apr, 2002) Normal myelopoiesis but abnormal T lymphocyte responses in vitamin D receptor knockout mice. J Clin Invest 109(8):1091–9
31. Yu S, Cantorna MT (Apr 1, 2008) The vitamin D receptor is required for iNKT cell development. Proc Natl Acad Sci USA 105(13):5207–12
32. Koeffler HP, Amatruda T, Ikekawa N, Kobayashi Y, DeLuca HF (Dec, 1984) Induction of macrophage differentiation of human normal and leukemic myeloid stem cells by 1,25-dihydroxyvitamin D3 and its fluorinated analogues. Cancer Res 44(12 Pt 1): 5624–8
33. Provvedini DM, Deftos LJ, Manolagas SC (1986) 1,25-Dihydroxyvitamin D3 promotes *in vitro* morphologic and enzymatic changes in normal human monocytes consistent with their differentiation into macrophages. Bone 7(1):23–8

11 Vitamin D and Hematologic Malignancies

34. Choudhuri U, Adams JA, Byrom N, McCarthy DM, Barrett J (1990) 1,25-Dihydroxyvitamin D3 induces normal mononuclear blood cells to differentiate in the direction of monocyte-macrophages. Haematologia (Budap) 23(1):9–19

35. Kreutz M, Andreesen R (Dec 15, 1990) Induction of human monocyte to macrophage maturation *in vitro* by 1,25-dihydroxyvitamin D3. Blood 76(12):2457–61

36. Paquette RL, Koeffler HP (June 1992) Differentiation therapy. Hematol Oncol Clin North Am 6(3):687–706

37. Grande A, Montanari M, Tagliafico E, Manfredini R, Marani TZ, Siena M, Tenedini E, Gallinelli A, Ferrari S (Apr, 2002) Physiological levels of lalpha, 25 dihydroxyvitamin D3 induce the monocytic commitment of CD34+ hematopoietic progenitors. J Leukoc Biol 71(4):641–51

38. Berer A, Stöckl J, Majdic O, Wagner T, Kollars M, Lechner K, Geissler K, Oehler L (May, 2000) 1,25-Dihydroxyvitamin D(3) inhibits dendritic cell differentiation and maturation *in vitro*. Exp Hematol 28(5):575–83

39. Tobler A, Gasson J, Reichel H, Norman AW, Koeffler HP (June 1987) Granulocyte-macrophage colony-stimulating factor. Sensitive and receptor-mediated regulation by 1,25-dihydroxyvitamin D3 in normal human peripheral blood lymphocytes. J Clin Invest 79(6):1700–5

40. D'Ambrosio D, Cippitelli M, Cocciolo MG, Mazzeo D, Di Lucia P, Lang R, Sinigaglia F, Panina-Bordignon P (Jan 1, 1998) Inhibition ofIL-12 production by 1,25-dihydroxyvitamin D3. Involvement ofNF-kappaB downregulation in transcriptional repression of the p40 gene. J Clin Invest 101(1):252–62

41. Abe E, Miyaura C, Sakagami H, Takeda M, Konno K, Yamazaki T, Yoshiki S, Suda T (Aug, 1981) Differentiation of mouse myeloid leukemia cells induced by 1 alpha,25-dihydroxyvitamin D3. Proc Natl Acad Sci USA 78(8):4990–4

42. Honma Y, Hozumi M, Abe E, Konno K, Fukushima M, Hata S, Nishii Y, DeLuca HF, Suda T (Jan, 1983) 1 alpha,25-Dihydroxyvitamin D3 and 1 alpha-hydroxyvitamin D3 prolong survival time of mice inoculated with myeloid leukemia cells. Proc Natl Acad Sci USA 80(1):201–4

43. Okamoto R, Akagi T, Koeffler HP (Jan, 2008) Vitamin D compounds and myelodysplastic syndrome. Leuk Lymphoma 49(1):12–3

44. Koeffler HP, Aslanian N, O'Kelly J (Nov, 2005) Vitamin D(2) analog (Paricalcitol; Zemplar) for treatment of myelodysplastic syndrome. Leuk Res 29(11):1259–62

45. Petrich A, Kahl B, Bailey H, Kim K, Turman N, Juckett M (Jan, 2008) Phase II study of doxercalciferol for the treatment of myelodysplastic syndrome. Leuk Lymphoma 49(1):57–61

46. Hu ZB, Ma W, Uphoff CC, Lanotte M, Drexler HG (Nov, 1993) Modulation of gene expression in the acute promyelocytic leukemia cell line NB4. Leukemia 7(11):1817–23

47. Elstner E, Linker-Israeli M, Le J, Umiel T, Michl P, Said JW, Binderup L, Reed JC, Koeffler HP (Jan 15, 1997) Synergistic decrease of clonal proliferation, induction of differentiation, and apoptosis of acute promyelocytic leukemia cells after combined treatment with novel 20-epi vitamin D3 analogs and 9-cis retinoic acid. J Clin Invest 99(2):349–60

48. Godyn JJ, Xu H, Zhang F, Kolla S, Studzinski GP (Jan, 1994) A dual block to cell cycle progression in HL60 cells exposed to analogues of vitamin D3. Cell Prolif 27(1):37–46

49. Kumagai T, O'Kelly J, Said JW, Koeffler HP (June 18, 2003) Vitamin D2 analog 19-nor-1,25-dihydroxyvitamin D2: antitumor activity against leukemia, myeloma, and colon cancer cells. J Natl Cancer Inst 95(12):896–905

50. Hisatake J, O'Kelly J, Uskokovic MR, Tomoyasu S, Koeffler HP (Apr 15, 2001) Novel vitamin D(3) analog, 21-(3-methyl-3-hydroxy-butyl)-19-nor D(3), that modulates cell growth, differentiation, apoptosis, cell cycle, and induction of PTEN in leukemic cells. Blood 97(8):2427–33

51. Koeffler HP (Oct, 1983) Induction of differentiation of human acute myelogenous leukemia cells; therapeutic implications. Blood 62(4):709–21

52. Mangelsdorf DJ, Koeffler HP, Donaldson CA, Pike JW, Haussler MR (Feb, 1984) 1,25-Dihydroxyvitamin D3-induced differentiation in a human promyelocytic leukemia cell line (HL-60): receptor-mediated maturation to macrophage-like cells. J Cell Biol 98(2):391–8

53. Koeffler HP, Hirji K, Itri L (Dec, 1985) 1,25-Dihydroxyvitamin D3: *in vivo* and *in vitro* effects on human preleukemic and leukemic cells. Cancer Treat Rep 69(12):1399–407

54. Tasaka T, Tokuda M, Taoka T, Itano T, Matsui H, Etoh S, Nishio H, Miyamoto O, Irino S, Hatase O (Jan, 1991) Mechanism of transient increase in intracellular concentration of free calcium ions in HL-60 cell differentiation induced by vitamin D3 and phorbol ester. Biochem Int 23(l):137–43

55. Nagler A, Merchav S, Fabian I, Tatarsky I, Weisman Y, Hochberg Z (Nov, 1987) Myeloid progenitors from the bone marrow of patients with vitamin D resistant rickets (type II) fail to respond to 1,25(OH)2D3. Br J Haematol 67(3):267–71

56. Koeffler HP, Bishop JE, Reichel H, Singer F, Nagler A, Tobler A, Walka M, Norman AW (Mar 26, 1990) Lymphocyte cell lines from vitamin D-dependent rickets type II show functional defects in the 1 alpha,25-dihydroxyvitamin D3 receptor. Mol Cell Endocrinol 70(1):1–11

57. Rots NY, Iavarone A, Bromleigh Y, Freedman LP (Apr 15, 1999) Induced differentiation of U937 cells by 1,25-dihydroxyvitamin D3 involves cell cycle arrest in G1 that is preceded by a transient proliferative burst and an increase in cyclin expression. Blood 93(8):2721–9

58. Campbell MJ, Drayson MT, Durham J, Wallington L, Siu-Caldera ML, Reddy GS, Brown G (Mar 25, 1999) Metabolism of 1alpha,25(OH)2D3 and its 20-epi analog integrates clonal expansion, maturation and apoptosis during HL-60 cell differentiation. Mol Cell Endocrinol 149(1–2):169–83

59. Liu M, Lee MH, Cohen M, Bommakanti M, Freedman LP (Jan 15, 1996) Transcriptional activation of the Cdk inhibitor p21 by vitamin D3 leads to the induced differentiation of the myelomonocytic cell line U937. Genes Dev 10(2):142–53

60. Schwaller J, Koeffler HP, Niklaus G, Loetscher P, Nagel S, Fey MF, Tobler A (Mar, 1995) Posttranscriptional stabilization underlies p53-independent induction of p21WAF1/CIP1/SD11 in differentiating human leukemic cells. J Clin Invest 95(3):973–9

61. Wang QM, Jones JB, Studzinski GP (Jan 15) Cyclin-dependent kinase inhibitor p27 as a mediator of the G1 -S phase block induced by 1,25-dihydroxyvitamin D3 in HL60 cells. Cancer Res 56(2):264–7

62. Xu HM, Tepper CG, Jones JB, Fernandez CE, Studzinski GP (Dec, 1993) 1,25-Dihydroxyvitamin D3 protects HL60 cells against apoptosis but down-regulates the expression of the bcl-2 gene. Exp Cell Res 209(2):367–74

63. Wu YL, Jiang XR, Lillington DM, Allen PD, Newland AC, Kelsey SM (Feb 15, 1998) 1,25-Dihydroxyvitamin D3 protects human leukemic cells from tumor necrosis factor-induced apoptosis via inactivation of cytosolic phospholipase A2. Cancer Res 58(4):633–40

64. Dalla-Favera R, Wong-Staal F, Gallo RC (Sep 2, 1982) Onc gene amplification in promyelocytic leukaemia cell line HL-60 and primary leukaemic cells of the same patient. Nature 299(5878):61–3

65. Obeid LM, Okazaki T, Karolak LA, Hannun YA (Feb 5, 1990) Transcriptional regulation of protein kinase C by 1,25-dihydroxyvitarnin D3 in HL-60 cells. J Biol Chem 265(4):2370–4

66. Reitsma PH, Rothberg PG, Astrin SM, Trial J, Bar-Shavit Z, Hall A, Teitelbaum SL, Kahn AJ (Dec 1–7, 1983) Regulation of myc gene expression in HL-60 leukaemia cells by a vitamin D metabolite. Nature 306(5942):492–4

67. Simpson RU, Hsu T, Begley DA, Mitchell BS, Alizadeh BN (Mar 25, 1987) Transcriptional regulation of the c-myc protooncogene by 1,25-dihydroxyvitamin D3 in HL-60 promyelocytic leukemia cells. J Biol Chem 262(9):4104–8

68. Zhou JY, Norman AW, Lübbert M, Collins ED, Uskokovic MR, Koeffler HP (Jul, 1989) Novel vitamin D analogs that modulate leukemic cell growth and differentiation with little effect on either intestinal calcium absorption or bone mobilization. Blood 74(1):82–93

69. Rowley PT, Farley B, Giuliano R, LaBella S, Leary JF (1992) Induction of the fms proto-oncogene product in HL-60 cells by vitamin D: a flow cytometric analysis. Leuk Res 16(4):403–10

70. Pan Q, Martell RE, O'Connell TD, Simpson RU (Oct, 1996) 1,25-Dihydroxyvitamin D3-regulated binding of nuclear proteins to a c-myc intron element. Endocrinology 137(10):4154–60

11 Vitamin D and Hematologic Malignancies

71. Pan Q, Simpson RU (Mar 26, 1999) c-myc intron element-binding proteins are required for 1,25-dihydroxyvitamin D3 regulation of c-myc during HL-60 cell differentiation and the involvement of HOXB4. J Biol Chem 274(13):8437–44
72. Rots NY, Liu M, Anderson EC, Freedman LP (Apr, 1998) A differential screen for ligand-regulated genes: identification of HoxA10 as a target of vitamin D3 induction in myeloid leukemic cells. Mol Cell Biol 18(4):1911–8
73. Du H, Daftary GS, Lalwani SI, Taylor HS (Sep, 2005) Direct regulation of HOXA 10 by 1,25-(OH)2D3 in human myelomonocytic cells and human endometrial stromal cells. Mol Endocrinol 19(9):2222–33
74. Koschmieder S, Agrawal S, Radomska HS, Huettner CS, Tenen DG, Ottmann OG, Berdel WE, Serve HL, Müller-Tidow C (Feb, 2007) Decitabine and vitamin D3 differentially affect hematopoietic transcription factors to induce monocytic differentiation. Int J Oncol 30(2):349–55
75. Kawagoe H, Potter M, Ellis J, Grosveld GC (Sep 1, 2004) TEL2, an ETS factor expressed in human leukemia, regulates monocytic differentiation of U937 Cells and blocks the inhibitory effect of TEL1 on ras-induced cellular transformation. Cancer Res 64(17):6091–100
76. Lasky SR, Iwata K, Rosmarin AG, Caprio DG, Maizel AL (Aug 25, 1995) Differential regulation of JunD by dihydroxycholecalciferol in human chronic myelogenous leukemia cells. J Biol Chem 270(34):19676–9
77. Gaynor R, Simon K, Koeffler P (June 15, 1991) Expression of c-jun during macrophage differentiation of HL-60 cells. Blood 77(12):2618–23
78. Urahama N, Ito M, Sada A, Yakushijin K, Yamamoto K, Okamura A, Minagawa K, Hato A, Chihara K, Roeder RG, Matsui T (Dec, 2005) The role of transcriptional coactivator TRAP220 in myelomonocytic differentiation. Genes Cells 10(12):1127–37
79. Ito M, Yuan CX, Okano HJ, Darnell RB, Roeder RG (Apr, 2000) Involvement of the TRAP220 component of the TRAP/SMCC coactivator complex in embryonic development and thyroid hormone action. Mol Cell 5(4):683–93
80. Ruthardt M, Testa U, Nervi C, Ferrucci PF, Grignani F, Puccetti E, Grignani F, Peschle C, Pelicci PG (Aug, 1997) Opposite effects of the acute promyelocytic leukemia PML-retinoic acid receptor alpha (RAR alpha) and PLZF-RAR alpha fusion proteins on retinoic acid signalling. Mol Cell Biol 17(8):4859–69
81. Puccetti E, Obradovic D, Beissert T, Bianchini A, Washburn B, Chiaradonna F, Boehrer S, Hoelzer D, Ottmann OG, Pelicci PG, Nervi C, Ruthardt M (Dec 1, 2002) AML-associated translocation products block vitamin D(3)-induced differentiation by sequestering the vitamin D(3) receptor. Cancer Res 62(23):7050–8
82. Ward JO, McConnell MJ, Carlile GW, Pandolfi PP, Licht JD, Freedman LP (Dec 1, 2001) The acute promyelocytic leukemia-associated protein, promyelocytic leukemia zinc finger, regulates 1,25-dihydroxyvitarnin D(3) induced monocytic differentiation of U937 cells through a physical interaction with vitamin D(3) receptor. Blood 98(12):3290–300
83. Bories D, Raynal MC, Solomon DH, Darzynkiewicz Z, Cayre YE (Dec 22, 1989) Down-regulation of a serine protease, myeloblastin, causes growth arrest and differentiation of promyelocytic leukemia cells. Cell 59(6):959–68
84. Savli H, Aalto Y, Nagy B, Knuutila S, Pakkala S (Sep, 2002) Gene expression analysis of 1,25(OH)2D3 dependent differentiation of HL-60 cells: a cDNA array study. Br J Haematol 118(4):1065–70
85. Suzuki T, Tazoe H, Taguchi K, Koyama Y, Ichikawa H, Hayakawa S, Munakata H, Isemura M (June 2006) DNA microarray analysis of changes in gene expression induced by 1,25-dihydroxyvitamin D3 in human promyelocytic leukemia HL-60 cells. Biomed Res 27(3):99–109
86. Suzuki T, Koyama Y, Ichikawa H, Tsushima K, Abe K, Hayakawa S, Kuruto-Niwa R, Nozawa R, Isemura M (2005) 1,25-Dihydroxyvitamin D3 suppresses gene expression of eukaryotic translation initiation factor 2 in human promyelocytic leukemia HL-60 cells. Cell Struct Funct 30(1):1–6
87. Williams CJB (1849) Cod liver oil in phthisis. Lond J Med 1:1–18

88. Martineau AR, Honecker FU, Wilkinson RJ, Griffiths CJ (Mar, 2007) Vitamin D in the treatment of pulmonary tuberculosis. J Steroid Biochem Mol Biol 103(3–5):793–8

89. Rook GA, Steele J, Fraher L, Barker S, Karmali R, O'Riordan J, Stanford J (Jan, 1986) Vitamin D3, gamma interferon, and control of proliferation of Mycobacterium tuberculosis by human monocytes. Immunology 57(1):159–63

90. Wang TT, Nestel FP, Bourdeau V, Nagai Y, Wang Q, Liao J, Tavera-Mendoza L, Lin R, Hanrahan JW, Mader S, White JH (Sep 1, 2004) Cutting edge: 1,25-dihydroxyvitamin D3 is a direct inducer of antimicrobial peptide gene expression. J Immunol 173(5): 2909–12

91. Gombart AF, Borregaard N, Koeffler HP (Jul, 2005) Human cathelicidin antimicrobial peptide (CAMP) gene is a direct target of the vitamin D receptor and is strongly up-regulated in myeloid cells by 1,25-dihydroxyvitamin D3. FASEB J 19(9):1067–77

92. Weber G, Heilborn JD, Chamorro Jimenez CI, Hammarsjo A, Törmä H, Stahle M (May, 2005) Vitamin D induces the antimicrobial protein hCAP18 in human skin. J Invest Dermatol 124(5):1080–2

93. Gombart AF, O'Kelly J, Saito T, Koeffler HP (Mar, 2007) Regulation of the CAMP gene by 1,25(OH)2D3 in various tissues. J Steroid Biochem Mol Biol 103(3–5):552–7

94. Koeffler HP, Bar-Eli M, Territo MC (Mar, 1981) Phorbol ester effect on differentiation of human myeloid leukemia cell lines blocked at different stages of maturation. Cancer Res 41(3):919–26

95. Tonetti M, Cavallero A, Botta GA, Niederman R, Eftimiadi C (Feb, 1991) Intracellular pH regulates the production of different oxygen metabolites in neutrophils: effects of organic acids produced by anaerobic bacteria. J Leukoc Biol 49(2):180–8

96. Pan Q, Granger J, O'Connell TD, Somerman MJ, Simpson RU (Oct 15, 1997) Promotion of HL-60 cell differentiation by 1,25-dihydroxyvitamin D3 regulation of protein kinase C levels and activity. Biochem Pharmacol 54(8):909–15

97. Berry DM, Meckling-Gill KA (Oct, 1999) Vitamin D analogs, 20-Epi-22-oxa-24a,26a,27a,-trihomo-1alpha,25(OH)2-vitamin D3, 1,24(OH)2–22-ene-24-cyclopropyl-vitamin D3 and 1alpha,25(OH)2-lumisterol3 prime NB4 leukemia cells for monocytic differentiation via nongenomic signaling pathways, involving calcium and calpain. Endocrinology 140(10):4779–88

98. Hmama Z, Nandan D, Sly L, Knutson KL, Herrera-Velit P, Reiner NE (Dec 6, 1999) 1alpha,25-dihydroxyvitamin D(3)-induced myeloid cell differentiation is regulated by a vitamin D receptor-phosphatidylinositol 3-kinase signaling complex. J Exp Med 190(11):1583–94

99. Marcinkowska E, Wiedlocha A, Radzikowski C (Sep-Oct, 1998) Evidence that phosphatidylinositol 3-kinase and p70S6K protein are involved in differentiation of HL-60 cells induced by calcitriol. Anticancer Res 18(5A):3507–14

100. Zhang Y, Zhang J, Studzinski GP (Feb, 2006) AKT pathway is activated by 1, 25-dihydroxyvitamin D3 and participates in its anti-apoptotic effect and cell cycle control in differentiating HL60 cells. Cell Cycle 5(4):447–51

101. Marcinkowska E, Wiedłocha A, Radzikowski C (Dec 18, 1997) 1,25-Dihydroxyvitamin D3 induced activation and subsequent nuclear translocation of MAPK is upstream regulated by PKC in HL-60 cells. Biochem Biophys Res Commun 241(2):419–26

102. Song X, Bishop JE, Okamura WH, Norman AW (Feb, 1998) Stimulation of phosphorylation of mitogen-activated protein kinase by 1alpha,25-dihydroxyvitamin D3 in promyelocytic NB4 leukemia cells: a structure-function study. Endocrinology 139(2):457–65

103. Ji Y, Kutner A, Verstuyf A, Verlinden L, Studzinski GP (Nov-Dec, 2002) Derivatives of vitamins D2 and D3 activate three MAPK pathways and upregulate pRb expression in differentiating HL60 cells. Cell Cycle 1(6):410–5

104. Pepper C, Thomas A, Hoy T, Milligan D, Bentley P, Fegan C (Apr 1, 2003) The vitamin D3 analog EB1089 induces apoptosis via a p53-independent mechanism involving p38 MAP kinase activation and suppression of ERK activity in B-cell chronic lymphocytic leukemia cells in vitro. Blood 101(7):2454–60

11 Vitamin D and Hematologic Malignancies

105. Wang X, Studzinski GP (2001) Activation of extracellular signal-regulated kinases (ERKs) defines the first phase of 1,25-dihydroxyvitamin D3-induced differentiation of HL60 cells. J Cell Biochem 80(4):471–82

106. Marcinkowska E (Jan-Feb, 2001) Evidence that activation of MEK1,2/erk1,2 signal transduction pathway is necessary for calcitriol-induced differentiation of HL-60 cells. Anticancer Res 21(1A):499–504

107. Wang X, Wang TT, White JH, Studzinski GP (Nov 9, 2006) Induction of kinase suppressor of RAS-1(KSR-1) gene by 1, alpha25-dihydroxyvitamin D3 in human leukemia HL60 cells through a vitamin D response element in the 5'-flanking region. Oncogene 25(53):7078–85

108. Wang X, Wang TT, White JH, Studzinski GP (Aug 15, 2007) Expression of human kinase suppressor of Ras 2 (hKSR-2) gene in HL60 leukemia cells is directly upregulated by 1,25-dihydroxyvitamin D(3) and is required for optimal cell differentiation. Exp Cell Res 313(14):3034–45

109. Wang X, Studzinski GP (Nov, 2006) Raf-1 signaling is required for the later stages of 1,25-dihydroxyvitamin D3-induced differentiation of HL60 cells but is not mediated by the MEK/ERK module. J Cell Physiol 209(2):253–60

110. Brown G, Bunce CM, Rowlands DC, Williams GR (May, 1994) All-trans retinoic acid and 1 alpha,25-dihydroxyvitamin D3 co-operate to promote differentiation of the human promyeloid leukemia cell line HL60 to monocytes. Leukemia 8(5):806–15

111. Bunce CM, Wallington LA, Harrison P, Williams GR, Brown G (Mar, 1995) Treatment of HL60 cells with various combinations of retinoids and 1 alpha,25 dihydroxyvitamin D3 results in differentiation toward neutrophils or monocytes or a failure to differentiate and apoptosis. Leukemia 9(3):410–8

112. Masciulli R, Testa U, Barberi T, Samoggia P, Tritarelli E, Pustorino R, Mastroberardino G, Camagna A, Peschle C (May, 1995) Combined vitamin D3/retinoic acid induction of human promyelocytic cell lines: enhanced phagocytic cell maturation and hybrid granulomonocytic phenotype. Cell Growth Differ 6(5):493–503

113. Elstner E, Linker-Israeli M, Umiel T, Le J, Grillier I, Said J, Shintaku IP, Krajewski S, Reed JC, Binderup L, Koeffler HP (Aug 1, 1996) Combination of a potent 20-epi-vitamin D3 analogue (KH 1060) with 9-cis-retinoic acid irreversibly inhibits clonal growth, decreases bcl-2 expression, and induces apoptosis in HL-60 leukemic cells. Cancer Res 56(15):3570–6

114. Taimi M, Chateau MT, Cabane S, Marti J (1991) Synergistic effect of retinoic acid and 1,25-dihydroxyvitamin D3 on the differentiation of the human monocytic cell line U937. Leuk Res 15(12):1145–52

115. Doré BT, Uskokovic MR, Momparler RL (Sep, 1993) Interaction of retinoic acid and vitamin D3 analogs on HL-60 myeloid leukemic cells. Leuk Res 17(9):749–57

116. Muto A, Kizaki M, Yamato K, Kawai Y, Kamata-Matsushita M, Ueno H, Ohguchi M, Nishihara T, Koeffler HP, Ikeda Y (Apr 1, 1999) 1,25-Dihydroxyvitamin D3 induces differentiation of a retinoic acid-resistant acute promyelocytic leukemia cell line (UF-1) associated with expression of p21(WAFl/CIPl) and p27(KIPl). Blood 93(7):2225–33

117. Bastie JN, Balitrand N, Guillemot I, Chomienne C, Delva L (Nov 1, 2005) Cooperative action of 1alpha,25-dihydroxyvitamin D3 and retinoic acid in NB4 acute promyelocytic leukemia cell differentiation is transcriptionally controlled. Exp Cell Res 310(2):319–30

118. Cellier MF, Taimi M, Chateau MT, Cannat A, Marti J (Aug, 1993) Thermal stress as an inducer of differentiation of U93 7 cells. Leuk Res 17(8):649–56

119. Kumagai T, Shih LY, Hughes SV, Desmond JC, O'Kelly J, Hewison M, Koeffler HP (Mar 15, 2005) 19-Nor-1,25(OH)2D2 (a novel, noncalcemic vitamin D analogue), combined with arsenic trioxide, has potent antitumor activity against myeloid leukemia. Cancer Res 65(6):2488–97

120. Bunce CM, French PJ, Durham J, Stockley RA, Michell RH, Brown G (Apr, 1994) Indomethacin potentiates the induction of HL60 differentiation to neutrophils, by retinoic acid and granulocyte colony-stimulating factor, and to monocytes, by vitamin D3. Leukemia 8(4):595–604

121. Sokoloski JA, Sartorelli AC (Feb, 1998) Induction of the differentiation of HL-60 promyelocytic leukemia cells by nonsteroidal anti-inflammatory agents in combination with low levels of vitamin D3. Leuk Res 22(2):153–61

122. Bhatia M, Kirkland JB, Meckling-Gill KA (Oct, 1994) M-CSF and 1,25 dihydroxy vitamin D3 synergize with 12-O-tetradecanoylphorbol-13-acetate to induce macrophage differentiation in acute promyelocytic leukemia NB4 cells. Leukemia 8(10):1744–9

123. Danilenko M, Wang X, Studzinski GP (Aug 15, 2001) Carnosic acid and promotion of monocytic differentiation of HL60-G cells initiated by other agents. J Natl Cancer Inst 93(16):1224–33

124. Danilenko M, Wang Q, Wang X, Levy J, Sharoni Y, Studzinski GP (Mar 15, 2003) Carnosic acid potentiates the antioxidant and prodifferentiation effects of 1alpha,25-dihydroxyvitamin D3 in leukemia cells but does not promote elevation of basal levels of intracellular calcium. Cancer Res 63(6):1325–32

125. Sharabani H, Izumchenko E, Wang Q, Kreinin R, Steiner M, Barvish Z, Kafka M, Sharoni Y, Levy J, Uskokovic M, Studzinski GP, Danilenko M (June 15, 2006) Cooperative antitumor effects of vitamin D3 derivatives and rosemary preparations in a mouse model of myeloid leukemia. Int J Cancer 118(12):3012–21

126. Wang X, Rao J, Studzinski GP (Aug 1, 2000) Inhibition of p38 MAP kinase activity up-regulates multiple MAP kinase pathways and potentiates 1,25-dihydroxyvitamin D(3)-induced differentiation of human leukemia HL60 cells. Exp Cell Res 258(2):425–37

127. Wang Q, Harrison JS, Uskokovic M, Kutner A, Studzinski GP (Oct, 2005) Translational study of vitamin D differentiation therapy of myeloid leukemia: effects of the combination with a p38 MAPK inhibitor and an antioxidant. Leukemia 19(10):1812–7

128. Torres R, Calle C, Aller P, Mata F (May, 2000) Etoposide stimulates 1,25-dihydroxyvitamin D3 differentiation activity, hormone binding and hormone receptor expression in HL-60 human promyelocytic cells. Mol Cell Biochem 208(1–2):157–62

129. Siwińska A, Opolski A, Chrobak A, Wietrzyk J, Wojdat E, Kutner A, Szelejewski W, Radzikowski C (May–June 2001) Potentiation of the anti proliferative effect *in vitro* of doxorubicin, cisplatin and genistein by new analogues of vitamin D. Anticancer Res 21(3B):1925–9

130. Ikezoe T, Bandobashi K, Yang Y, Takeuchi S, Sekiguchi N, Sakai S, Koeffler HP, Taguchi H (Aug, 2006) HIV-1 protease inhibitor ritonavir potentiates the effect of 1,25-dihydroxyvitamin D3 to induce growth arrest and differentiation of human myeloid leukemia cells via down-regulation of CYP24. Leuk Res 30(8):1005–11

131. Göttlicher M, Minucci S, Zhu P, Krämer OH, Schimpf A, Giavara S, Sleeman JP, Lo Coco F, Nervi C, Pelicci PG, Heinzel T (Dec 17, 2001) Valproic acid defines a novel class of HDAC inhibitors inducing differentiation of transformed cells. EMBO J 20(24):6969–78

132. Siitonen T, Timonen T, Juvonen E, Terävä V, Kutila A, Honkanen T, Mikkola M, Hallman H, Kauppila M, Nyländen P, Poikonen E, Rauhala A, Sinisalo M, Suominen M, Savolainen ER, Koistinen P, Pirjo Koistinen for the Finnish Leukemia Group (Aug, 2007) Valproic acid combined with13-cis retinoic acid and 1,25 dihydroxyvitamin D3 in the treatment of patients with myelodysplastic syndromes. Haematologica 92(8):1119–22

133. Mehta AB, Kumaran TO, Marsh GW, McCarthy DM (Sep 29, 1984) Treatment of advanced myelodysplastic syndrome with alfacalcidol. Lancet 2(8405):761

134. Motomura S, Kanamori H, Maruta A, Kodama F, Ohkubo T (Sep, 1991) The effect of 1-hydroxyvitamin D3 for prolongation of leukemic transformation-free survival in myelodysplastic syndromes. Am J Hematol 38(1):67–8

135. Frazão JM, Elangovan L, Maung HM, Chesney RW, Acchiardo SR, Bower JD, Kelley BJ, Rodriguez HJ, Norris KC, Robertson JA, Levine BS, Goodman WG, Gentile D, Mazess RB, Kyllo DM, Douglass LL, Bishop CW, Coburn JW (Sep, 2000) Intermittent doxercalciferol (1alpha-hydroxyvitamin D(2)) therapy for secondary hyperparathyroidism. Am J Kidney Dis 36(3):550–61

136. Mellibovsky L, Díez A, Aubia J, Nogues X, Pérez-Vila E, Serrano S, Recker RR (Dec, 1993) Long-standing remission after 25–OH D3 treatment in a case of chronic myelomonocytic leukaemia. Br J Haematol 85(4):811–2

11 Vitamin D and Hematologic Malignancies

137. Binderup L, Bramm E (Mar 1, 1988) Effects of a novel vitamin D analogue MC903 on cell proliferation and differentiation *in vitro* and on calcium metabolism *in vivo*. Biochem Pharmacol 37(5):889–95
138. Müller K, Svenson M, Bendtzen K (Apr, 1988) 1 alpha,25-Dihydroxyvitamin D3 and a novel vitamin D analogue MC 903 are potent inhibitors of human interleukin 1 *in vitro*. Immunol Lett 17(4):361–5
139. Müller K, Heilmann C, Poulsen LK, Barington T, Bendtzen K (Mar-Apr, 1991) The role of monocytes and T cells in 1,25-dihydroxyvitamin D3 mediated inhibition of B cell function *in vitro*. Immunopharmacology 21(2):121–8
140. Rebel VI, Ossenkoppele GJ, van de Loosdrecht AA, Wijermans PW, Beelen RH, Langenhuijsen MM (1992) Monocytic differentiation induction of HL-60 cells by MC 903, a novel vitamin D analogue. Leuk Res 16(5):443–51
141. Jung SJ, Lee YY, Pakkala S, de Vos S, Elstner E, Norman AW, Green J, Uskokovic M, Koeffler HP (June 1994) 1,25(OH)2-16ene-vitamin D3 is a potent antileukemic agent with low potential to cause hypercalcemia. Leuk Res 18(6):453–63
142. Hisatake J, Kubota T, Hisatake Y, Uskokovic M, Tomoyasu S, Koeffler HP (Aug 15, 1999) 5,6-trans-16-ene-vitamin D3: a new class of potent inhibitors of proliferation of prostate, breast, and myeloid leukemic cells. Cancer Res 59(16):4023–9
143. Pakkala S, de Vos S, Elstner E, Rude RK, Uskokovic M, Binderup L, Koeffler HP (Jan, 1995) Vitamin D3 analogs: effect on leukemic clonal growth and differentiation, and on serum calcium levels. Leuk Res 19(1):65–72
144. Brown AJ, Dusso A, Slatopolsky E (Mar, 1994) Selective vitamin D analogs and their therapeutic applications. Semin Nephrol 14(2):156–74
145. Zhou JY, Norman AW, Chen DL, Sun GW, Uskokovic M, Koeffler HP (May, 1990) 1,25-Dihydroxy-16-ene-23-yne-vitamin D3 prolongs survival time of leukemic mice. Proc Natl Acad Sci USA 87(10):3929–32
146. Campbell MJ, Reddy GS, Koeffler HP (Sep 1, 1997) Vitamin D3 analogs and their 24-oxo metabolites equally inhibit clonal proliferation of a variety of cancer cells but have differing molecular effects. J Cell Biochem 66(3):413–25
147. Shiohara M, Uskokovic M, Hisatake J, Hisatake Y, Koike K, Komiyama A, Koeffler HP (Apr 15, 2001) 24-Oxo metabolites of vitamin D3 analogues: disassociation of their prominent antileukemic effects from their lack of calcium modulation. Cancer Res 61(8):3361–8
148. Binderup L, Latini S, Binderup E, Bretting C, Calverley M, Hansen K (Sep 27, 1991) 20-epi-vitamin D3 analogues: a novel class of potent regulators of cell growth and immune responses. Biochem Pharmacol 42(8):1569–75
149. Elstner E, Lee YY, Hashiya M, Pakkala S, Binderup L, Norman AW, Okamura WH, Koeffler HP (Sep 15, 1994) 1 alpha,25-Dihydroxy-20-epi-vitamin D3: an extraordinarily potent inhibitor of leukemic cell growth *in vitro*. Blood 84(6):1960–7
150. Llach F, Keshav G, Goldblat MV, Lindberg JS, Sadler R, Delmez J, Arruda J, Lau A, Slatopolsky E (Oct, 1998) Suppression of parathyroid hormone secretion in hemodialysis patients by a novel vitamin D analogue: 19-nor- 1,25-dihydroxyvitamin D2. Am J Kidney Dis 32(2 Suppl 2):S48–54
151. Martin KJ, González EA, Gellens M, Hamm LL, Abboud H, Lindberg J (Aug, 1998) 19-Nor-l-alpha-25 dihydroxyvitamin D2 (Paricalcitol) safely and effectively reduces the levels of intact parathyroid hormone in patients on hemodialysis. J Am Soc Nephrol 9(8):1427–32
152. Molnár I, Kute T, Willingham MC, Powell BL, Dodge WH, Schwartz GG (Jan, 2003) 19-nor-lalpha,25 dihydroxyvitamin D(2) (paricalcitol): effects on clonal proliferation, differentiation, and apoptosis in human leukemic cell lines. J Cancer Res Clin Oncol 129(1):35–42
153. Molnár I, Kute T, Willingham MC, Schwartz GG (May, 2004) 19-Nor-lalpha,25-dihydroxyvitamin D2 (paricalcitol) exerts anticancer activity against HL-60 cells *in vitro* at clinically achievable concentrations. J Steroid Biochem Mol Biol 89–90(1–5):539–43

154. Norman AW, Manchand PS, Uskokovic MR, Okamura WH, Takeuchi JA, Bishop JE, Hisatake JI, Koeffler HP, Peleg S (Jul 13, 2000) Characterization of a novel analogue of 1alpha,25(OH)(2)-vitamin D(3) with two side chains: interaction with its nuclear receptor and cellular actions. J Med Chem 43(14):2719–30
155. O'Kelly J, Uskokovic M, Lemp N, Vadgama J, Koeffler HP (Aug, 2006) Novel Gemini-vitamin D3 analog inhibits tumor cell growth and modulates the Akt/mTOR signaling pathway. J Steroid Biochem Mol Biol 100(4–5):107–16
156. Mellibovsky L, Díez A, Pérez-Vila E, Serrano S, Nacher M, Aubía J, Supervía A, Recker RR (Mar, 1998) Vitamin D treatment in myelodysplastic syndromes. Br J Haematol 100(3):S16–20
157. Molnár I, Stark, Lovato J, Powell BL, Cruz J, Hurd DD, Mathieu JS, Chen TC, Holick MF, Cambra S, McQuellon RP, Schwartz GG (May, 2007) Treatment of low-risk myelodysplastic syndromes with high-dose daily oral cholecalciferol (2000–4000 IU vitamin D(3)). Leukemia 21(5):1 089–92

Chapter 12
The Vitamin D Signaling Pathway in Mammary Gland and Breast Cancer

Glendon M. Zinser, Carmen J. Narvaez, and JoEllen Welsh

Abstract Epidemiologic data have demonstrated that breast cancer incidence is inversely correlated with indices of vitamin D status, including UV exposure, which enhances epidermal vitamin D synthesis. The vitamin D receptor (VDR) is expressed in mammary epithelial cells, suggesting that vitamin D may directly influence sensitivity of the gland to transformation. Consistent with this concept, in vitro studies have demonstrated that the VDR ligand, 1,25-dihydroxyvitamin D (1,25D), exerts negative growth regulatory effects on mammary epithelial cells that contribute to maintain the differentiated phenotype and protection of the genome. Mammary cells also have the ability to internalize the major circulating vitamin D metabolite, 25-hydroxyvitamin D (25D), and convert it to 1,25D. Furthermore, deletion of the VDR gene in mice alters the balance between proliferation and apoptosis in the mammary gland which ultimately enhances its susceptibility to carcinogenesis. Dietary supplementation with vitamin D, or chronic treatment with synthetic VDR agonists, reduces the incidence of carcinogen-induced mammary tumors in rodents. Collectively, these observations have reinforced the need to further define the human requirement for vitamin D and the molecular actions of the VDR in relation to prevention of breast cancer.

Keywords Vitamin D receptor • Breast cancer • Mammary gland • Vitamin D • Prevention

---

J. Welsh (✉)
GenNYsis Center for Excellence in Cancer Genomics, University of Albany, Rensselaer, NY 12144, USA
e-mail: jwelsh@albany.edu

D.L. Trump and C.S. Johnson (eds.), *Vitamin D and Cancer*,
DOI 10.1007/978-1-4419-7188-3_12, © Springer Science+Business Media, LLC 2011

Abbreviations

1,25D	1,25 Dihydroxyvitamin D
DBP	D binding protein
DMBA	Dimethylbenzanthracene
EGF	Epidermal growth factor
HME	Human mammary epithelial
25D	25 Hydroxyvitamin D
IGF-1	Insulin like growth factor 1
KGF	Keratinocyte growth factor
SV40	Simian virus 40
TGFβ	Transforming growth factor beta
UV	Ultraviolet
VDR	Vitamin D receptor
WT	Wild type

12.1 Introduction to Vitamin D and Breast Cancer

Although originally identified based on its ability to prevent the bone disease rickets, it is now recognized that $1\alpha,25$ dihydroxyvitamin D_3 (1,25D), the biologically active form of vitamin D_3, is a global regulator of gene expression and signal transduction in virtually every tissue. In breast cells, the vitamin D receptor (VDR) and its ligand 1,25D contribute to maintenance of the quiescent, differentiated phenotype, providing defense against cancer development. The presence of functional VDR in the majority of human breast tumors (initially discovered over 25 years ago) suggested that this receptor might represent a target for breast cancer therapy. Since that time, multiple studies have confirmed the antiproliferative effects of vitamin D on breast cancer cells in vitro and rodent tumors in vivo. Dozens of synthetic vitamin D structural analogs have been tested for efficacy and side effects in animal models of cancer, individually and in combination with standard therapies such as anti-estrogens, chemotherapeutic drugs and radiation. While these studies have generally been supportive of targeting the vitamin D pathway in breast cancer therapy, issues of dosing, toxicity and efficacy (particularly against various tumor sub-types) remain to be resolved.

More recently, studies have focused on characterization of the expression and function of the vitamin D pathway in normal mammary tissue and the possible role of the vitamin D pathway in breast cancer prevention. In particular, data from the VDR knockout mouse model have indicated that complete abrogation of vitamin D signaling alters glandular morphology and susceptibility to cancer development. In this review, we will summarize the currently available data generated from both in vitro and in vivo studies, with an emphasis on the cellular and molecular mechanisms by which vitamin D may contribute to breast cancer prevention.

12 The Vitamin D Signaling Pathway in Mammary Gland and Breast Cancer

12.2 Vitamin D and Breast Cancer Links in Populations

12.2.1 Diet, Sunlight Exposure and Breast Cancer Risk

Population studies on vitamin D in relation to chronic diseases such as breast cancer are complicated by difficulties in accurately assessing dietary sources (confounders include natural foods versus fortified foods, supplement use, intake of D_2 versus D_3, and calcium status) and estimating the amount of vitamin D generated through sunlight exposure (confounders include lifestyle, latitude, pollution, sunscreen, skin pigmentation and age). Despite this caveat, the cumulative population data support an inverse correlation between vitamin D sources (diet or sunlight exposure) and relative risk of breast cancer. Due to space constraints, only a few of these studies are highlighted here. An evaluation of the Nurse's Health Study found that intakes of dairy products, dairy calcium and vitamin D were inversely associated with breast cancer risk in premenopausal, but not postmenopausal, women [1]. Similarly, a prospective analysis of breast cancer incidence in relation to vitamin D intake for over 30,000 participants in the Women's Health Study indicated that higher intake of vitamin D was moderately associated with a lower risk of pre- but not post- menopausal breast cancer [2]. These data are consistent with reports of inverse associations between vitamin D status and mammographic density in premenopausal women [3, 4]. In addition to dietary vitamin D, [5] demonstrated that sunlight exposure was associated with reduced risk of breast cancer, and that this association was dependent on region of residence. A recent follow-up analysis indicated that the beneficial effect of sunlight exposure on risk was dependent on skin pigmentation, with significant correlations demonstrated only in women with fair skin [6]. In larger international studies, a significant inverse correlation between incident solar radiation and breast cancer rates was confirmed [7–9].

12.2.2 Serum 25-Hydroxyvitamin D and Breast Cancer Risk

Studies which have assessed serum parameters as indicators of vitamin D status in relation to breast cancer risk have been more consistent. Although confounders remain (dietary calcium, serum PTH, seasonal influences and assay methodology), multiple studies have reported significant inverse relationships between relative risk of breast cancer and serum 25D, the most accurate indicator of vitamin D status [10–12]. A pooled analysis of data on serum 25D in relation to breast cancer demonstrated that the highest quintile of serum 25D was associated with a 50% reduction in breast cancer risk [13]. These data suggested that serum 25D concentrations above 100 nM may be required to optimize vitamin D signaling in mammary tissue. This serum 25D concentration is considerably higher than that necessary for prevention of rickets (approximately 50 nM), suggesting that prolonged sub-optimal vitamin D status (rather than overt deficiency) is associated

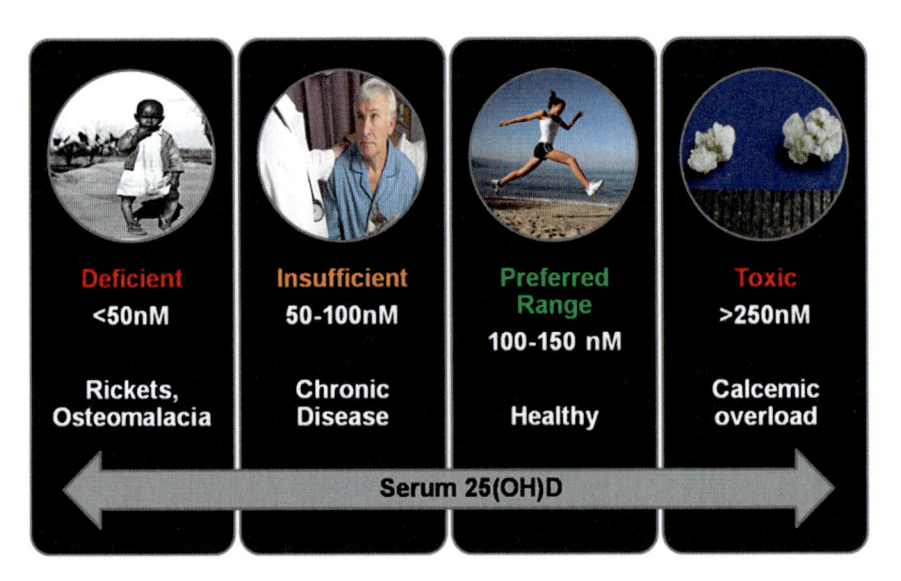

Fig. 12.1 Proposed stages of vitamin D deficiency and sufficiency according to serum 25-hydroxyvitamin D. Vitamin D deficiency leading to rickets in children and osteomalacia in adults is associated with circulating concentrations of 25-hydroxyvitamin D (*25D*) below 50 nM. In contrast, data from epidemiological studies suggest that serum 25D concentrations between 50 and 100 nM are associated with an increased risk of chronic diseases, including cancer. Toxicity, which is associated with soft tissue calcification and renal stones, occurs when 25D increases above 250 nM. Thus, the preferred range of 25D for optimal health is likely between 100 and 150 nM, which may not be attainable without dietary supplementation (2,000 IU/day or higher in individuals with limited sun exposure)

with increased risk for breast cancer. Similar data generated on serum 25D in relation to other chronic diseases support the concept that a healthy 25D range for adults is between 100 and 150 nM, which is well below the toxic range associated with calcemic overload (above 250 nM). A summary of the proposed relationship between serum 25D, health and disease is provided in Fig. 12.1.

12.2.3 Prevalence of Vitamin D Insufficiency

Unfortunately, it is difficult for most people to maintain serum 25D in the proposed healthy range (100–150 nM) from dietary sources alone due to the low amounts of vitamin D in natural foods [14]. Particularly relevant to the possible relationship between vitamin D and breast cancer, vitamin D deficiency has been reported in a high percentage of women, including during adolescence, pregnancy/lactation and after menopause, even in sunny climates [15–17]. The amount of vitamin D usually present in over the counter supplements (400 IU) is too low to significantly elevate serum 25D [18]. Supplementation studies suggest that 2,000 IU/day (and possibly

12   The Vitamin D Signaling Pathway in Mammary Gland and Breast Cancer

up to 4,000 IU/day for individuals with limited sun exposure) is needed to maintain serum 25D at 100 nM, but further research is needed on vitamin D supplementation in relation to chronic disease. One small, double blinded intervention study of healthy post-menopausal women indicated that daily supplementation with 1,100 IU vitamin D_3 reduced cancer risk at all sites [19]. Collectively, these observations provide initial evidence that vitamin D may reduce breast cancer incidence, and emphasize the need for re-evaluation of public health recommendations regarding sun exposure, vitamin D intake, food fortification and supplement use in relation to vitamin D status and breast cancer.

12.3   Mechanistic Links Between Vitamin D and Breast Cancer

12.3.1   General Effects of VDR Agonists in Breast Cancer Cells

In response to the initial identification of VDR in cancer cells, numerous studies examined the effects of 1,25D on breast cancer cells. Furthermore, a large number of structural analogs of vitamin D, developed by pharmaceutical companies and academic researchers, have been used to probe the mechanisms of vitamin D mediated growth inhibition. In general, the effects of VDR agonists on breast cancer cells are similar to those reported in other cancer cell types: modulation of key cell cycle regulators to arrest the cycle at either G0/G1 or G2/M (depending on cell type), induction of differentiation markers, and/or activation of cell death (via apoptosis or autophagy). Mechanisms have recently been reviewed in detail [20, 21], and thus are briefly discussed here. Notably, studies with cells derived from VDR null mice has definitely established that the VDR is required for the antiproliferative and proapoptotic effects of 1,25D in transformed mammary cells in vitro [22, 23].

12.3.2   Cellular and Molecular Targets of VDR in Breast Cancer Cells

Screening for molecular changes induced by 1,25D or vitamin D analogs in various breast cancer cells has identified scores of VDR regulated genes and proteins in diverse pathways, indicating a broad range of downstream targets [24–26]. The induction of cell cycle arrest in both estrogen receptor positive and negative breast cancer cells by 1,25D results from alterations in key cell cycle regulators including cyclin D1, the cyclin dependent kinase inhibitors p21 and/or p27 and the retinoblastoma protein. 1,25D also blocks mitogenic signaling, including that of estrogen, EGF, IGF-1 and KGF and up-regulates negative growth factors such as TGFβ [27–29]. In many breast cancer cell lines, 1,25D mediated growth arrest is associated with the induction of differentiation markers such as casein, lipid droplets, and adhe-

sion proteins [30, 31]. In some transformed breast cells, 1,25D induces apoptotic cell death via generation of reactive oxygen species, dissipation of the mitochondrial membrane potential and cytochrome c release in association with down regulation of bcl-2 and activation of bax [32, 33]. The role of proteases in 1,25D mediated cell death appears to be cell type dependent, with caspases, cathepsins and calpains being activated under different contexts [22, 34]. Notably, 1,25D exerts additive or synergistic effects in combination with other triggers of apoptosis, such as ionizing radiation and chemotherapeutic agents [35–37]. Collectively, these studies indicate that a wide variety of different signaling pathways, cell cycle and apoptotic regulatory proteins and proteases may contribute to the antiproliferative, prodifferentiating and apoptotic effects of 1,25D depending on the specific cell type and/or context.

12.3.3 Emerging Role of Vitamin D in Cellular Stress Responses

Normal cells continuously sense and respond to a variety of stresses, including DNA damage, oxidative stress, endoplasmic reticulum overload, unfolded proteins and others. One of the major sensors of cellular stress is the tumor suppressor protein p53, a transcription factor that integrates the response to DNA damage. Germline mutations in p53 that disable its transcriptional activity strongly predispose to the breast to cancer [38]. It has recently been demonstrated that p53 transcriptionally up-regulates the human VDR via direct binding to conserved intronic p53 response elements [39]. Other studies have implicated the p53-related family members p63 and p73 in regulation of the VDR gene [40]. Exposure of cells to DNA damaging agents such as doxorubicin, etoposide or ionizing radiation resulted in up-regulation of VDR expression [39, 41, 77]. Notably, VDR and p53 mediate similar biological effects (growth arrest, apoptosis, DNA repair) via common target genes (p21, bax, GADD45). On the p21 promoter, both independent and overlapping VDR and p53 binding sites have been characterized [42]. These studies suggest that VDR regulated pathways may contribute to the tumor suppressive effects of the p53 family, particularly those involved in cellular stress response. Other studies have implicated c-jun in the control of VDR expression and activity in response to arsenic stress, suggesting that VDR signaling may also protect against nongenotoxic cellular damage via p53 independent pathways [43]. Further studies to clarify how p53, c-jun and other stress responsive pathways regulate VDR signaling, and how VDR activation in turn modulates cellular responses, are needed.

12.3.4 Preclinical Studies of VDR Agonists in Animal Models of Breast Cancer

Although therapeutic use of 1,25D is precluded by dose-limiting calcemic toxicity, synthetic analogs of 1,25D with low calcemic potency have provided proof of principle that VDR agonists can inhibit growth and induce regression of mammary

12 The Vitamin D Signaling Pathway in Mammary Gland and Breast Cancer   285

tumors in animal models [44, 45]. The effects of vitamin D analogs were comparable
to standard anti-estrogen therapies such as tamoxifen, and additive effects were
observed in combination studies (with tamoxifen, ionizing radiation, and chemotherapeutic drugs). Vitamin D analog therapy was effective in both estrogen receptor positive and estrogen receptor negative xenografts [45–47]. Under the conditions utilized
in these in vivo studies, the vitamin D analogs did not cause weight loss or hypercalcemia, but the therapeutic window for most of these compounds is narrow. Of particular interest, studies on xenografts derived from WT and VDR null cells indicated that
expression of functional VDR in tumor epithelial cells (rather than in accessory cells
such as fibroblasts, immune cells or endothelial cells) is necessary for the antitumor
effects of the vitamin D analog EB1089 and UV generated vitamin D *in vivo* [48].

12.4 Evidence for Breast Cancer Prevention by Vitamin D

12.4.1 VDR Expression in Normal Mammary Cells

The link between vitamin D and breast cancer prevention is based on the concept
that the 1,25D-VDR complex promotes or maintains the differentiated phenotype in
normal mammary cells. Consistent with this concept, the VDR is expressed in normal mammary epithelial tissue in vivo and in nontransformed human mammary
epithelial (HME) cells in vitro [49–51]. In mouse mammary gland, VDR is localized
predominantly in differentiated epithelial cells, and its expression increases 100-fold
during the course of pregnancy and lactation [51, 52]. The mammary epithelial cells
are organized in ducts which are contained within an extensive adipose rich stromal
compartment (the mammary fat pad) which includes fibroblasts, adipocytes,
endothelial cells, immune cells, and extracellular matrix proteins. VDR is also
expressed in the stromal fibroblasts, the mammary adipocytes, and the infiltrating
immune cells, indicating potential cross-talk between compartments. A potential
role of vitamin D signaling in the mammary adipocytes is supported by the demonstration that lipophilic vitamin D steroids are stored in fat [53] and by the presence
of both VDR and Cyp27B1 expression and activity in human adipocytes (Zinser,
unpublished). A working model for the cross-talk between cells in the epithelial,
stromal and adipose compartments of the breast is provided in Fig. 12.2.

12.4.2 Vitamin D Metabolites Mediate Growth Inhibition in Normal Mammary Cells

The function of the vitamin D pathway in nontransformed HME cells has recently
been evaluated. The effects of 1,25D on HME cells include growth arrest and
induction of differentiation markers such as E-cadherin, but apoptosis has not been
observed [50]. Notably, mammary cells express Cyp27B1, can generate 1,25D

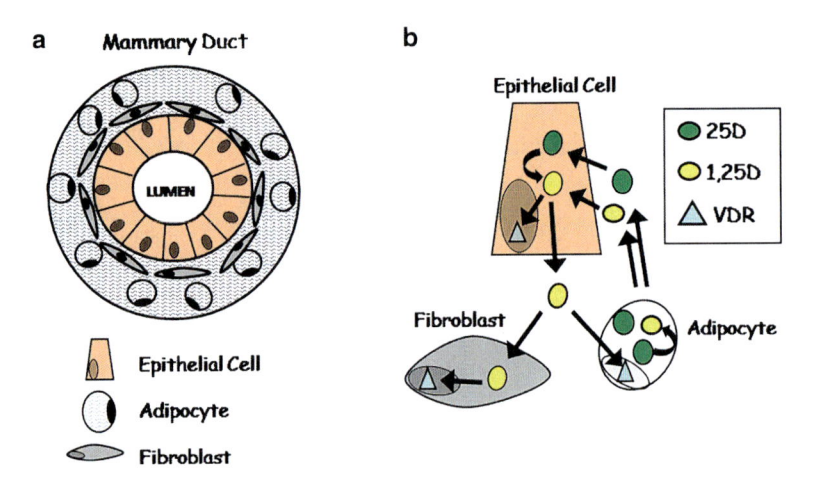

Fig. 12.2 (**a**) Cellular organization in mammary gland. Glands are composed of epithelial cells arranged in ducts around a central lumen, which are embedded in an adipocyte rich mammary fat pad/stroma containing fibroblasts, immune cells, endothelial cells and extracellular matrix proteins. (**b**) Model for cell type specific vitamin D activation and function in mammary gland. In mouse mammary gland, the three major cell types (epithelial cells, fibroblasts and adipocytes) express VDR and have the ability to respond to 1,25D. Mammary epithelial cells and adipocytes express CYP27B1 and can generate 1,25D from 25D. We propose that like adipose tissue elsewhere in the body, the mammary fat pad acts as a storage depot for 25D. This model predicts that optimal vitamin D signaling in the epithelial, stromal and adipose compartments is required for maintenance of differentiation, genomic stability and protection against breast cancer

from 25D, and are growth inhibited by physiological concentrations of 25D [50, 54]. These data suggest that 25D may be the most biologically relevant metabolite in the mammary gland, but one caveat to these studies is that the mechanisms by which 25D is taken up by mammary cells have yet to be identified. It is well known that circulating 25D is bound to serum DBP, therefore it is likely that 25D is delivered to the mammary gland in complex with DBP. However, whether 25D dissociates from the 25D-DBP complex or whether the 25-DBP complex is internalized intact by mammary cells is unclear. Recent studies have demonstrated that both murine and human mammary cells express megalin and cubilin, proteins required for the endocytic uptake of DBP in kidney. Furthermore, uptake of DBP occurred in mammary cells in vitro and was correlated with 25D mediated transactivation of VDR [55], however, further studies are necessary to determine whether endocytosis of the 25D-DBP complex occurs in mammary tissue *in vivo*.

12.4.3 Prevention of Mammary Carcinogenesis by VDR Agonists in Animal Models

Animal studies also support the concept that vitamin D signalling reduces breast cancer development. Rodents fed western style diets (low in vitamin D

and calcium, high in saturated fat) exhibited hyperproliferation in the mammary gland and developed significantly more mammary tumors when treated with 7,12-dimethylbenzanthracence (DMBA) compared to rats fed adequate calcium and vitamin D (reviewed by [56]). In mouse mammary gland organ culture, 1,25D and the synthetic analog $1\alpha(OH)D_5$ reduced the incidence of preneoplastic lesions in response to DMBA during both the initiation and the promotion stages, demonstrating that vitamin D compounds exert *direct* antineoplastic effects on mammary gland at multiple steps [57]. Prevention of N-methyl-N-nitrosourea-induced mammary tumors with vitamin D analogs, including Ro24–5531 (1,25-dihydroxy-16-ene-23-yne-26–27-hexafluorocholecalciferol) and $1\alpha(OH)D_5$ provided further support that the vitamin D pathway may protect against breast cancer in vivo [58, 59].

12.5 Mammary Gland Development and Carcinogenesis in VDR Null Mice

Mice lacking the VDR demonstrate excess proliferation and branching as well as impaired apoptosis during the reproductive cycle compared to their normal counterparts [51, 52]. Organ culture experiments indicated that 1,25D blocked the growth stimulatory effects of estrogen and progesterone in glands from wild-type mice but was without effect in glands from VDR null mice, indicating that the VDR acts in a ligand dependent manner to mediate negative growth signaling directly in mammary tissue. Comparison of gene expression in normal and VDR null mice has identified cyclin D1, p21, clusterin, β-catenin and TGFβ1 as potential VDR target genes in the mammary gland in vivo (Zinser, Matthews and Welsh, unpublished data). Demonstration that VDR ablation alters growth regulatory pathways in mammary gland raised the possibility that VDR null mice might display enhanced risk for cancer development in this tissue. Indeed, the incidence of mammary hyperplasias and development of ER negative tumors in response to the carcinogen DMBA was higher in VDR null mice than their WT counterparts [60]. Furthermore, on the MMTV-neu transgenic background (a model of her2 positive human breast cancer), VDR heterozygote mice demonstrated higher incidence of mammary tumors than did WT mice [61]. Notably, differences in cancer susceptibility were not limited to the mammary gland, as VDR null mice displayed increased sensitivity to tumors in the lymph nodes and skin in response to DMBA compared to WT mice [60, 62]. These *in vivo* studies have provided the most direct evidence that VDR signaling can protect against cancer development. Collectively, these and other animal studies have confirmed that the effects of vitamin D signalling observed *in vivo* translate to effects on cell proliferation, differentiation and apoptosis *in vivo* that are of sufficient magnitude to impact on the carcinogenic process.

12.6 Vitamin D Resistance Pathways

Some transformed breast cells display reduced sensitivity to 1,25D, suggesting that the vitamin D pathway may become deregulated during cancer development. Multiple mechanisms have been identified that contribute to 1,25D resistance, including loss of VDR expression, alterations in transcriptional co-regulators and overexpression of Cyp24, the enzyme that catabolizes 1,25D. Stable expression of the antiapoptotic protein bcl-2 rendered cancer cells resistant to 1,25D mediated apoptosis, and expression of certain oncogenes (including ras and SV40 large T antigen) abrogated VDR signaling [32, 63, 64]. Amplification of the Cyp24 gene was reported in human breast tumors, and higher Cyp24 expression was detected in tumors compared to adjacent normal tissue [54, 65]. De-sensitization of breast cancer cells to growth inhibition by VDR ligands has also been associated with changes in nuclear receptor co-repressors via epigenetic mechanisms, which are potentially reversible [66–68]. Sub-clones of the MCF-7 breast cancer cell line selected for resistance to 1,25D in vitro have been independently developed and characterized [69, 70]. These cell lines retain low level expression of transcriptionally active VDR but exhibit changes in protein expression that alter redox status, favor autonomous growth signaling, and down-regulate the apoptotic pathway [24, 26, 71]. One of these 1,25D resistant MCF-7 cell lines was tested in a xenograft model and retained resistance to the antitumor effects of the vitamin D analog EB1089, providing an in vivo model for the study of vitamin D resistance [45]. Notably, despite deregulation of multiple signaling pathways, the MCF-7 cells selected for 1,25D resistance are cross-resistant to structurally related vitamin D analogs but retain sensitivity to other growth inhibitory agents, including retinoids and anti-estrogens. Uncovering the molecular basis for selective vitamin D resistance will be critical in design and implementation of new vitamin D analogs for clinical use.

Kemmis and Welsh [72] used a series of isogenic, progressively transformed HME cell lines [73] to study the effects of transformation per se on the vitamin D pathway. In this model, HME cells transduced with SV40 large T antigen and oncogenic ras undergo the epithelial-mesenchymal transition (loss of E-cadherin) and acquire tumorigenic potential. VDR expression in HME cells expressing SV40 or ras was reduced more than 70% compared to the nontransformed HME cells from which they were derived. Loss of VDR may be associated with up-regulation of transformation-associated corepressor proteins such as slug or snail, which have been shown to directly repress the human VDR promoter in breast and colon cancer cells [74, 75]. In the HME series, oncogenic transformation was also associated with reduced Cyp27B1 expression and activity (as measured by 125D synthesis), but the underlying molecular mechanisms for this change remain unknown. The reductions in VDR and Cyp27B1 in the oncogene-transformed HME cells were of sufficient magnitude to reduce cellular sensitivity to growth inhibition by both 1,25D and 25D approximately 100-fold. These studies provide evidence that disruption of the vitamin D signaling pathway may occur early in the cancer development process, and that cancer cells employ multiple mechanisms to evade the negative growth regulatory effects of the vitamin D signaling pathway.

12.7 Conclusions and Directions for Future Research

In summary, the VDR is expressed in normal mammary epithelial cells, where it regulates proliferation, apoptosis & differentiation via distinct targets at different stages of development. In mice, deficiency of the VDR alters glandular growth during puberty, pregnancy and aging, and enhances risk for mammary gland transformation. 1,25D and numerous synthetic vitamin D analogs effectively inhibit growth and induce apoptosis in breast cancer cells & tumors, and these effects require the VDR. VDR agonists also inhibit growth of normal human mammary epithelial cells, and evidence suggests that autocrine bio-activation of vitamin D precursors can occur within mammary cells. Thus, data from both human tissues and animal models support the concept that the VDR and its ligand induce a program of gene expression that contributes to maintenance of the differentiated phenotype in breast cells, a concept which is consistent with a role for vitamin D in both prevention and treatment of breast cancer. However, the specific mechanisms by which the 1,25D-VDR complex elicits such diverse changes in cell behavior, in particular the relative importance of genomic versus nongenomic mechanisms (Fig. 12.3), have yet to be fully elucidated. Since emerging evidence indicates that aggressive cancer cells can develop deregulation of VDR and Cyp27B1, clarifying the pathways by which vitamin D signaling contributes to breast cancer prevention is of critical importance.

Although a tentative relationship between serum 25D and health outcomes was proposed in Fig. 12.1, the amount of vitamin D (either from diet or endogenous synthesis) needed to optimize growth inhibitory signaling through the VDR in vivo

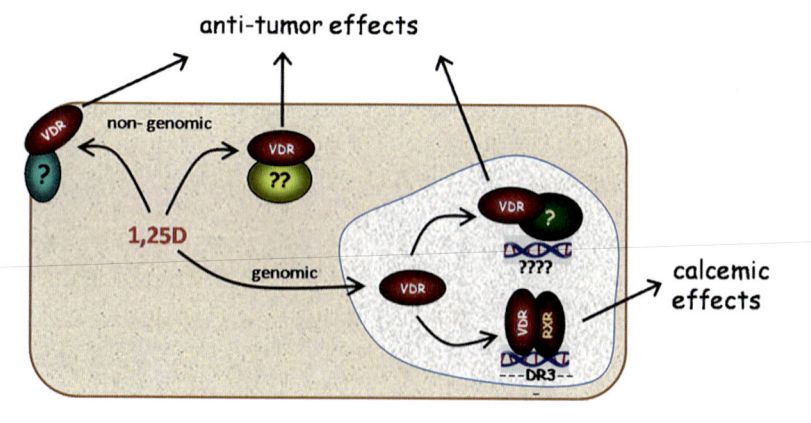

Fig. 12.3 Potential pathways for vitamin D action in mammary cells. The vitamin D receptor (VDR) is required for the antitumor effects of 1,25D, but the intracellular mechanisms may include nongenomic actions at the membrane or the cytosol (i.e., via interactions with signal transduction pathways) and/or genomic actions via heterodimerization with RXR on well characterized vitamin D response elements known to be involved in calcium metabolism (i.e., direct repeat 3 (DR3) sites) or novel elements in association with other transcription factors

is currently undefined, and further studies are needed before guidelines or requirements for human populations can be established. Collectively, studies to date have confirmed that multiple components of the vitamin D signaling system are present in normal mammary epithelial cells, but have also emphasized the need for additional research on regulation and function of these proteins in intact mammary tissue in vivo, particularly in relation to maintenance of the differentiated phenotype.

In addition to gaps in knowledge of vitamin D signaling in mammary epithelial cells, little is known about the in vivo compartmentalization of the metabolic enzymes, transport proteins and receptor for vitamin D in the gland. Still unresolved as well are the molecular mechanisms for cellular uptake, storage and intracellular transport of the various vitamin D metabolites in mammary tissue. Use of targeted mouse models with cell type specific ablation of VDR, Cyp27B1 and other candidate genes involved in vitamin D signaling should be highly informative in clarifying some of the relationships postulated in our working model.

References

1. Knekt P, Jarvinen R, Seppanen R et al (1996) Intake of dairy products and the risk of breast cancer. Br J Cancer 73:687–691
2. Lin J, Manson JE, Lee IM et al (2007) Intakes of calcium and vitamin D and breast cancer risk in women. Arch Intern Med 167:1050–1059
3. Berube S, Diorio C, Verhoek-Oftedahl W et al (2004) Vitamin D, calcium and mammographic breast densities. Cancer Epidemiol Biomarkers Prev 13:1466–1472
4. Brisson J, Berube S, Diorio C et al (2007) Synchronized seasonal variations of mammographic breast density and plasma 25-hydroxyvitamin D. Cancer Epidemiol Biomarkers Prev 16:929–933
5. John EM, Schwartz GG, Dreon DM et al (1999) Vitamin D and breast cancer risk: the NHANES I Epidemiologic follow-up study, 1971–1975 to 1992. Cancer Epidemiol Biomarkers Prev 8:399–406
6. John EM, Schwartz GG, Koo J et al (2007) Sun exposure, vitamin D receptor gene polymorphisms, and breast cancer risk in a multiethnic population. Am J Epidemiol 166: 1409–1419
7. Giovannucci E (2005) The epidemiology of vitamin D and cancer incidence and mortality: a review (United States). Cancer Causes Control 16:83–95
8. Grant WB, Garland CF (2006) The association of solar ultraviolet B (UVB) with reducing risk of cancer: multifactorial ecologic analysis of geographic variation in age-adjusted cancer mortality rates. Anticancer Res 26:2687–2699
9. Mohr SB, Garland CF, Gorham ED et al (2008) Relationship between low ultraviolet B irradiance and higher breast cancer risk in 107 countries. Breast J 14:255–260
10. Abbas S, Linseisen J, Slanger T et al (2008) Serum 25-hydroxyvitamin D and risk of postmenopausal breast cancer-results of a large case-control study. Carcinogenesis 29:93–99
11. Bertone-Johnson ER, Chen WY, Holick MF et al (2005) Plasma 25-hydroxyvitamin D and 1, 25-dihydroxyvitamin D and risk of breast cancer. Cancer Epidemiol Biomarkers Prev 14:1991–1997
12. Lowe LC, Guy M, Mansi JL et al (2005) Plasma 25-hydroxy vitamin D concentrations, vitamin D receptor genotype and breast cancer risk in a UK Caucasian population. Eur J Cancer 41:1164–1169

13. Garland CF, Gorham ED, Mohr SB et al (2007) Vitamin D and prevention of breast cancer: pooled analysis. J Steroid Biochem Mol Biol 103:708–711
14. Glerup H, Mikkelsen K, Poulsen L et al (2000) Commonly recommended daily intake of vitamin D is not sufficient if sunlight exposure is limited. J Intern Med 247:260–268
15. Bodnar LM, Simhan HN, Powers RW et al (2007) High prevalence of vitamin D insufficiency in black and white pregnant women residing in the northern United States and their neonates. J Nutr 137:447–452
16. Gonzalez G, Alvarado JN, Rojas A et al (2007) High prevalence of vitamin D deficiency in Chilean healthy postmenopausal women with normal sun exposure: additional evidence for a worldwide concern. Menopause 14:455–461
17. Siddiqui AM, Kamfar HZ (2007) Prevalence of vitamin D deficiency rickets in adolescent school girls in Western region, Saudi Arabia. Saudi Med J 28:441–444
18. Vieth R, Bischoff-Ferrari H, Boucher BJ et al (2007) The urgent need to recommend an intake of vitamin D that is effective. Am J Clin Nutr 85:649–650
19. Lappe JM, Travers-Gustafson D, Davies KM et al (2007) Vitamin D and calcium supplementation reduces cancer risk: results of a randomized trial. Am J Clin Nutr 85:1586–1591
20. Bouillon R, Eelen G, Verlinden L et al (2006) Vitamin D and cancer. Steroid Biochem Mol Biol 102:156–162
21. Deeb KK, Trump DL, Johnson CS (2007) Vitamin D signalling pathways in cancer: potential for anticancer therapeutics. Nat Rev Cancer 7:684–700
22. Valrance ME, Brunet AH, Acosta A et al (2007) Dissociation of growth arrest and CYP24 induction by VDR ligands in mammary tumor cells. J Cell Biochem 101:1505–1519
23. Zinser GM, McEleney K, Welsh JE (2003) Characterization of mammary tumor cell lines from wild type and vitamin D(3) receptor knockout mice. Mol Cell Endocrinol 200:67–80
24. Byrne B, Welsh JE (2007) Identification of novel mediators of vitamin D signaling and 1, 25(OH)2D3 resistance in mammary cells. J Steroid Biochem Mol Biol 103:703–707
25. Swami S, Raghavachari N, Muller UR et al (2003) Vitamin D growth inhibition of breast cancer cells: gene expression patterns assessed by cDNA microarray. Breast Cancer Res Treat 80:49–62
26. Towsend K, Trevino V, Falciani F et al (2006) Identification of VDR-responsive gene signatures in breast cancer cells. Oncology 71:111–123
27. Cordero JB, Cozzolino M, Lu Y et al (2002) 1, 25-Dihydroxyvitamin D down-regulates cell membrane growth- and nuclear growth-promoting signals by the epidermal growth factor receptor. J Biol Chem 277:38965–38971
28. Kawata H, Kamiakito T, Takayashiki N et al (2006) Vitamin D3 suppresses the androgen-stimulated growth of mouse mammary carcinoma SC-3 cells by transcriptional repression of fibroblast growth factor 8. J Cell Physiol 207:793–799
29. Wu Y, Craig TA, Lutz WH et al (1999) Identification of 1 alpha, 25-dihydroxyvitamin D3 response elements in the human transforming growth factor beta 2 gene. Biochemistry 38:2654–2660
30. Lazzaro G, Agadir A, Qing W et al (2000) Induction of differentiation by 1alpha-hydroxyvitamin D(5) in T47D human breast cancer cells and its interaction with vitamin D receptors. Eur J Cancer 36:780–786
31. Pendas-Franco N, Gonzalez-Sancho JM, Suarez Y et al (2007) Vitamin D regulates the phenotype of human breast cancer cells. Differentiation 75:193–207
32. Mathiasen IS, Lademann U, Jaattela M (1999) Apoptosis induced by vitamin D compounds in breast cancer cells is inhibited by Bcl-2 but does not involve known caspases or p53. Cancer Res 59:4848–4856
33. Narvaez CJ, Welsh J (2001) Role of mitochondria and caspases in vitamin D-mediated apoptosis of MCF-7 breast cancer cells. J Biol Chem 276:9101–9107
34. Mathiasen IS, Sergeev IN, Bastholm L et al (2002) Calcium and calpain as key mediators of apoptosis-like death induced by vitamin D compounds in breast cancer cells. J Biol Chem 277:30738–30745

35. Chaudhry M, Sundaram S, Gennings C et al (2001) The vitamin D3 analog, ILX-23–7553, enhances the response to adriamycin and irradiation in MCF-7 breast tumor cells. Cancer Chemother Pharmacol 47:429–436

36. Posner GH, Crawford KR, Peleg S et al (2001) A non-calcemic sulfone version of the vitamin D(3) analogue seocalcitol (EB 1089): chemical synthesis, biological evaluation and potency enhancement of the anticancer drug adriamycin. Bioorg Med Chem 9:2365–2371

37. Sundaram S, Gewirtz DA (1999) The vitamin D3 analog EB 1089 enhances the response of human breast tumor cells to radiation. Radiat Res 152:479–486

38. Petitjean A, Mathe E, Kato S et al (2007) Impact of mutant p53 functional properties on TP53 mutation patterns and tumor phenotype: lessons from recent developments in the IARC TP53 database. Hum Mut 28:622–629

39. Maruyama R, Aoki F, Toyota M et al (2006) Comparative genome analysis identifies the vitamin D receptor gene as a direct target of p53-mediated transcriptional activation. Cancer Res 66:4574–4583

40. Kommagani R, Caserta TM, Kadakia MP (2006) Identification of vitamin D receptor as a target of p63. Oncogene 25:3745–3751

41. Roy D, Calaf G, Hei TK (2003) Role of vitamin D receptor gene in radiation-induced neoplastic transformation of human breast epithelial cell. Steroids 68:621–627

42. Saramaki A, Banwell CM, Campbell MJ et al (2006) Regulation of the human p21(waf1/cip1) gene promoter via multiple binding sites for p53 and the vitamin D3 receptor. Nucleic Acids Res 34:543–554

43. Li QP, Qi X, Pramanik R et al (2007) Stress-induced c-jun-dependent vitamin D receptor (VDR) activation dissects the non-classical VDR pathway from the classical VDR activity. J Biol Chem 282:1544–1551

44. James SY, Mercer E, Brady M et al (1998) EB1089, a synthetic analogue of vitamin D, induces apoptosis in breast cancer cells in vivo and in vitro. Br J Pharmacol 125:953–962

45. VanWeelden K, Flanagan L, Binderup L et al (1998) Apoptotic regression of MCF-7 xenografts in nude mice treated with the vitamin D3 analog, EB1089. Endocrinol 139:2102–2110

46. Flanagan L, Packman K, Juba B et al (2003) Efficacy of Vitamin D compounds to modulate estrogen receptor negative breast cancer growth and invasion. J Steroid Biochem Mol Biol 84:181–192

47. Nolan E, Donepudi M, VanWeelden K et al (1998) Dissociation of vitamin D3 and anti-estrogen mediated growth regulation in MCF-7 breast cancer cells. Mol Cell Biochem 188:13–20

48. Valrance ME, Brunet AH, Welsh JE (2007) Vitamin D receptor-dependent inhibition of mammary tumor growth by EB1089 and ultraviolet radiation in vivo. Endocrinol 148:4887–4894

49. Eisman JA, Macintyre I, Martin TJ et al (1980) Normal and malignant breast tissue is a target organ for 1, 25-(OH)2 vitamin D3. Clin Endocrinol 13:267–272

50. Kemmis CM, Salvador SM, Smith KM et al (2006) Human mammary epithelial cells express CYP27B1 and are growth inhibited by 25-hydroxyvitamin D-3, the major circulating form of vitamin D3. J Nutr 136:887–892

51. Zinser G, Packman K, Welsh JE (2002) Vitamin D(3) receptor ablation alters mammary gland morphogenesis. Development 129:3067–3076

52. Zinser GM, Welsh J (2004) Accelerated mammary gland development during pregnancy and delayed postlactational involution in vitamin D3 receptor null mice. Mol Endocrinol 18:2208–2223

53. Brouwer DA, van Beek J, Ferwerda H et al (1998) Rat adipose tissue rapidly accumulates and slowly releases an orally-administered high vitamin D dose. Br J Nutr 79:527–532

54. Townsend K, Banwell CM, Guy M et al (2005) Autocrine metabolism of vitamin D in normal and malignant breast tissue. Clin Cancer Res 11:3579–3586

55. Rowling MJ, Kemmis CM, Taffany DA et al (2006) Megalin-mediated endocytosis of vitamin D binding protein correlates with 25-hydroxycholecalciferol actions in human mammary cells. J Nutr 136:2754–2759

56. Lipkin M, Newmark HL (1999) Vitamin D, calcium and prevention of breast cancer: a review. J Am Coll Nutr 18:392S–397S

57. Mehta RG, Moriarty RM, Mehta RR et al (1997) Prevention of preneoplastic mammary lesion development by a novel vitamin D analogue, 1alpha-hydroxyvitamin D5. J Natl Cancer Inst 89:212–218

58. Anzano M, Smith J, Uskokovic M et al (1994) 1α-Dihydroxy-16-ene-23-yne-26, 27-hexafluorocholcalciferol (Ro24–5531), a new deltanoid (vitamin D analog) for prevention of breast cancer of breast cancer in the rat. Cancer Res 54:1653–1656

59. Mehta R, Hawthorne M, Uselding L et al (2000) Prevention of N-methyl-N-nitrosourea-induced mammary carcinogenesis in rats by 1α-hydroxyvitamin D5. J Natl Cancer Inst 92:1836–1840

60. Zinser GM, Suckow M, Welsh JE (2005) Vitamin D receptor (VDR) ablation alters carcinogen-induced tumorigenesis in mammary gland, epidermis and lymphoid tissues. J Steroid Biochem Mol Biol 97:153–164

61. Zinser GM, Welsh JE (2004) Vitamin D receptor status alters mammary gland morphology and tumorigenesis in MMTV-neu mice. Carcinogenesis 25:2361–2372

62. Zinser GM, Sundberg JP, Welsh JE (2002) Vitamin D(3) receptor ablation sensitizes skin to chemically induced tumorigenesis. Carcinogenesis 23:103–109

63. Agadir A, Lazzaro G, Zheng Y et al (1999) Resistance of HBL100 human breast epithelial cells to vitamin D action. Carcinogenesis 20:577–582

64. Solomon C, Kremer R, White JH et al (2001) Vitamin D resistance in RAS-transformed keratinocytes: mechanism and reversal strategies. Radiat Res 155:156–162

65. Albertson DG, Ylstra B, Segraves R et al (2000) Quantitative mapping of amplicon structure by array CGH identifies CYP24 as a candidate oncogene. Nat Genet 25:144–146

66. Abedin SA, Banwell CM, Colston KW et al (2006) Epigenetic corruption of VDR signalling in malignancy. Anticancer Res 26:2557–2566

67. Banwell CM, MacCartney DP, Guy M et al (2006) Altered nuclear receptor corepressor expression attenuates vitamin D receptor signaling in breast cancer cells. Clin Cancer Res 12:2004–2013

68. Malinen M, Saramäki A, Ropponen A et al (2008) Distinct HDACs regulate the transcriptional response of human cyclin-dependent kinase inhibitor genes to Trichostatin A and 1alpha, 25-dihydroxyvitamin D3. Nucleic Acids Res 36:121–132

69. Hansen CM, Rohde L, Madsen MW et al (2001) MCF-7/VD(R): a new vitamin D resistant cell line. J Cell Biochem 82:422–436

70. Narvaez CJ, Vanweelden K, Byrne I et al (1996) Characterization of a vitamin D3-resistant MCF-7 cell line. Endocrinol 137:400–409

71. Narvaez CJ, Byrne BM, Romu S et al (2003) Induction of apoptosis by 1, 25-dihydroxyvitamin D(3) in MCF-7 Vitamin D(3)-resistant variant can be sensitized by TPA. J Steroid Biochem Mol Biol 84:199–209

72. Kemmis CM, Welsh JE (2008) Mammary epithelial cell transformation is associated with deregulation of the vitamin D pathway. J Cell Biochem Sept 2 [Epub ahead of print]

73. Elenbaas B, Spirio L, Koerner F et al (2001) Human breast cancer cells generated by oncogenic transformation of primary mammary epithelial cells. Genes Dev 15:50–65

74. Mittal MK, Myers JN, Misra S et al (2008) In vivo binding to and functional repression of the VDR gene promoter by SLUG in human breast cells. Biochem Biophys Res Commun 372:30–34

75. Pálmer HG, Larriba MJ, García JM et al (2004) The transcription factor SNAIL represses vitamin D receptor expression and responsiveness in human colon cancer. Nat Med 10:917–919

76. Binkley N, Novotny R, Krueger D et al (2007) Low vitamin D status despite abundant sun exposure. J Clin Endocrinol Metab 92:2130–2135

77. Kommagani R, Payal V, Kadakia MP. Differential regulation of vitamin D receptor (VDR) by the p53 Family: p73-dependent induction of VDR upon DNA damage. J Biol Chem. 2007 Oct 12:282(41);29847–54. Epub 2007 Aug 23

Chapter 13
Vitamin D and Colorectal Cancer

Marwan Fakih, Annette Sunga, and Josephia Muindi

Abstract An inverse association between sunlight exposure and colon cancer mortality has been previously described. This protective effect has been attributed to increased vitamin D synthesis. Indeed, vitamin D deficiency has been repeatedly associated with an increased risk of adenomatous polyp recurrence and increased colorectal cancer incidence in case–control studies, supporting a direct role for this vitamin against colorectal carcinogenesis. Despite the supporting epidemiological evidence, the Women Health Initiative (WHI) prevention trial failed to demonstrate any reduction in colorectal cancer with 400 IU/day of vitamin D.

We show that dosing at or in excess of 2,000 IU/day of vitamin D3 may be required to achieve optimal serum levels. Prospective studies of such doses need to be investigated to adequately test vitamin D in colorectal cancer prevention. We also review the status of vitamin D in patients with metastatic disease where we demonstrate severe insufficiency and decreased response to vitamin D supplementation, supporting the need of a more aggressive approach in this population.

Keywords Vitamin D • Colorectal cancer • Polyps • Prevention • Chemotherapy

13.1 Epidemiology of Vitamin D and Colorectal Cancer

13.1.1 Sunlight and Colorectal Cancer

An inverse association between sunlight exposure and the risk of certain cancers has long been recognized. One of the first reports to indirectly suggest the association between sunlight and a decreased incidence of non-skin cancer was in 1936 when Peller reported an inverse association between a higher incidence of skin cancer and

---

M. Fakih (✉)
Roswell Park Cancer Institute, Elm & Carlton Streets, Buffalo, NY 14263, USA
e-mail: marwan.fakih@roswellpark.org

D.L. Trump and C.S. Johnson (eds.), *Vitamin D and Cancer*,
DOI 10.1007/978-1-4419-7188-3_13, © Springer Science+Business Media, LLC 2011

decreased mortality from other cancers [1]. Peller attributed this reverse association to a protective effect of skin cancer against the development of other cancers rather than on the protective effects of sunlight [1]. In an attempt to confirm the correlation between skin cancer and protection against non-skin cancers, Apperly studied skin cancer mortality between 1934 and 1938 in the USA [2]. He noted a decreased risk of skin cancer in States with mean annual temperatures $< 42°F$. His subsequent investigations confirmed an inverse association between solar radiation and general cancer rates. Apperly attributed the solar protective effects to the induction of anticancer immunity [2]. It was in 1980 that Garland and Garland first reported the link between vitamin D deficiency, as a result of limited sun exposure, and an increased risk of colorectal cancer [3]. Garland noted an increased rate of colorectal cancer in states with low levels of solar radiation as well as in large cities where population life style limits sunlight exposure [3]. Garland pointed to a parallel increase in the risk of rickets and low vitamin D levels in low solar exposure areas and drew attention to a potential association between vitamin D levels and risk of colorectal cancer [3]. Other supporting data for the protective role of sunlight come from Grant's ecological study on ultra violet (UV-B) light exposure and risk of cancer between 1950 and 1994 [4]. Higher exposure to UV-B protected against colorectal cancer in both White and African Americans [4]. Grant estimated that more than 10% of the deaths from colorectal cancer were premature and related to inadequate UV-B exposure [4]. Case–control studies also strongly support an inverse association between solar radiation and colon cancer. Freedman et al. conducted a death certificate based case–control study of five different types of cancers including colon cancer [5]. Cases were identified as cancer deaths in 24 states in the USA between 1984 and 1995. Controls were frequency matched by 5-year age groups and excluded death from cancer and other neurological illnesses linked with residential sunlight exposure. The risk of colorectal cancer in the highest residential exposure areas was 0.73 (95% CI 0.71–0.74) suggesting a protective effect of sunlight exposure against colorectal cancer mortality [5].

13.1.2 Vitamin D Status and Risk of Colorectal Neoplasia

13.1.2.1 Vitamin D Metabolism

The two universally accepted prerequisites for eliciting vitamin D_3 antitumor effects are the tissue expression of the vitamin D receptor (VDR) and adequate supply of vitamin D_3. Current data suggest that response to vitamin D_3 therapy is highly dose dependent and exhibits substantial inter-patient variability. Furthermore, the physiological range of serum vitamin D_3 metabolite levels required for healthy bones may be different from that required for cell growth inhibition, differentiation and programmed cell death.

The biological basis for the variable vitamin D_3 status in cancer patients could stem from an inadequate supply of vitamin D_3 precursors and inter-patient

13   Vitamin D and Colorectal Cancer

differences in vitamin D_3 gastrointestinal absorption and/or metabolism. With the exception of fish, eggs and vitamin D fortified foods, the human diet is not an vital source of vitamin D_3 [6, 7]. Because more than 90% of vitamin D_3 is produced by exposure of the skin to sunlight, inadequate exposure to sunlight is the leading cause of vitamin D_3 deficiency in humans [6–9]. In humans, ultraviolet light catalyzes the conversion of 7-dehydroxycholesterol to cholecalciferol (vitamin D_3). Vitamin D_3 is then sequentially metabolized in the liver by a number cytochrome P450 enzymes ($cyp27A1$, cyp 2J3, cyp 2R1 and $cyp3A4$) to 25-hydroxyvitamin D_3 (25-D_3) and by 1α-hydroxylase (cyp 27B1) in the kidney to form calcitriol, the biologically most active form of vitamin D_3. Renal 24-hydroxylase (24-OHase, cyp 24A 1), is the major vitamin D_3 inactivating enzyme [10–13]. Simplified vitamin D_3 activation and inactivation oxidative metabolism pathways are shown in Fig. 13.1 These vitamin D_3 activating and inactivating cytochrome P450 enzymes show wider tissue distribution than previously reported. In addition to the classical tissues (gastrointestinal mucosa, liver and kidney), substantial variations in vitamin D_3 activating and inactivating cytochrome P450 enzymes have been reported in a variety of human lung, colon, breast and prostate cancer cell lines and in tissue samples derived from healthy volunteers and cancer patients [14–17]. Recent reports have also identified other non classical vitamin D_3 metabolizing cytochrome P450 enzymes that contribute to the 1α-hydroxylation and 24-hydroxylation of vitamin D_3 hydroxylation [18–20].

The contribution of imbalances in cytochrome P450 enzyme activities that activate and inactivate vitamin D_3 in the pathogenesis of vitamin D_3 deficiency and the responses to vitamin D_3-based therapies in cancer patients has not been fully investigated.

13.1.2.2   Assessment of Vitamin D_3 Status

The serum 25-Hydroxy D_3 (25-D_3) level is the generally accepted and the best indicator of vitamin D_3 status in humans [21, 22]. The utility of 25-D_3 level in assessing vitamin D_3 status is based on its long serum half life (ranging from 2 to 6 weeks), because its synthesis is unregulated, and that serum 25-D_3 levels reflect the overall supply of vitamin D_3 metabolic precursors [23]. There is no universally accepted optimal serum 25-D_3 level. The most widely accepted classification of vitamin D status based on serum 25-D_3 measurement in humans consists of six categories [24]: (i) vitamin D_3 deficiency (serum 25-D_3 levels <20 ng/mL), (ii) vitamin D_3 insufficiency (serum 25-D_3 20–32 ng/mL), (iii) vitamin D_3 sufficiency ≥ 32–100 ng/mL, (iv) vitamin D_3 excess >100 ng/mL, and (v) vitamin D_3 intoxication (serum 25-D_3 >150 ng/mL). High performance liquid chromatography (HPLC) with UV detection method is accepted as the gold standard for measuring serum 25-D_3 levels [25–27]. The HPLC assay is, however, time consuming, often requires large sample volumes and is not free of inaccuracies in serum 25-D_3 quantitation. The three FDA approved and most commonly used analytical assays for measuring serum 25-D_3 levels are: Nichols Diagnostics fully automated

VITAMIN D₃ OXIDATIVE METABOLISM PATHWAYS

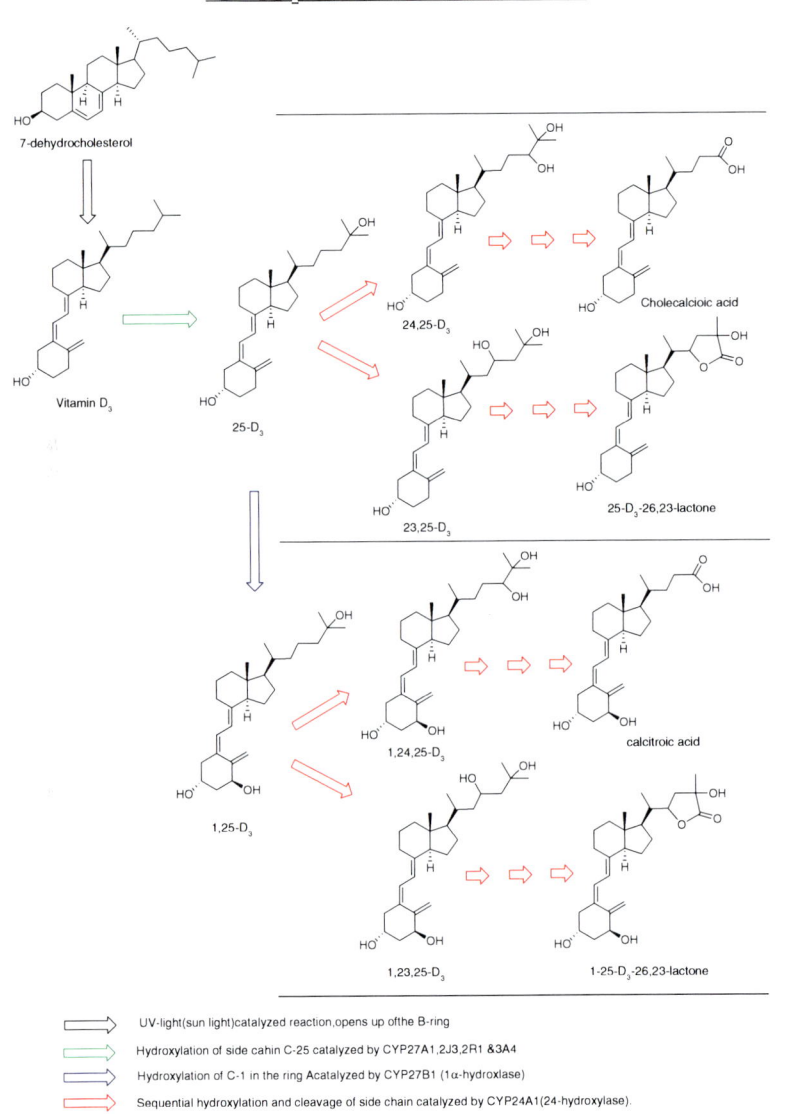

Fig. 13.1 Simplified vitamin D₃ activation and inactivation oxidative metabolism pathways

chemoluminescence ADVANTAGE 25(OH)-D assay system, the DiaSorin LIAISON 25(OH)-D radio immuno assay (RIA) and the immunodiagnostic systems (IDS). The specificity, precision, limitations as well as accuracy of these assays have been extensively documented [28–31]. These assays tend to overestimate the basal levels and greatly underestimate the exogenously added 25- D₃ levels [32]. Serum 25-D₃ measurements are more reliable when performed in

laboratories experienced and dedicated to performing these assays when compared to results obtained from standard hospital clinical chemistry laboratories. A number of RIA are also available for serum 1, 25-D$_3$ measurements and have extensive utilities in the management of patients with chronic renal diseases and more recently in cancer patients on calcitriol Phase I/II clinical trials.

More comprehensive and simultaneous analysis of the various serum vitamin D3 metabolites profiles will be needed as our knowledge of the impact of vitamin D3 status on a number of important chronic human diseases expands. The use of new analytical technologies such as atmospheric pressure chemical ionization (APCI) with positive ion mode LC/MS/MS method is likely to improve the specificity and accuracy of the analysis of the serum vitamin D$_3$ metabolites. At the same time, this new technology can provide comprehensive serum vitamin D$_3$ metabolites profiles including serum 24,25-D$_3$ levels that have not been reported in cancer. Our study which utilizes APCI with positive ion mode LC/MS/MS and DiaSorin RIA to measure serum 25-D$_3$ levels in colorectal patients receiving 400 and 2,000 IU of oral cholecalciferol daily have confirmed the dose dependency and biphasic characteristics of the serum 25-D$_3$ pharmacokinetics. The initial phase of increase in serum 25-D$_3$ levels is approximately 2 months long while the second phase is characterized by the attainment of a steady state (plateau) serum 25-D$_3$ levels that lasts as long as cholecalciferol therapy is continued (Fig. 13.2, Panel A). These results also show that plateau serum 25-D$_3$ levels of >32 ng/mL are attained in patients receiving 2,000 IU of cholecalciferol but not in patients receiving 400 IU. Correlation of serum 25-D$_3$ levels measured by both LC/MS/MS and RIA in these samples is shown in Fig. 13.2, Panel B. The results show that serum 25-D$_3$ levels measured by RIA are higher than those measured by LC/MS/MS. The RIA overestimation of serum 25-D$_3$ levels could be attributed to the cross- reaction with other hydroxylated vitamin D metabolites including 25-D$_2$ and 24,25-D$_3$. Vitamin D3 dose effect

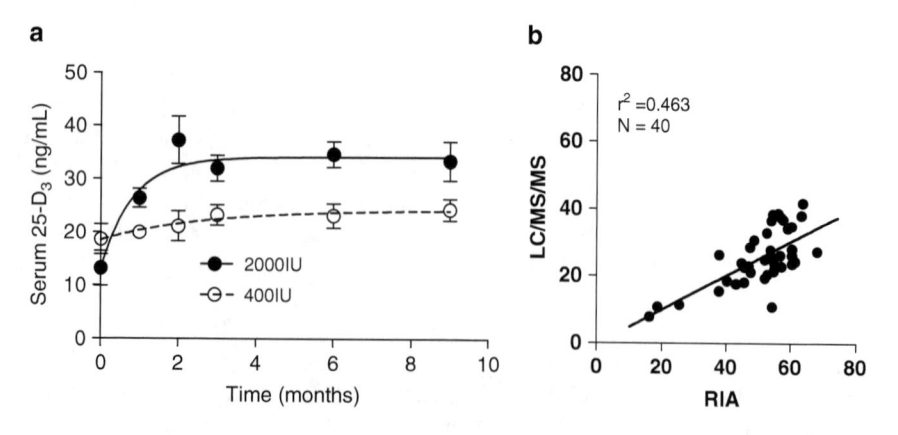

Fig. 13.2 Cholecalciferol dose effect and time course of the changes in serum 25-D$_3$ levels (panel A, presented as mean ± SEM). Correlation of serum 25-D$_3$ levels measured by radio immuno assay (*RIA*) and atmospheric pressure chemical ionization (*APCI*) in positive ion mode LC/MS/MS assay (panel B)

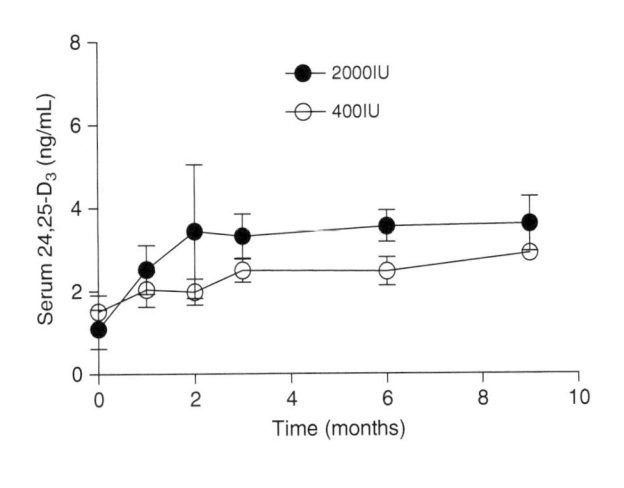

Fig. 13.3 Cholecalciferol dose effect and time course of changes in serum 24, 25-D$_3$ levels in colorectal cancer patients. Serum 24,25-D$_3$ levels determined by atmospheric pressure chemical ionization (*APCI*) in positive ion mode LC/MS/MS. Data = mean ± SEM

and time course of the changes in serum 24, 25-D$_3$ in this cohort of colorectal cancer patients is shown in Fig. 13.3. These results suggest that comprehensive profiles of serum vitamin D$_3$ metabolite in cancer patients are now achievable.

13.1.2.3 Vitamin D Status and Adenomatous Polyps

Several studies suggest a correlation between vitamin D intake or 25-hydroxy vitamin D (25-D$_3$) status and the risk of adenomatous polyps (Table 13.1). Levine et al. conducted a case–control study where 473 patients with a finding of at least one adenoma on initial sigmoidoscopy were compared to controls without any polyps on sigmoidoscopy or without any prior history of adenoma [33]. Plasma 25-D$_3$ was assayed by a competitive binding assay. Increasing plasma levels of 25-D$_3$ was associated with a decreased risk of adenoma (OR = 0.74 for the highest quartile compared to lowest; CI 0.49–1.09). The benefit from higher serum 25-D$_3$ was more pronounced in the population with lower calcium intake. In another case–control study, 222 patients with newly diagnosed adenomas on colonoscopy were compared to 479 controls who had adenoma-free colonoscopies [34]. One hundred and eleven cases and 238 controls had available serum for 25-D$_3$ assay by enzyme immunoassay. A significant association was present between the highest tertile of 25-D$_3$ and a lower risk of adenoma in comparison to the lowest tertile (OR 0.51; CI 0.27–0.98). Contrary to the findings by Levine, the benefit noted on this study seemed more pronounced in the population with a higher calcium intake. In the third study, 239 patients with colonic adenomas diagnosed by sigmoidoscopy were compared to 228 controls with an adenoma-free sigmoidoscopy [35]. 25-D$_3$, assayed by RIA, was found protective against adenoma formation, with a risk reduction by 26% for each 10 ng/mL increase in serum levels. Only one study failed to show a clear association between 25-D$_3$ and polyps [36]. In this study, cases and controls were drawn from the Nurse's Health Study. Cases were diagnosed to have at least one adenoma by endoscopy and were compared with

13   Vitamin D and Colorectal Cancer   301

Table 13.1 Case–control studies (2000–2007): vitamin D status and adenomatous polyps

Year	Author	Population	OR (CI)
2000	Platz [36]	Nurses Health Study	0.34 (0.16–0.75) (1st vs 4 quintile)
2001	Peters [35]	Nat Naval Med Center	0.74 (0.60–0.92) (10 ng/mL inc)
2001	Levine [33]		0.74 (0.49–1.09)
2003	Lieberman [39]	13 VA	0.94 (0.90–0.99)
2004	Hartman [38]	Polyp Prevention trial	0.82 (0.68–0.99) (supplemental Vit D)
2007	Oh [37]	Nurses Health Study	0.79 (0.63–0.99)
2007	Miller [34]	Diet and Health Study III	0.51 (0.27–0.98)

adenoma-free controls. Blood samples, in contrast to the other three trials, were collected several years after the endoscopic procedure. No difference in the median levels of $25\text{-}D_3$ by RIA were seen between cases and controls. However, in subjects with a consistent vitamin D intake across the years, an inverse association between $25\text{-}D_3$ and risk of adenomatous polyps was noted (OR of 0.64, 0.41, and 0.34 for 2nd, 3rd, and 4th quartiles when compared to the 1st quartile).

These case–control studies suggest a potential protective effect of higher levels of plasma $25\text{-}D_3$ against polyp formation. These findings are supported further by several other epidemiological studies associating an increased dietary vitamin D with a lower risk of colorectal adenomas [35–39].

13.1.2.4   Vitamin D Status and Colorectal Cancer

Vitamin D insufficiency, assessed by $25\text{-}D_3$ serum levels, has been associated with an increased risk of colorectal cancer in several case-control studies (Table 13.2). Garland et al. performed a case–control study based on a volunteer population with donated blood samples in 1974 who were subsequently followed for eight years [40]. Thirty-four colorectal cancer cases were matched to 67 controls by age, race, sex, and month of blood draw. $25\text{-}D_3$ serum levels were assayed by HPLC. The risk of colorectal cancer was reduced by 75% in the third quintile and by 80% in the fourth quintiles of serum $25\text{-}D_3$. The odds of getting colorectal cancer was 70% less for patients with $25\text{-}D_3$ levels ≥ 20 ng/mL compared to <20 ng/mL. These results were not confirmed in another case–control study from the same base population [41]. A Finnish case–control study matched 146 newly diagnosed colon cancer cases to 292 non-cancer controls by clinic, age, and date of blood draw. Participants were selected from the Alpha-Tocopherol, Beta-carotene Cancer Prevention Study (ATBC Study) [42]. Pre-diagnosis $25\text{-}D_3$ serum levels were determined by RIA. Increasing levels of $25\text{-}D_3$ were associated with a reduction in the risk of colorectal cancer. The highest risk reduction was seen in the highest quartile, with more than 40% risk reduction encountered in this group. A nested case-control study of $25\text{-}D_3$ and risk of colorectal cancer was conducted within the Health Professionals Follow-up Study

Table 13.2 Case–control studies (1989–2007): vitamin D status and colorectal cancer

Year	Author	Population	Risk reduction
1989	Garland [40]	Men in Maryland	~80%
1995	Braun [41]	ATBC Study	~40%
1997	Tangrea [42]	Finnish Study	40%
2004	Feskanich [44]	NHS	47%
2006	WaktawskiWende [45]	WHI	60%
2007	Wu [43]	HPHS plus NHS	34%
2007	Otani [46]	JPHC	Neg

ATBC Alpha-Tocopherol, Beta-Carotene Cancer Prevention Study, *JPHC* Japan Public Health Centre-base Prospective Study, *WHI* Women's Health Initiative, *NHS* Nurses Health Study, *HPHS* Health Professionals Health Study

(HPFS) [43]. One hundred and seventy-nine of the patients enrolled on the HPFS study were diagnosed with colorectal cancer between 1993 and 2002. These cases were matched to 356 controls by age and by month and year of blood collection. Blood was collected between 1993 and 1995 pre-diagnosis. Serum 25-D_3 was assayed by RIA. An inverse association between higher levels of 25-D_3 and risk of colorectal cancer was noted in the case-control population but did not reach statistical significance. However, the association was highly significant when the analysis was limited to colon cancer (OR 0.46; CI 0.24–0.89). A nested case-control study was also performed in the Nurse's Health Study (NHS) [44]. One-hundred and ninety-three cases of colorectal cancer were identified within 11 years of their initial blood draw. Three-hundred and fifty-six controls were selected in a 2:1 ratio from the same cohort as the case. Controls had to be cancer free at the time of diagnosis and were matched to cases by age and month of blood draw. 25-D_3 was assayed by RIA. A significant inverse association was noted between 25-D_3 and risk of colorectal cancer ($p = 0.02$). The risk reduction in colorectal cancer for the highest 25-D_3 quintile when compared to the lowest quintile was 47%, close to statistical significance (OR = 0.53; CI 0.27–1.04). Another nested case-control study, from the Women's Health Initiative (WHI), matched 317 women with colorectal cancer to 317 non-cancer controls by age, center, race or ethnic group, and date of blood sampling [45]. 25-D_3 was assayed using chemiluminescent RIA. A significant inverse association between 25-D_3 and risk of colorectal cancer was confirmed ($p = 0.02$). The risk of colorectal cancer among the lowest quartile of 25-D_3 was 2.53-fold higher than the highest quartile (CI 1.49–4.32). A recent Japanese case-control study from two large male and female cohorts failed to support the above findings [46]. Three-hundred and seventy-five colorectal cancer cases were identified within 11.5 years from blood collection. Two controls were matched to each case by age, sex, study area, date of blood draw, and fasting time. 25-D_3 was assayed by a competitive protein-binding assay. Although no association was found between 25-D_3 levels and risk of colorectal cancer, low 25-D_3 levels were associated with a statistically significant increased risk of rectal cancer in both males and females.

13   Vitamin D and Colorectal Cancer   303

In summary, most case-control studies support an inverse association between 25-D$_3$ levels and colorectal cancer. Indeed, a combined analysis of five of the studies listed above shows a strong statistically significant inverse correlation between 25-D$_3$ and risk of colorectal cancer [40–42, 44, 45, 47]. The odds ratio for colorectal cancer was 1, 0.82, 0.66, and 0.46 ($p_{trend} < 0.0001$) from the lowest to the highest quartiles of 25-D$_3$. Individuals in the highest quartile had less than half the risk of colorectal cancer of those in the lowest quartile. The combined analysis projected a 50% reduction in risk of colorectal cancer with levels of 34 ng/mL and higher. This is consistent with recent data from the National Health and Nutrition Examination Survey (NHANES) that support an association between a replete vitamin D status and colorectal cancer mortality [48]. Between 1988 and 1994, the NHANES enrolled US individuals aged 17 years and older and included non-Hispanic whites, non-Hispanic blacks, and Mexican Americans with oversampling from the latter two. Among all enrolled patients, 16,818 (95%) had a baseline 25-D$_3$ level by RIA and a known mortality status. Follow-up of this cohort continued until the last day of the year 2000. Sixty-six patients died of colorectal cancer during the follow-up period. An inverse association was present between 25-D$_3$ levels and colorectal cancer mortality ($p = 0.02$). Patients with 25-D$_3$ levels exceeding 32 ng/ mL had a lower risk of colorectal cancer mortality compared to patients with levels lower than 20 ng/mL (risk ratio $= 0.28$, 95% CI 0.11–0.68) [48].

13.1.2.5   Vitamin D Intake and the Risk of Colorectal Cancer

Several case–control and cohort studies have evaluated the effect of dietary vitamin D on the risk of colorectal polyps or cancer. Garland first evaluated vitamin D intake and risk of subsequent colorectal cancer incidence based on 28 day dietary intake diaries completed between 1957 and 1959 in 1954 men [49]. The incidence of colorectal cancer decreased from the lowest to the highest quartiles of vitamin D intake [49]. Another analysis of 35,216 women on the Iowa Women's Health Study investigated the association between baseline dietary questionnaires and the risk of subsequent colorectal cancer [50]. Females with the highest quintile of vitamin D intake had a 32% lower risk of colorectal cancer compared to the lowest quintile; this did not reach statistical significance [50]. The Health Professionals Follow-up Study consisted of 51,529 male professionals who had provided baseline information about dietary habits. Six year follow-up to assess colorectal cancer incidence and death was obtained by mail (response rate 94%) [51]. A higher intake of vitamin D was associated with a slight decrease in the risk of colorectal cancer (relative risk [RR] = 0.88; CI: 0.54–1.42) on multivariate analysis [51]. A larger cohort was evaluated from the Nurses' Health Study [52]. Among study participants, 89,448 respondents to dietary questionnaires and who were free of cancer were followed for colorectal cancer incidence. An inverse association between dietary vitamin D intake and risk of colorectal cancer was noted. The relative risk (RR) for colorectal cancer was 0.84 for the highest quintile of vitamin D when compared to the lowest quintile (CI: 0.63–1.13). The relative risk reduction was more pronounced when

females with subsequent variations in milk intake were excluded (RR = 0.59; CI: 0.3–1.16) [52]. Several other studies confirmed similar findings of inverse association between vitamin D intake and colorectal cancer [53–62]. Indeed, in an analysis of 14 observational studies that investigated oral intake of vitamin D and subsequent incidence of colorectal cancer, Gorham et al. identified the median Effective Dose in preventing 50% of the colorectal cancer cases (ED_{50}) to be 1,000 IU/day when compared to a reference of 100 IU/day [63].

13.1.2.6 Vitamin D Receptor Polymorphism and Colorectal Cancer

There is ample and well documented evidence suggesting that low serum $25\text{-}D_3$ level is associated with an increased risk of developing colorectal cancer. The anti carcinogenesis effects of vitamin D_3 is generally thought to be mediated via 1,25-hydroxyvitamin D_3, the most biologically active form of vitamin D_3, which interacts with VDR to activate key antiproliferative, pro-apoptotic, pro-differentiating and anti-angiogenesis genes in the colorectal mucosa. Down regulation of VDR expression and increased cyp24A1 expression in neoplastic colorectal epithelial cells (when compared to normal colonic epithelial cells) could potentially augment dysregulation of vitamin D_3 homeostasis at the target tissue and thus perpetuate colorectal carcinogenesis.

There are conflicting reports on the association of genetic polymorphisms in VDR gene and the risk of developing colorectal adenoma and cancers in humans. Several studies suggest an association between certain VDR polymorphisms and risk for colorectal adenomas and cancer. A study of 26 patient colorectal cancer patients and 52 controls found the VDR *TtFf* or *TTFf* genotypes to be protective against colorectal carcinogenesis [64]. Another study of 373 colorectal adenoma patients and 394 controls demonstrated that VDR *Fok*1 genotype was associated with large adenomas in patients on low dietary calcium and vitamin D_3 intake [65]. However, another study of 239 colorectal adenoma cases and 228 controls reported that VDR *Fok*1 polymorphism was not significantly associated with colorectal adenoma and did not modify the effect of either calcium or vitamin D3 [35]. VDR *Taq*1 genotype has similarly not been associated with increased risk of developing colorectal adenomas [66].

A more recent study of 170 colorectal cancer patients and 122 healthy controls reported significant down regulation of VDR expression on colonic cancer tissue compared to normal mucosa. However, this study found no differences in VDR *Bsm*1 genotypes in colonic tumor tissues and normal colonic mucosa [67]. Similarly genotyping studies of VDR *Cdx*2, *Fok*1, *Bsm*1, *Apa*1 and *Taq*1 polymorphisms in 546 patients with colorectal adenomas showed that these VDR polymorphisms had no direct effect on the colorectal adenoma recurrence risk [68]. In summary, current literature show no clear cut association between VDR polymorphisms and the risk of developing colorectal adenoma and adenocarcinoma. These reports suggest that the role of VDR polymorphism on colorectal carcinogenesis may be dependent on other factors including the dietary vitamin D_3 and calcium intake.

13.2 Colorectal Cancer Prevention with Vitamin D Supplementation

13.2.1 Pathological Basis for Vitamin D Supplementation

Colonic normal, pre-cancerous, and cancerous epithelium may be targets to vitamin D through a direct effect on vitamin D receptors (VDR) [69–73]. VDR expression increases in the progression from normal mucosa to pre-malignant or malignant tissue (aberrant crypt foci [ACF], polyps, and differentiated adenocarcinoma) [71, 73, 74]. Vitamin D 1α-hydroxylase, the enzyme responsible for the transformation of 25-D_3 to the active form 1, 25-D_3, is expressed in colon tissue. The expression appears equally prominent in normal, ACF, polyps, and differentiated colonic adenocarcinomas [73]. Recent reports, however, suggest that high grade tumors lose VDR and vitamin D 1α-hydroxylase suggesting the importance of VDR and its activation in maintaining normal tissue differentiation [75, 76]. While this may lessen the enthusiasm to investigate vitamin D compounds as antitumor agents in advanced colon cancer, it suggests a window of potential opportunity for vitamin D compounds from the early pre-ACF stage to the development of colon cancer (Fig. 13.4).

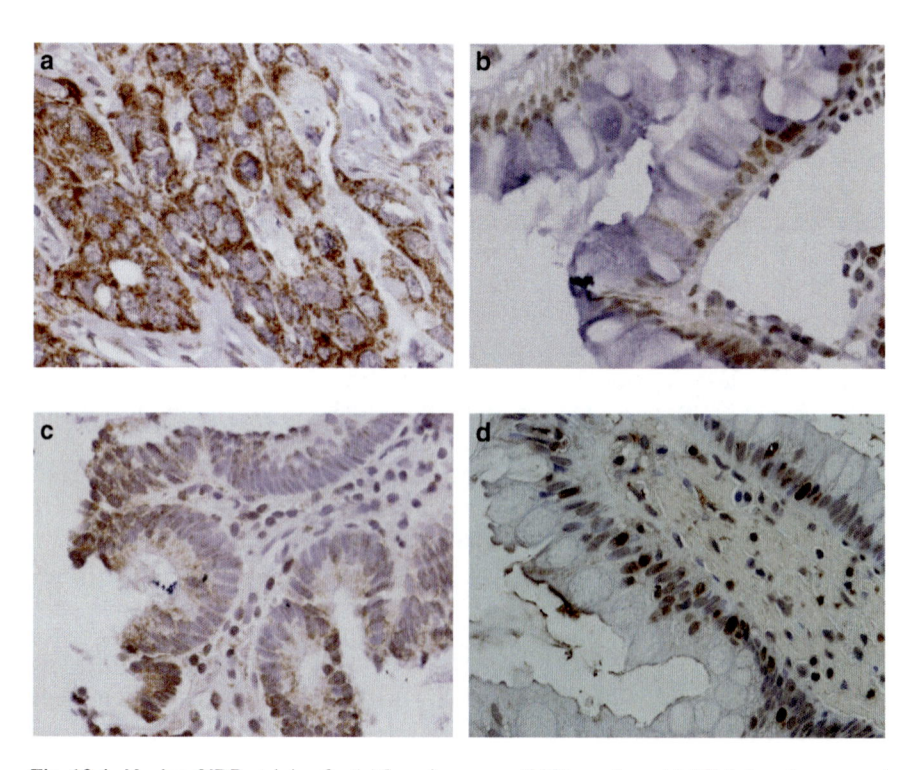

Fig. 13.4 Nuclear VDR staining for (**a**) Invasive cancer (**b**) Normal crypt (**c**) Tubular adenoma and (**d**) Aberrant crypt foci

The importance of VDR activation in the prevention of colorectal carcinogenesis has been demonstrated in several pre-clinical models. An inverse association between cellular proliferation and VDR expression was demonstrated in mice colon [75]. Furthermore, complete loss of VDR (knockout) was associated with an increased proliferation and increased oxidative DNA stress, which may promote carcinogenesis [75, 77]. Vitamin D antitumor activity was also documented in an APC^{min} mouse model [78]. Vitamin D may also induce detoxification through VDR-induced expression of cyp3A, a cytochrome P450 enzyme that detoxifies the secondary bile acid, lithocolic acid [79].

An association between vitamin D levels and supplementation and intestinal mucosal proliferation has also been proposed. Holt et al. have shown an inverse association between colonic epithelial proliferation and increasing levels of 25-D$_3$ [80]. In a subsequent study of daily 400 IU of cholecalciferol and three-times daily 1,500 mg of calcium carbonate, the same investigator showed a decrease in proliferation in both normal and polyp tissue after 6 months of replacement [81].

13.2.2 Clinical Studies with Vitamin D Supplementation

Only one large randomized study evaluated and reported on the effect of vitamin D supplementation on the risk of colorectal cancer [45]. Participants on the Women's Health Initiative (WHI) study were randomized to receive daily vitamin D (400 IU) and elemental calcium (1,000 mg) or placebo. Study participated were post-menopausal females with an age ranging between 50 and 79 years. Among WHI participants, 18,176 were randomized to receive vitamin D plus calcium and 18,106 were randomized to the placebo arm. The study population was followed for outcome after an average of 7 years of treatment. The incidence of colorectal cancer did not differ significantly between the vitamin D and placebo arms (168 cases in the vitamin D arm vs 154 in the placebo; HR = 1.08 (95% CI: 0.86–1.34)) [45]. While the study shows no beneficial effect of low dose vitamin D on the risk of colorectal cancer, several considerations should be kept in mind regarding this study design and its limitations. First, the dose of vitamin D used on this study was likely inadequate to test for a protective role for vitamin D against colorectal cancer. Most epidemiological studies suggest that if a benefit is derived with higher 25-D$_3$ levels, this benefit is typically limited to the highest quartile or quintiles of the population. This suggests that to derive a benefit from vitamin D supplementation, we would need to aim for 25-D$_3$ concentration considerably in excess of 30 ng/mL. The median 25-D$_3$ levels from a nested case-control from the WHI study was noted to be 17 ng/mL. Since, 400 IU/day of vitamin D is expected to raise 25-D$_3$ levels by only 3–4 ng/mL, it would be unlikely that the dose selected for this study would have resulted in any significant shift in 25-D$_3$ levels towards the favorable protective range. Second, the follow-up period on this study may have been insufficient. The carcinogenesis process for colorectal cancer may span a

course of decades. As such, a 7 year treatment period may be insufficient to detect a protective effect for vitamin D, especially if vitamin D effects are limited to the earlier steps in the carcinogenesis process. Finally, there was no limitation or control process on the population enlisted on the WHI study in regards to their vitamin D intake. Indeed, the average vitamin D intake in this population was estimated at 367 IU/day at the time of enrollment and rose further during the conduct of the study. Vitamin D intake on the WHI study was estimated as double the national average [45]. It is possible that the increased baseline vitamin D intake on the WHI study washed out any potential protective effects of the study supplementation.

It is fair to conclude from the WHI study that a low dose of vitamin D supplementation is not effective in preventing colorectal cancer. However, the effectiveness of higher doses of vitamin D in preventing colorectal cancer is still an open question. Based on our studies and those of others in the literature, a dose of 2,000 IU/day or higher may be needed to adequately investigate a role for vitamin D protection in colorectal cancer (see below).

13.3 Vitamin D Status in Advanced Colorectal Cancer

Little is known about vitamin D status in patients with advanced colorectal cancer. One study has assessed 25-D_3 levels across a small number of patients with stages I-IV colon cancer who had not received any chemotherapy treatment. No significant difference was noted in 25-D_3 levels across different stages [82]. We have evaluated 25-D_3 levels in more than 300 patients with colorectal cancer with stage II-IV disease and in various stages of treatment. Patients were stratified according to age, body mass index (BMI), season of blood draw, location of their primary tumor, stage of disease (I-III vs. IV), and chemotherapy status (no chemotherapy, or chemotherapy within 3 months from 25-D_3 level draw) [83]. Vitamin D deficiency was common among patients with colorectal cancer with a median 25-D_3 level of 21.3 ng/mL. On multivariate analysis, only primary site of disease and chemotherapy status were associated with very low 25-D_3 levels ($\leq$15 ng/mL). Chemotherapy was associated with a fourfold increase in risk of severe vitamin D deficiency while a rectal primary was associated with a 2.6-fold increase. This suggests that chemotherapy may increase the risk of vitamin D deficiency in patients with colorectal cancer. The etiology of the increased 25-D_3 deficiency with chemotherapy is under current investigation. Possible etiologies include decreased vitamin D absorption secondary to chemotherapy-induced gastrointestinal toxicity or modulation of 25-D_3 hydroxylation. These retrospective findings, if validated in prospective settings, suggest that patients with colorectal cancer may require more aggressive vitamin D supplementation in the setting of chemotherapy in comparison to a prevention setting. This may be particularly important given recent suggestions that vitamin D status impacts the overall survival of patients with established colorectal cancer [84, 85].

13.4 Vitamin D Status and Colorectal Cancer Outcome

At least two reports have recently associated vitamin D status and patient outcome after a diagnosis of colorectal cancer. The first study is an ecological study from Norway [84]. Norway was divided into three geographical regions based on solar exposure and vitamin D intake based on regional differences in fish consumption. Diagnosis of colorectal cancer was stratified per season (Winter: December–February; Spring: March–May; Summer: June–August; Autumn: September–November). Collected samples from various outpatient clinics on more than 14,000 individuals showed that 25-D_3 levels were lowest in winter and spring. Data regarding colorectal cancer diagnosis and mortality were obtained from the cancer registry for the period of 1964–1992. Data on 12,823 men and 14,922 women with colorectal cancer were analyzed. Colorectal cancer mortality was higher when the diagnosis of colon cancer was made during winter or spring (low 25-D_3) compared to summer and autumn (higher 25-D_3) [84].

The second study analyzed the outcome of colorectal cancer patients who had a baseline vitamin D level (RIA) at least 2 years prior to cancer diagnosis through a retrospective analysis of the NHS and HPFS studies [85]. Three hundred and four colorectal cancer cases were identified. Stages I–IV of colorectal cancer were equally distributed among all four quartiles of 25-D_3. Yet, the mortality rate was the lowest in the highest quartile of 25-D_3. Compared to the lowest quartile, the highest quartile had an adjusted HR for overall mortality of 0.52 (95% CI: 0.29–0.94). The HR for colorectal cancer mortality was 0.61 (95% CI: 0.31–1.19) for the highest 25-D_3 quartile compared to the lowest [85]. This study strongly suggests a correlation between vitamin D status and the risk of death from colorectal cancer. Whether this association is a cause – effect association or signifies a common association between a more replete vitamin D status and other factors that positively impact colorectal cancer outcome remains to be determined.

13.5 Vitamin D Replacement Strategies and Recommended Dosing in Colorectal Cancer

The recommended daily dose of cholecalciferol in the general population is a subject of debate, as is the dosing for the purpose of colorectal cancer prevention. The Institute of Medicine recommends 400 IU of cholecalciferol/day for the population older than 50 years [86]. However, this dose has been regarded by many experts in this field to be inadequate, especially in a vitamin D insufficient population. Indeed, a dose of 400 IU of cholecalciferol/day is estimated to raise 25-D_3 levels by a modest 2.8 ng/mL [87]. It is thus no surprise that clinical trials evaluating doses of 400 IU of cholecalciferol/day failed to show any benefit against osteoporotic fractures while higher doses did [88]. In a review of randomized studies of cholecalciferol vs. placebo and fracture prevention, only the study treatment arms achieving a 25-D_3

mean concentrations of 30 ng/mL or higher resulted in a reduction in parathyroid hormone levels (PTH) and in the risk of fractures [89]. Similarly, no reduction in colorectal cancer has been noted with 400 IU of cholecalciferol and calcium supplementation [45]. However, a combined analysis of five case–control studies supports a 50% risk reduction in patients with 25-D_3 levels exceeding 34 ng/mL [47]. Furthermore, colorectal cancer related mortality has been shown to be inversely associated with 25-D_3 levels with significant benefits noted in the population with levels exceeding 32 ng/mL compared to those <20 ng/mL [48].

These data strongly suggest that if a protective effect for vitamin D supplementation exists, it would likely be achieved with a cholecalciferol dose resulting in 25-D_3 levels in excess of 30 ng/mL. Given that the majority of the US population has 25-D_3 concentrations below 30 ng/mL and that up to 36% of normal healthy population has concentrations below 20 ng/mL [90], it becomes evident that cholecalciferol supplementation doses considerably higher than 400 IU/day would be needed for prevention purposes. Data from healthy volunteers receiving cholecalciferol at 1,000–10,000 IU/day suggest that a dose of 1,700 IU/day is required to achieve the optimal 32 ng/mL concentration [87]. Other data suggest the need for a cholecalciferol dose of 4,000 IU/day to achieve an average steady state concentration of 38 ng/mL [91]. Therefore, a conservative dose of cholecalciferol of 2,000 IU/day is suggested for the goal of achieving the optimal 25-D_3 concentrations exceeding 30 ng/mL.

It is important to point that the current epidemiological and prospective data support an *association* between low levels of 25-D_3 and increased incidence of colorectal cancer and increased colorectal mortality. This does not necessitate a cause effect relationship between vitamin D deficiency and colorectal cancer. It is possible that other biological factors or life style practices predispose subjects to both colorectal cancer and vitamin D deficiency. It is therefore essential that prospective randomized clinical trials with higher doses of cholecalciferol vs. placebo are conducted to determine if vitamin D status plays a significant role in colorectal carcinogenesis and mortality.

References

1. Peller S (1936) Carcinogenesis as a means of reducing cancer mortality. Lancet 2:552–556
2. Apperly F (1941) The relation of solar radiation to cancer mortality in North America. Cancer Res 1:191–195
3. Garland CF, Garland FC (1980) Do sunlight and vitamin D reduce the likelihood of colon cancer? Int J Epidemiol 9(3):227–231
4. Grant WB (2002) An estimate of premature cancer mortality in the US due to inadequate doses of solar ultraviolet-B radiation. Cancer 94(6):1867–1875
5. Freedman DM, Dosemeci M, McGlynn K (2002) Sunlight and mortality from breast, ovarian, colon, prostate, and non-melanoma skin cancer: a composite death certificate based case–control study. Occup Environ Med 59(4):257–262
6. DeLuca HF (2004) Overview of general physiologic features and functions of vitamin D. Am J Clin Nutr 80(6 Suppl):1689S–1696S

7. Holick MF (2004) Sunlight and vitamin D for bone health and prevention of autoimmune diseases, cancers, and cardiovascular disease. Am J Clin Nutr 80(6 Suppl):1678S–1688S
8. Hollis BW (2005) Circulating 25-hydroxyvitamin D levels indicative of vitamin D sufficiency: implications for establishing a new effective dietary intake recommendation for vitamin D. J Nutr 135(2):317–322
9. Heaney RP (2004) Functional indices of vitamin D status and ramifications of vitamin D deficiency. Am J Clin Nutr 80(6 Suppl):1706S–1709S
10. Prosser DE, Jones G (2004) Enzymes involved in the activation and inactivation of vitamin D. Trends Biochem Sci 29(12):664–673
11. Okuda K, Usui E, Ohyama Y (1995) Recent progress in enzymology and molecular biology of enzymes involved in vitamin D metabolism. J Lipid Res 36(8):1641–1652
12. Beckman MJ et al (1996) Human 25-hydroxyvitamin D3-24-hydroxylase, a multicatalytic enzyme. Biochemistry 35(25):8465–8472
13. Omdahl JL, Morris HA, May BK (2002) Hydroxylase enzymes of the vitamin D pathway: expression, function, and regulation. Annu Rev Nutr 22:139–166
14. Miller GJ et al (1995) Vitamin D receptor expression, 24-hydroxylase activity, and inhibition of growth by 1alpha, 25-dihydroxyvitamin D3 in seven human prostatic carcinoma cell lines. Clin Cancer Res 1(9):997–1003
15. Jones G et al (1999) Expression and activity of vitamin D-metabolizing cytochrome P450s (CYP1alpha and CYP24) in human nonsmall cell lung carcinomas. Endocrinology 140(7):3303–3310
16. Anderson MG et al (2006) Expression of VDR and CYP24A1 mRNA in human tumors. Cancer Chemother Pharmacol 57(2):234–240
17. Mimori K et al (2004) Clinical significance of the overexpression of the candidate oncogene CYP24 in esophageal cancer. Ann Oncol 15(2):236–241
18. Gupta RP et al (2004) CYP3A4 is a human microsomal vitamin D 25-hydroxylase. J Bone Miner Res 19(4):680–688
19. Xu Y et al (2006) Intestinal and hepatic CYP3A4 catalyze hydroxylation of 1alpha, 25-dihydroxyvitamin D(3): implications for drug-induced osteomalacia. Mol Pharmacol 69(1):56–65
20. Zhou C et al (2006) Steroid and xenobiotic receptor and vitamin D receptor crosstalk mediates CYP24 expression and drug-induced osteomalacia. J Clin Invest 116(6):1703–1712
21. Holick MF (1990) The use and interpretation of assays for vitamin D and its metabolites. J Nutr 120(Suppl 11):1464–1469
22. Zerwekh JE (2004) The measurement of vitamin D: analytical aspects. Ann Clin Biochem 41(Pt 4):272–281
23. Wootton AM (2005) Improving the measurement of 25-hydroxyvitamin D. Clin Biochem Rev 26(1):33–36
24. Grant WB, Holick MF (2005) Benefits and requirements of vitamin D for optimal health: a review. Altern Med Rev 10(2):94–111
25. Lensmeyer GL et al (2006) HPLC method for 25-hydroxyvitamin D measurement: comparison with contemporary assays. Clin Chem 52(6):1120–1126
26. Binkley N et al (2008) Correlation among 25-hydroxy-vitamin D assays. J Clin Endocrinol Metab 93(5):1804–1808
27. Roth HJ et al (2008) Accuracy and clinical implications of seven 25-hydroxyvitamin D methods compared with liquid chromatography-tandem mass spectrometry as a reference. Ann Clin Biochem 45(Pt 2):153–159
28. Insert P (2001) Nichol ADVANTAGE 25-hydroxyvitamin D assay. Nichols Institute Diagnosis, San Juan Capistrano, CA
29. Product insert (2004) LIAISON chemoluminescence 25-hydroxyvitamin D assay. DiaSorin Corporation, Still Water, MN
30. Vitamin D External Quality Assessment Scheme (DEQAS) 25(OH)D report (2002). Charing Cross Hospital London, UK
31. Vitamin D External Quality Assessment Scheme (DEQAS) 25(OH)D report (2004) Charing Cross Hospital, London UK

32. Hollis BW (2004) Editorial: the determination of circulating 25-hydroxyvitamin D: no easy task. J Clin Endocrinol Metab 89(7):3149–3151
33. Levine AJ et al (2001) Serum 25-hydroxyvitamin D, dietary calcium intake, and distal colorectal adenoma risk. Nutr Cancer 39(1):35–41
34. Miller EA et al (2007) Calcium, dietary, and lifestyle factors in the prevention of colorectal adenomas. Cancer 109(3):510–517
35. Peters U et al (2001) Vitamin D, calcium, and vitamin D receptor polymorphism in colorectal adenomas. Cancer Epidemiol Biomarkers Prev 10(12):1267–1274
36. Platz EA et al (2000) Plasma 1, 25-dihydroxy- and 25-hydroxyvitamin D and adenomatous polyps of the distal colorectum. Cancer Epidemiol Biomarkers Prev 9(10):1059–1065
37. Oh K et al (2007) Calcium and vitamin D intakes in relation to risk of distal colorectal adenoma in women. Am J Epidemiol 165(10):1178–1186
38. Hartman TJ et al (2005) The association of calcium and vitamin D with risk of colorectal adenomas. J Nutr 135(2):252–259
39. Lieberman DA et al (2003) Risk factors for advanced colonic neoplasia and hyperplastic polyps in asymptomatic individuals. Jama 290(22):2959–2967
40. Garland CF et al (1989) Serum 25-hydroxyvitamin D and colon cancer: eight-year prospective study. Lancet 2(8673):1176–1178
41. Braun MM et al (1995) Colon cancer and serum vitamin D metabolite levels 10–17 years prior to diagnosis. Am J Epidemiol 142(6):608–611
42. Tangrea J et al (1997) Serum levels of vitamin D metabolites and the subsequent risk of colon and rectal cancer in Finnish men. Cancer Causes Control 8(4):615–625
43. Wu K et al (2007) A nested case control study of plasma 25-hydroxyvitamin D concentrations and risk of colorectal cancer. J Natl Cancer Inst 99(14):1120–1129
44. Feskanich D et al (2004) Plasma vitamin D metabolites and risk of colorectal cancer in women. Cancer Epidemiol Biomarkers Prev 13(9):1502–1508
45. Wactawski-Wende J et al (2006) Calcium plus vitamin D supplementation and the risk of colorectal cancer. N Engl J Med 354(7):684–696
46. Otani T et al (2007) Plasma vitamin D and risk of colorectal cancer: the Japan Public Health Center-Based Prospective Study. Br J Cancer 97(3):446–451
47. Gorham ED et al (2007) Optimal vitamin D status for colorectal cancer prevention: a quantitative meta analysis. Am J Prev Med 32(3):210–216
48. Freedman DM et al (2007) Prospective study of serum vitamin D and cancer mortality in the United States. J Natl Cancer Inst 99(21):1594–1602
49. Garland C et al (1985) Dietary vitamin D and calcium and risk of colorectal cancer: a 19-year prospective study in men. Lancet 1(8424):307–309
50. Bostick RM et al (1993) Relation of calcium, vitamin D, and dairy food intake to incidence of colon cancer among older women. The Iowa Women's Health Study. Am J Epidemiol 137(12):1302–1317
51. Kearney J et al (1996) Calcium, vitamin D, and dairy foods and the occurrence of colon cancer in men. Am J Epidemiol 143(9):907–917
52. Martinez ME et al (1996) Calcium, vitamin D, and the occurrence of colorectal cancer among women. J Natl Cancer Inst 88(19):1375–1382
53. Pritchard RS, Baron JA, Gerhardsson de Verdier M (1996) Dietary calcium, vitamin D, and the risk of colorectal cancer in Stockholm, Sweden. Cancer Epidemiol Biomarkers Prev 5(11):897–900
54. La Vecchia C et al (1997) Intake of selected micronutrients and risk of colorectal cancer. Int J Cancer 73(4):525–530
55. Marcus PM, Newcomb PA (1998) The association of calcium and vitamin D, and colon and rectal cancer in Wisconsin women. Int J Epidemiol 27(5):788–793
56. Jarvinen R et al (2001) Prospective study on milk products, calcium and cancers of the colon and rectum. Eur J Clin Nutr 55(11):1000–1007
57. McCullough ML et al (2003) Calcium, vitamin D, dairy products, and risk of colorectal cancer in the Cancer Prevention Study II Nutrition Cohort (United States). Cancer Causes Control 14(1):1–12

58. Kampman E et al (2000) Calcium, vitamin D, sunshine exposure, dairy products and colon cancer risk (United States). Cancer Causes Control 11(5):459–466
59. Terry P et al (2002) Dietary calcium and vitamin D intake and risk of colorectal cancer: a prospective cohort study in women. Nutr Cancer 43(1):39–46
60. Ferraroni M et al (1994) Selected micronutrient intake and the risk of colorectal cancer. Br J Cancer 70(6):1150–1155
61. Peters RK et al (1992) Diet and colon cancer in Los Angeles County, California. Cancer Causes Control 3(5):457–473
62. Pietinen P et al (1999) Diet and risk of colorectal cancer in a cohort of Finnish men. Cancer Causes Control 10(5):387–396
63. Gorham ED et al (2005) Vitamin D and prevention of colorectal cancer. J Steroid Biochem Mol Biol 97(1–2):179–194
64. Yaylim-Eraltan I et al (2007) Investigation of the VDR gene polymorphisms association with susceptibility to colorectal cancer. Cell Biochem Funct 25(6):731–737
65. Ingles SA et al (2001) Vitamin D receptor polymorphisms and risk of colorectal adenomas (United States). Cancer Causes Control 12(7):607–614
66. Peters U et al (2004) Circulating vitamin D metabolites, polymorphism in vitamin D receptor, and colorectal adenoma risk. Cancer Epidemiol Biomarkers Prev 13(4):546–552
67. Parisi E et al (2008) Vitamin D receptor levels in colorectal cancer Possible role of BsmI polymorphism. J Steroid Biochem Mol Biol [Epub ahead of print]
68. Hubner RA et al (2008) Dairy products, polymorphisms in the vitamin D receptor gene and colorectal adenoma recurrence. Int J Cancer 123(3):586–593
69. Thomas MG et al (1999) Vitamin D receptor expression in colorectal cancer. J Clin Pathol 52(3):181–183
70. Meggouh F et al (1990) Evidence of 1, 25-dihydroxyvitamin D3-receptors in human digestive mucosa and carcinoma tissue biopsies taken at different levels of the digestive tract, in 152 patients. J Steroid Biochem 36(1–2):143–147
71. Cross HS et al (1996) Vitamin D receptor and cytokeratin expression may be progression indicators in human colon cancer. Anticancer Res 16(4B):2333–2337
72. Evans SR et al (1998) Vitamin D receptor expression as a predictive marker of biological behavior in human colorectal cancer. Clin Cancer Res 4(7):1591–1595
73. Matusiak D et al (2005) Expression of vitamin D receptor and 25-hydroxyvitamin D3-1{alpha}-hydroxylase in normal and malignant human colon. Cancer Epidemiol Biomarkers Prev 14(10):2370–2376
74. Murillo G et al (2007) Chemopreventive efficacy of 25-hydroxyvitamin D3 in colon cancer. J Steroid Biochem Mol Biol 103(3–5):763–767
75. Kallay E et al (2002) Vitamin D receptor activity and prevention of colonic hyperproliferation and oxidative stress. Food Chem Toxicol 40(8):1191–1196
76. Bises G et al (2004) 25-Hydroxyvitamin D3-1alpha-hydroxylase expression in normal and malignant human colon. J Histochem Cytochem 52(7):985–989
77. Kallay E et al (2001) Characterization of a vitamin D receptor knockout mouse as a model of colorectal hyperproliferation and DNA damage. Carcinogenesis 22(9):1429–1435
78. Huerta S et al (2002) 1Alpha, 25-(OH)(2)-D(3) and its synthetic analogue decrease tumor load in the Apc(min) Mouse. Cancer Res 62(3):741–746
79. Makishima M et al (2002) Vitamin D receptor as an intestinal bile acid sensor. Science 296(5571):1313–1316
80. Holt PR et al (2002) Colonic epithelial cell proliferation decreases with increasing levels of serum 25-hydroxy vitamin D. Cancer Epidemiol Biomarkers Prev 11(1):113–119
81. Holt PR et al (2006) Calcium plus vitamin D alters preneoplastic features of colorectal adenomas and rectal mucosa. Cancer 106(2):287–296
82. Niv Y et al (1999) In colorectal carcinoma patients, serum vitamin D levels vary according to stage of the carcinoma. Cancer 86(3):391–397
83. Sunga A et al (2008) Chemotherapy is linked to severe vitamin D deficiency in patients with colorectal cancer. Gastrointestinal Cancers Symposium, Abstract 297

84. Moan J et al (2007) Colon cancer: prognosis for different latitudes, age groups and seasons in Norway. J Photochem Photobiol B 89(2–3):148–155
85. Ng K et al (2008) Circulating 25-hydroxyvitamin d levels and survival in patients with colorectal cancer. J Clin Oncol 26(18):2984–2991
86. Standing Committee on the Scientific Evaluation of Dietary Reference Intakes Food and Nutrition Board Institute of Medicine (1999) Dietary reference intakes for calcium, phosphorous, magnesium, vitamin D, and fluoride. Vitamin D. National Academy Press, Washington, DC, pp 250–287
87. Heaney RP et al (2003) Human serum 25-hydroxycholecalciferol response to extended oral dosing with cholecalciferol. Am J Clin Nutr 77(1):204–210
88. Bischoff-Ferrari HA et al (2006) Estimation of optimal serum concentrations of 25-hydroxyvitamin D for multiple health outcomes. Am J Clin Nutr 84(1):18–28
89. Dawson-Hughes B, Bischoff-Ferrari HA (2007) Therapy of osteoporosis with calcium and vitamin D. J Bone Miner Res 22(Suppl 2):V59–V63
90. Nesby-O'Dell S et al (2002) Hypovitaminosis D prevalence and determinants among African American and white women of reproductive age: third National Health and Nutrition Examination Survey, 1988–1994. Am J Clin Nutr 76(1):187–192
91. Vieth R, Chan PC, MacFarlane GD (2001) Efficacy and safety of vitamin D3 intake exceeding the lowest observed adverse effect level. Am J Clin Nutr 73(2):288–294

Chapter 14
Unique Features of the Enzyme Kinetics for the Vitamin D System, and the Implications for Cancer Prevention and Therapeutics

Reinhold Vieth

Abstract An inadequate vitamin D supply per se does not fully explain the role of vitamin D in the prevention of cancer. The paradigm for the vitamin D system differs from the rest of endocrinology because the enzymes that metabolize 25-hydroxyvitamin D [25(OH)D] behave according to first-order reaction kinetics in vivo. Perpetually fluctuating 25(OH)D in the circulation forces perpetually adaptive adjustments to the enzymes, CYP27B1 and CYP24, that respectively synthesize and catabolize 1,25-dihydroxyvitamin D [1,25(OH)2D] in various tissues. Low levels of 1,25(OH)2D within tissues such as breast and prostate are thought to increase propensity toward cancer. This chapter details the hypothesis that during the times when 25(OH)D levels are declining, such as during fall and winter, concentrations of 1,25(OH)2D within tissues cannot be maintained at any cellular set point for optimal cellular biology. If higher latitude increases the risk of cancer, then vitamin D supplementation will raise and stabilize serum 25(OH)D concentrations, and this will lessen the adverse effects of seasonal fluctuations in serum 25(OH)D.

Keywords Latitude • Seasonality • Enzyme kinetics • Pharmacokinetics • Feedback control • Regulation • Dosage interval • Cholecalciferol • Paracrine

14.1 Introduction

Although environmental ultraviolet light (UVB) is associated with fewer internal cancers, there is no direct experimental evidence that exposure of a person or an animal to light prevents or moderates an internal cancer. The vitamin D system is regarded as one mechanism by which lower latitude and/or higher UVB exposure lower cancer risk or improve prognosis. But aside from colorectal cancer,

---

R. Vieth (✉)

Departments of Nutritional Sciences, and Laboratory Medicine and Pathobiology,
University of Toronto, 600 University Ave,
Toronto, Ontario, M5G 1X5, Canada
e-mail: rvieth@mtsinai.on.ca

D.L. Trump and C.S. Johnson (eds.), *Vitamin D and Cancer*,
DOI 10.1007/978-1-4419-7188-3_14, © Springer Science+Business Media, LLC 2011

case-control studies have generally failed to demonstrate that a higher prediagnostic serum 25(OH)D level lowers risk of cancer. The vitamin D relationship with cancer is not a simple one, where more is better. As latitude increases, so does the seasonal variability in UVB exposure and serum 25(OH)D of populations. Humans are a tropical species for whom large seasonal fluctuations in serum 25(OH)D may not be something for which their biology has been adapted. Metabolism within tissues responsive to the paracrine synthesis of 1,25(OH)$_2$D needs to adapt to prevailing 25(OH)D levels, through adjustments to the CYP27B1 and CYP24 that respectively synthesize and catabolize 1,25(OH)$_2$D. These enzymes are unique in endocrinology, because their activity in vivo is a first-order relationship with substrate. Consequently, so long as serum 25(OH)D levels are in a phase of decline, as they are during winters, there will be a relative excess in tissue catabolism of 1,25(OH)$_2$D, lowering tissue 1,25(OH)$_2$D levels, and potentially affecting cancer risk and prognosis.

14.1.1 Relationship Between Vitamin D and Prostate Cancer

Increased exposure to UVB light is associated with lower risk of internal cancers [1–3], but those benefits are at the cost of higher risk of skin cancer [4, 5]. The benefits of a high-UV environment are widely attributed to the vitamin D produced as a result of UVB light. But for prostate cancer as well as cancer of the pancreas, there is much controversy about whether higher vitamin D status (measured as serum 25-hydroxyvitamin D [25(OH)D]) is beneficial or harmful [6–8]. One alternative to the vitamin D hypothesis is that melatonin or lighting cycles themselves can moderate cellular biology to prevent cancer [9, 10]; however, this speculation is not supported by direct evidence.

As an example, prostate cancer cells possess *both* of the enzymes needed to convert vitamin D into the active paracrine hormone, 1,25-dihydroxyvitamin D3 (1,25(OH)$_2$D; calcitriol) [11]. In cultured prostate epithelial cells, a physiological level of vitamin D$_3$ (the simple product of UV-exposed skin) inhibits growth, induces differentiation, and up-regulates VDR, RXRs, and androgen receptors, suggesting that the observed effects are receptor-mediated [11]. The vitamin D system targets many genes that can play a cancer-preventive role, including genes involved in protection from oxidative stress, and cell–cell and cell–matrix interactions [12, 13]. Anti-inflammatory effects include the inhibition of tumor angiogenesis, invasion, and metastasis [14]. Calcitriol inhibits stromal invasion of prostate cancer cells by modulation of protease activity [15]. In xenograft mouse models, calcitriol and its analogs suppress the nuclear proliferation marker MIB-1 (or Ki-67) in ovarian [16] and breast cancers [17]. Prostate cancer LNCaP cells respond to calcitriol and its analogs with decreased MIB-1 expression [13, 18, 19] and when implanted into in vivo mouse models, they respond to calcitriol with greatly diminished growth [20, 21]. In a mouse cancer model, the calcitriol precursor, calcidiol, slowed tumor growth and improved differentiation of ras-transformed keratinocytes, confirming that in vivo, extrarenal 1-hydroxylase plays an important role in paracrine/autocrine control of growth and differentiation [22].

The cellular mechanisms that explain why and how vitamin D can affect prostate cancer have been studied thoroughly, but the practical question of whether vitamin

14   Unique Features of the Enzyme Kinetics for the Vitamin D System   317

Table 14.1 Conceptual issues complicating the vitamin D hypothesis for cancer prevention

- How can latitude and environmental ultraviolet light be associated with increased risk of prostate cancer [3, 25, 26], and pancreatic cancer [27], yet not be a significant contributor to the lower average 25(OH)D concentrations theorized to be the key component of the mechanism that relates latitude to cancer risk [7]?
- Except for gastrointestinal cancer [28], efforts to relate serum 25(OH)D to cancer risk prospectively have not been prospectively associated with cancer risk
- A U-shaped risk curve has been reported for prostate cancer in relation to serum 25(OH)D concentrations, suggesting that higher serum 25(OH)D is not necessarily a good thing [29, 30]
- The rate of rise in prostate-specific antigen (PSA) slower in summer than in other seasons [31] and vitamin D supplementation appears to slow the rate of rise in PSA [32], yet in epidemiological studies, serum 25(OH)D levels are not related to lower cancer risk
- In regions of the United States where environmental UVB is low, is there a positive association between pancreatic cancer versus serum 25(OH)D, but in regions where UVB is high (presumably providing even higher serum 25(OH)D levels), is there no relationship with · 25(OH)D [33]

D supplementation or a UV-light environment can do anything to prevent an internal cancer or to improve prognosis has never been addressed with an in vivo experimental model. Furthermore, an incomplete understanding of the relationships between vitamin D and cancer has impeded any substantial adjustment in policy to take advantage the potential role for vitamin D. The World Health Organization, through its International Agency for Research in Cancer (IARC) published a major review of cancer and vitamin D [7]. The authors of the IARC report found no compelling reason to change existing public advice about vitamin D. However, the IARC has joined the National Institutes of Health in calling for randomized clinical trials to address vitamin D treatment and cancer prevention [7, 23].

There are many arguments against the "vitamin D hypothesis" and cancer. Serum 25(OH)D levels are similar or even higher in northern Europeans than they are in the south [7, 24]. An inadequate vitamin D supply per se does not explain for the positive latitudinal correlation with prostate cancer incidence. Table 14.1 lists some difficulties that need to be resolved before the vitamin D hypothesis for cancer prevention can be more widely accepted. The rest of this paper describes how an understanding of the enzymology of the vitamin D system may help to resolve the apparently contradictory issues surrounding the roles of vitamin D, latitude, and ultraviolet light in the context of certain cancers.

14.2   Vitamin D Hydroxylase Enzyme Kinetics

There are several reasons why the paradigm for the vitamin D system is very different from the rest of endocrinology (Table 14.2). Metabolism in the vitamin D system behaves according to enzyme-kinetic principles that are very different from those underlying other hormone control systems. The hydroxylase enzymes that metabolize 25(OH)D in vivo behave according to first-order reaction kinetics. In essence, a doubling in availability of substrate to the enzyme results in a transient doubling in the rate of product (i.e. $1,25(OH)_2D$) synthesis. After a time, an increase in 25(OH)D

Table 14.2 Similarities and Differences between the vitamin D system and the classic hormone systems

	Similarities to conventional hormones	Differences from the rest of the endocrine system
Signalling molecule	Endocrine 1,25(OH)$_2$D, which is released into circulation	Paracrine 1,25(OH)$_2$D, which is not normally released into circulation
Site of synthesis of signaling molecule	Endocrine gland for the vitamin D system is the kidney	Breast, prostate, many cell types
What is regulated	Calcium absorption at the intestine	Cell cycle, proliferation, differentiation, many genes affected.
Feedback via	Serum calcium and parathyroid hormone, respectively suppressing and stimulating secretion of 1,25(OH)$_2$D	Autocatabolism by CYP24 (24-hydroxylase) induced by both the substrate 25(OH)D and the product, 1,25(OH)$_2$D.
Substrate availability	(for hormones in general, substrate supply is not rate limiting in the context of ability to produce a hormone)	Supply of vitamin D and 25(OH)D depend on UV light or food sources that were once in UV light, and can range from deficiency to over 200 nmol/L (80 ng/mL) without supplements
Enzyme kinetics	(for hormones in general, substrate supply is abundant and not rate limiting)	For CYP27B1 and CYP24, substrate concentrations are below the Km of the enzyme. Hence the enzyme activity is in a first-order relationship with the substrates, 25(OH)D and 1,25(OH)$_2$D
Effect of season	(for hormones in general, there is no seasonality in substrate supply)	For most who live at temperate latitudes, there is seasonality in substrate supply to produce 1,25(OH)$_2$D

produces an increase in the rate of catabolism, by inducing 1,25(OH)$_2$D-24-hydroxylase (CYP24). Tissue levels of both 1-hydroxylase [CYP27B1] and 24-hydroxylase need to be balanced according to the prevailing supply of 25(OH)D. The inverse relationship between these enzymes has been shown in vivo in rats [34].

14.2.1 Vitamin D Metabolism and Points of Regulation

The metabolism of vitamin D behaves in a manner consistent with the model illustrated in Fig. 14.1, in which a molecule of vitamin D can flow through a series of virtual compartments as represented for each metabolite. Flow is regulated at several steps in the system. At the level of 25-hydroxylase in the liver, metabolism of the vitamin D substrate is relatively automatic and unregulated. Passage of 25(OH)D at the kidney into the hormone, 1,25(OH)$_2$D, is regulated tightly, depending on the need for calcium. At peripheral tissues where its role in the prevention of cancer becomes

14 Unique Features of the Enzyme Kinetics for the Vitamin D System   319

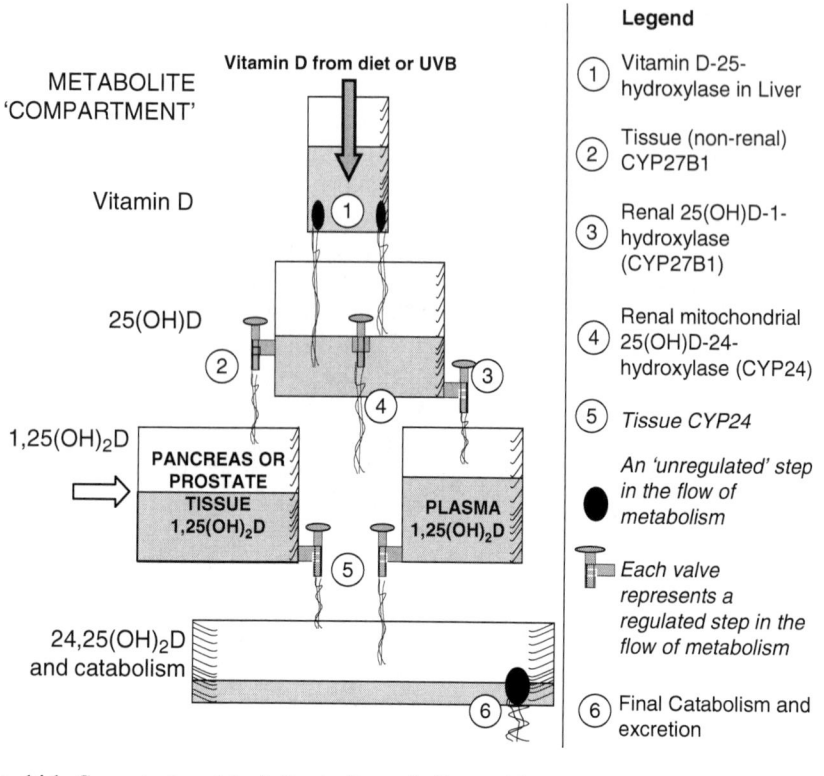

Fig. 14.1 Conceptual model of vitamin D metabolism and its points of regulation. The vessels represent virtual body compartments for vitamin D and its major metabolites. The height of material in the shaded portion of each vessel represents the relative concentration of metabolite. Open passages represent stages at which the pertinent enzymes are relatively unregulated. Valves represent stages at which there is regulation of flow at the enzyme level. A higher supply of 25(OH)D leads to down-regulation of CYP27B1 and an up-regulation of CYP24. The net effect of this model is to maintain tissue 1,25(OH)$_2$D at the set-point level indicated by the block *arrows*

relevant, the regulation of 1,25(OH)$_2$D production is poorly understood. The 1,25(OH)$_2$D generated in peripheral tissues is not normally released into the circulation, and tissue levels of 1,25(OH)$_2$D are very difficult to measure. In Fig. 14.1, the valves represent the stages at which hydroxylases of the vitamin D system need to be regulated. In both the circulation and peripheral tissues, the concentration of 1,25(OH)$_2$D needs to be regulated according to serum 25(OH)D concentration. At the endocrine kidney, there are multiple regulatory mechanisms to moderate circulating 1,25(OH)$_2$D quickly. In comparison peripheral tissues represent a black box in terms of regulating 1,25(OH)$_2$D locally. The control of 1,25(OH)$_2$D is a classic engineering problem of feedback control. A basic concern for systems is the time it takes for a system to sense a change in input, to initiate the appropriate response, and for the response mechanism to fully complete the necessary correction.

14.2.2 Vitamin D Cellular Adaptation

In the fields of biochemistry and cellular biology, the time required for an enzyme to respond to a change in environment (e.g. a change in vitamin D supply) has been assumed to be so fast that the duration of disequilibrium insignificant. Few publications have addressed the rate of adaptation of the vitamin D hydroxylases to changes in vitamin D supply [34–37]. What we know is that endocrine adjustments to $1,25(OH)_2D$ in response to calcium or to changes in 25(OH)D take about 3 days [34, 35, 37]. However, the endocrine secretion of $1,25(OH)_2D$ (i.e. what we measure in serum or plasma) is regulated at the kidney by at least three mechanisms: by plasma calcium, parathyroid hormone [PTH], and through direct feedback by the product, $1,25(OH)_2D$. In contrast, regulation of paracrine, non-renal $1,25(OH)_2D$ production is poorly understood. Outside the kidney, there is no regulation of $1,25(OH)_2D$ production by calcium or PTH [38]. Because they lack the multiple systems to regulate CYP27B1 and CYP24, the prostate and pancreas probably do take longer than the kidney to adapt to altered vitamin D supply.

14.2.3 Vitamin D Modulation of Hydroxylases

If the concentration of $1,25(OH)_2D$ within cells beyond the kidneys is mediated by the ratio between 25(OH)D-1-hydroxylase and $1,25(OH)_2D$-24-hydroxylase (CYP27B1/CYP24 ratio), then the negative impact of higher CYP24 could be described as the product of an "oncogene" [39–41]. A relative excess of CYP24 lowers the tissue concentration of $1,25(OH)_2D$ that promotes cellular differentiation and reduces replication [42, 43]. Conversely, CYP27B1 could be described as "a tumor suppressor" [44]. Prostate cancer cells, both primary cultured cells and cell lines, possess lower CYP27B1 activity than normal cells from the prostate, making them partly resistant to the tumor suppressor activity of circulating 25(OH)D [45–47]. If CYP27B1 and CYP24 need to be maintained in a ratio that compensates for changes in circulating 25(OH)D levels, then the reportedly lower cellular CYP27B1 within prostate cancer cell lines suggests that those cells have lost some of their ability to adapt to low 25(OH)D concentrations (Fig. 14.2).

14.3 Vitamin D and Cancer Risk: Sun Exposure and Levels of 25(OH)D

If prostate and pancreas are particularly slow to adapt to declining 25(OH)D concentrations, then rates of these types of cancer could increase with latitude despite average 25(OH)D concentrations that may not necessarily trend downwards with

14 Unique Features of the Enzyme Kinetics for the Vitamin D System   321

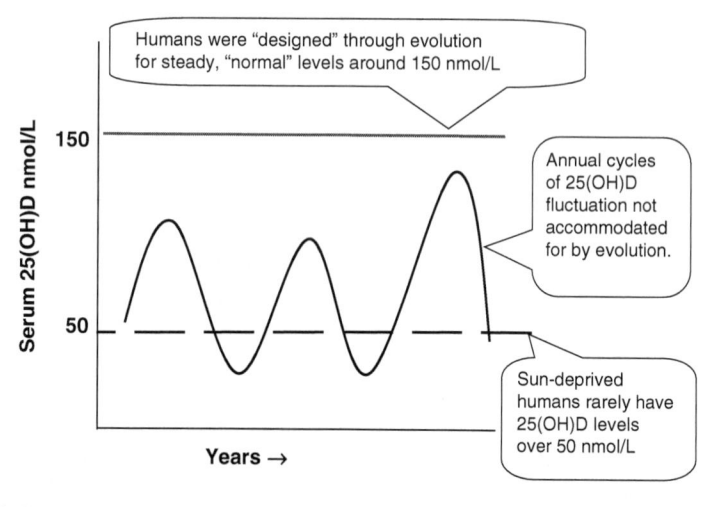

Fig. 14.2 Long-term patterns of 25(OH)D levels in modern populations, showing two patterns for modern humans that would have been unlikely to have existed during our evolution. Populations at temperate latitudes who avoid exposing skin to sunshine exhibit perpetually low serum 25(OH)D concentrations. Populations that sunbathe during summer will exhibit annual cycles of rising and falling serum 25(OH)D concentrations. The prevalent view is that low 25(OH)D may not be ideal, but cyclic patterns in serum 25(OH)D may also have adverse consequences even though average levels may appear to be comparatively high. Cycles of rising and falling 25(OH)D would force the system of enzymes represented in Fig. 14.1 to adapt, and during the declining phase CYP24 would be in relative excess, causing insufficient tissue levels of 1,25(OH)$_2$D

latitude. At latitudes distant from the equator, persons who exhibit the highest serum 25(OH)D concentrations during the summer should as a consequence suffer the largest absolute and relative declines in 25(OH)D through the "vitamin D winter," when at high latitudes, the sun does not reach high enough in the sky to deliver vitamin D-forming UVB to the earth's surface [48, 49]. Those who avoid exposing skin to summer sunlight will exhibit the smallest amplitude fluctuations in serum 25(OH)D. In other words, at the level of tissues like the prostate, serum 25(OH)D levels that are actively declining may be just as bad as very low levels of 25(OH)D, because the tissue level of the catabolic enzyme, 1,25(OH)$_2$D-24-Ohase are relatively excessive during declining phases in serum 25(OH)D. Near the equator there seasonal variability in UVB radiation is minimal, but with increasing latitude, the variability in environmental UVB increases dramatically [50]. Serum 25(OH)D concentrations cycle in a pattern and amplitude that lags by about 3 months the fluctuations in UVB light throughout the seasons. We humans are the hairless primates, with a biology suited for an environment where the dermal vitamin D factory is exposed throughout the year. We have been designed through evolution to be optimized for tropical latitudes where serum 25(OH)D concentrations do remain high and stable all year. Consequently, it is reasonable to infer that perpetually fluctuating inputs of vitamin D may pose a risk to certain aspects of our biology.

14.3.1 Adaptation of Vitamin D Hydroxylases and Cancer Risk

What needs to be established is whether a slow rate of adaptation of the vitamin D hydroxylases can be enough of a problem to affect cancer risk. It has recently been shown that risk of pancreatic cancer in the US north increases with rising 25(OH) D levels measured in summer, but in the US south, there is no such relationship [33]. This would be expected based on the hypothesis proposed here. The present hypothesis is also logically consistent with the evidence that some antineoplastic drugs suppress expression of CYP24 [51].

Not all vitamin D-responsive tissues are likely to behave in the manner proposed here for the prostate. Colon cancer has been well validated epidemiologically as being protected against by higher 25(OH)D concentrations [2, 7, 52]. However, the epidemiology of prostate [6, 29] and pancreatic cancers [8, 33] suggests that these tissues are inefficient at adapting to seasonal UV light and the seasonal cycles in serum 25(OH)D.

So long as serum 25(OH)D concentrations are in a phase of decline, there can be no full achievement of tissue $1,25(OH)_2D$ to match its ideal set-point concentration. No matter how small the true increment below the set-point may prove to be, it is by definition, a sub-optimal concentration. This may not be harmful as a single event in an individual, but over many lifetimes, annual cycles of below set-point phases in tissue $1,25(OH)_2D$ will have an adverse effect on the risk of promotion or progression of certain types of cancer.

14.4 Vitamin D Hypothesis for Cancer Prevention

The hypothesis presented here integrates with the vitamin D hypothesis for cancer prevention in a manner that accounts for the apparent contradictions outlined in Table 14.1. This hypothesis is based on the unusual, first-order in vivo enzyme kinetics of the vitamin D system. The key prediction based on this hypothesis was published in 2004 was in relation to prostate cancer [30], and has been confirmed subsequently at least once, in the context of cancer of the pancreas as shown in Fig. 14.1 [33]. The hypothesis is testable in experimental models, such as the TRAMP mouse model of prostate cancer, as well as with epidemiologic data. The prediction is not tenable as a primary study outcome for human clinical trials, because it predicts an increased risk of cancer of prostate and pancreas in individuals given large doses of vitamin D at dosing intervals of more than 2 months.

14.4.1 Implications of the Model

A major problem for clinical research is poor adherence to medication, which results in negative findings [53, 54]. One way to improve adherence is to give vitamin D less frequently, but at larger doses [55, 56]. A general guideline for a dosing interval

14 Unique Features of the Enzyme Kinetics for the Vitamin D System

for a drug is an interval at about the half-life of the drug [57]. For vitamin D, the effective half-life for the decline in 25(OH)D after a dose of vitamin D_3 is approximately 2 months [58]. However, during the first month after a dose of vitamin D, serum 25(OH)D concentrations are relatively stable [37, 59]. In contrast, with vitamin D the total serum 25(OH)D concentration after the subsequent month ends up even lower than the baseline level before the first dose was given [59]. The phenomenon of a total serum 25(OH)D falling to below its initial level a month after a dose of vitamin D_2 is clinical support for the present contention that the rate of adaptation of metabolic clearance is too slow to respond to fluctuations in vitamin D supply. A key implication of the theory described here is that clinical trials using vitamin D at intermittent doses should avoid vitamin D_2 and they should avoid dosing intervals of any form of vitamin D that go beyond 1 or 2 months. Vitamin D_3 given on a once weekly or once monthly may be an optimal, because less frequent dosing improves adherence compared to daily dosing [60, 61] while minimizing fluctuations in serum 25(OH)D concentration that would occur with semi-annual or annual doses.

An inherent benefit to moderately higher 25(OH)D concentrations makes a lot of sense in many respects [62–64], but it has not been the purpose of this chapter to deal with those aspects. The perspective presented here provides an explanation that can account for many of the things that led to the reservations IARC has expressed against broad advice to increase vitamin D as a way to prevent cancer [7]. The present perspective helps to justify vitamin D supplementation for situations in which latitude appears to increase the risk of cancer even if the population 25(OH) D concentrations might average higher than for populations at lower latitudes. Supplementation raises overall serum 25(OH)D concentrations, and it will lessen the role of seasonal fluctuations in serum 25(OH)D.

References

1. Kricker A, Armstrong B (2006) Does sunlight have a beneficial influence on certain cancers? Prog Biophys Mol Biol 92(1):132–139
2. Knight JA, Lesosky M, Barnett H, Raboud JM, Vieth R (2007) Vitamin D and reduced risk of breast cancer: a population-based case-control study. Cancer Epidemiol Biomarkers Prev 16(3):422–429
3. Grant WB (2002) An estimate of premature cancer mortality in the US due to inadequate doses of solar ultraviolet-B radiation. Cancer 94(6):1867–1875
4. John EM, Schwartz GG, Koo J, Van Den BD, Ingles SA (2005) Sun exposure, vitamin D receptor gene polymorphisms, and risk of advanced prostate cancer. Cancer Res 65(12):5470–5479
5. Rukin NJ, Zeegers MP, Ramachandran S, Luscombe CJ, Liu S, Saxby M et al (2007) A comparison of sunlight exposure in men with prostate cancer and basal cell carcinoma. Br J Cancer 96(3):523–528
6. Ahn J, Peters U, Albanes D, Purdue MP, Abnet CC, Chatterjee N et al (2008) Serum vitamin D concentration and prostate cancer risk: a nested case-control study. J Natl Cancer Inst 100(11):796–804
7. IARC Working Group on Vitamin D (2008) Vitamin D and cancer. WHO Press, Geneva, Switzerland, Report No.: 5

8. Stolzenberg-Solomon RZ, Vieth R, Azad A, Pietinen P, Taylor PR, Virtamo J et al (2006) A prospective nested case-control study of vitamin D status and pancreatic cancer risk in male smokers. Cancer Res 66(20):10213–10219
9. Oh EY, Ansell C, Nawaz H, Yang CH, Wood PA, Hrushesky WJ (2010) Global breast cancer seasonality. Breast Cancer Res Treat (Pub Online First) Feb 4
10. Holdaway IM, Mason BH, Gibbs EE, Rajasoorya C, Lethaby A, Hopkins KD et al (1997) Seasonal variation in the secretion of mammotrophic hormones in normal women and women with previous breast cancer. Breast Cancer Res Treat 42(1):15–22
11. Tokar EJ, Webber MM (2005) Chemoprevention of prostate cancer by cholecalciferol (vitamin D3): 25-hydroxylase (CYP27A1) in human prostate epithelial cells. Clin Exp Metastasis 22(3):265–273
12. Krishnan AV, Peehl DM, Feldman D (2003) Inhibition of prostate cancer growth by vitamin D: Regulation of target gene expression. J Cell Biochem 88(2):363–371
13. Chen TC, Holick MF (2003) Vitamin D and prostate cancer prevention and treatment. Trends Endocrinol Metab 14(9):423–430
14. Krishnan AV, Feldman D (2010) Molecular pathways mediating the anti-inflammatory effects of calcitriol: implications for prostate cancer chemoprevention and treatment. Endocr Relat Cancer 17(1):R19–R38
15. Bao BY, Yeh SD, Lee YF (2006) 1{alpha}, 25-dihydroxyvitamin D3 inhibits prostate cancer cell invasion via modulation of selective proteases. Carcinogenesis 27(1):32–42
16. Zhang X, Jiang F, Li P, Li C, Ma Q, Nicosia SV et al (2005) Growth suppression of ovarian cancer xenografts in nude mice by vitamin D analogue EB1089. Clin Cancer Res 11(1):323–328
17. Sundaram S, Sea A, Feldman S, Strawbridge R, Hoopes PJ, Demidenko E et al (2003) The combination of a potent vitamin D3 analog, EB 1089, with ionizing radiation reduces tumor growth and induces apoptosis of MCF-7 breast tumor xenografts in nude mice. Clin Cancer Res 9(6):2350–2356
18. Elstner E, Campbell MJ, Munker R, Shintaku P, Binderup L, Heber D et al (1999) Novel 20-epi-vitamin D3 analog combined with 9-cis-retinoic acid markedly inhibits colony growth of prostate cancer cells. Prostate 40(3):141–149
19. Polek TC, Stewart LV, Ryu EJ, Cohen MB, Allegretto EA, Weigel NL (2003) p53 Is required for 1, 25-dihydroxyvitamin D3-induced G0 arrest but is not required for G1 accumulation or apoptosis of LNCaP prostate cancer cells. Endocrinology 144(1):50–60
20. Vegesna V, O'Kelly J, Said J, Uskokovic M, Binderup L, Koeffle HP (2003) Ability of potent vitamin D3 analogs to inhibit growth of prostate cancer cells in vivo. Anticancer Res 23(1A):283–289
21. Peleg S, Khan F, Navone NM, Cody DD, Johnson EM, Pelt CS et al (2005) Inhibition of prostate cancer-mediated osteoblastic bone lesions by the low-calcemic analog 1alpha-hydroxymethyl-16-ene-26, 27-bishomo-25-hydroxy vitamin D(3). J Steroid Biochem Mol Biol 97(1–2):203–211
22. Huang DC, Papavasiliou V, Rhim JS, Horst RL, Kremer R (2002) Targeted disruption of the 25-hydroxyvitamin D3 1alpha-hydroxylase gene in ras-transformed keratinocytes demonstrates that locally produced 1alpha, 25-dihydroxyvitamin D3 suppresses growth and induces differentiation in an autocrine fashion. Mol Cancer Res 1(1):56–67
23. Bouillon R, Moody T, Sporn M, Barrett JC, Norman AW (2005) NIH deltanoids meeting on Vitamin D and cancer. Conclusion and strategic options. J Steroid Biochem Mol Biol 97(1–2):3–5
24. Travis RC, Crowe FL, Allen NE, Appleby PN, Roddam AW, Tjonneland A et al (2009) Serum Vitamin D and risk of prostate cancer in a case-control analysis nested within the European Prospective Investigation into Cancer and Nutrition (EPIC). Am J Epidemiol 169(10):1223–1232
25. Luscombe CJ, French ME, Liu S, Saxby MF, Jones PW, Fryer AA et al (2001) Prostate cancer risk: associations with ultraviolet radiation, tyrosinase and melanocortin-1 receptor genotypes. Br J Cancer 85(10):1504–1509
26. Schwartz GG, Hanchette CL (2006) UV, latitude, and spatial trends in prostate cancer mortality: all sunlight is not the same (United States). Cancer Causes Control 17(8):1091–1101

14 Unique Features of the Enzyme Kinetics for the Vitamin D System

27. Neale RE, Youlden DR, Krnjacki L, Kimlin MG, van der Pols JC (2009) Latitude variation in pancreatic cancer mortality in Australia. Pancreas 38(4):387–390
28. Jenab M, Bueno-de-Mesquita HB, Ferrari P, van Duijnhoven FJ, Norat T, Pischon T et al (2010) Association between pre-diagnostic circulating vitamin D concentration and risk of colorectal cancer in European populations:a nested case-control study. BMJ 340:b5500
29. Tuohimaa P, Tenkanen L, Ahonen M, Lumme S, Jellum E, Hallmans G et al (2004) Both high and low levels of blood vitamin D are associated with a higher prostate cancer risk: a longitudinal, nested case-control study in the Nordic countries. Int J Cancer 108(1):104–108
30. Vieth R (2004) Enzyme kinetics hypothesis to explain the U-shaped risk curve for prostate cancer vs 25-hydroxyvitamin D in nordic countries. Int J Cancer 111(3):468
31. Vieth R, Choo R, Deboer L, Danjoux C, Morton G, Klotz L (2006) Rise in prostate-specific antigen in men with untreated low-grade prostate cancer is slower during spring-summer. Am J Ther 13(5):394–399
32. Woo TC, Choo R, Jamieson M, Chander S, Vieth R (2005) Pilot study: potential role of vitamin D (cholecalciferol) in patients with PSA relapse after definitive therapy. Nutr Cancer 51(1):32–36
33. Stolzenberg-Solomon RZ, Hayes RB, Horst RL, Anderson KE, Hollis BW, Silverman DT (2009) Serum vitamin D and risk of pancreatic cancer in the prostate, lung, colorectal, and ovarian screening trial. Cancer Res 69(4):1439–1447
34. Vieth R, McCarten K, Norwich KH (1990) Role of 25-hydroxyvitamin D3 dose in determining rat 1, 25- dihydroxyvitamin D3 production. Am J Physiol 258(5 Pt 1):E780–E789
35. Vieth R, Milojevic S, Peltekova V (2000) Improved cholecalciferol nutrition in rats is noncalcemic, suppresses parathyroid hormone and increases responsiveness to 1, 25-dihydroxycholecalciferol. J Nutr 130(3):578–584
36. Vieth R, Milojevic S (1995) Moderate vitamin D3 supplementation lowers serum 1, 25-dihydroxy-vitamin D3 in rats. Nutr Res 15(5):725–731
37. Ish-Shalom S, Segal E, Salganik T, Raz B, Bromberg IL, Vieth R (2008) Comparison of daily, weekly, and monthly vitamin D3 in ethanol dosing protocols for two months in elderly hip fracture patients. J Clin Endocrinol Metab 93(9):3430–3435
38. Young MV, Schwartz GG, Wang L, Jamieson DP, Whitlatch LW, Flanagan JN et al (2004) The prostate 25-hydroxyvitamin D-1 alpha-hydroxylase is not influenced by parathyroid hormone and calcium: implications for prostate cancer chemoprevention by vitamin D. Carcinogenesis 25(6):967–971
39. Albertson DG, Ylstra B, Segraves R, Collins C, Dairkee SH, Kowbel D et al (2000) Quantitative mapping of amplicon structure by array CGH identifies CYP24 as a candidate oncogene. Nat Genet 25(2):144–146
40. Mimori K, Tanaka Y, Yoshinaga K, Masuda T, Yamashita K, Okamoto M et al (2004) Clinical significance of the overexpression of the candidate oncogene CYP24 in esophageal cancer. Ann Oncol 15(2):236–241
41. Farhan H, Wahala K, Cross HS (2003) Genistein inhibits vitamin D hydroxylases CYP24 and CYP27B1 expression in prostate cells. J Steroid Biochem Mol Biol 84(4):423–429
42. Chen TC, Holick MF, Lokeshwar BL, Burnstein KL, Schwartz GG (2003) Evaluation of vitamin D analogs as therapeutic agents for prostate cancer. Recent Results Cancer Res 164:273–288
43. Peehl DM, Shinghal R, Nonn L, Seto E, Krishnan AV, Brooks JD et al (2004) Molecular activity of 1, 25-dihydroxyvitamin D3 in primary cultures of human prostatic epithelial cells revealed by cDNA microarray analysis. J Steroid Biochem Mol Biol 92(3):131–141
44. Chen TC (2008) 25-Hydroxyvitamin D-1 alpha-hydroxylase (CYP27B1) is a new class of tumor suppressor in the prostate. Anticancer Res 28(4A):2015–2017
45. Ma JF, Nonn L, Campbell MJ, Hewison M, Feldman D, Peehl DM (2004) Mechanisms of decreased Vitamin D 1alpha-hydroxylase activity in prostate cancer cells. Mol Cell Endocrinol 221(1–2):67–74
46. Whitlatch LW, Young MV, Schwartz GG, Flanagan JN, Burnstein KL, Lokeshwar BL et al (2002) 25-Hydroxyvitamin D-1alpha-hydroxylase activity is diminished in human prostate cancer cells and is enhanced by gene transfer. J Steroid Biochem Mol Biol 81(2):135–140

47. Hsu JY, Feldman D, McNeal JE, Peehl DM (2001) Reduced 1alpha-hydroxylase activity in human prostate cancer cells correlates with decreased susceptibility to 25-hydroxyvitamin D3-induced growth inhibition. Cancer Res 61(7):2852–2856
48. Harris SS, Dawson-Hughes B (1998) Seasonal changes in plasma 25-hydroxyvitamin D concentrations of young American black and white women. Am J Clin Nutr 67(6):1232–1236
49. Bolland MJ, Grey AB, Ames RW, Mason BH, Horne AM, Gamble GD et al (2007) The effects of seasonal variation of 25-hydroxyvitamin D and fat mass on a diagnosis of vitamin D sufficiency. Am J Clin Nutr 86(4):959–964
50. Kimlin MG (2008) Geographic location and vitamin D synthesis. Mol Aspects Med 29(6):453–461
51. Tan J, Dwivedi PP, Anderson P, Nutchey BK, O'Loughlin P, Morris HA et al (2007) Antineoplastic agents target the 25-hydroxyvitamin D3 24-hydroxylase messenger RNA for degradation: implications in anticancer activity. Mol Cancer Ther 6(12 Pt 1):3131–3138
52. Giovannucci E (2005) The epidemiology of vitamin D and cancer incidence and mortality: a review (United States). Cancer Causes Control 16(2):83–95
53. Tang BM, Eslick GD, Nowson C, Smith C, Bensoussan A (2007) Use of calcium or calcium in combination with vitamin D supplementation to prevent fractures and bone loss in people aged 50 years and older: a meta-analysis. Lancet 370(9588):657–666
54. Bischoff-Ferrari HA (2007) How to select the doses of vitamin D in the management of osteoporosis. Osteoporos Int 18(4):401–407
55. Vieth R (1999) Vitamin D supplementation, 25-hydroxyvitamin D concentrations, and safety. Am J Clin Nutr 69(5):842–856
56. Trivedi DP, Doll R, Khaw KT (2003) Effect of four monthly oral vitamin D3 (cholecalciferol) supplementation on fractures and mortality in men and women living in the community: randomised double blind controlled trial. BMJ 326:469–475
57. Buxton ILO (2006) Pharmacokinetics and pharmacodynamics: the dynamics of drug absorption, distribution, action, and elimination. In: Brunton LL (ed) Goodman & Gilman's the pharmacological basis of therapeutics, 11th edn. McGraw-Hill, Medical Publishing Division, New York, pp 1–41
58. Vieth R (2007) Vitamin D toxicity, policy, and science. J Bone Miner Res 22(Suppl 2):V64–V68
59. Armas LA, Hollis BW, Heaney RP (2004) Vitamin D2 is much less effective than vitamin D3 in humans. J Clin Endocrinol Metab 89(11):5387–5391
60. Kruk ME, Schwalbe N (2006) The relation between intermittent dosing and adherence: preliminary insights. Clin Ther 28(12):1989–1995
61. Rossini M, Viapiana O, Gatti D, James G, Girardello S, Adami S (2005) The long term correction of vitamin D deficiency: comparison between different treatments with vitamin D in clinical practice. Minerva Med 96(2 Suppl 2):1–7
62. Barreto AM, Schwartz GG, Woodruff R, Cramer SD (2000) 25-Hydroxyvitamin D3, the prohormone of 1, 25-dihydroxyvitamin D3, inhibits the proliferation of primary prostatic epithelial cells. Cancer Epidemiol Biomarkers Prev 9(3):265–270
63. Skinner HG, Michaud DS, Giovannucci E, Willett WC, Colditz GA, Fuchs CS (2006) Vitamin D intake and the risk for pancreatic cancer in two cohort studies. Cancer Epidemiol Biomarkers Prev 15(9):1688–1695
64. Schwartz GG, Eads D, Rao A, Cramer SD, Willingham MC, Chen TC et al (2004) Pancreatic cancer cells express 25-hydroxyvitamin D-1 alpha-hydroxylase and their proliferation is inhibited by the prohormone 25-hydroxyvitamin D3. Carcinogenesis 25(6):1015–1026

Chapter 15
Assessment of Vitamin D Status in the 21st Century

Bruce W. Hollis

Abstract The field of Vitamin D assay technology has progressed significantly over the past 4 decades. Further, the clinical utility of these measurements has moved from esoteric into mainstream clinical diagnosis. This movement has been fueled by the realization that Vitamin D is involved in bodily systems beyond skeletal integrity. The clinical assay techniques for circulating 25(OH)D and 1,25(OH)$_2$D have progressed away from competitive protein-binding assay (CPBAs) that utilize tritium reporters to radioimmunoassay (RIAs) that utilize both I125 and chemiluminescent reporters. These advances have allowed direct serum analysis of 25(OH)D in an automated format that provides a huge sample throughput. Detection of circulating 25(OH)D can also be achieved utilizing direct high-performance liquid chromatographic (HPLC) or liquid chromatography coupled with mass spectrometry (LC-MS) techniques. These methods are accurate, however, they require expensive equipment and restrict sample throughput in the large clinical laboratory. Direct serum detection of 1,25(OH)$_2$D is unlikely to occur for many reasons as a sample pre-purification will always be required. However, a semi-automated chemiluminescent detection system with automated sample preparation is in final development for the determination of circulating 1,25(OH)$_2$D. These advances will allow both 25(OH)D and 1,25(OH)$_2$D to be detected in an accurate, rapid fashion to meet the clinical demands we see emerging.

Keywords Vitamin D assay • 25-hydroxyvitamin D • 1,25-dihydroxyvitamin D

Disclosure Dr. Hollis is an academic consultant to the DiaSorin Corp.

B.W. Hollis (✉)
Department of Pediatrics,
Darby Children's Research Institute, Medical University of South Carolina,
173 Ashley Ave., Room 313, Charleston, SC 29425, USA
e-mail: holisb@musc.edu

D.L. Trump and C.S. Johnson (eds.), *Vitamin D and Cancer,*
DOI 10.1007/978-1-4419-7188-3_15, © Springer Science+Business Media, LLC 2011

15.1 Introduction

In 1971, Haddad and Chyu published a seminal paper in *The Journal of Clinical Endocrinology and Metabolism* that described a competitive protein-binding assay (CPBA) for the determination of circulating 25-hydroxycalfierol [25(OH)D] in human subjects [1]. In this paper they also presented limited patient data for definition of "normal" circulating 25(OH)D levels in humans (Table 15.1). Their "normal" subjects were basically asymptomatic for rickets or osteomalacia and thus were considered "normal" for 25(OH)D status. Their study also presented a group of lifeguards that had circulating 25(OH)D levels 2.5 times that of "normals." Countless similar studies have been performed in the ensuing decades, reiterating the same conclusion. I, however, interpret the original Haddad differently; I suggest that the 25(OH)D levels in the lifeguards are normal and the Haddad "normals" were actually vitamin D deficient. Fortunately, many others now agree with this idea and as a result "normal" circulating 25(OH)D levels, from a clinical standpoint, are 30–100 ng/mL [2]. Because of this newly defined "normal" range a great many patients are deficient in circulating 25(OH)D when tested by their physician. As a result, clinical testing of circulating 25(OH)D has literally exploded in the past 5 years and almost every clinical laboratory wants to perform the test as it is very profitable to do so. I will review the methods currently utilized to perform this testing as well as those for $1,25(OH)_2D$ testing.

15.2 Vitamin D Structure and Chemistry

Vitamin D is a 9,10-seco steroid and exists in two distinct forms: vitamin D_2 and vitamin D_3. Vitamin D_2 is a 28-carbon molecule derived from the plant sterol ergosterol, while vitamin D_3 is a 27-carbon derivative of cholesterol. Vitamin D_2 differs from vitamin D_3 in that it contains an extra methyl group and a double bond between carbons 22 and 23.

The most important aspect of vitamin D chemistry centers on its *cis*-triene structure. This unique structure makes vitamin D and related metabolites susceptible to oxidation, ultraviolet (UV) light-induced conformational changes, heat-induced conformational changes, and attacks by free radicals. Most of these transformation

Table 15.1 Original assessment of nutritional vitamin D status circa 1971 (From [1])

Group	n	Age year	Weekly consumption of vitamin D IU	Weekly exposure to sunlight h	Plasma 25(OH)D nmol
Normal volunteers	40	30.2 ± 12.9	$2,230 \pm 1,041$	8.8 ± 6.1	68.3 ± 29.5
Biliary cirrhosis	4	1.5–55	2,500 (est.)	–	16 ± 6.5*
Lifeguards	8	18.5 ± 2.0	$2,895 \pm 677$	53.0 ± 10.3	161 ± 21.8*

Values are means $\pm$ SD

*$P < 0.001$

products have less biological activity than vitamin D. Research has now demonstrated that vitamin D_2 is much less bioactive than vitamin D_3 in humans [3, 4] although a recent study disputes this finding [5]. The parent compounds, vitamins D_2 and D_3 are sometimes referred to as calciferol.

Hydroxylation reactions at both carbon 25 of the side chain and, subsequently, carbon 1 of the A ring result in metabolic activation of vitamin D. Metabolic inactivation of vitamin D takes place primarily through a series of oxidative reactions at carbons 23, 24, and 26 of the molecule's side chain. This metabolic activation and inactivation are well characterized and result in a plethora of vitamin D metabolites [6]. Of these metabolites, only 25(OH)D and 1,25-dihydroxyvitamin D provide any clinically relevant information. $25(OH)D_2$ and $25(OH)D_3$ are commonly known as calcifediol and the $1,25(OH)_2D$ metabolites as calcitriol. The assay of these vitamin D metabolites will be discussed in this chapter.

15.3   Methods of 25(OH)D Quantitation

The assessment of circulating 25(OH)D started its journey approximately 4 decades ago with the advent of the competitive protein-binding assay (CPBA) [1]. From that early time to the present we have progressed to radioimmunoassay (RIA), high-performance liquid chromatography (HPLC) and liquid chromatography coupled with mass spectrometry (LC/MS). A detailed procedural description of these methods can be reviewed in a recent publication [7]. I will provide a brief description of each technique in this text.

15.3.1   Competitive Protein-Binding Assay

A major factor responsible for the explosion of information on vitamin D metabolism and its relation to clinical disease was the introduction of a CPBA for 25(OH)D. John Haddad, Jr., introduced this CPBA almost 4 decades ago [1]. The assay assessed circulating 25(OH)D concentrations using the vitamin D-binding protein (DBP) as a primary binding agent and ^3H-$25(OH)D_3$ as a reporter. Although this CPBA was valid, it was also relatively cumbersome. Technicians had to extract the sample with organic solvent, dry it under nitrogen, and purify it using column chromatography. This assay was suitable for the research laboratory but did not meet the requirements of a high-throughput clinical laboratory.

The major difficulty in measuring 25(OH)D is attributable to the molecule itself. 25(OH)D is probably the most hydrophobic compound measured by protein-binding assay (PBA), which constitutes either CPBA or radioimmunoassay (RIA). The fact that the molecule exists in two forms, $25(OH)D_2$ and $25(OH)D_3$, compounds the difficulties with its quantitation by PBA. 25(OH)D's lipophilic nature renders it

especially vulnerable to the matrix effects of any PBA. Anything present in the sample assay vessel that is not present in the calibrator assay vessel can cause matrix effects. These matrix effect substances are usually lipid but in the newer direct assays, they could be anything contained in the serum or plasma sample. These matrix factors change the ability of the binding agent, antibody or binding protein to associate with 25(OH)D in the sample or standard in an equal fashion. When this occurs, it markedly diminishes the assay's validity. Experience has demonstrated that the DBP is more susceptible to these matrix effects than antibodies [8]. The original Haddad procedure overcame the matrix problem by using chromatographic sample purification before CPBA [1].

Researchers had a strong desire to simplify this cumbersome CPBA for 25(OH) D, so Belsey and colleagues developed a streamlined CPBA in 1974 [9]. The goal of this second-generation CPBA was to eliminate chromatographic sample purification as well as individual sample recovery using ^{3}H-25(OH)D$_3$. However, after several years of trying, researchers were unable to validate the Belsey assay due to matrix problems originating from ethanolic sample extraction [10].

The 25(OH)D CPBA's did have the advantage of being co-specific for 25(OH) D$_2$ and 25(OH)D$_3$ and thus provided a "total" 25(OH)D value if the assay was valid. The DBP's binding co-specificity for 25(OH)D$_2$ and 25(OH)D$_3$, as well as its stability, made it an attractive candidate for incorporation into automated direct chemiluminescent assays. In fact, Nichols Institute Diagnostics used this approach when its researchers developed the Advantage® 25(OH)D Assay. The U.S. Food and Drug Administration (FDA) approved this assay for clinical use but Nichols ultimately withdrew it from the market place due to its propensity to overestimate total circulating 25(OH)D concentrations and its surprising inability to detect circulating 25(OH)D$_2$ [11, 12]. Although never described, these problems were probably linked to the DBP's inability to resolve the matrix problems associated with direct sample assay. Currently, the CPBA for 25(OH) D is rarely used. Also, one cannot accurately compare most CPBA results for circulating 25(OH)D concentrations from the past with values from current methods because many of the matrix interferences were not linear in the old CPBA's.

15.3.2 Radioimmunoassay

In the early 1980s, my group decided that a non-chromatographic RIA for circulating 25(OH)D would be the best approach to measuring the substance. We therefore designed an antigen that would generate an antibody that was co-specific for 25(OH)D$_2$ and 25(OH)D$_3$ [13]. In addition, we designed a simple extraction method that allowed simple non-chromatographic quantification of circulating 25(OH)D. In 1985 Immunonuclear Corp., now known as DiaSorin, introduced this ^{3}H-based RIA as a kit on a commercial basis. This RIA was further modified in 1993 to incorporate a ^{125}I-labeled reporter and calibrators (standards) in a serum matrix [14].

15   Assessment of Vitamin D Status in the 21st Century   331

This modification finally made mass assessment of circulating 25(OH)D possible. In that same year this assay became the first FDA-approved device for the clinical diagnosis of nutritional vitamin D deficiency. Further, during these past 23 years, these DiaSorin tests have been utilized in the vast majority of large clinical studies worldwide to define "normal" circulating 25(OH)D levels in a variety of disease states. This test still remains today the only RIA-based assay that provides a "total" 25(OH)D value.

15.3.3   Random-Access Automated Instrumentation

DiaSorin Corporation, Roche Diagnostics, and the now defunct Nichols Institute Diagnostics all introduced methods for the direct (no extraction) quantitative determination of 25(OH)D in serum or plasma using competitive protein assay chemiluminescence technology [15]. These assays appear quite similar on the surface but they are not.

In 2001, Nichols Diagnostics introduced the fully automated chemiluminescence Advantage® 25(OH)D assay system. In this assay system, non-extracted serum or plasma was added directly into a mixture containing human DBP, acridinium-ester labeled anti-DBP, and $25(OH)D_3$-coated magnetic particles. Note that the primary binding agent was human DBP. Thus, this assay was a CPBA, much like the manual procedure introduced in 1974 by Belsey et al. [9]. The major difference between these procedures was that Belsey depotenized the sample with ethanol before assaying it. The calibrators for the Belsey assay were in ethanol. In the Advantage assay, the calibrators were in a serum-based matrix, and its developers assumed that this matrix would replicate the serum or plasma sample introduced directly into the assay system. In the end, the 1974 Belsey assay never worked and neither did the Advantage 25(OH)D Assay. The company removed the assay from the market in 2006.

In 2004, the DiaSorin Corporation introduced the fully automated chemiluminescence Liaison® 25(OH)D Assay System [15]. This assay is very similar to the late Advantage assay, with one major difference – the Liaison assay uses an antibody as a primary binding agent as opposed to the human DBP in the Advantage system. Thus, the Liaison is a true RIA method. Details on this procedure are available elsewhere [15]. The Liaison 25(OH)D assay is co-specific for $25(OH)D_2$ and $25(OH)D_3$, so it reports a "total" 25(OH)D concentration. DiaSorin recently introduced a second-generation Liaison 25(OH)D assay. This new version has increased functional sensitivity and much improved assay precision. The Liaison 25(OH)D assay is the single most widely used 25(OH)D assay in the world for clinical diagnosis.

The most recent addition to the automated 25(OH)D assay platforms is from Roche Diagnostics. Their test is an RIA called vitamin D_3(25-OH) and it can be performed on their Elecsys and Cobas systems. Roche only released this assay in 2007, so very little information on it is available. However, the assay can only

detect 25(OH)D$_3$, so it will not be a viable product in countries in which vitamin D$_2$ is used clinically, including the United States.

15.3.4 Direct Physical Detection Methods

Direct detection methodologies for determining circulating 25(OH)D include both HPLC and LC/MS procedures [16–20]. The HPLC methods separate and quantitate circulating 25(OH)D$_2$ and 25(OH)D$_3$ individually. HPLC followed by UV detection is highly repeatable and, in general, most people consider it the gold standard method. However, these methods are cumbersome and require a relatively large sample as well as an internal standard. Sample throughout is slow and is not suited to a high demand clinical laboratory processing up to 10,000 25(OH)D assays per day.

Researchers have recently revitalized LC/MS as a viable method to assess circulating 25(OH)D [17–20]. As with HPLC, LC/MS quantitates 25(OH)D$_2$ and 25(OH)D$_3$ separately. When performed properly, LC/MS is a very accurate testing method. However, the equipment is very expensive and its overall sample throughput when performed properly and ease of operation cannot match that of the automated instrumentation format. As a methodology, LC/MS can compare favorable with RIA techniques [18, 19]. One unique problem with LC/MS is its relative inability to discriminate between 25(OH)D$_3$ and its inactive isomer 3-epi-25(OH)D$_3$. This problem has been especially noticeable in the circulation of newborn infants [17]. Next to the DiaSorin assays, LC/MS is the next most utilized procedure for the clinical assessment of circulating 25(OH)D.

15.4 Clinical Reporting of Circulating 25(OH)D Concentrations

As highlighted earlier, all DiaSorin 25(OH)D assays are approved by the FDA for clinical utility. Thus, the diagnostic 25(OH)D tests sold by DiaSorin and IDS Diagnostics (Fountain Hills, AZ) are under strict FDA control and monitoring for assay performance and reliability. In what I consider a distributing trend, many clinical reference laboratories are replacing these FDA-approved test with "homebrew" LC/MS methods that are diverse and not under FDA scrutiny. The reasons for this switch in utilization are the "perceived" advantages of LC/MS technology being more accurate, precise, specific, cost effective, and providing the separate determination of 25(OH)D$_2$ and 25(OH)D$_3$. First, with respect to accuracy and precision, the DiaSorin and IDS RIA methods perform at least as well as LC/MS methods according to the Vitamin D External Quality Assessment Scheme (DEQAS) operated out of London, UK. As far as specificity goes, the DiaSorin tests appear more specific than LC/MS methodology in that the DiaSorin assays do not detect the inactive 3-epimer of 25(OH)D$_3$ [17]. Finally, LC/MS assays are

marketed on their ability to separately measure $25(OH)D_2$ and $25(OH)D_3$ in a blood sample. Clinically, however, there is no advantage to this separate measurement claim. Not a single scientific publication exists that demonstrates separate $25(OH)$ D_2 and $25(OH)D_3$ measurements are superior to a "total" $25(OH)D$ value as supplied by the DiaSorin tests. In fact, this separate reporting has been shown to confuse the clinician [21]. The truth is, LC/MS laboratories report separate values because that is how LC/MS technology has to report the data [17–20] and is not a reason to "spin" it into a clinical advantage. The fact is, this individual quantitation has been going on for the past 3 decades utilizing HPLC detection and no one claimed it to be clinically advantageous. Some LC/MS laboratories have actually billed inappropriate CPT codes to enhance return for these separate reported values. I consider this practice to be abusive and fraudulent and feel it must end. Further, 99% of all patient samples assayed will not contain any $25(OH)D_2$.

Replacement of FDA-controlled devices such as the DiaSorin and IDS assays with "home-brew" LC/MS assays from a clinical diagnostic standpoint is, again, disturbing. It is disturbing because the DiaSorin assays have and continue to be the standard of clinical $25(OH)D$ assessment. I can say this because the "normal" range of circulating $25(OH)D$ is almost entirely based on clinical studies using the DiaSorin tests. In fact, Labcorp (Burlington, NC) uses a publication by Hollis [2] on which to base their clinical range of $25(OH)D$ levels. In turn, this publication is based on DiaSorin assay-based clinical studies so unless a given LC/MS method is calibrated against the DiaSorin methods, this reference range should not be reported against.

Many years and clinical studies have gone into establishing the DiaSorin reference range and as we stated earlier, this consists of thousands of scientific publications. To prove my point we have selected some large significant clinical studies on which the "normal" circulating level of $25(OH)D$ is based, most of which utilized DiaSorin and some IDS assays as their method of analysis. I have not included any LC/MS clinical studies because basically none exist, which is my point exactly.

The DiaSorin RIA has been used to generate all of the $25(OH)D$ data from the third National Health and Nutrition Examination Survey (NHANES III). I have included selected references on this topic to validate my claim [22–51]. Many more studies from NHANES exist with respect to vitamin D and all use the DiaSorin RIA. Studies from the huge NIH sponsored Women's Health Initiative (WHI) used the DiaSorin LIAISON assay for the first two major publications [49, 50] with others to follow.

The Harvard-based studies, the Health Professionals' Follow-up Study (HPFS) and the Nurses' Health Study (NHS) have been used to establish much of the information in the last decade with regard to the relationship of circulating $25(OH)D$ levels and various disease states such as cancer, autoimmune, cardiovascular and renal. All of these studies again utilized DiaSorin-based assays [29–48]. Of course, we cannot forget the relationship of vitamin D status, PTH and skeletal integrity. Hundreds of papers have been published on this topic; most using DiaSorin assays none using LC/MS testing.

What then should LC/MS laboratories do? If they are going to use the current DiaSorin-based reference range [2] they had better target their values to that of the

334   B.W. Hollis

DiaSorin test. In fact, this is basically how the FDA approves new devices for 25(OH)
D assessment through the 510 K process since the DiaSorin RIA was the first device
approved in 1993. The alternative is that each LC/MS site establish their own reference range which will take years of clinical study since a normal Gaussian distribution is useless in establishing a normative 25(OH)D range. In fact, this "normalization"
of values is quite common between other 25(OH)D assays and DiaSorin testing as
recent articles demonstrate [51]. For instance, if a recently published LC/MS article
was used for diagnosis, the levels reported would have to be increased by 13% if the
DiaSorin reference range is to be used for clinical diagnosis [19].

Finally, clinical reference laboratories should simply use a single reference
range to report circulating 25(OH)D levels as does Labcorp, 32–100 ng/mL.
Compare this to the Mayo Clinic which reports four different "classes" of 25(OH)
D status. This type of reporting is confusing and should be discontinued.

15.5 Methods of 1,25(OH)$_2$D Quantitation

Of all the steroid hormones, 1,25(OH)$_2$D represented the most difficult challenge
to the analytical biochemist with respect to quantitation. 1,25(OH)$_2$D circulates
at picomole (pmol) levels. The development of a simple, rapid assay for this
compound has proven to be a daunting task.

15.5.1 Radioreceptor Assay

The first radioreceptor assay (RRA) for 1,25(OH)$_2$D was introduced in 1974 [52].
Although this initial assay was extremely cumbersome, it did provide invaluable
information with respect to Vitamin D homeostasis. This initial RRA required a
20 mL serum sample, which was extracted using Bligh-Dyer organics. The extract
had to be purified by three successive chromatographic systems, and chickens had
to be sacrificed and Vitamin D receptor (VDR) harvested from their intestines. By
1977, the volume requirement for this RRA had been reduced to a 5 mL sample and
sample pre-purification had been modified to include HPLC [53]. However, the
sample still had to be extracted using a modified Bligh-Dyer procedure and then
pre-purified on Sephadex LH-20. Chicken intestinal VDR was still utilized as a
binding agent.

A major advancement occurred in 1984 with the introduction of a radically new
concept for the RRA determination of circulating 1,25(OH)$_2$D [54]. This new RRA
utilized solid phase extraction of 1,25(OH)$_2$D from serum along with silica car-
tridge purification of 1,25(OH)$_2$D. As a result, the need for HPLC sample pre-
purification was eliminated. Also, this assay utilized VDR isolated from calf
thymus, which proved to be quite stable and thus had to be prepared only periodi-
cally. Further, the volume requirement was reduced to 1 mL of serum or plasma.
This assay opened the way for any laboratory to measure circulating 1,25(OH)$_2$D.
This procedure also resulted in the production of the first commercial kit for

1,25(OH)$_2$D measurement. This RRA was further simplified in 1986 by decreasing the required chromatographic purification steps [55]. This major improvement has recently become a citation classic [56].

As good as the calf thymus RRA for 1,25(OH)$_2$D was, it still possessed two serious shortcomings. First, VDR had to be isolated from thymus glands. Second, because the VDR is so specific for its ligand, only ^{3}H-1,25(OH)$_2$D$_3$ could be used as a reporter, eliminating the use of a ^{125}I or chemiluminescent reporter. This was a major handicap, especially for the commercial laboratory.

15.5.2 Radioimmunoassay

In 1978, the first RIA for 1,25(OH)$_2$D was introduced [57]. Although it was an advantage not to have to isolate the VDR as a binding agent, this RIA was relatively nonspecific, so the cumbersome sample preparative steps were still required. Over the next 18 years all RIAs developed for 1,25(OH)$_2$D suffered from the same shortcomings. In 1996, we developed the first significant advance in 1,25(OH)$_2$D quantification in a decade [58]. This RIA incorporated and ^{125}I-reporter, as well as standards in an equivalent serum matrix, so individual sample recoveries were no longer required. The sample purification procedure is the same one previously used for the rapid RRA procedure [55]. The assay has 100% cross-reactivity between 1,25(OH)$_2$D$_2$ and 1,25(OH)$_2$D$_3$ and is FDA-approved for clinical diagnosis in humans.

Another ^{125}I-based RIA for 1,25(OH)$_2$D is also commercially available from IDS Ltd. The basis of this kit is a selective immunoextraction of 1,25(OH)$_2$D from serum or plasma with a specific monoclonal antibody bound to a solid support. This antibody is directed toward the 1α-hydroxylated A ring of 1,25(OH)$_2$D [59]. This assay procedure has never been published in detail so critical evaluation is difficult. I concluded that this immunoextraction procedure was highly specific for the 1-hydroxylated forms of Vitamin D. However, I also believe that this procedure overestimates circulating 1,25(OH)$_2$D levels. Evidence of this overestimation is evident in a recent publication which shows a correlation of circulating 25(OH)D and 1,25(OH)$_2$D at physiologic levels [60] indicating that 25(OH)D may be interfering with the assay.

ELISAs for circulating 1,25(OH)$_2$D determinations do exist commercially from Immunodiagnostik and IDS. However, their performance has never been published in detail. Further, no automated platforms or LC/MS methods yet exist for the assay of circulating 1,25(OH)$_2$D.

References

1. Haddad JG, Chyu K (1971) Competitive protein-binding radioassay for 25-hydroxy cholecalciferol. J Clin Endocrinol Metab 33:992–995
2. Hollis BW (2005) Circulating 25-hydroxyvitamin D levels indicative of vitamin D sufficiency: implications for establishing a new effective dietary intake recommendation for vitamin D. J Nutr 135:317–322

3. Houghton LA, Vieth R (2006) The case against ergocalciferol (vitamin D_2) as a vitamin supplement. Am J Clin Nutr 84:694–697
4. Aramas LA, Hollis BW, Heaney RP (2004) Vitamin D_2 is much less than vitamin D_3 in humans. J Clin Endocrinol Metab 89:5387–5391
5. Holick MF, Biancusso RM, Chen TC et al (2008) Vitamin D_2 is as effective as vitamin D_3 in maintaining circulating concentrations of 25-hydroxyvitamin D. J Clin Endocrinol Metab 93:677–681
6. Bouillon R, Okamura WH, Norman AW (1995) Structure-function relationships in the vitamin D endocrine system. Endocrine Rev 16:200–257
7. Hollis BW (2005) Detection of vitamin D and its major metabolites. In: Feldman D, Pike J, Glorieux F (eds) Vitamin D, 2nd edn. Elsevier, San Diego, CA
8. Bouillion R, Auwerx J, Dekeyser et al (1984) Two direct (nonchromatographic) assays for 25-hydroxyvitamin D. Clin Chem 30:1731–1736
9. Belsey R, DeLuca HF, Potts JT (1974) A rapid assay 25-OH-vitamin D_3 without preparative chromatography. J Clin Endocrinol Metab 38:1046–1051
10. Dorantes LM, Arnaud SB, Arnaud CD (1978) Importance of the isolation of 25-hydroxyvitamin D before assay. J Lab Clin Med 91:791–796
11. Leventis HC, Garrison L, Sibley M et al (2005) Underestimation of serum 25-hydroxyvitamin D by the Nichols Advantage Assay in patients receiving vitamin D_2 replacement therapy. Clin Chem 51:1072–1074
12. Carter GD, Jones JC, Berry JL (2007) The anomalous behaviour of exogenous 25-hydroxyvitamin D in competitive binding assays. J Steroid Biochem Mol Biol 103:480–482
13. Hollis BW, Napoli JL (1985) Improved radioimmunoassay for vitamin D and its use in assessing vitamin D status. Clin Chem 31:1815–1819
14. Hollis BW, Kamerud JQ, Selvaag SR et al (1993) Determination of vitamin D status by radioimmunoassay with an ^{125}I-labeled tracer. Clin Chem 39:529–533
15. Ersfeld DL, Rao DS, Body JJ et al (2004) Analytical and clinical validation of the 25 OH vitamin D assay for the LIAISON automated analyzer. Clin Biochem 37:867–874
16. Lensmeyer GL, Wiebe DA, Binkley N, Drezner MK (2006) HPLC measurement for 25-hydroxyvitamin D measurement: comparison with contemporary assays. Clin Chem 52:1120–1126
17. Singh RJ, Taylor RL, Reddy GS, Grebe SK (2006) C-3 epimers can account for asignificant proportion of total circulating 25(OH)D in infants, complicating accurate measurement and interpretation of vitamin D status. J Clin Endocrinol Metab 91:3055–3061
18. Maunsell Z, Wright DJ, Rainbow SJ (2005) Routine Isotope-dilution liquid chromatography-tandem mass spectrometry assay for simultaneous measurement of the 25-hydroxy metabolites of vitamins D_2 and D_3. Clin Chem 51:1683–1690
19. Chen H, McCoy LF, Schleicher RL, Pfeiffer CM (2008) Measurement of 25(OH)D_2 and 25(OH)D_3 in human serum using liquid chromatography-tandem mass spectrometry and it comparisons to a radioimmunoassay method. Clin Chem Acta 391:6–12
20. Saenger AK, Laha TJ, De B, Sadzadeh MH (2006) Quantification of serum 25-hydroxyvitamin D_2 and D_3 using HPLC-tandem mass spectrometry and examination of reference intervals for diagnosis of vitamin D deficiency. Am J Clin Pathol 125:914–920
21. Binkley N, Drezner MK, Hollis BW (2006) Laboratory reporting of 25-hydroxyvitamin D results: potential for clinical misinterpretation. Clin Chem 52:2124–2125
22. Gloth FM, Gundberg CM, Hollis BW et al (1995) Vitamin D deficiency in homebound elderly persons. JAMA 274:1683–1686
23. Vieth R, Ladaky WPG (2003) Age-related changes in the 25(OH)D versus parathyroid hormone relationship suggest a different reason why older adults require more vitamin D. J Clin Endocrinol Metab 88:185–191
24. Heaney RP, Dowell MS, Hale CA, Bendich A (2003) Calcium absorption varies within the reference range for serum 25-hydroxyvitamin D. J Am Coll Nutr 22:142–146
25. Looker AC, Mussolino ME (2008) Serum 25-hydroxyvitamin D and hip fracture risk in older white adults. J Bone Min Res 23:143–150

26. Dawson-Hughes B (2006) Calcium plus vitamin D and the risk of fractures. N Engl J Med 354:2285–2287
27. Hollis BW, Wagner CL (2005) Normal serum vitamin D levels. N Engl J Med 352:515–516
28. Abbas S, Linseisen J, Slanger T et al (2008) Serum 25-hydroxyvitamin D and risk of post-menopausal breast cancer – results of a large case-control study. Carcinogenesis 29:93–99
29. Betone-Johnson ER, Chen WY, Holick MF et al (2005) Plasma 25-hydroxyvitamin D and 1,25-dihydroxyvitamin D and risk of breast cancer. Cancer Epidemiol Biomarkers Prev 14:1991–1997
30. Feskanich D, Ma J, Fuchs CS et al (2004) Plasma vitamin D metabolites and risk of colorectal cancer in women. Cancer Epidemiol Biomarkers Prev 13:1502–1508
31. Giovannucci E, Liu Y, Rimm EB et al (2006) Prospective study of predictors of vitamin D status and cancer incidence and mortality in men. J Natl Cancer Inst 98:451–459
32. Zhou W, Heist RS, Liu G et al (2007) Circulating 25-Hydroxyvitamin D levels predict survival in early-state non-small-cell lung cancer patients. J Clin Oncol 25:474–485
33. Tworoger SS, Lee IM, Buring JE et al (2007) Plasma 25-hydroxyvitamin D and 1, 25-dihydroxyvitamin D and risk of incident ovarian cancer. Cancer Epidemiol Biomarkers Prev 16:783–788
34. Mikhak B, Hunter DJ, Speigelman D et al (2007) Vitamin D receptor (VDR) gene polymorphisms and haplotypes, interactions with plasma 25-hydroxyvitamin D and 1, 25-dihydroxyvitamin D, and prostate cancer risk. Prostate 67:911–923
35. Wu K, Feskanich D, Fuchs CS et al (2007) A nested case control study of plasma 25-hydroxyvitamin D concentrations and risk of colorectal cancer. J Natl Cancer Inst 99:1120–1129
36. Wactawski-Wende J, Kotchen JM, Anderson GL et al (2006) Calcium plus vitamin D supplementation and the risk of colorectal cancer. N Engl J Med 354:684–696
37. Liu PT, Sterdger S, Li H et al (2006) Toll-like receptor triggering of a vitamin D-mediated human antimicrobial response. Science 311:1770–1773
38. Zasloff M (2006) Fighting infections with vitamin D. Nat Med 12:388–390
39. Dietrich T, Nunn M, Dawson-Hughes B, Bischoff-Ferrari HA (2005) Association between serum concentrations of 25-hydroxyvitamin D and gingival inflammation. Am J Clin Nutr 82:575–580
40. Dietrich T, Joshipura KJ, Dawson-Hughes B, Bischoff-Ferrari HA (2004) Association between serum concentrations of 25-hydroxyvitamin D and periodontal disease in the US population. Am J Clin Nutr 80:108–113
41. Forman JP, Giovannucci E, Holmes MD et al (2007) Plasma 25-hydroxyvitamin D levels and risk of incident hypertension. Hypertension 49:1063–1069
42. Martins D, Wolf M, Pan D et al (2007) Prevalence of cardiovascular risk factors and the serum levels of 25-hydroxyvitamin D in the United States. Arch Intern Med 167:1159–1165
43. Wang TJ, Pencina MJ, Booth SL et al (2008) Vitamin D deficiency and risk of cardiovascular disease. Circulation 117:503–511
44. Giovannucci E, Liu Y, Hollis BW, Rimm EB (2008) A prospective study of 25-hydroxyvitamin D and risk of myocardial infarction in men. Arch Intern Med 168(11):1174–1180
45. Chiu KC, Chu A, Go V, Saad MF (2004) Hypovitaminosis D is associated with insulin resistance and beta cell dysfunction. Amer J Clin Nutr 79:820–825
46. Zipitis CS, Abodeng AK (2008) Vitamin D supplementation in early childhood and risk of type 1 diabetes: a systematic review and meta-analysis. Arch Dis Child 93:512–517
47. Chonchol M, Scragg R (2007) 25-hydroxyvitamin D, insulin resistance and kidney function in the third national health and nutrition examination survey. Kidney Int 71:134–139
48. Munger KL, Levin LI, Hollis BW et al (2006) Serum 25-hydroxyvitamin D levels and risk of multiple sclerosis. JAMA 20:2832–2838
49. Jackson RD, LaCroix AZ, Gass M et al (2006) Calcium plus vitamin D supplementation and the risk of fractures. N Engl J Med 354:669–683
50. Nesby-O'Dell S, Scanlon KS, Cogswell ME et al (2002) Hypovitaminosis D prevalence and determinants among African American and white women of reproductive age: third national health and nutrition examination survey, 1988–1994. Am J Clin Nutr 76:187–192

51. Rovner AJ, Stallings VA, Schall JI et al (2007) Vitamin D insufficiency in children, adolescents, and young adults with cystic fibrosis despite routine oral supplementation. Amer J Clin Nutr 86:1694–1699
52. Brumbaugh PF, Haussler DH, Bursac KM, Haussler MR (1974) Filter assay for 1, 25(OH)D3: utilization of the hormones target tissue chromatin receptor. Biochem J 13:4091–4097
53. Eisman JA, Shepard RM, DeLuca HF (1977) Determination of 25-hydroxyvitamin D_2 and 25-hydroxyvitamin D_3 in human plasma using high pressure liquid chromatography. Anal Biochem 80:298–305
54. Reinhardt TA, Horst RL, Orf JW, Hollis BW (1984) A microassay for 1, 25-dihydroxyvitamin D not requiring high performance liquid chromatography: application to clinical studies. J Clin Endocrinol Metab 58:91–98
55. Hollis BW (1986) Assay of circulating 1, 25-dihydroxyvitamin D involving a novel single-cartridge extraction and purification procedure. Clin Chem 32:2060–2063
56. Hollis BW (2008) Phase switching SPE for faster 1, 25(OH)2D analysis. Clin Chem 54:446–447
57. Clemens TL, Hendy GN, Graham EG et al (1978) A radioimmunoassay for 1,25-dihydroxy-cholecalciferol. Clin Sci Mol Med 54:329–332
58. Hollis BW, Kamerud JQ, Kurkowski A et al (1996) Quantification of circulating 1, 25-dihydroxyvitamin D by radioimmunoassay with 125I-labeled tracer. Clin Chem 42:586–592
59. Fraser WD, Durham BH, Berry JL, Mawer EB (1997) Measurement of plasma 1, 25 dihydroxyvitamin D using a novel immunoextraction technique and immunoassay with iodine labeled vitamin D tracer. Ann Clin Biochem 34(pt 6):632–637
60. El-Hajj Fuleihan G, Nabulsi M, Tamim H et al (2006) Effect of vitamin D replacement on musculoskeletal parameters in school children: a randomized controlled trial. J Clin Endocrinol Metab 91:405–412

Index

A

Activating protein 1 (AP-1), 28, 148, 153, 156–158, 163, 254, 259

Acute myeloid leukemia (AML), 155, 156, 158, 162, 163, 256–258, 263, 264, 267

Acute promyelocytic leukemia (APL), 35, 145, 260, 263

Adenomatous polyposis coli (APC), 56, 147, 177, 306

1a,25dihydroxyvitaminD3 (1a,25(OH)$_2$D$_3$), 2, 26, 27, 29, 32–41, 43

AKT. *See* Serine/threonine-specific protein kinase B

Alkaline phosphatase (Alk Pase), 146, 154

All-trans retinoic acid (ATRA), 105, 145, 263

AML. *See* Acute myeloid leukemia

Androgen (A), 9, 151, 152, 159, 179, 180

Androgen receptor (AR), 9, 29, 31, 148, 151, 152, 316

Angiogenesis, 54, 55, 57, 58, 74, 99–107, 232, 258, 304, 316

Anti-inflammatory actions, 14, 54–64

B

Bacis helix loop helix (bHLH), 31

Bone morphogenetic protein (BMP), 151

Breast cancer (BC), 8–9, 37–41, 55, 74, 83–87, 89, 91, 102, 145, 149–150, 176, 180–182, 185, 193, 194, 232, 279–290, 316

C

Calcitriol, 53–64, 107, 125, 131, 187, 196, 233–240, 257, 297, 299, 316, 329

Calcium (Ca), 4, 59, 90, 91, 147, 163, 177, 180, 181, 199–201, 210, 230, 258, 268

Calcium-sensing surface receptor (CaR), 14, 153, 200, 205

Cancer prevention, 15, 18, 63, 64, 86, 180, 186–187, 208, 209, 280, 285–287, 289, 295, 301, 302, 305–308, 315–323

Cardiovascular (CV), 59, 89, 106, 107, 115–132, 187, 333

CCAAT/enhancer binding protein (C/EBP), 31, 151, 157, 162–163, 259, 260

Cdk5. *See* Cyclin-dependent kinase 5

Chemoprevention, 36, 54, 59, 63, 64, 145, 146, 175–187, 207, 209

Chemoprevention and treatment, 186

Chemotherapy, 59, 103, 145, 232, 237–240, 262–264, 280, 284, 285, 307

Cholecalciferol (D$_3$), 76, 90, 178, 184–185, 195–196, 257, 297, 299, 300, 306, 308, 309

CI. *See* Confidence interval

9 cis retinoic acid (9 cRA), 105, 183, 263

Co-activator (CoA), 27, 29, 30, 39, 147, 148, 154, 196–198, 206, 224, 260, 268

Colorectal cancer (CRC), 4, 5, 7, 10, 13–15, 55, 56, 63, 75, 78–81, 89, 91, 106, 147, 176–178, 180, 185, 186, 295–309, 315

Combination therapy, 16, 17, 59, 60, 268, 280

Confidence interval (CI), 78–81, 83–91, 121, 125–128, 185, 186, 238, 239, 296, 300–304, 306, 308

C-reactive protein (CRP), 130, 131

CV. *See* Cardiovascular

Cyclin-dependent kinase 5 (Cdk5), 157, 160–163

"Cyclin-like" neuronal Cdk5 activator (Nck5a), 157, 160

D

1,25D$_3$ (calcitriol), 103–107, 298
1,25(OH)$_2$ D$_3$, 2–10, 12–17, 30, 130, 177–184, 193, 194, 196–201, 203, 205–209, 212, 253–269, 316, 335
D$_2$. *See* Ergocalciferol
25D (25 hydroxyvitamin D), 282, 283, 286, 288, 289, 297
25(OH)D. *See* 25-Hydroxyvitamin D
D binding protein (DBP), 26, 119, 196, 268, 286, 329–331
Differentiation, 2–7, 16, 29, 32–37, 41, 42, 55, 56, 74, 143–163, 177, 187, 199–201, 204–206, 212, 252–256, 258–264, 268, 283–287, 289, 296, 305, 316, 318, 320
1,25-Dihydroxyvitamin D$_3$ (1,25(OH)$_2$D$_3$), 2, 54, 122, 123, 193, 199–206, 329
Dimethylbenzanthracene (DMBA), 180–182, 287
Dosage interval, 322, 323
Downstream regulatory element antagonist modulator (DREAM), 29

E

Early growth response protein 1(EGR-1), 161
Early progenitor (EP), 57, 58, 162, 253
Endothelial cells, 11, 57–58, 100–104, 106, 107, 122, 232, 285, 286
Enzyme kinetics, 315–323
Epidemiology, 73–92, 121, 184, 226–231, 240, 295–304, 322
Epidermal growth factor (EGF), 5, 106
Epidermal growth factor receptor (EGFR), 5, 17, 36, 102, 147, 148, 150, 154, 225
Ergocalciferol (D$_2$), 76, 194, 195, 240, 256, 257, 263, 265–267, 281, 299, 322, 323, 328–333, 335
Estrogen receptor (ER), 7, 8, 10, 14, 15, 150, 152, 283, 285, 287
Estrogens, 7–11, 34, 151, 152, 159, 283, 287
Expression of extrarenal vitamin D hydroxy-lases, 2
Extracellular-signal regulated kinase (ERK), 147, 152, 153, 156, 158–162, 168, 197–199, 209, 259, 261–263

F

Farnesoid X-activated receptor (FXR), 42
Feedback control, 259, 318
Flow cytometry (FC), 104, 155

G

Growth factor (GF), 17, 37, 56, 100–103, 152, 156, 157, 159, 224, 252, 253, 283
Growth factor receptor (GFR), 148, 152, 226

H

Heat shock protein (HSP), 29, 263
Hematopoiesis, 29, 155, 160, 162, 251–253, 255–256, 258, 260, 261, 268
Histone deacetylase (HDAC), 11, 13, 27, 28, 31, 35, 36, 41, 197, 198, 264
Histone deacetylase inhibitor (HDACi), 18
Human mammary epithelial (HME), 9, 285, 288, 289
Human osteocalcin (hOC), 81, 156
Human osteopontin (hOP), 156
24-hydroxylase (24OHase), 2, 6, 104, 156, 196, 223–225, 268, 297, 298, 318–320
25-Hydroxyvitamin D (25(OH)D), 36, 38, 74–76, 78–82, 84–92, 119–120, 193, 281–282, 298, 315–323, 329–335

I

IGF binding protein-5 (IBP-5), 154
Inflammatory cytokines, 55, 61–64, 107, 130
Inositol triphosphate (IP$_3$), 153, 199, 205
Insulin like growth factor 1 (IGF-1), 283
Insulin-like growth factor binding protein-3 (IGFBP-3), 36, 62, 151, 225, 232
Interleukin (IL), 61, 62, 102, 105–107, 130, 131, 151, 163, 225, 226, 252–256, 266
International unit (IU), 32, 76, 79, 81, 84–87, 90, 117, 118, 120, 122, 123, 128–130, 176–178, 180, 181, 183, 185–187, 211, 282–283, 299, 304, 306–309, 328

J

Jun N-terminal kinase (JNK), 61, 147, 150, 153, 154, 156–158, 207, 208, 263

K

Keratinocyte growth factor (KGF), 283
Keratinocytes, 2, 29, 33, 34, 152–154, 194, 200, 205–207, 316
Kinase suppressor of Ras-1 (KSR-1), 156, 157, 159, 259, 262
Kruppel-like factor 4 (KLF-4), 153

Index

L

Latitude, 3, 32, 74, 76, 79, 82, 91, 106, 120, 121, 176, 202, 227, 228, 281, 315, 317, 318, 320, 321, 323

Leukemia, 3, 34, 35, 145, 146, 149, 155, 156, 158–163, 254, 256, 259, 260, 262–268

Ligand-dependent nuclear receptor corepressor (LCOR), 26

Lipopolysaccharides (LPS), 155, 263

Lithocholic acid (LCA), 14, 41, 42

Liver X receptor (LXR), 42

M

Matrix-assisted laser desorption/ionization time-of-flight mass spectrometry (MALDI-TOFMS), 147

Matrix-metallo-protease (MMP), 62, 100–102, 105, 129–132

Mitogen activated protein kinase (MAPK), 6, 16, 61, 148, 154, 156–161, 197–199, 206, 207, 224, 225, 254, 259, 261–263

Mitogen-activated protein kinase phos-phatase 5 (MKP5), 55, 60–61, 63, 64

Molecular mechanisms, 38, 177, 184, 209, 211, 258–262, 280, 288, 290

Monocyte-specific esterase ("non-specific" esterase) (MSE), 155

N

Nanograms per milliliter (ng/mL), 79, 81, 90, 119, 121, 126–128, 176, 178, 229, 231, 236, 297, 299–301, 303, 306–309, 318, 328, 334

Nanomoles per liter (nmol/L), 78–79, 81, 82, 86, 88–91, 119, 121, 123, 126–129, 131, 186, 210, 318, 321

National Health and Nutrition Examination Survey (NHANES), 126–128, 303, 333

NCOR2/SMRT. See Silencing mediator of retinoid and thyroid hormone receptors/ Nuclear receptor co-repressor 2

Nitroblue tetrazolium (NBT), 155, 160, 258

Non-Hodgkin lymphoma (NHL), 88–89, 92, 193, 255, 265, 266

Non-specific esterase (NSE), 155, 157, 160, 161, 258

Nuclear factor-kB (NF-kB), 37, 105, 201, 263

Nuclear receptor (NR), 2, 148, 151, 152, 159, 163, 204, 253, 260

Nuclear receptor co-repressor 1 (NCOR1), 27, 28, 36, 40–41

P

Paracrine, 2, 5, 12, 57, 194, 204, 205, 223, 315, 316, 318, 319

Parathyroid hormone (PTH), 2, 4, 31, 106, 107, 122–124, 128, 200, 201, 210, 223–225, 240, 266, 281, 309, 318–320, 333

PCa. See Prostate cancer

Peroxisome proliferator activated receptor (PPAR), 151, 153

PGs. See Prostaglandins

Pharmacokinetics, 234, 235, 299

Phosphatidylinositol 3-kinase (PI3K), 107, 152–154, 158–159, 197, 199, 254, 259, 261, 262

Phosphatidylinositol 3,4,5-triphosphate (PIP$_3$), 152, 153, 199

Phospholipase C gamma-1 ((PLC-g1), 153

Photoprotection, 194, 207–209, 212

PKC. See Protein kinase C

Polyps, 56, 177, 185, 300–301, 303, 305, 306

PPAR. See Peroxisome proliferator activated receptor

Prevention, 180, 184

Progenitor (P), 33, 100, 101, 146, 162, 251–253, 255, 256, 260

Prostaglandins (PGs), 15, 54–60, 102, 233

Prostate cancer (PCa), 6, 7, 9, 13, 15–17, 40, 54–64, 74, 75, 80–84, 90–92, 105, 106, 146, 150–152, 176, 179–180, 221–241, 267, 297, 316–317, 320, 322

Prostate Cancer & Lung Cancer, 176, 182–183, 187, 227

Prostate specific antigen (PSA), 59, 60, 64, 82, 150–151, 233–235, 237–240, 317

Protein kinase C (PKC), 30, 153, 197, 199, 224, 225, 259, 261

PTH. See Parathyroid hormone

R

Randomized controlled trial (RCT), 78, 81, 90–91, 127, 186

Reactive oxygen species (ROS), 5, 36–38, 154, 155, 284

Regulation, 2, 28, 55, 148, 194, 223, 253, 254, 284, 304, 317

Regulation of colonic vitamin D synthesis, 14

Relative risk (RR), 78–81, 83–91, 126, 127, 186, 227, 228, 281, 303–304

Retinoblastoma protein (Rb), 104, 183, 184, 283

Retinoic acid receptor (RAR), 29, 35, 148, 260
Retinoid X receptor (RXR), 27, 30, 42, 154, 156, 157, 159, 196, 197, 206, 207, 212, 224, 232, 253, 254, 258, 268, 289
Retinoid X receptor alpha (RXRa), 154
Ribosomal s6 kinase (MAPK-activated protein kinase-1) (p90RSK), 156, 157, 159

S

Seasonality, 75, 92, 106, 120, 121, 123, 185, 193, 202, 227, 281, 307, 308, 315, 317, 318, 320, 322, 323
Serine/threonine-specific protein kinase B (AKT), 104, 152, 158, 159, 199, 209, 225, 232, 259, 261, 262
Silencing mediator of retinoid and thyroid hormone receptors/Nuclear receptor co-repressor 2 (NCOR2/SMRT), 27, 28, 31, 40
Simian virus 40 (SV40), 280
Skin cancer, 37, 74, 80, 88, 90, 176, 183, 187, 191–212, 295–296, 316
Solar UV radiation, 120, 195, 201
Specificity protein 1(Sp-1), 149
Squamous cell carcinoma (SCC), 17, 88, 103, 104, 152–154, 195, 201, 202, 206, 209, 228, 232
SRA stem loop-interacting RNA-binding protein (SLIRP), 28
Steroid receptor co-activator (SRC), 29, 30, 154, 197, 198, 206, 233

T

T-cell transcription factor 4, 149
Thyroid hormone receptor interactor 2 (TRIP2/DRIP205), 30
Thyroid hormone receptor interactor 15 (TRIP15/COPS2/Alien), 28
Transforming growth factor beta (TGFb), 101, 283, 287
TRIP15/COPS2/Alien. *See* Thyroid hormone receptor interactor 15

TRIP2/DRIP205. *See* Thyroid hormone receptor interactor 2
Tumor necrosis factor (TNF), 61, 102, 105, 107, 130–131, 225, 226

U

Ultraviolet (UV), 32, 36, 37, 41, 83, 89, 120, 121, 128, 156, 157, 193–196, 201, 203, 207–209, 211, 212, 226–228, 285, 297, 298, 316–318, 322, 328, 332
Ultraviolet B light (UV-B), 3, 36, 74–76, 79, 82, 91, 106, 120, 128, 176, 183–184, 193, 195, 207–209, 222, 228, 296, 315–317, 319, 320

V

Vascular smooth muscle cells (VSMCs), 103, 106–107
Vasculature, 100, 103, 107
Vitamin D, 1–18, 73–92, 99–107, 115–132, 143–163, 175–187, 191–212, 221–241, 251–269, 279–290, 295–309, 315–323, 327–335
Vitamin D_3, 3, 15–16, 26, 32, 41, 107, 120, 122, 123, 128, 176–178, 180–186, 195–201, 257, 280, 283, 296–300, 304, 316, 322, 323, 328–329, 331
Vitamin D analogues, 7, 16, 17, 104, 105, 150, 177, 180, 181, 209, 211, 212, 231, 256, 261–268, 283–285, 287–289
Vitamin D assay, 14, 103, 119, 120, 155, 161, 177, 281, 297–302, 328–335
Vitamin D_3 receptor (VDR), 2, 26, 62, 103, 147, 178, 193, 222, 253, 280, 296, 316, 334
Vitamin D_3 response element (VDRE), 30, 41, 60, 148, 154, 156, 157, 159, 196–198, 206, 224, 225, 260–262, 268

W

Wild type (WT), 37, 56, 100, 104, 178, 179, 183, 205, 207, 255, 267, 285, 287
Wingless-related MMTV integration site (Wnt), 6, 33, 34, 39, 147–149